Possibilities and Limitations in Current Translational Stem Cell Research

Edited by Diana Kitala

Published in London, United Kingdom

Possibilities and Limitations in Current Translational Stem Cell Research
http://dx.doi.org/10.5772/intechopen.100951
Edited by Diana Kitala

Contributors
Patricia Semedo-Kuriki, Niels Olsen Saraiva Camara, Danilo Cândido de Almeida, Gabriel Pereira, Ganna Bilousova, Igor Kizub, Andrii Rozhok, Hong-Thuy Bui, Nhat-Thinh Nguyen, Phuong Truc Lam Do, Anh My Le Ba, Nguyen Van Thuan, Udaykumar Kolkundkar, Suresh Kannan, Pawan Kumar Gupta, Swaroop Bhagwat, Sonali Rawat, Sujata Mohanty, Yashvi Sharma, Misba Majood, Manuela Cristina Russu, Neil H. Hugh Riordan, Luis Gerardo Jiménez Arias, Ramón Coronado, Kuldip Sidhu, Devashree Vakil, Riddhesh Doshi, Flynn Mckinnirey, Arnaud Martino Capuzzo, Riccardo Ossanna, Alessandro Negri, Andrea Sbarbati, Sheila Veronese, Anita Conti, Lindsey Alejandra Quintero Sierra, Federica Virla, Ponpandian Samuel, Shenbagamoorthy Sundarraj, D.N. P.Sudarmani, Fatemeh Sadeghian-Nodushan, Leila Dehghani, Amir Hossein Kheirkhah, Arsalan Jalili, Arman Saadati Partan, Habib Nikukar, Hossein Motedayyen, Reza ArefNezhad, Zorica A Becker-Kojic, José Manuel García-Verdugo, Vicente Herranz-Pérez, Anne-Kathrin Schott, Ivan Zipančić, Vicente Hernández-Rabaza, Shin Mukai

First published in London, United Kingdom, 2023 by IntechOpen
IntechOpen is the global imprint of INTECHOPEN LIMITED, registered in England and Wales, registration number: 11086078, 5 Princes Gate Court, London, SW7 2QJ, United Kingdom
Printed in Croatia

British Library Cataloguing-in-Publication Data
A catalogue record for this book is available from the British Library

Additional hard and PDF copies can be obtained from orders@intechopen.com

Possibilities and Limitations in Current Translational Stem Cell Research
Edited by Diana Kitala
p. cm.

This title is part of the Biochemistry Book Series, Volume 44
Topic: Cell and Molecular Biology
Series Editor: Miroslav Blumenberg
Topic Editor: Rosa María Martínez-Espinosa

Print ISBN 978-1-83768-032-0
Online ISBN 978-1-83768-033-7
eBook (PDF) ISBN 978-1-83768-034-4
ISSN 2632-0983

For EU product safety concerns:
IN TECH d.o.o., Prolaz Marije Krucifikse Kozulić 3, 51000 Rijeka, Croatia,
info@intechopen.com or visit our website at intechopen.com.

IntechOpen Book Series
Biochemistry
Volume 44

Aims and Scope of the Series

Biochemistry, the study of chemical transformations occurring within living organisms, impacts all of the life sciences, from molecular crystallography and genetics, to ecology, medicine and population biology. Biochemistry studies macromolecules - proteins, nucleic acids, carbohydrates and lipids –their building blocks, structures, functions and interactions. Much of biochemistry is devoted to enzymes, proteins that catalyze chemical reactions, enzyme structures, mechanisms of action and their roles within cells. Biochemistry also studies small signaling molecules, coenzymes, inhibitors, vitamins and hormones, which play roles in the life process. Biochemical experimentation, besides coopting the methods of classical chemistry, e.g., chromatography, adopted new techniques, e.g., X-ray diffraction, electron microscopy, NMR, radioisotopes, and developed sophisticated microbial genetic tools, e.g., auxotroph mutants and their revertants, fermentation, etc. More recently, biochemistry embraced the 'big data' omics systems. Initial biochemical studies have been exclusively analytic: dissecting, purifying and examining individual components of a biological system; in exemplary words of Efraim Racker, (1913 –1991) "Don't waste clean thinking on dirty enzymes." Today, however, biochemistry is becoming more agglomerative and comprehensive, setting out to integrate and describe fully a particular biological system. The 'big data' metabolomics can define the complement of small molecules, e.g., in a soil or biofilm sample; proteomics can distinguish all the proteins comprising e.g., serum; metagenomics can identify all the genes in a complex environment e.g., the bovine rumen.
This Biochemistry Series will address both the current research on biomolecules, and the emerging trends with great promise.

Meet the Series Editor

Miroslav Blumenberg, Ph.D., was born in Subotica and received his BSc in Belgrade, Yugoslavia. He completed his Ph.D. at MIT in Organic Chemistry; he followed up his Ph.D. with two postdoctoral study periods at Stanford University. Since 1983, he has been a faculty member of the RO Perelman Department of Dermatology, NYU School of Medicine, where he is codirector of a training grant in cutaneous biology. Dr. Blumenberg's research is focused on the epidermis, expression of keratin genes, transcription profiling, keratinocyte differentiation, inflammatory diseases and cancers, and most recently the effects of the microbiome on the skin. He has published more than 100 peer-reviewed research articles and graduated numerous Ph.D. and postdoctoral students.

Meet the Volume Editor

Diana Kitala graduated with degrees in Biotechnology and Biomedical Engineering and completed postgraduate studies in clinical research, biostatistics, laboratory diagnostics, LEAN manufacturing, and Six Sigma management. She also obtained an MBA in Healthcare. She previously worked at a tissue bank and advanced therapy medicinal product (ATMP) facility. She was also a university lecturer. Currently, Dr. Kitala works for a Medical Research Agency in Poland. In 2019, she was awarded an EWMA (European Wound Management Association) grant for her project, "Theory of constraints (TOC) and LEAN management for wider application of amniotic mesenchymal stem cells in group of patients with chronic wounds." In addition, she was nominated for the Golden Order for Merits of the Silesian Voivodeship.

Contents

Preface

The need for effective translation in the field of stem cells is demonstrated by the small number of products on the market. This means a constant need to carry out basic research and to manipulate the properties of stem cells in such a way as to maximize their therapeutic effect while maintaining an appropriate level of safety of therapy. Thus, there is a serious disproportion between the demand and pressure from the patient and medical community and the ability of the market to provide safe and effective stem cell-based products. Mesenchymal stem cells (MSCs) are currently the most used type of stem cells in clinical trials. MSCs, also known as mesenchymal stromal cells [1] or medical signal cells [2], are adult stem cells with regenerative properties such as self-renewal, secretion of trophic factors, and ability to differentiate into other mesodermal lines. MSCs can also migrate and colonize the damaged site and secrete a number of trophic factors, modulating the immune system, influencing the microenvironment around damaged tissues, and improving endogenous tissue repair, thus offering a wide perspective in cell therapies. Therefore, MSCs are widely used in clinical trials. However, to date, several constraints have been identified in the development of clinical trials in stem cells [3]. The lack of knowledge, which led to the premature termination of several studies, can be considered obvious. However, there are many smaller but equally important obstacles. First, the quality of the obtained stem cells is important, which is influenced by many factors, such as selected isolation and culture methods. According to current data, MSCs are heterogeneous fractions that may contain a different fraction of "regenerative" cells. In addition, individual variability (e.g., sex, age, medical history, etc.) plays an important role here. Therefore, when choosing cells for a clinical trial, all parameters should be considered. In the clinicaltrials.gov database, there are three types of common sources of MSCs used in clinical trials, with almost 500 studies on bone marrow mesenchymal cells, more than 400 studies on adipose-derived mesenchymal cells, and almost 400 studies on umbilical cord mesenchymal cells. In addition, there are studies on mesenchymal cells isolated from the amniotic membrane [4]. Although the topic of mesenchymal cells isolated from hair follicles is extremely popular, there is not a single study on their clinical use in the database. MSCs can be isolated from virtually any tissue, however, for clinical use, a source is needed that provides a sufficient amount that is not defined by top-down standards, which makes MSC research very heterogeneous in terms of the number of cells applied. More than 800 studies on MSCs have been registered as phase I studies, although these calculations are subject to error in the form of registration of some studies in the phase I/II scheme. However, the number of studies does not translate to the number of products available to patients. This illustrates the constant need for research on stem cells and their shifting and reprogramming properties, which we cover in this book.

This book also examines issues of preparation and preservation of MSCs in translational research. Chapters describe organoids, nanotechnology, and exosomes. Exosomes isolated from MSCs have biological functions like those of MSCs, thus leading to tissue regeneration by encapsulating and transferring active biomolecules such as peptides, proteins, and RNA to diseased cells/tissues [5]. However, the use of MSC exosomes in clinical trials is limited due to the lack of established cell culture conditions and optimal protocols. Organoids from MSCs provide the appropriate functions and functionalities of the original native organs, which is essential for therapeutic application and drug screening [6]. Chapters also cover such topics as female germline stem cells, which are of particular importance in the face of growing fertility problems [7].

In the section on clinical applications, chapters discuss novel indications for stem cells and examine ethical concerns in stem cell research. They discuss induced pluripotent stem cells and targeting cancer stem cells. The clinical introduction of stem cells in pelvic floor disorders is presented as an example of practical application of stem cell therapy in an area that cannot be covered by other therapies.

This edited volume covers a wide diversity of problems, solutions, and translational issues that may accelerate future stem cell research.

Diana Kitala
Medical Research Agency,
Warsaw, Poland

References

[1] Wu X, Jiang J, Gu Z, et al. Mesenchymal stromal cell therapies: Immunomodulatory properties and clinical progress. Stem Cell Research and Therapy. 2020;**11**:345. DOI: 10.1186/s13287-020-01855-9

[2] Hosseiniyan Khatibi SM, Kheyrolahzadeh K, Barzegari A, Rahbar Saadat Y, Zununi Vahed S. Medicinal signaling cells: A potential antimicrobial drug store. Journal of Cellular Physiology. Nov 2020;**235**(11):7731-7746. DOI: 10.1002/jcp.29728

[3] Musiał-Wysocka A, Kot M, Majka M. The Pros and Cons of Mesenchymal Stem Cell-Based Therapies. Cell Transplant. Jul 2019;**28**(7):801-812. DOI: 10.1177/0963689719837897

[4] https://clinicaltrials.gov/

[5] Nikfarjam S, Rezaie J, Zolbanin NM, et al. Mesenchymal stem cell derived-exosomes: A modern approach in translational medicine. Journal of Translational Medicine. 2020;**18**:449. DOI: 10.1186/s12967-020-02622-3

[6] Kim W, Gwon Y, Park S, Kim H, Kim J. Therapeutic strategies of three-dimensional stem cell spheroids and organoids for tissue repair and regeneration. Bioact Mater. 4 Apr 2022;**19**:50-74. DOI: 10.1016/j.bioactmat.2022.03.039

[7] Hong W, Wang B, Zhu Y, Wu J, Qiu L, Ling S, et al. Female germline stem cells: Aging and anti-aging. Journal of Ovarian Research. 4 Jul 2022;**15**(1):79. DOI: 10.1186/s13048-022-01011-2

Section 1

Stem Cells Properties Shifting and Reprogramming

Chapter 1

Membrane-to-Nucleus Signaling in Human Blood Progenitor Cells Reveals an Efficient GM-Free Reprogramming to Pluripotency

Zorica A Becker-Kojić, José Manuel García-Verdugo, Anne-Kathrin Schott, Vicente Herranz-Pérez, Ivan Zipančić and Vicente Hernández-Rabaza

Abstract

The generation of induced pluripotent stem cells (iPSCs) by forced expression of defined transcription factors has revolutionized regenerative medicine. These cells have similar features to embryonic stem cells (ESCs) regarding self-renewal and their ability to differentiate into any cell type in the body. In spite of many improvements, in using nonviral delivery reprogramming methods, there are still challenges to overcome regarding safety before patient-made iPSCs can be used in regular clinical practice. We have recently reported about a gene manipulation-free method of generating human pluripotent stem cells (PSCs), based on activation of the novel human GPI-linked glycoprotein ACA. The process of dedifferentiation of blood progenitor cells that leads to the generation of blood-derived pluripotent stem cells (BD-PSCs) is initiated upon cross-linking of this protein *via* activation of PLCγ/PI3K/Akt pathway. These cells are mortal, express pluripotent markers, and redifferentiate *in vitro* into cells of all three germ layers. The ultrastructural analysis of BD-PSCs, by means of electron microscopy, revealed them similar to human ESCs with large dense nucleolus and scarce cytoplasm. BD-PSCs are autologous stem cells and while nonteratogenic offer a new alternative that overcomes immunological, ethical, and safety concerns and opens up a new avenue in treating contemporarily intractable diseases and generally in human therapeutics.

Keywords: blood progenitor cells, membrane signaling, dedifferentiation, pluripotency, redifferentiation

1. Introduction

Human induced pluripotent stem cell (iPSC) technology has paved the way for new possibilities to investigate and potentially cure diseases. The iPSCs derived from patients can be used in at least two ways: regenerative medicine and drug discovery, for example, screening chemicals, natural compounds, and derivatives to identify drug candidates. This new technology promises to provide a powerful tool for modeling human pathology that allows for investigation and understanding of the underlying mechanisms and causes of various human diseases. Particularly, disease-specific iPSCs are of great potential for disease modeling and therapeutic benefits [1, 2].

iPSCs have characteristics of human embryonic stem cells (ESCs), including pluripotency and potentially unlimited self-renewal. During the last decade, patient-made iPSCs have been differentiated into a variety of functional cell types *in vitro* and are expected to reconstruct disease phenotypes, as already demonstrated in several animal disease models [3].

Originally, iPSCs were generated by retroviral transduction of four specific transcription factors, Oct3/4, Sox2, Klf4, and cMyc or Oct3/4 Sox2/Nanog/LIN, using retroviral or lentiviral vectors [4, 5]. Later, lentivirus was the preferred delivery method, since, unlike retrovirus can infect proliferating and nondividing cells. Viral vectors for iPSC generation are very effective for integrating exogenous genes into the genome of somatic cells; however, they could be permanently integrated into the cell's genome, which generates serious concerns about changes in cell behavior and therefore, limiting their use in patients [6].

Despite the fact that iPSCs, as well as ESCs, are being proclaimed to have a great advantage as a source of stem cells that can be used in regenerative medicine, the ultimate goal to use them in clinical practice has not been achieved.

Cells convert one kind of signal into another through a process called signal transduction. This mechanism comprises the coupling of a ligand-receptor interaction to many intracellular events. These events include phosphorylation by tyrosine kinases and/or serine/threonine kinases. Differential localization of protein that participates in signaling pathways is essential for cells to respond efficiently to changes in their environment. In the plasma membrane, such compartmentalization is performed through lipid rafts [7] that are enriched in cholesterol, sphingolipids, and GPI-anchored proteins. During the signaling processes, various lipids are phosphorylated, recruited, and activated by different signaling components, which are essential for the regulation of cell survival and growth.

PI3Ks are a family of intracellular signal transducer enzymes involved in a variety of cellular functions like growth proliferation and differentiation. These enzymes are capable of phosphorylating position 3 of the inositol ring of phosphatidylinositol (PtdIns), and this lipid modification initiates a set of events that leads to cell activation and growth [8].

We have recently shown that human GPI-linked glycoprotein ACA, expressed in all adult stem cells, including hESCs, is involved in developmental process, which shapes the human embryo and controls adult stem cell compartments. Activation by this protein on the membrane of human blood progenitor cells drives membrane-to-nucleus signaling pathways, thus regulating pluripotency [9].

We investigate here the signaling machinery behind the antibody cross-linking activation of GPI-linked membrane glycoprotein ACA that drives the immature blood progenitor cells to pluripotency and the capacity of these cells to redifferentiate into cells belonging to different germ layers.

The newly generated human blood-derived (BD-PSCs) as well as their redifferentiated progeny was assessed by the methods of immunocytochemistry (ICC), flow cytometry, and electron microscopy (EM). Signaling competence of ACA receptor was analyzed by studying the phosphorylation pattern and real-time analysis of developmentally relevant genes, such as *NOTCH, WNT, CTNNB, C-KIT,* and others.

2. Materials and methods

2.1 Cell cultures

Human mononuclear cells (MNCs) were isolated from peripheral blood (PB) samples obtained from healthy donors. All of the human blood samples were used after obtaining written informed consent from the donors. PBMNCs were isolated by density gradient centrifugation using Ficoll and activated by specific antibody cross-linking as described elsewhere. Briefly, 6×10^6 MNCs in a 15 ml polystyrol tube were incubated for 30 min with antibody (30 μg/mL in 1% PBS/BSA) to GPI-linked membrane protein ACA and further cultured and maintained in suspension in Iscove's modified Dulbecco's medium (IMDM) supplemented with 10% FBS (Gibco Life Technology, Grand Island, NY), 0.1 mM nonessential amino acids (NEAA), 1 mM L-glutamine (all from Invitrogen, Carlsbad CA). The cells were taken at different time points for immunophenotyping (IP) by flow cytometry, ICC, and western blot (WB) analysis.

For growing of human ESCs line H9 (Wi Cell Inc., Medison, Wi) on feeder cells, mouse embryonic fibroblasts (MEFs), obtained from mouse strain CF-1 (American Type Culture Collection Manassas, VA, USA) mitotically inactivated by radiation were prepared according to standard protocol and approved (07-WO25) by the ethics committee established at Príncipe Felipe Centro de Investigación (CIPF) in Valencia (Spain), where these experiments were conducted. Human ES cells were placed on a freshly prepared MEF layer and further cultivated in ESC medium consisting of Knockout-DMEM (Invitrogen, Carlsbad CA), 100 μM β-mercaptoethanol (Sigma, St Lois), 1 mM L-glutamine (Invitrogen), 1% penicillin/streptomycin (Invitrogen) and 8 ng/ml bFGF (Invitrogen), or in condition media (see below). The medium was changed every other day. Recovery of pluripotent phenotype of the differentiated H9 cells line was done by cross-linking the membrane of these cells with a GPI protein-specific antibody. Cells were maintained in IMDM supplemented with 10% FBS.

The culture of BD-PSCs on feeder-free culture dishes was performed to assess the expression of hESC markers by means of immunofluorescence. For that purpose, BD-PSCs were grown in MEF-conditioned media supplemented with 8 ng/ml bFGF on culture dishes coated with Matrigel 1:30 (BD Bioscience, Franklin Lakes, NJ).

MNCs isolated from PB were cultured in Iscove's media supplemented with 10% FBS, preincubated with the inhibitors Et-18-CH_3 (ET) at 50 μM, LY 294002 (LY) at 20 μM or ET + LY purchased from Calbiochem (USA) for 1 h, and after activation in the presence of inhibitors cultured at 37°C, as specified. After labeling with CD34APC, CD45FITC (BD Pharmingen), and CD14PE (eBioscience), the generation of CD34 cells was daily assessed by multiple flow cytometry analyses. The nonviable cells were excluded after performing a propidium iodide (PI) assay. Conjugated isotype-matched irrelevant antibodies were used as controls.

Antibodies to SSEA-4 and TRA-1-81 purchased by Chemicon (Temecula, CA) labeled with phycoerythrin (PE) were used for phenotypic analysis of pluripotent markers expressed on BD-stem cells by means of flow cytometry. Gating was done with matched isotype control monoclonal antibodies. Conjugated isotype-matched irrelevant mAbs were used as controls.

2.2 BD-PSCs differentiation toward neuronal and hepatic cells

Differentiation to neuronal lineages was adapted from previously described protocol [10]. Briefly, BD-PSCs were grown on glass coverslips coated with 1:5 diluted poly-L-ornithin/laminin for 2 days in neuronal initiating medium N2, followed by neuronal differentiation medium (Neurobasal medium, L-glutamin, B27 supplement, nonessential amino acids (NEAA), recombinant human glial-derived neurotrophic factor (GDNF), recombinant human brain-derived neurotrophic factor (BDNF), and ascorbic acid solution. The cells were grown at 37°C, 5% CO_2 for 1–30 days. Cells were taken at D8, D20, and D30 for ICC analysis.

Differentiation of BD-PSCs toward endoderm/hepatocytes was performed on Biolaminin 111 (Biolamina Sundbyberg, Sweden) treated glass coverslips in KSR/DMSO media consisting of 80% of knockout DMEM media (KO-DMEM), 20% knockout serum replacement, 0.5% NEAA (all from Invitrogen), 0.1% β-mercaptoethanol, 1% DMSO (Sigma), and 1% penicillin–streptomycin for 7 days, followed by culturing cells in HepatoZYME maturation medium, 1% glutamax, 10 μM hydrocortisone 21-hemisuccinate sodium salt (HCC) (Biomol, Hamburg, Germany), supplemented with 10 ng/mL hepatocyte growth factor (HGF) (Life Technology) and 2 ng oncostatin M (OSM) (Biotechne), for additional 2 weeks. The cells were taken at D7 and D21 for ICC analysis.

2.3 Cell culture, inhibition, and western blot analysis

MNCs isolated from PB were cultured in Iscove's media supplemented with 10% FBS, preincubated with the inhibitors ET at 50 μM, LY at 20 μM, and PD098059 (PD) at 10 μM, or (ET + LY + PD) purchased from Calbiochem (USA) for 1 h, and after activation in the presence of inhibitors cultured at 37°C, 5% CO_2. Cell-free extracts of these cells were subjected to western blot analysis. Nonactivated MNCs were used as controls.

2.4 Western immunoblotting

Cell lysis, protein extraction, and western blot analysis of ACA-activated PBMNCs vs. nontreated samples were performed as described elsewhere [9, 11]. Briefly, cells were lysed in Triplex buffer (50 mM Tris HCl pH 8, 120 mM NaCl, 0.1% SDS, 1% Nonidet P-40, and 0.54% deoxycholate), 300 μg of protein extracts were submitted to electrophoresis by using 10% SDS-PAGE. Immunodetection was performed by using appropriate primary antibodies followed by incubation with HPR-conjugated secondary antibodies (all purchased by cell Signaling technology, GAPDH antibody by Santa Cruz biotechnology). ECL Western blotting substrate (Pierce) was used for the detection of proteins on PVDF.

2.5 Quantitative PCR analysis

Total RNA was extracted from cells using TRIzol (Invitrogen) and transcribed into cDNA using oligo (dT) 16 and ReverTra Ace reverse transcriptase. PCR reactions were carried out by mixing 1 μL of cDNA template, 250 nM of each primer, 200 μM dNTP mixture, and 1 U of Taq DNA polymerase in a total volume of 20 μL. Samples were amplified in a thermocycler. For qPCR, each sample was analyzed in triplicate with GADH as the internal control. Amplification data were collected using ABI PRISM 7900 and analyzed using the sequence detection system 2.0 software. The primers and TaqMan probes used in this experiment are presented in **Table 1**.

Gene	Assay identification	Exon boundary	Assay location	NCBI reference sequence	Amplicon length
hGAPDH	Hs99999905_m1	3–3	157	NM_002046.3	122
hc-KIT	Hs00174029_m1	1–2	158	NM_001093772.1	64
hHOXB4	Hs00256884_m1	1–2	526	NM_024015.4	54
hCTNNB1	Hs00170025_m1	7–8	1351	NM_001098210.1	88
hWNT10B	Hs00559664_m1	3–4	684	NM_003394.2	63
hNOTCH	Hs00413187_m1	4–5	745	NM_017617.3	95
hBCL2	Hs00153350_m1	2–3	1079	NM_000633.2	96
hBMI	Hs00180411_m1	3–4	719	NM_005180.5	105
hTGFβ	Hs99999918_m1	4–5	1583	NM_000660.3	125

Table 1.
*List of primers and TaqMan probes used in **Figure 1d**.*

2.6 Immunocytochemistry (ICC)

Immunofluorescence analysis of cells growing on Matrigel (Corning CA) was first performed to evaluate the presence of pluripotent stem cell markers, such as SSEA-4, TRA-1-60, and TRA-1-81. All antibodies were purchased by chemicon. Secondary PE or Alexa Fluor 488 labeled antibody (Life Technology, Carlsbad CA) was used to reveal the expression of pluripotent markers on BD-PSCs.

Activated PBMNCs growing in suspension were taken at different time points during culture time period from D1—D14, transferred on glass coverslips, coated with Poly-L-lysine (Sigma-Aldrich St. Louis, USA), and the acquisition of pluripotent markers, TRA-1-60, SOX2, NANOG, and OCT3/4 monitored by means of ICC. Cells were fixed in 4% paraformaldehyde (PFA) for 15 min at RT followed by permeabilization with 0.1% Triton X-100 in PBS for 30 min in phosphate buffer saline (PBS) containing 3% BSA. All directly labeled primary antibodies were diluted in the same blocking buffer and incubated with samples overnight at 4°C. The nuclei were stained with DAPI (Sigma-Aldrich) for 3 min at RT. All images were acquired with an inverted Olympus IX71 Microscope. All antibodies used in this experiment are presented in **Table 2**.

ICC of human neuronal cells generated from BD-PSCs was performed using antibodies to Nestin, GFAP, MAP2, Neun, and Tuj1.

Cells were fixed with prewarmed fixative (PBS, PFA, $MgCL_2$, EGTA, and sucrose) for 15 min, then treated with 0.3% Triton X-100 in PBS containing 3% BSA for 5 min as previously described [10]. Appropriate dilution of antibodies was prepared in PBS containing 1% BSA and incubated for 1.5 h, at RT, washed three times with PBS, and anti-chicken, anti-rabbit, and anti-mouse fluorochrome-conjugated antibodies were used to reveal the expression of specific neuronal markers (antibodies used in this experiment are presented in **Tables 3** and **4**). DAPI was used for staining the nuclei of cells. All images were acquired with an inverted Olympus IX71 microscope.

BD-PSC Differentiation to human endoderm/hepatocytes was assessed by means of ICC using antibodies to alpha-fetoprotein (AFP), anti-transthyretin (TTR) (endoderm), and anti-Albumin (ALB), anti-Hepatocyte Nuclear Factor 4 alpha (HNF4 α) (hepatocytes) and their relevant fluorescent-labeled secondary antibodies (all antibodies used are presented in **Tables 5** and **6**). Cells were fixed with 4% PFA for

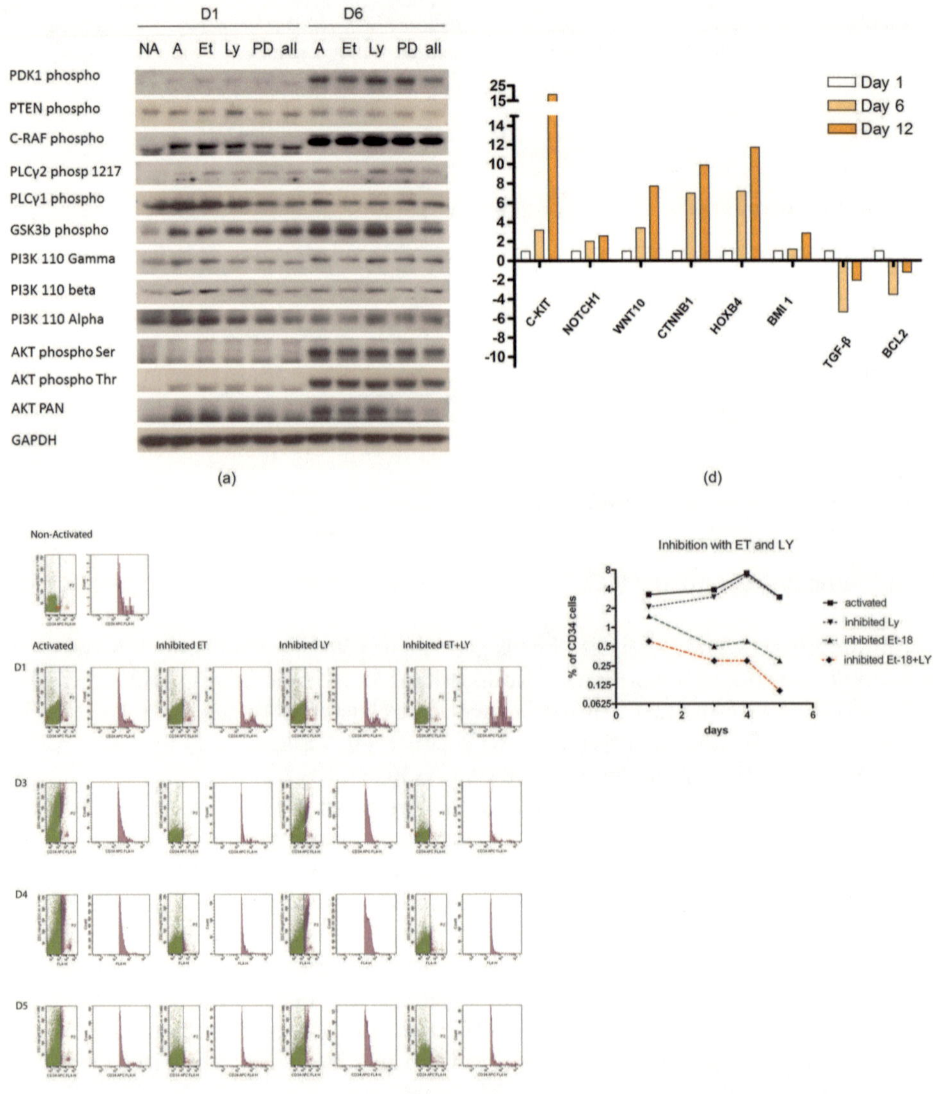

Figure 1.
Relative protein expression and phosphorylation status of mediators of ACA-signaling and its inhibition with pharmacological inhibitors. The PBMNCs activated cells were cultured either in presence or absence of inhibitors ET, LY, PD, or ET + LY. Relative protein expression and phosphorylation status of mediators of this signaling were determined by western blot analysis. Nonactivated mononuclear cells isolated from PB were used as controls. (a): Western blotting of cell lysate derived from activated PBMNCs expression and phosphorylation status of proteins involved in the signaling mechanisms. (b): The PB-generated progenitor/HSCs were cultured either in presence or absence of inhibitors ET, LY, PD, or ET + LY. Nonactivated MNCs isolated from PB were used as controls. ET, LY, and ET + LY inhibited membrane protein antibody cross-linking induced generation of CD34 cells. Flow cytometry shows the generation of CD34 cells in the presence or absence of ET, LY, or ET + LY respectively in a 1–5 days' culture time period. (c): The total number of viable CD34 cells in cultures stimulated with or without ET, LY or ET + LY was estimated by flow cytometry. Data are presented as the mean ± SEM for three independent experiments. (d): Regulation of gene expression after GPI-anchored protein-induced generation of BD-PSCs. Living cells were isolated before (D1) and after activation at D6 and D12. The total RNA of these cells was reverse transcribed, and the expression of the following genes was studied by real-time analysis: CTNNB1 HOXB4, C-KIT, NOTCH1, WNT10, BMI1, TGFB1, and BCL2. Represented is the average of triplicate gene-expression changes measured by TaqMan as described in Methods.

antibody	Clone	Isotype	Fluorochrome
anti-human Nanog	N31–355	Mouse IgG1, ƙ	Alexa Fluor® 488
anti-Oct3/4	40/oct-3	Mouse IgG1, ƙ	Alexa Fluor® 488
anti-Sox2	245,610	Mouse IgG2a	Alexa Fluor® 647
anti-human TRA-1-60 antigen	TRA-1-60	Mouse (BALB/c) IgM, ƙ	Alexa Fluor® 488

Table 2.
Directly labeled antibodies from BD Pharmingen (California, US) are used in ***Figure 2****.*

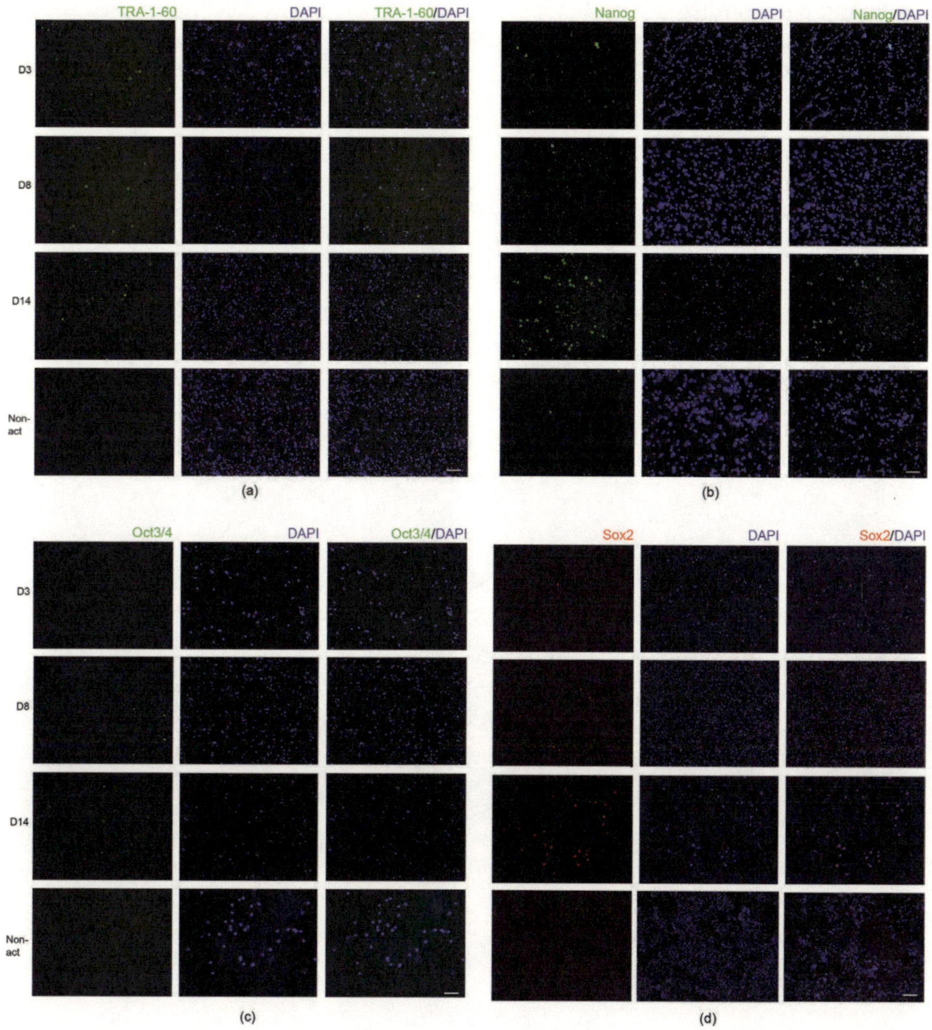

Figure 2.
Inducing pluripotency in BD-progenitor/stem cells throughout cell culture time. (a-d): PBMNCs were activated by antibody-specific cross-linking and cultured in Iscove's medium for 14 days. Aliquots were taken at D3, 8, and 14, incubated with fluorescent-labeled antibodies to pluripotent markers, and subjected to immunofluorescence microscopy. Immunofluorescence microscopy of cell cultures for 14 days showing the expression in vitro *of the pluripotency markers, (a) TRA-1-60, (b) Nanog, (c) Oct3/4, and (d) Sox2, throughout the cell culture time. Nuclei were stained with DAPI. Scale bars 50 μm.*

Primary Antibodies	Clone	Isotype	Company
anti-Nestin	10C2	Mouse, IgG1, monoclonal	Antibodies online
anti-GFAP		Rabbit, polyclonal	Dako
anti-MAP2	Ap20	Mouse, IgG1, monoclonal	BD Pharmingen
anti-NeuN	1B7	Mouse, IgG2a, monoclonal	Antibodies online
anti-Tuj1		Chicken, polyclonal	Antibodies online

Table 3.
*Primary antibodies used in **Figure 3**.*

10 min and permeabilized with 0.1% Triton-X-100 for 3 min. DAPI was used for nuclear staining. Expressions of these markers were visualized with an inverted Olympus Microscope CKX53.

2.7 Electron microscopy

Cells were seeded at 6.25×10^5cells/cm^2 in 8-well Lab-Tek chamber slides (Nalgene Nunc International, Naperville, IL). Cells were fixed in 3.5% glutaraldehyde for 1 h at 37°C, postfixed in 2% OsO_4 for an additional 1 h at RT, and stained in 2% uranyl acetate in the dark at 4°C for 2 h. Finally, cells were rinsed with distilled water, dehydrated in ethanol, and embedded in Durcupan (Fluka) epoxy resin overnight. Following resin hardening, embedded cultures were detached from the chamber slide and glued to araldite blocks. Serial semi-thin sections (1.5 μm) were cut with an

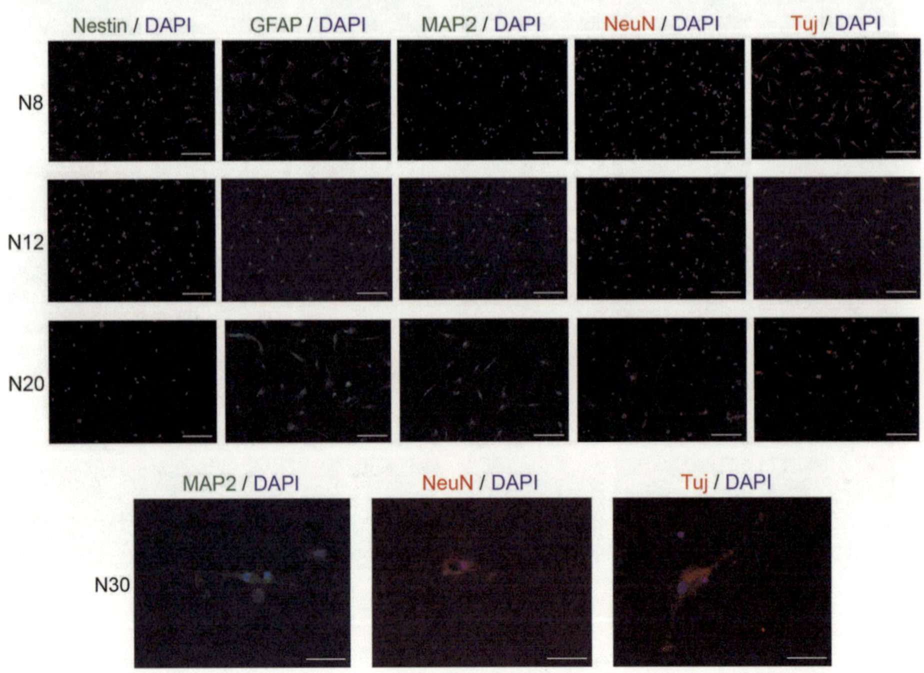

Figure 3.
Ectodermal/neural differentiation of BD-PSCs in adherent monolayer from N8-N30. Immunocytochemical and immunophenotypic profiling of BD-neuronal cells throughout the culture time period from N8 to N30. Scale bars 100 μm (N8-N20) and 50 μm (N30).

Secondary Antibodies	Isotype	Fluorochrome	Company
Goat anti-Mouse	IgG, polyclonal	DyLight 488	Antibodies online
Goat anti-Mouse	IgG, polyclonal	Alexa Fluor 647	Life Technologies
Goat anti-Rabbit	IgG, polyclonal	Alexa Fluor 488	Invitrogen
Rabbit anti-Chicken	IgG, polyclonal	Texas Red	Antibodies online

Table 4.
Secondary antibodies used in ***Figure 3****.*

Ultracut UC-6 (Leica, Heidelberg, Germany), mounted onto glass slides, and lightly stained with 1% toluidine blue. Selected semi-thin sections were glued with Super-Glue-3 Loctite (Henkel, Düsseldorf, Germany) to resin blocks and detached from the glass-slide by repeated freezing (in liquid nitrogen) and thawing. Ultrathin sections (60–80 nm) were prepared with an ultramicrotome and contrasted with lead citrate. Finally, photomicrographs were obtained at 80 kV using an FEI Tecnai G^2 Spirit transmission electron microscope (FEI Europe, Eindhoven, and Netherlands) equipped with a Morada CCD digital camera (Olympus Soft Image Solutions GmbH, Münster, Germany).

3. Results

3.1 Membrane-to-nucleus signaling induced by ACA activation

To investigate a phosphorylation pattern across the plasma membrane, we activated GPI-linked protein by means of ACA-antibody cross-linking. The cell-free extracts were prepared from peripheral blood (PB) cells after Ficoll gradient centrifugation before and 6 days after activation. The evaluation of the expression and phosphorylation status of the proteins, presumably involved in this signaling network, was assessed by western blot analysis.

As shown in **Figure 1(a)**, antibody cross-linking induces PI3K activation that phosphorylates and activates the known members of this pathway described in [12].

The phosphoinositol-phospholipase γ (PLCγ) is a member of family of PLC enzymes consisting of various isoforms with different cellular functions. PLCγ is linked to tyrosine kinase signaling pathways with its primary function to catalyze the hydrolysis of phosphatidylinositol-4,5-bisphosphat (PIP2) to generate inositol (1,4,5)-triphosphate (PIP3) and 1,2-diacylglycerol (DAG). PIP3 initiates an increase in intracellular, whereas DAG activates protein kinase C, and the control over this important

Primary Antibodies	Clone	Isotype	Company
anti-human alpha-1-Fetoprotein		Rabbit, polyclonal	DakoCytomation
anti-Transthyretin		Chicken, polyclonal	Sigma-Aldrich
anti-Albumin		Chicken, polyclonal	Sigma-Aldrich
anti-HNF-4alpha	4C19	ZooMab Rabbit, monoclonal	Merck

Table 5.
Primary antibodies used in ***Figure 4****.*

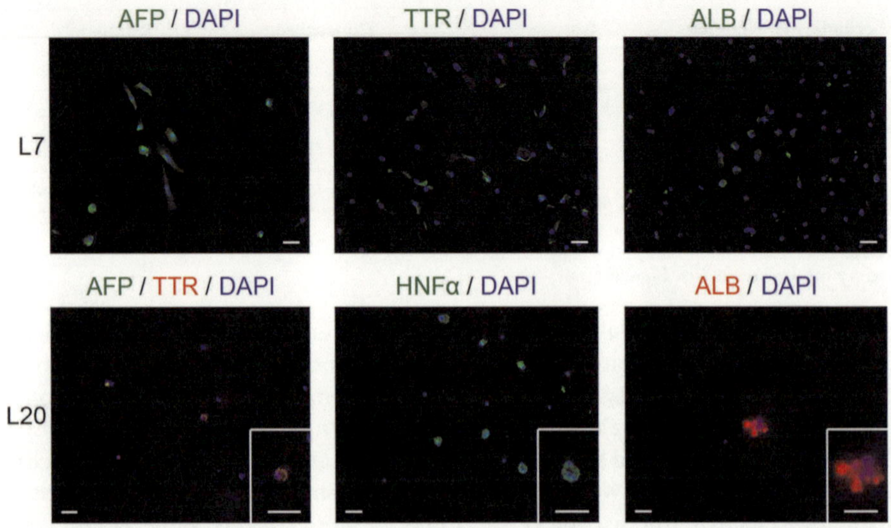

Figure 4.
Endodermal/hepatocytes differentiation of BD-PSC in culture time period from L7 to L20. Shown is immunocytochemical and immunophenotypic analysis of endoderm/hepatocyte-specific markers through culture time period from L7 to L20. Scale bars 50 and 20 μm (L20 ALB/DAPI).

second messenger pathway is a key to changes in cellular activity and function [13]. Phosphorylation of PLCγ and its activation seemed to be the earliest event in ACA-initiated specific signaling network, leaving the hydrolytic products of this protein, like inositol phosphate and a diacyl-myristate, the latter known as the most powerful activator of PI3K.

To evaluate the function and role of PLC in the signaling pathway, we used the specific inhibitors of PLCγ, ET-18-0-CH_3 (ET), as well as the inhibitors of various kinases involved in this signaling like LY 294002 (LY) inhibitor of PI3K, and the MAP kinase inhibitor, PD098059 (PD). Peripheral blood mononuclear cells (PBMNCs) were preincubated with these inhibitors and cultured for 1–6 days after activation. Relative protein expression and phosphorylation status of various mediators of the initiated signaling were assessed by western blot analysis. Nonactivated PBMNCs were used as controls. The inhibitor of PLCγ caused a partial suppression of expression as well as reduced phosphorylation extent of various kinases involved in this signaling, whereas specific inhibitors of PI3K or MAP kinase alone had no effect on the initialization of the signaling (data not shown). Conversely, preincubation with all three inhibitors caused significant suppression of protein expression and phosphorylation extent of all participants (**Figure 1(a)**).

Secondary Antibodies	Isotype	Fluorochrome	Company
Goat-anti-Chicken	IgY(H + L),polyclonal	CF 488A	Sigma Aldrich
Rabbit anti-Chicken	IgG, polyclonal	Texas Red	Antibodies online
Goat anti-Rabbit	IgG, polyclonal	Alexa Fluor 488	Invitrogen

Table 6.
*Secondary antibodies used in **Figure 4**.*

We further assessed the role of PLCγ and PI3K in membrane-to-nucleus-induced signaling, which first leads to generation of hematopoietic progenitor cells. This is an inevitable intermediate step on the described route to pluripotency [11]. Cells expressing CD34 protein are normally found in umbilical cord blood as well as in bone marrow cells, and antibodies to this protein are often used clinically to quantify the number of HSCs used in HSC transplantations [14]. $CD34^+$ cells were generated in the presence or absence of their specific inhibitors, such as ET, LY, and ET + LY. The culture conditions were used to follow the effect of inhibitors on generation of $CD34^+$ cells for the culture time period of 5 days.

PBMNCs were preincubated with ET, LY, and ET + LY and upon activation; newly generated cells were assessed by multiple flow cytometry analyses using antibodies to CD34, CD45, and CD14. The growing population of $CD34^+$ during culture time, from D1–D5, was monitored. Upon inhibition with ET, we observed a slight decrease in the number of newly generated $CD34^+$ cells, whereas LY alone had no effect on induced hematopoiesis. Most significantly, a dramatic decrease in fluorescence intensity occurred when both inhibitors were used together. Inhibition of PLCγ alone as well as inhibition of PLCγ and PI3K with their specific inhibitors ET + LY induced significant changes regarding *de novo* generation of $CD34^+$ cells, whereas no changes were observed when IP3K was inhibited by using LY alone see **Figure 1(b)**.

The same experiment was performed with PBMNCs from various donors confirming the previous findings (**Figure 1(c)**).

Herewith, we conclude that phosphorylation and activation of PLCγ are indispensable for initiation of the signaling cascade. The changes induced by ET during the generation of $CD34^+$ cells are highly likely to be responsible for the initial events started by GPI-linked protein stimulation. This implies that phosphorylation and activation of PLCγ is a crucial event in this mediated signaling, while PI3K and AKT represent downstream activators, essential for induced route to pluripotency *via* generation of human hematopoietic progenitor cells.

3.2 Antibody activation upregulates the expression of developmentally relevant genes

In order to study these specific signal transduction networks, we used quantitative RT-PCR (TaqMan) analysis to determine the gene-expression pattern of the molecules potentially involved in this signaling.

We activated the GPI-linked protein at the surface of blood progenitor cells as described above and analyzed the way in which the initiated signaling machinery regulates the expression of genes known to play a role in human embryonic development *via* specific protein phosphorylation as an important regulatory mechanism in cellular processes.

We compared the gene expression profile of the PBMNCs before and after activation to assess a transcript level for candidate molecules, the most important among them NOTCH and WNT/CTNNB1 (**Figure 1(d)**). The signaling pathways linked to these genes are developmentally conserved and play a significant role in embryonic development as well as in the regulation of adult cell compartments [15].

Dysregulation of Wnt and Notch pathways due to their involvement in the key functions of human cells is a reason for their implication in many human diseases [16].

As shown in **Figure 1(d)**, we demonstrated that GPI-linked glycoprotein upregulates both Notch and Wnt signaling pathways, thus acting in a hierarchical manner in the relationship to both signaling pathways. In consistency with the previous report about β-catenin as a downstream regulator of the canonical Wnt pathway [17], our results clearly show the upregulation of β-catenin as a consequence of this activation.

The involvement and significance of this GPI-linked protein in the signaling process regarding development and dedifferentiation of the human (ES) cell line H9 is shown in **Figure 5**, in which a spontaneously differentiated colony of ESCs was restored (dedifferentiated) to its primordial state, upon specific antibody cross-linking and culturing for 1 day in Iscove's medium supplemented with 10% FBS.

3.3 BD-PSCs are generated from unmanipulated steady-state PB

A PB sample (30 mL) of healthy donors was collected after obtaining informed consent. Mononuclear cells (MNCs) were isolated after Ficoll centrifugation and activated at the membrane by antibody cross-linking using specific antibodies. **Figure 6** shows the steady growing new population of cells in time course modus from day 5 to 14, while the nonactivated PB cells, under identical culture conditions, show gradual deterioration of cell structure and function, which leads to disappearance and death of the majority of the cells from day 5 to 14 of culture time period.

3.4 Expression of ESC markers on BD-PSCs

As shown, membrane-to-nucleus signaling network initiates *via* PLCγ/PI3K/Akt/mTor/PTEN a process of de-differentiation of blood cells that leads *via* generation of HSCs to PSCs. By means of immunofluorescence, we analyzed the expression of pluripotent markers that BD-PSCs share with ESCs.

Human PSCs are characterized by specific cell surface markers, such as the glycolipid antigens SSEA-3 and SSEA-4, as well as the glycoprotein antigens TRA-1-60 and TRA-1-81. Stage-specific embryonic antigen (SSEA-4) is a glycosphingolipid expressed in early human embryonic development and PSCs and acts as a mediator of cell adhesion as well as a modulator of signal transduction. The expression of human SSEA-4 decreases following differentiation of ESCs [18]. Glycoprotein antigens TRA-1-60 and TRA-1-81 are expressed in early human embryonic development and PSCs; they also mark cells of the inner cell mass of preimplantation embryos [19].

Activated PBMNCs were grown in MEF-conditioned media in Matrigel-coated cell culture dishes for 2 weeks and immunocytochemistry (ICC) analysis was performed using antibodies to SSEA-4, TRA-1-60, and TRA-1-81. As shown in **Figure 7(a–e)**, cell surface protein marker analysis demonstrated that BD-PSCs express the pluripotent markers SSEA-4 as well as TRA-1-60 and TRA-1-81.

Unlike ESCs and PSCs, BD-PSCs can grow in suspension in Iscove's medium supplemented with 10% FBS without any addition of cytokines or growth factors. Flow cytometry expression analysis of BD-PSCs grown in suspension revealed the expression of ESCs specific markers SSEA-4 and TRA-1-81 on BD-PSCs (**Figure 7(f–m)**).

3.5 Expression of the factors that maintain pluripotency

To further investigate the expression of pluripotency markers on BD-PSCs at the protein level, we extended the immunofluorescence analysis to the transcription

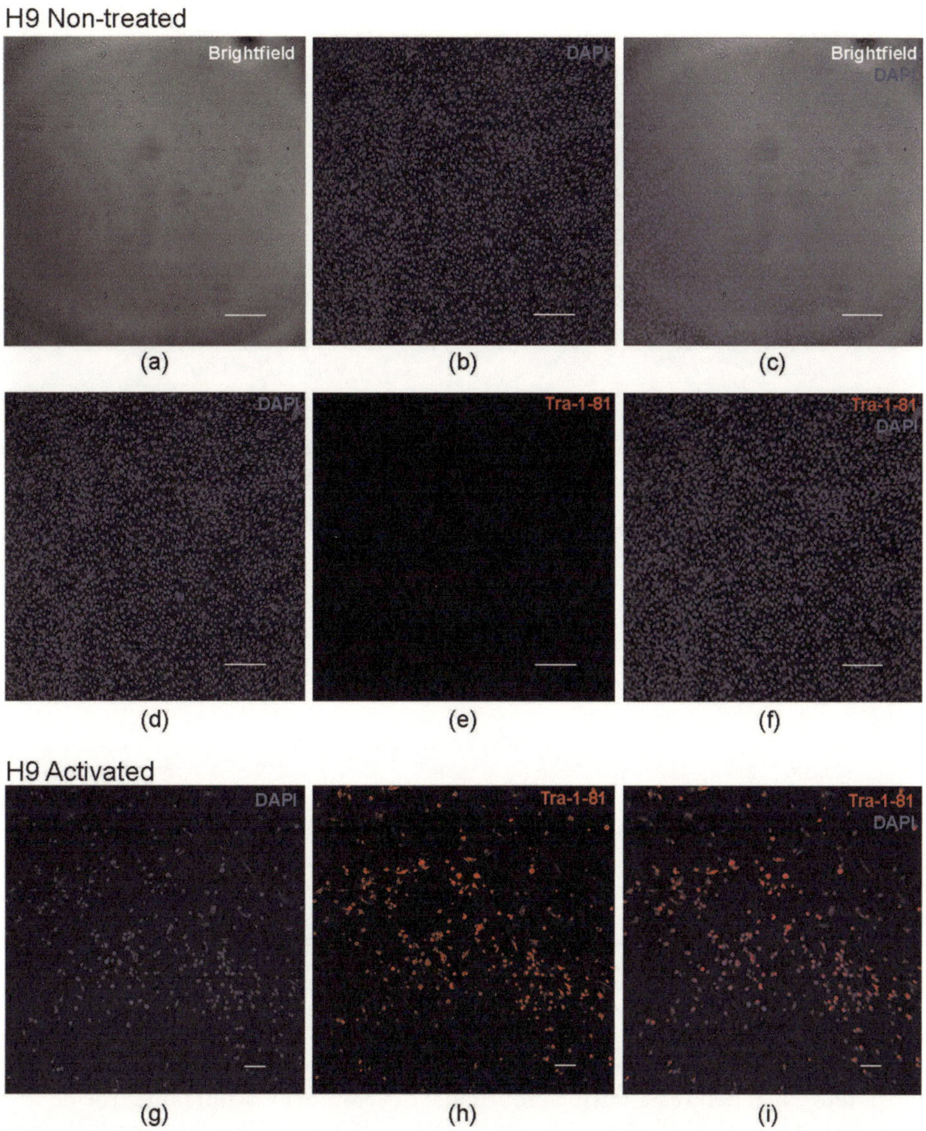

Figure 5.
Immunofluorescence analysis during redifferentiation process of spontaneously differentiated human ESCs initiated after membrane antibody cross-linking activation of human GPI-linked protein. A spontaneously differentiated ESCs colony was activated by antibody cross-linking incubated with TRA-1-81 and DAPI and micrographs were taken by confocal microscopy. (a): Brightfield micrograph of spontaneously differentiated ESC colony. (b): Nuclei stained with DAPI. (c): Brightfield/DAPI (merged). (d): Spontaneously differentiated ESC colony stained with DAPI. (e): Spontaneously differentiated ESC colony stained with antibody to TRA-1-81. (f): Spontaneously differentiated ESC colony stained with TRA-1-81/DAPI (merged). Scale bars 200 μm. Fluorescence images of restored (activated) pluripotency in spontaneously differentiated ESC colony. (g): Activated ESC colony stained with DAPI. (h): Activated ESC colony stained with antibody to TRA-1-81. (i): Activated ESC colony stained with TRA-1-81/DAPI (merged). Scale bars 100 μm.

factors Nanog, Sox2, and Oct3/4 that reside at the core of pluripotency network, where they can regulate their own expression and interact with a number of other pluripotency factors. Nanog, also called a pluripotency master molecule, is a

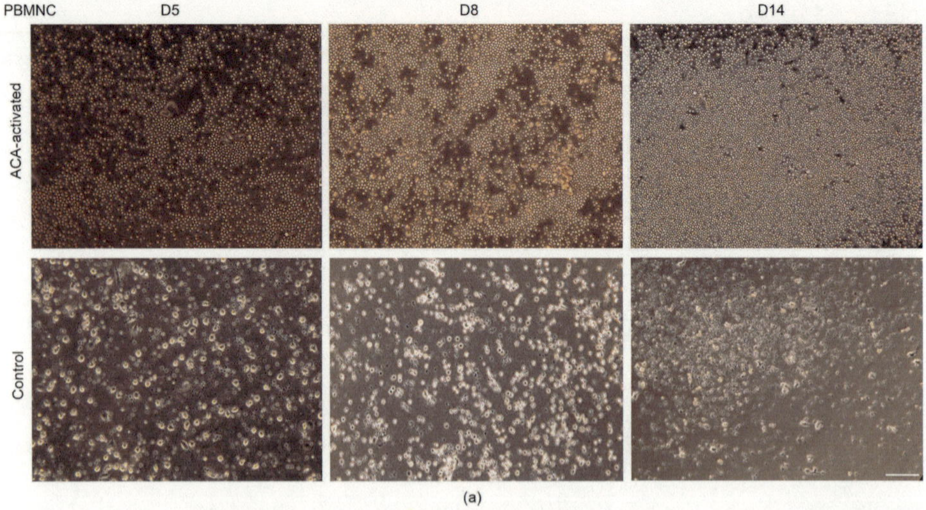

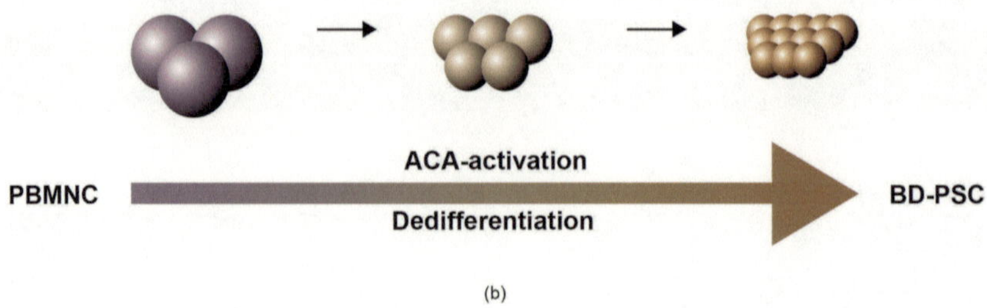

Figure 6.
Reprogramming of PBMNCs after activation. PBMNCs isolated by Ficoll gradient centrifugation were activated by specific antibody cross-linking and cultured in Iscove's medium supplemented with 10% FBS. The brightfield images were taken at D5, 8, and 14. Nonactivated PBMNCs were used as controls. Scale bars 100 μm.

unique homeobox transcription factor that is critical in regulating the cell fate of the pluripotent inner cell mass during embryonic development maintaining the pluripotency and blocking the differentiation of PSCs [20]. Activated preparations of PBMNCs were grown in suspension in a time-course manner from day 3 to 14. The newly generated cells were plated on Poly-L-lysine coated glass coverslips and ICC was performed using appropriate antibodies for specific pluripotency markers on BD-PSCs.

Expression of TRA-1-60, Sox2, Nanog, and Oct3/4 was induced following day 3, showing a rising trend from day 8 and had been completed at day 14.

Immunofluorescence analysis of the newly generated cells at different time points showed the gradual enhancement of the expression of the pluripotency markers on activated PBMNCs cultures. In contrast, nonactivated PBMNCs cultures showed no expression of these markers. These data are presented in **Figure 2(a–d)**.

Karyotype analysis was conducted on BD-PSCs by a qualified service provider (CELL Line Service Heidelberg, Germany) using standard G-banding methods as described elsewhere [21]. Analyses showed that BD-PSCs maintained a normal karyotype (data not shown).

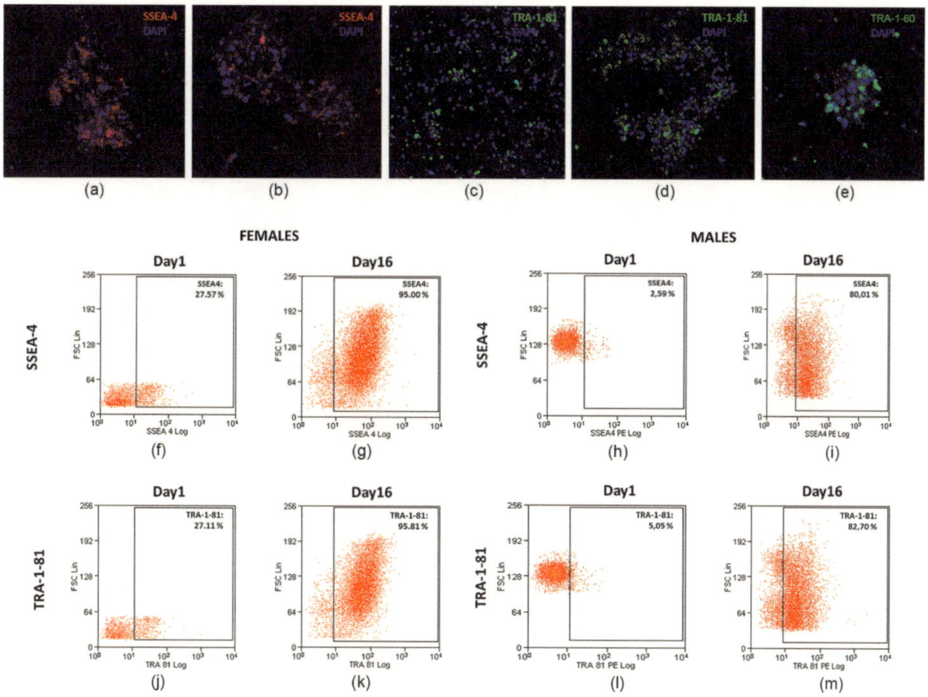

Figure 7.
BD-PSCs express PSC markers. (a–e): Activated PBMNCs were grown in MEF-conditioned media on Matrigel-coated culture dishes. Immunofluorescence analysis was performed with characteristic pluripotent markers, including SSEA-4, TRA-1-81, and TRA-1-60. Immunofluorescence images were taken by confocal microscopy. DAPI was used to stain nuclei. Generation and culture of BD-PSCs in suspension. (f-m): Activated PBMNCs cultures were grown in suspension in Iscove's medium supplemented with 10% FBS for 16 days, and flow cytometry analysis for pluripotency marker was performed using antibodies to SSEA-4 and TRA-1-8. Percentages were determined relative to appropriate isotype control.

3.6 Ultrastructural studies

Various blood donors provided blood samples for this study showed no differences among them in terms of cell morphology or ultrastructure. *In vitro* cell culture studies, after 1 day of dedifferentiation (D1), upon activation, resulted in the observation of a large cell population comprised of approximately 60% of agranular mononuclear cells and 40% of granular mononuclear cells (**Figure 8A–C**). Granular cells displaying deeply invaginated nuclei, scarce cytoplasm, and abundant primary and secondary granules disappeared from the culture after 5 days. Furthermore, red blood cells and platelets were detected. The population of interest for this study are agranular cells, morphologically characterized by small size and a nucleus with condensed chromatin (**Figure 8B–C**). These cells showed rounded shapes with slender filipodia-like cytoplasmic expansions. Their cytoplasm was electron-dense and contained few organelles, highlighting the presence of small dictiosomes, some mitochondria, and rough endoplasmic reticulum cisterns. The nucleus occasionally showed deep invaginations with the emphasis on large nucleoli and condensed chromatin, preferentially associated with the nuclear membrane. On day 5 (D5), most cells in the culture were classified as agranulocytes, which exhibited a clear decrease in the number of cytoplasmic organelles (**Figure 8D**). The nuclear/cytoplasmic ratio of these cells was high,

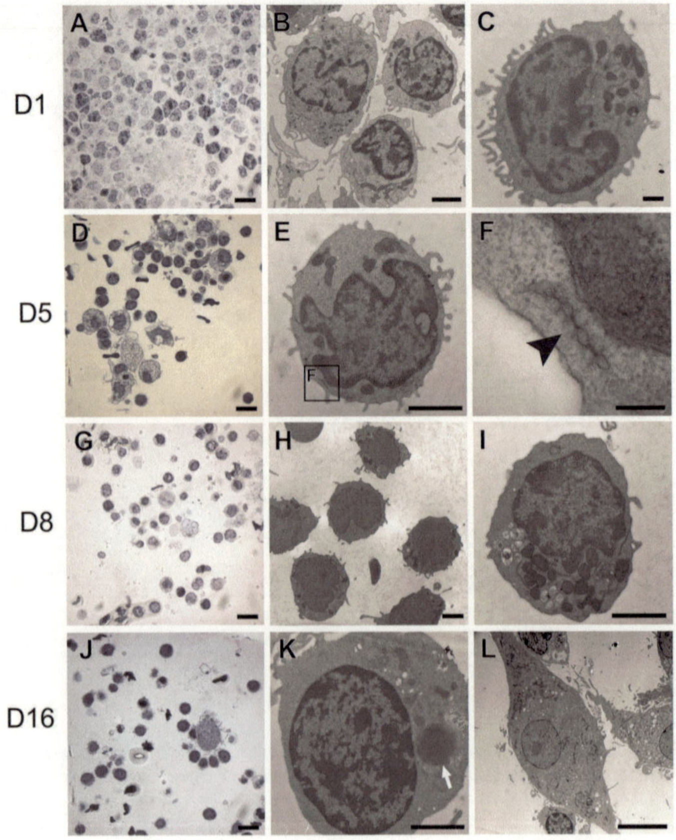

Figure 8.
Blood cells progressively showed morphological de-differentiation features upon activation. A, B: One day (D1) after activation in appropriate culture medium, typical blood cell types could be detected in the samples. Red blood cells, platelets, granulocytes, and agranulocytes could be found in the sample. C–E: After 5 days (D5), most cells could be classified as agranulocytes, showing some nondifferentiation characteristics, such as scarce cytoplasm, low number of organelles, and the presence of annulate lamellae (E, arrowhead). F-H: The homogeneity of the culture peaked 8 days upon activation (D8), showing similar ultrastructural characteristics as defined in D5. This state was maintained in D12 (not shown). I–K: After 16 days (D16), cells slowly started to show some differentiation signs, such as lipid drops (K, arrow). A small population of cells started to change their morphology and appeared as bigger elongated cells (L), with increased cytoplasm and more organelles. Photomicrographs in the first column correspond to toluidine blue-stained semithin (1.5 µm) sections. The center and right columns are transmission electron microscopy (TEM) images. Scale bars: A, C, F, I, K = 10 µm; B, D, G, H, J = 2 µm; E = 200 nm.

similar to ESCs (**Figure 8D–E**). We could observe small dictiosomes, some mitochondria, and rough endoplasmic reticulum cisterns. On the other side, polyribosomes and filamentous structures were abundant. Moreover, some of these cells occasionally presented annulate lamellae (**Figure 8E–F**). Regarding the nucleus, we observed nuclear invaginations with abundant condensed chromatin (heterochromatin). On the cell surface, some short cytoplasmic expansions were appreciated (**Figure 8E**). At this stage of the cellular culture, we could occasionally see another subpopulation of large cells as well, containing a wide heterogeneity of cellular structures and highlighting the presence of lysosomes (**Figure 8D**). On day 16 (D16), reprogrammed cells underwent morphological changes as the culture medium was gradually changed to neuronal medium, showing signs of differentiation with the presence of lipid drops, bigger cytoplasm, and more organelles. Their shape changes to larger cells with elongated appearance resembling cells of neuroectodermal origin.

3.7 Redifferentiation of BD-PSCs to the cells belonging to different germ layers

The capability of BD-PSCs to redifferentiate in neuroectodermal layer was demonstrated by growing these cells on laminin-ornithin coated plates in N2 medium to initiate the differentiation toward neuronal cells and further cultivation in neuronal differentiation media containing B27 supplement BDNF and GDNF as described in Material and Methods. The conditions described above enable for redifferentiation of BD-PSCs toward various neuronal lineages. Depicted are different populations of cells expressing the specific markers (**Figure 3**). Specific neuronal lineages from BD-PSCs can be generated by slightly modifying time and culture conditions. In neuronal differentiation culture time period from D8 to D30, we observed a clear decrease in the expression of neuroepithelial stem cell protein Nestin, which is a major intermediate filament (IF) protein of embryonic central nervous system also known as neuronal progenitor marker, while the expression of MAP2, a member of neuron-specific microtubule-associated protein family, neuronal nuclear antigen NeuN, a common neuron marker, and class III β-tubulin element of tubulin family, Tuj 1 a specific marker for human neurons, significantly increases during neuronal differentiation.

Glial fibrillary acidic protein GFAP is a type of IF expressed in various cells belonging to central nervous system, such as glial cells and astrocytes. These cells are mainly expressed in the central nervous system, such as brain and spinal cord, contributing to astrocytes-neuron interactions as well as cell–cell communication. Using antibodies to GFAP, we confirmed that such structures are recognized in newly generated BD-neuronal cells confirming the feature of BD-PSCs to redifferentiate to neuroectoderm [22].

Capacity BD-PSCs to redifferentiate into endoderm/hepatocytes was assessed by growing the cells in appropriate medium as described in Material and Methods. Following initial differentiation into endoderm in KSR/DMSO medium, as confirmed by ICC using antibodies to AFP and TTR, in the second phase by using hepatocytes maturation medium cells turned to mature hepatocytes like cells expressing their specific marker ALB and HNF4α, (**Figure 4**), recapitulating liver development *in vivo* [23].

The membrane activation of human glycoprotein ACA initiates a dedifferentiation process, consecutively generating more primitive cells until the final stage of this process is reached. BD-PSCs capable of redifferentiation into all three germ layers are the final product of this dedifferentiation process initiated by the membrane glycoprotein ACA [9–11] depicted in **Figure 9**.

4. Discussion

4.1 Signaling

Antibody cross-linking of a GPI-linked protein ACA initiates, *via* PLCγ/PI3K/Akt mTor/PTEN up-regulation of Wnt, Notch, c-Kit, and/or HoxB4 genes, among others. Signaling network linked to these genes induces dedifferentiation of blood progenitor cells leading to generation of BD-PSCs [9, 11]. Briefly, PI3K activation phosphorylates and activates Akt localizing it at the plasma membrane. Akt is a serine/threonine-specific protein kinase that plays a key role in multiple cellular processes like cell proliferation, transcription, and apoptosis [25]. The components of the PI3K/Akt Pathway, such as α, β, and γ p110 catalytic subunits, as well as subunits of 3-

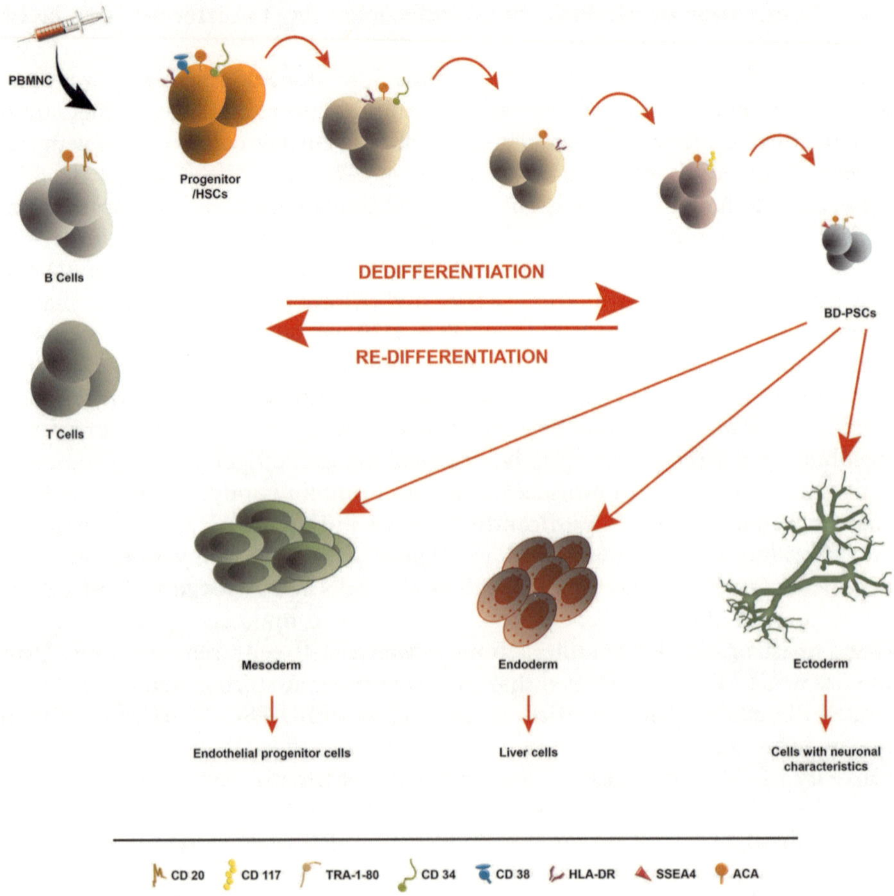

Figure 9.
Schematic presentation of dedifferentiation process that starts with human blood progenitor cells expressing CD34 via HSCs, following side population (SP) cells mainly not expressing CD34, (low HLA) and ending up with the generation of BD-PSCs expressing pluripotency marker (SSEA-4) [24].

phosphoinositide dependent kinase 1 (PDK1), which is the major transducer of IP3 Kinase, likewise the downstream effector proteins like glycogen kinase-3 (GSK-3), which plays a central role in the regulation of the stability and synthesis of proteins involved in the cell cycle entry; C-Raf, a serine/threonine kinase, whose main role in cells is the phosphorylation and activation of the MAP kinases MEK1 and MEK2, and the mTOR complex that controls translation and acts as a critical regulator of protein synthesis, are regulated in ACA-dependent manner.

Most importantly, a lipid-protein phosphatase PTEN, also called an anti-tumor agent, which is the natural inhibitor of PI3K/Akt signaling pathway that regulates p53 protein level and activity as well. PTEN works by dephosphorylating phosphatidylinositol (3,4,5)-trisphosphate (PIP3) to phosphatidylinositol-4, 5-bisphosphat (PIP2), which limits Akt's ability to bind to the membrane, decreasing its activity. Deletion of the PTEN tumor-suppressor gene in adult hematopoietic stem cells (HSCs) leads to myeloproliferative diseases, and recent studies showed the inactivation of PTEN in human T-ALL cell lines as well as primary cells [26–28].

PTEN is a downstream target of phosphorylation at this specific route to pluripotency. Upregulation of PTEN indicates that a proliferative control of the process that leads to generation of self-renewable BD-PSCs is tightly regulated in a GPI-linked protein-dependent manner. Our results indicate that this protein promotes the generation of self-renewing cells by activating PI3K/AKT pathway, but one of the best-conserved functions of AKT in promoting growth in this induced signaling cascade appears to be under the control of activated PTEN by preventing oncogenic outgrow.

Canonical Wnt/ß-catenin pathway is involved in the regulation of various functions like embryonic development, proliferation, survival, cell polarity, migration, and maintenance of somatic stem cells in many tissues, modulating a delicate balance between stemness and differentiation. Binding of Wnt proteins to their receptors inhibits the phosphorylation of β-catenin resulting in stabilization and accumulation of β-catenin in the cytosol and its nuclear translocation followed by transcriptional regulation of target genes [29]. Notch protein is a hetero-oligomer type of transmembrane receptor and the pathway linked to his protein is highly conserved and functionally involved in various processes from development to cell growth and cell death. Among them, the most important, together with other signaling pathways, such as Wnt, are the regulation of stem cell self-renewal, maintenance of homeostasis, cell–cell communication, regulation of cell-fate decision, neuronal function and development, and expansion of HSCs [30].

Tyrosine-protein kinase (c-Kit) is a receptor tyrosine kinase that belongs to the type III family of kinases. When the ligand binds to this receptor, the dimer is formed, which activates intrinsic kinase activity that initiates the phosphorylation of various signal-transducing molecules and propagates the signals within the cells.

C-Kit signaling is involved in various mechanisms, such as differentiation, cell survival, and proliferation. It is expressed in various cell types, most importantly in HSCs, and binding on its ligand SCF causes blood progenitor cells to grow [31].

Further downstream partner genes in the initiated signaling pathways, such as HOXB4 and BMI1, belong to the Homeobox and/or Polycomb group (PcG) family of genes involved in the development. HOXB4 gene encodes a nuclear protein with a homeobox DNA-binding domain. Ectopic expression of this protein expands HSCs and progenitor cells *in vivo* and *in vitro,* making it a potential candidate for therapeutic stem cell expansion [32]. BMI1, as a member of the PcG family of transcriptional repressors, is involved in the control of development by regulating cell growth and differentiation. It is also expressed in HSCs proven to be essential for generation of self-renewing HSCs [33].

Transforming growth factor beta (TGF-β) is a multifunctional cytokine that belongs to the transforming growth superfamily that includes endogenous growth-inhibiting proteins [34]. Activation by ACA down-regulates TGF-β, which is one of the most potent inhibitors of HSC growth *in vitro.* One of the features of HSCs is their relative quiescence and given the strong inhibitory properties of TGF-β, it has been proposed to be the main regulator of quiescence *in vivo* [35]. Anti-apoptotic BCL-2 family proteins represent a family of evolutionary conserved cytoplasmic proteins that are known for their regulation of programmed cell death and survival. In response to intracellular damage, signals initiate the proteolytic cascade that disintegrates the cells. In our findings, these genes are down-regulated compared to unmanipulated PBMNCs, indicating that apoptosis represents an important regulatory factor in the maintenance of stem cells and is a part of the molecular mechanisms regulated by GPI-anchored membrane protein.

ACA signaling network *via* PLCγ/IP3K/Akt/mTOR/PTEN up-regulates the critical genes that are involved in the signaling pathways that regulate human development, such as NOTCH and WNT. Moreover, due to hierarchy among them, upregulation of these genes remains under the control of tumor suppressor gene PTEN, which is also upregulated in an ACA-specific manner. Apoptosis through downregulation of BCL-2 is an additional mechanism that regulates growth and proliferation. Finally, tumor suppressor gene P53 remains constant during reprogramming by ACA (data not shown).

Notably, the highest extent of upregulation of target genes is reached with c-Kit receptor tyrosine kinase [36], which is the gene critical for proliferation and survival of HSCs, indicative of a direct link that exists between these two proteins. C-Kit, a receptor tyrosine kinase type II, activates signaling through second messengers, such as cyclic adenosine monophosphate (cAMP), which are membrane-associated and diffuse from the plasma membrane into intermembrane space, where they can reach and regulate other membrane proteins. This reaction is probably the key to molecular mechanisms regulated by GPI-anchored membrane glycoprotein.

4.2 Reprogramming by dedifferentiation

IPSCs appear to represent the greatest promise for regenerative medicine without the ethical and immunological concerns incurred by the use of ESCs. They are pluripotent and have high replicative capability. Furthermore, iPSCs have the potential to generate all the tissues of the human body and provide researchers with patient- and disease-specific cells, which can recapitulate the disease *in vitro,* allowing for specific drug discovery. The iPSC technology provides an opportunity to generate cells with characteristics of ESCs, including pluripotency and potentially unlimited self-renewal.

Although methods have been improved from viral integration to integration-free, there are still challenges down the road to achieving their clinical application in humans.

The use of iPSCs in autologous cell-based therapy represents an ideal approach for regenerative medicine since the patients do not require long-term immunosuppressive drugs. The derivation of iPSCs over a decade ago has been raising high expectations and enthusiasm that iPSC technology can deliver autologous cell-based therapeutics to treat a high number of degenerative diseases, but actually, autologous therapy is related to the high cost and long period of time which should be spent in the manufacturing process that includes generation, characterization, differentiation into relevant cell types, scale up, and careful validation of the generated cell product. In order to reduce production time and costs, iPSCs therapies are moving toward allogeneic approaches by establishing clinical-grade iPSC banking [37, 38].

Banking of iPSCs from healthy donors would throw the iPSC reprogramming strategy once claimed as advantageous when compared to hESCs, while autologous at its beginning background. Therefore, the reprogramming strategies entirely free of DNA-based vectors could lead to solving the problems regarding genetic induced pluripotency.

4.3 Blood-derived pluripotent stem cells

Stem cell therapy is the ultimate goal of personalized medicine and individual care for many degenerative diseases, such as Alzheimer's disease, Parkinson, diabetes, and others. It has already been shown that human PB cells can be successfully reprogrammed into blood cells using the Yamanaka factors [39]. Blood is one of the most easily accessible sources of patient cells for reprogramming because there is no need to maintain cell cultures extensively prior to reprogramming experiments. Therefore, it is a potentially unlimited and safe source of cells.

Our own work showed recently that blood cells can be reprogrammed to PSCs without any genetic manipulation [9–11]. The present study shows that the signaling network activated by human GPI-linked protein ACA is sufficient to generate cells from circulating blood that is pluripotent, according to their morphology, pluripotent marker proteins, and differentiation potential. In fact, it is possible to reprogram adult progenitor cells that can be obtained from PB through protein activation, by means of antibody cross-linking, making them return to a similar state to that of ESCs.

Immunophenotyping of BD-PSCs by using antibodies to pluripotency markers by means of flow cytometry and immunofluorescence analysis, revealed the expression of SSEA-4, TRA-1-60, TRA-1-81, NANOG, SOX2, and OCT3/4, indicating that newly generated cells possess the properties of ES cells. Electron microscopy analysis showed the morphological changes during the culture time period from D1-D16. A scarce cytoplasm and decreased number of organelles indicate that undifferentiated characteristics appeared through culture time. Most importantly, the appearance of annulate lamellae, stacked sheets of membranes embedded with pore complexes which are frequently found in cells with high proliferative activity, such as oocytes, embryonic, and tumor cells [40], suggest that BD-PCSs correspond to an actively dividing cell population. In addition, when neuronal differentiation medium was added to the cell culture system the morphology of these cells changed to larger elongated cells with more organelles and increased cytoplasm, supporting the notion that they are able to redifferentiate [10].

Activation of a membrane protein ACA at the surface of blood progenitor cells by cross-linking at the membrane of blood cells with its specific antibody and analysis of the mode of how this signaling machinery regulates the expression of genes known to play a role in human development *via* specific protein phosphorylation as an important regulatory mechanism in the cellular processes related to its signaling competence showed its involvement in the processes that determine the cell type, its fate, and identity.

Our results confirm the previously published data that the initiation of GPI-linked protein ACA upon activation of PBMNCs described here is sufficient to induce signaling machinery that leads to generation of self-renewing PSCs. Moreover, it ensures the maintenance of pluripotency in ESCs as well, indicating the involvement of this protein in pluripotency signaling network in humans.

Dedifferentiation process initiated upon membrane activation follows exactly the opposite way that is known for differentiation of PB progenitor cells. It is generally accepted that the proliferation capacity is higher by hematopoietic progenitor cells and declined by more primitive cells like HSCs and SP cells. The process initiated by ACA is due to activation of tumor suppressor genes, under strict proliferative control, leading to generation of BD-PSCs and explaining the lack of teratogenicity of these cells resulting in advantage for their application in cell tissue replacement.

BD-PSCs generated through dedifferentiation process are capable of redifferentiate into cells belonging to all three germ layers.

Most importantly, an antibody that acts at the surface of PB progenitor cells initiating membrane-to-nucleus signaling pathways may have numerous potential advantages regarding clinical safety for application of these cell products in regenerative medicine.

5. Conclusions

So far today, no iPSCs-based therapy is implemented into routine clinical use [41]. The hurdles regarding use of iPSCs in clinical practice are related primarily to genetic instability of these cells that may cause mutations leading to tumor formation and cancer. Another safety concern, when it comes to their application in humans, is the presence of residual undifferentiated iPSCs, also linked to tumorigenicity [42].

Allogenic approaches by establishing clinical-grade iPSC banking must consider the populations with heterogeneous genetic background, which might be very challenging [43]. Therefore, it is very important to further improve the current iPSCs technology to minimize the possible side effects and genetic and epigenetic differences between reprogrammed cells and donors. Human ES cells derived from blastocyst imply its destruction that cause serious ethical concern. In addition, these cells are by their nature nonautologous and may cause graft-versus-host disease. They are mostly used for studying early human development using currently available hES cell lines, but they also have limited potential in medicine due to restrictions related to ethical and immunological issues [44].

Great effort is made to assure safe clinical applications using stem cell therapies. The international stem cell banking initiative (ISCB) published guidelines for the development of pluripotent stem cell stocks for clinical applications [45]. Due to complex nature of the cells to be used for therapies in regenerative medicine compared to drug therapies, standard regulations must include purity of the cells, sterility, viability, genomic stability, specific gene expression profile, functional evaluation of reprogrammed, and differentiated cells, absence of infectious pathogens and tumorigenicity [46]. The greatest attention is taken to ensure the safety of transgenic cells when compared to genetically unmodified cells, even more so when it comes to the use of more modern technology like CRISPR/Cas9 for removing randomly inserted foreign genes into human genome during reprogramming process because there is the possibility of off target mutagenesis [47, 48].

Our results reveal insight into the molecular events regulating cellular reprogramming and indicate that pluripotency may be controlled *in vivo* through the binding of soluble ligand(s) to ACA-protein and initiating the cascade of already known and partly characterized signaling pathways. The process of reprogramming is short (10–12 days), the source is easily accessible unmanipulated peripheral blood and no use of growth factors is necessary. BD-PSCs are autologous, capable of generating *in vitro* cell types of all three layers exhibiting neuronal, liver, or hematopoietic characteristics [9–11]. Due to their differentiation capacity, they could be potentially utilized to regenerate any type of tissue, and thus treat neurological and immune disorders, as well as injuries to critical organs, such as the heart and brain. Moreover, due to lack of teratogenicity, BD-PSCs can be used *in situ* without the necessity to be differentiated before their application [11] as is also shown in the wound healing experiment currently ongoing in our laboratory. Due to tight proliferation control, BD-PSCs do not form cell lines and therefore must be freshly prepared.

It can be expected that the standard regulations for the use of BD-PSCs would be similar to that of genetically unmodified cells implying fewer hurdles compared to iPSCs ensuring a fast and safe application of these cells in routine clinical practice.

The potential application of BD-PSCs in regenerative stem cell therapies is innovative and promising. Additional studies are underway in order to determine *in vivo* therapeutic potential and to ensure a safe platform for translation of basic research to new clinical therapies.

Our report provides a practical and efficient way to generate patient-specific PSCs. This will also be valuable for the generation of clinical-grade PSCs for future therapeutic applications so that the possibility to develop a truly personalized medicine becomes more realistic.

Acknowledgements

Dedicated to the memory of Dr. Rainer Saffrich.

The authors express unending gratitude to Professor Gennady T. Sukhikh for his encouragement and permanent support. We gratefully acknowledge the technical assistance provided by Patricia García-Tárraga in Valencia and by John and Oksana Greenacre in Heidelberg.

This work was supported by the Prometeo Grant for Excellence Research Groups [PROMETEO/2019/075] given to José Manuel García-Verdugo; and private funding from ACA CELL Biotech GmbH Heidelberg, Germany.

Conflict of interest

The corresponding author declares that she is a patent holder related to Novel Human GPI-linked Protein ACA, she also cofounded and works in ACA CELL Biotech GmbH. The other authors declare that there is no conflict of interest.

Abbreviations

AFP	alpha-feto-protein
ALB	albumin
APC	allophycocyanin
BCL-2	B-cell lymphoma 2
BDNF	brain-derived neurotrophic factor
BD-PSCs	blood-derived pluripotent stem cells
bFGF	basic fibroblast growth factor
BMI1	B lymphoma Mo-MLV insertion region 1 homolog
BSA	bovine serum albumin
cAMP	cyclic adenosine monophosphate
CTNNB	catenin (cadherin-associated protein), beta 1
DAG	1,2-diacylglycerol
DAPI	4′,6-Diamidin-2-phenylindol
DMEM	dulbecco's Modified Eagle Medium
DMSO	dimethyl sulfoxide
ECL	enhanced chemiluminescence
EGTA	ethylene glycol-bis(β-aminoethyl ether)-*N*,*N*,*N′*,*N′*-tetraacetic acid
EM	electron microscopy
ESCs	embryonic stem cells
ET	Et-18-CH_3
FBS	fetal bovine serum
FITC	fluorescein isothiocyanate
GAPDH	glyceraldehyde 3-phosphate dehydrogenase
GDNF	glial-derived neurotrophic factor
GFAP	glial fibrillary acidic protein
GPI	glycosylphosphatidylinositol
GSK-3	glycogen kinase-3
HCC	hydrocortisone 21-hemisuccinate sodium salt

HGF	hepatocyte growth factor
HNF4 α	hepatocyte Nuclear Factor 4 alpha
HoxB4	homeobox protein B4
HPR	horseradish peroxidase
HSCs	hematopoietic stem cells
ICC	immunocytochemistry
IF	intermediate filament
IMDM	Iscove's modified Dulbecco's medium
IP	immunophenotyping
iPSCs	induced pluripotent stem cells
Klf4	kruppel-like factor 4
KSR	knockout serum replacement
LY	LY 294002
MAP2	microtubule-associated protein 2
MEFs	mouse embryonic fibroblasts
MEK1	mitogen-activated protein kinase kinase 1
MEK2	mitogen-activated protein kinase kinase 2
MNCs	mononuclear cells
mTor	mammalian target of rapamycin
NEAA	nonessential amino acids
Nestin	neuroepithelial stem cell protein
Oct3/4	octamer-binding transcription factor 3/4
OSM	oncostatin M
PB	peripheral blood
PBS	phosphate buffer saline
PcG	polycomb group
PD	PD098059
PDK1	3-phosphoinositide dependent kinase 1
PE	phycoerythrin
PFA	paraformaldehyde
PI	propidium iodide
PI3Ks	phosphoinositide 3-kinases
PIP2	phosphatidylinositol-4,5-bisphosphat
PIP3	phosphatidylinositol (3,4,5)-trisphosphate
PLCγ	phosphoinositol-phospholipase C γ
PtdIns	phosphatidylinositol
PTEN	phosphatase and tensin homolog
PVDF	polyvinylidene difluoride
SCF	stem cell factor
SDS-PAGE	sodium dodecyl sulfate-polyacrylamide gel electrophoresis
SSEA-4	stage-specific embryonic antigen
TGF-β	transforming growth factor beta
TTR	transthyretin
Tuj1	class III β-tubulin
WB	western blot

Author details

Zorica A Becker-Kojić[1]*, José Manuel García-Verdugo[2], Anne-Kathrin Schott[1], Vicente Herranz-Pérez[2], Ivan Zipančić[3] and Vicente Hernández-Rabaza[3]

1 ACA CELL Biotech GmbH, Heidelberg, Germany

2 Laboratory of Comparative Neurobiology, Cavanilles Institute of Biodiversity and Evolutionary Biology, University of Valencia, Paterna, Spain

3 Department of Biomedical Sciences, Cardenal Herrera-CEU University, Valencia, Spain

*Address all correspondence to: z.becker-kojic@aca-cell.de

References

[1] Takahashi K, Yamanaka S. Induction of pluripotent stem cells from mouse embryonic and adult fibroblast cultures by defined factors. Cell. 2006;**126**(4): 663-676. DOI: 10.1016/j.cell.2006.07.024

[2] Jaenisch R, Young R. Stem cells, the molecular circuitry of pluripotency and nuclear reprogramming. Cell. 2008;**132**(4):567-582. DOI: 10.1016/j.cell.2008.01.015

[3] Cibelli J, Emborg ME, Prockop DJ, Roberts M, Schatten G, Rao M, et al. Strategies for improving animal models for regenerative medicine. Cell Stem Cell. 2013;**12**(3):271-274. DOI: 10.1016/j.stem.2013.01.004

[4] Takahashi K, Tanabe K, Ohnuki M, Narita M, Ichisaka T, Tomoda K, et al. Induction of pluripotent stem cells from adult human fibroblasts by defined factors. Cell. 2007;**131**(5):861-872. DOI: 10.1016/j.cell.2007.11.019

[5] Yu J, Vodyanik MA, Smuga-Otto K, Antosiewicz-Bourget J, Frane JL, Tian S, et al. Induced pluripotent stem cell lines derived from human somatic cells. Science. 2007;**318**(5858):1917-1920. DOI: 10.1126/science.1151526

[6] Malik N, Rao MS. A review of the methods for human iPSC derivation. Methods in Molecular Biology. 2013;**997**: 23-33. DOI: 10.1007/978-1-62703-348-0_3

[7] Edidin M. The state of lipid rafts: From model membranes to cells. Annual Review of Biophysics and Biomolecular Structure. 2003;**32**:257-283. DOI: 10.1146/annurev.biophys.32.110601.142439

[8] Cantley LC. The phosphoinositide 3-kinase pathway. Science. 2002;**296**(5573):1655-1657. DOI: 10.1126/science.296.5573.1655

[9] Becker-Kojić ZA, Ureña-Peralta JR, Zipančić I, Rodriguez-Jiménez FJ, Rubio MP, Stojković P, et al. Activation by ACA induces pluripotency in human blood progenitor cells. Bulletin of Experimental Biology and Medicine. 2013;**155**(4):552-567. DOI: 10.1007/s10517-013-2196-z

[10] Becker-Kojić ZA, Schott AK, Zipančić I, Hernández-Rabaza V. GM-free generation of blood-derived neuronal cells. Journal of Visualized Experiments. 2021;(168):1-10. DOI: 10.3791/61634

[11] Becker-Kojić ZA, Ureña-Peralta JR, Saffrich R, Rodriguez-Jiménez FJ, Rubio MP, Rios P, et al. A novel human glycoprotein ACA is an upstream regulator of human hematopoiesis. Bulletin of Experimental Biology and Medicine. 2013;**155**(4):536-551. DOI: 10.1007/s10517-013-2195-0

[12] Engelman JA, Luo J, Cantley LC. The evolution of phosphatidylinositol 3-kinases as regulators of growth and metabolism. Nature Reviews Genetics. 2006;7(8):606-619. DOI: 10.1038/nrg1879

[13] Cocco L, Follo MY, Manzoli L, Suh PG. Phosphoinositide-specific phospholipase C in health and disease. Journal of Lipid Research. 2015;**56**(10): 1853-1860. DOI: 10.1194/jlr.R057984

[14] Sidney LE, Branch MJ, Dunphy SE, Dua HS, Hopkinson A. Concise review: Evidence for CD34 as a common marker for diverse progenitors. Stem Cells. 2014; **32**(6):1380-1389. DOI: 10.1002/stem.1661

[15] Ratajczak MZ, Ratajczak J, Suszynska M, Miller DM, Kucia M, Shin DM. A novel view of the adult stem cell

compartment from the perspective of a quiescent population of very small embryonic-like stem cells. Circulation Research. 2017;**120**(1):166-178. DOI: 10.1161/CIRCRESAHA.116.309362

[16] Zhan T, Rindtorff N, Boutros M. Wnt signaling in cancer. Oncogene. 2017;**36**(11):1461-1473. DOI: 10.1038/onc.2016.304

[17] Reya T, Duncan AW, Ailles L, Domen J, Scherer DC, Willert K, et al. A role for Wnt signalling in self-renewal of haematopoietic stem cells. Nature. 2003; **423**(6938):409-414. DOI: 10.1038/nature01593

[18] Kossack N, Meneses J, Shefi S, Nguyen HN, Chavez S, Nicholas C, et al. Isolation and characterization of pluripotent human spermatogonial stem cell-derived cells. Stem Cells. 2009;**27**(1):138-149. DOI: 10.1634/stemcells.2008-0439

[19] Henderson JK, Draper JS, Baillie HS, Fishel S, Thomson JA, Moore H, et al. Preimplantation human embryos and embryonic stem cells show comparable expression of stage-specific embryonic antigens. Stem Cells. 2002;**20**(4):329-337. DOI: 10.1634/stemcells.20-4-329

[20] Pan G, Thomson JA. Nanog and transcriptional networks in embryonic stem cell pluripotency. Cell Research. 2007;**1**:42-49. DOI: 10.1038/sj.cr.7310125

[21] Baghbaderani BA, Syama A, Sivapatham R, Pei Y, Mukherjee O, Fellner T, et al. Detailed characterization of human induced pluripotent stem cells manufactured for therapeutic applications. Stem Cell Reviews and Reports. 2016;**12**(4):394-420. DOI: 10.1007/s12015-016-9662-8

[22] Nasonkin I, Mahairaki V, Xu L, Hatfield G, Cummings BJ, Eberhart C, et al. Long-term, stable differentiation of human embryonic stem cell-derived neural precursors grafted into the adult mammalian neostriatum. Stem Cells. 2009;**27**(10):2414-2426. DOI: 10.1002/stem.177

[23] Fu GB, Huang WJ, Zeng M, Zhou X, Wu HP, Liu CC, et al. Expansion and differentiation of human hepatocyte-derived liver progenitor-like cells and their use for the study of hepatotropic pathogens. Cell Research. 2019;**29**(1):8-22. DOI: 10.1038/s41422-018-0103-x

[24] Becker-Kojić ZA, Ureña-Peralta JR, Zipančić I, Rodriguez-Jiménez FJ, Rubio MP, Stojković P, et al. Available from: https://globalmedicaldiscovery.com/key-scientific-articles/activation-aca-induces-pluripotency-human-blood-progenitor-cells/

[25] Manning BD, Cantley LC. AKT/PKB signaling: Navigating downstream. Cell. 2007;**129**(7):1261-1274. DOI: 10.1016/j.cell.2007.06.009

[26] Yilmaz OH, Valdez R, Theisen BK, Guo W, Ferguson DO, Wu H, et al. Pten dependence distinguishes haematopoietic stem cells from leukaemia-initiating cells. Nature. 2006; **441**(7092):475-482. DOI: 10.1038/nature04703

[27] Zhang J, Grindley JC, Yin T, Jayasinghe S, He XC, Ross JT, et al. PTEN maintains haematopoietic stem cells and acts in lineage choice and leukaemia prevention. Nature. 2006;**441**(7092):518-522. DOI: 10.1038/nature04747

[28] Gutierrez A, Sanda T, Grebliunaite R, Carracedo A, Salmena L, Ahn Y, et al. High frequency of PTEN, PI3K, and AKT abnormalities in T-cell acute lymphoblastic leukemia. Blood. 2009;

114(3):647-650. DOI: 10.1182/blood-2009-02-206722

[29] Clevers H, Nusse R. Wnt/β-catenin signaling and disease. Cell. 2012;**149**(6): 1192-1205. DOI: 10.1016/j.cell.2012.05.012

[30] LaFoya B, Munroe JA, Mia MM, Detweiler MA, Crow JJ, Wood T, et al. Notch: A multi-functional integrating system of microenvironmental signals. Developmental Biology. 2016;**418**(2): 227-241. DOI: 10.1016/j.ydbio.2016.08.023

[31] Lennartsson J, Rönnstrand L. Stem cell factor receptor/c-Kit: From basic science to clinical implications. Physiological Reviews. 2012;**92**(4):1619-1649. DOI: 10.1152/physrev.00046.2011

[32] Antonchuk J, Sauvageau G, Humphries RK. HOXB4-induced expansion of adult hematopoietic stem cells ex vivo. Cell. 2002;**109**(1):39-45. DOI: 10.1016/s0092-8674(02)00697-9

[33] Park IK, Qian D, Kiel M, Becker MW, Pihalja M, Weissman IL, et al. Bmi-1 is required for maintenance of adult self-renewing haematopoietic stem cells. Nature. 2003;**423**(6937):302-305. DOI: 10.1038/nature01587

[34] Massagué J. TGFβ signalling in context. Nature Reviews. Molecular Cell Biology. 2012;**13**(10):616-630. DOI: 10.1038/nrm3434

[35] Sitnicka E, Ruscetti FW, Priestley GV, Wolf NS, Bartelmez SH. Transforming growth factor beta 1 directly and reversibly inhibits the initial cell divisions of long-term repopulating hematopoietic stem cells. Blood. 1996;**88** (1):82-88

[36] Edling CE, Hallberg B. c-Kit–a hematopoietic cell essential receptor tyrosine kinase. The International Journal of Biochemistry & Cell Biology. 2007;**39**(11):1995-1998. DOI: 10.1016/j.biocel.2006.12.005

[37] Stacey GN, Crook JM, Hei D, Ludwig T. Banking human induced pluripotent stem cells: Lessons learned from embryonic stem cells? Cell Stem Cell. 2013;**13**(4):385-388. DOI: 10.1016/j.stem.2013.09.007

[38] Turner M, Leslie S, Martin NG, Peschanski M, Rao M, Taylor CJ, et al. Toward the development of a global induced pluripotent stem cell library. Cell Stem Cell. 2013;**13**(4):382-384. DOI: 10.1016/j.stem.2013.08.003

[39] Loh YH, Agarwal S, Park IH, Urbach A, Huo H, Heffner GC, et al. Generation of induced pluripotent stem cells from human blood. Blood. 2009;**113**(22):5476-5479. DOI: 10.1182/blood-2009-02-204800

[40] Kessel RG. Annulate lamellae: A last frontier in cellular organelles. International Review of Cytology. 1992; **133**:143-120. DOI: 10.1016/s0074-7696 (08)61858-6

[41] Deinsberger J, Reisinger D, Weber B. Global trends in clinical trials involving pluripotent stem cells: A systematic multi-database analysis. NPJ. Regenerative Medicine. 2020;**5**(15):1-13. DOI: 10.1038/s41536-020-00100-4

[42] Yoshihara M, Oguchi A, Murakawa Y. Genomic Instability of iPSCs and challenges in their clinical applications. Advances in Experimental Medicine and Biology. 2019;**1201**:23-47. DOI: 10.1007/978-3-030-31206-0_2

[43] de Rham C, Villard J. Potential and limitation of HLA-based banking of human pluripotent stem cells for cell therapy. Journal of Immunology

Research. 2014;**2014**:518135. DOI: 10.1155/2014/518135

[44] Schwartz SD, Regillo CD, Lam BL, Eliott D, Rosenfeld PJ, Gregori NZ, et al. Human embryonic stem cell-derived retinal pigment epithelium in patients with age-related macular degeneration and Stargardt's macular dystrophy: Follow-up of two open-label phase 1/2 studies. Lancet. 2015;**385**(9967):509-516. DOI: 10.1016/S0140-6736(14)61376-3

[45] Andrews PW, Baker D, Benvinisty N, Miranda B, Bruce K, Brüstle O, et al. Points to consider in the development of seed stocks of pluripotent stem cells for clinical applications: International Stem Cell Banking Initiative (ISCBI). Regenerative Medicine. 2015;**10**(2 Suppl):1-44. DOI: 10.2217/rme.14.93

[46] Cao J, Hao J, Wang L, Tan Y, Tian Y, Li S, et al. Developing standards to support the clinical translation of stem cells. Stem Cells Translational Medicine. 2021;**10**(Suppl 2):S85-S95. DOI: 10.1002/sct3.13035

[47] Wang X, Rivière I. Clinical manufacturing of CAR T cells: Foundation of a promising therapy. Molecular Therapy — Oncolytics. 2016;**3**:16015. DOI: 10.1038/mto.2016.15

[48] Moradi S, Mahdizadeh H, Šarić T, Kim J, Harati J, Shahsavarani H, et al. Research and therapy with induced pluripotent stem cells (iPSCs): Social, legal, and ethical considerations. Stem Cell Research & Therapy. 2019;**10**(1):341. DOI: 10.1186/s13287-019-1455-y

Chapter 2

Activation and Metabolic Shifting: An Essential Process to Mesenchymal Stromal Cells Function

Patricia Semedo-Kuriki, Gabriel Pereira,
Danilo Cândido de Almeida and Niels Olsen Saraiva Camara

Abstract

To elucidate the basal metabolism of Mesenchymal Stromal Cells (MSCs), as well as knowing how they are activated, can bring important clues to a successful cell-based therapy. Naive MSCs, in their niche, mainly keep the local homeostasis and the pool of tissue stem cells. Once activated, by an injury, MSCs' response leads to a lot of physiological differences in its metabolism that are responsible for its healing process. Since endogenous MSC seems to be ineffective in pathologic and aging conditions, cell-based therapy using MSC is focused on administration of exogenous MSC in patients to exert its healing functions. From quiescent to activated state, this "Metabolic Shifting" of MSC interferes directly in its secretion and cellular-derived particle generation. We will address here the differences between the MSCs activation phases and how they can modify the MSCs metabolism and its function. Moreover, understanding MSC in their niche and its damped function in pathologic and aging processes can improve stem cell-based therapies.

Keywords: mesenchymal stromal cell, metabolism, MSC activation, MSC niche, cell therapy

1. Introduction

Stem cells research has brought great insight in regenerative medicine. Currently, over 1700 clinical trials are registered at Clinicaltrials.gov (clinicaltrials.gov "mesenchymal stem cell OR mesenchymal stromal cell", August 2022), with ten approved MSC therapies worldwide [1].

Besides efforts to promote standardization of procedures and classifications for MSCs, the translation of promising preclinical results to human clinical trials has not matched full desired effects. Such variability may come from differences among species or source-tissues of MSCs in both *in vivo* or *in vitro* preclinical studies [1].

Initially, MSCs therapeutic potential was associated with engraftment of MSCs into tissues and to a contact-dependent cell communication. Advances in the field now confirm that paracrine mechanisms are the primary effector of MSCs for tissue regeneration, angiogenesis and modulatory effects on inflammation, apoptosis and fibrosis. These effects may be achieved by the secretion of biologically active molecules by MSC, such as cytokines and chemokines, growth factors, extracellular matrix and extracellular vesicles. Indeed, the use of secreted factors in the medicine and research fields lead to a cell-free approach, which can overcome major adversities found in the use of allogeneic or even autologous MSCs therapy [2, 3].

In fact, for some regenerative approaches, no additional cell is necessary, and nowadays, beyond adult stem cells; there are other stem cells-based products such as: (i) conditioned medium, (ii) concentrated supernatant, (iii) lyophilized secretome, (iv) cellular particles (i.e. exosomes, microvesicles, small body particles), and (v) small regulatory molecules (i.e. lncRNAs, microRNAs, ceRNAs, circRNAs). All together, these approaches are new fields to be explored in stem cell technologies and cellular-based therapies [4, 5].

Several questions can be formulated regarding the MSC paracrine mechanism of repair: How did MSC become so secretive? How is MSC activated to secrete those molecules responsible for its regenerative mechanism? How is MSC in its quiescent state?

Currently, the metabolism and cell activation of MSC has been the focus of study of many researchers worldwide. Recent reports have provided evidence that stem cells have a metabolic/activation signature which is distinct and specific to each tissue to maintain the homeostasis. Regarding therapy, the choice of the MSC origin and thus how it is MSC activated directly regulates therapy performance, since MSC metabolism is crucial to the paracrine effect. MSC activation is also controlled by the microenvironment, for instance, a metabolically activated MSC can interact with other cells in their niches and they are able to sense and to adapt to dietary changes, exercise, aging, epigenetics changes etc. [5, 6].

Thus, in this chapter, we will attempt to elucidate the importance of MSCs activation/metabolism in its therapeutic function. More specifically, we will describe here the impact of MSCs activation in its metabolism and function. In addition, we will discuss how this "Metabolic/activation Shifting" can interfere directly in the MSC secretory function and in its cellular-derived particle generation. Moreover, stem cell dysfunction and disabilities will also be discussed. Hence, understanding these basic steps about naïve and activated MSCs should improve the establishment of new stem cell-based therapies and other associated approaches around MSC technologies, expanding its use and resources for future implementation as a translational and effective therapy.

2. Mesenchymal stromal cells definitions

Isolated from a huge number of tissues, MSC has been used in several clinicals trials, despite its basic studies are still ongoing. Since MSC therapy leads to amelioration of pathologic state, it causes a frenesi in clinicals trials and cell-based therapies. This frenesi creates inconsistencies, for instance at MSC's characterization, nomenclature, culture parameters, etc. Lacking the principle of reproductibile and quality control, several works and clinical trials have still been done improperly [7, 8]. How are the MSCs defined? A brief historic event of its discovery may help to elucidate it.

Described in the early 1970 by Friedenstein and colleagues [9], they observed that bone marrow cells, in a cell culture condition, generated attached cells in culture plates. These cells showed fibroblastic shape that started growing in this condition. Moreover, they observed that these cells induce osteogenesis in an experimental model. In 1990, Caplan first used the name Mesenchymal Stem Cell to describe these cells with differentiation properties, depending on local (niche concept) and genetics factors [10]. In 1999, Pittenger studies had flourished in the MSC area. Pittenger et al. showed the isolation of MSC from human bone marrow, listing some criteria to define them, such as (i) adherent culture cells and (ii) differentiation capabilities under specific stimulation [11].

Then, the mess comes ... everybody, everywhere, every tissue could generate MSC. However each culture condition was different from each other, with different techniques of isolation, with different patterns of characterization [12]. And this confusion is used indiscriminately by some clinicals to sell cell therapies treatment [13]. Placing order to it, in 2006, the International Society for Cellular Therapy (ISCT) defined the minimal criteria to MSC [14]. Those criteria were upgraded in 2019 [15], where ISCT defines:

1. Terminology: MSC means Mesenchymal Stromal Cells. The terms Mesenchymal Stem Cell, Medicinal Signaling Cells, Multipotent Stromal Cells are not recommended. The term "stem" can be used if there is evidence for self-renewal and differentiation properties.

2. Tissue of origin must be described. If the cell is from bone marrow, it will be called bone marrow mesenchymal stromal cell.

3. Mesenchymal stromal cells are used to describe the heterogeneous populations of adherent cells. Characterization using several functional assays to define and exclude some cells must be done. It enrolls RNA analyses of selected genes, immunophenotyping assay, protein analysis of MSC secretome and IFNg activation assays. It requires attention that those assays are to be informed by the intended therapeutic mode of actions.

The most cited MSCs are the Bone marrow-MSC (BM-MSC) and adipose tissue-MSC (AT-MSC). Other sites are also well known such as Umbilical cord MSC (UC-MSC) and Wharton's Jelly MSC (WJ-MSC). Each one has different characterization patterns but all have paracrine effects and immunomodulatory properties, however with different amounts of molecules secreted by each one. For instance, AT-MSC shows a higher pro-angiogenic pattern than BM-MSC and WJ-MSC. WJ-MSC shows an increased expression of inflammatory cytokines and chemokines than BM and AT-MSC [16]. MSC secretome not only includes molecules secreted by them but also extracellular vesicles (EV) productions that reflect in its internal content the same pattern of MSC from origin. For clinicals trials, it's a quite exciting way to treat with MSC without MSC *per se*. In this sense, there are several clinicals trials ongoing using EV from MSC.

But again, regarding the use of EV at clinicals trials, the cell culture protocol standardization, as well as detailed description of isolating methods, requires more attention [13, 17]. For immune regulation capabilities of MSC, ISCT describes assays to standardize the protocols for clinicals trials. Several researchers and groups summarize three assays that must be followed by all clinicals trials: real time PCR of selected gene products, immunophenotyping assays by flow cytometry and secretome assays [18]. In addition, clinicals parameters such as time to administer MSC, dosage, delivery, homing, fresh or frozen MSC, autologous or allogeneic transplantation, etc., all these can generate different responses for patient's' treatment. Thus, more quality control to clinicals trials must be done [19].

3. Healing mechanisms of mesenchymal stromal cells

The physiological and clinical properties of MSCs include not only differentiation potential but also maintenance of tissue homeostasis, immunomodulation, secretion of particles and molecules, and of course, tissue regeneration/healing [20].

Initially, it was believed that MSCs could act directly in the tissue repair and regeneration through migration and engraftment to the site of injury, differentiating into functional local cells and promoting regeneration to the damaged tissue. However, it is now understood that MSCs major effects are promoted largely through secretion of modulatory factors (paracrine activity) and less due to its tissue replacement [21].

In this sense, the ability of regeneration and healing of tissue depends on multiple factors. In the aspect of wound healing, for example, different cell types are involved, including platelets, macrophages, fibroblasts and MSCs. Thus, the balance among proinflammatory M1 macrophages, transformation to anti-inflammatory M2 macrophages and fibroblast extracellular matrix production are crucial to the process of healing. For instance, Adipose-tissue derived MSCs (AT-MSC), as well as its derived exosomes, have been reported to induce M2 macrophage phenotype, modulating the inflammatory process and to enhance the proliferation and migration of fibroblast, contributing to the wound healing process [22].

MSC paracrine signaling can act as anti-inflammatory, anti-fibrotic and pro-angiogenic effects leading to tissue healing and regeneration. In this case, MSCs have been shown to promote accelerated peptic ulcer healing leading to higher proliferative cells population over the ulcer margin, by increasing vascularity in the site of lesion with increased expression of interleukin-10, an anti-inflammatory cytokine, resulting in ulcer healing, such as reepithelization, angiogenesis, and reduced inflammation [23].

Furthermore, this triad process of healing of MSCs based on its anti-inflammatory, anti-fibrotic and pro-angiogenic effects was observed in many studies confirming the pleiotropic effect of these cells during therapeutic process. Briefly, the use of MSCs in ischemic diseases have also been explored. In this scenario, transplantation of MSCs induced angiogenesis with reported differentiation of MSCs into endothelial cells to compose new blood vessels in the infarcted cardiac tissue. Classically, MSCs have been used in Graft versus host disease (GVHD) and autoimmune diseases and have presented decreasing of global inflammatory process with modulation of inflammatory cells (lymphocytes, NK cells, macrophages) and expanded survival or reduced the use of corticoids by transplanted patients [24]. MSCs paracrine secretion of extracellular vesicles or soluble factors may also contribute to angiogenic or immunomodulatory activity in the ischemic heart and brain, even leading to activation of endogenous cardiac stem cells responsible for myocardial regeneration [25].

4. Mesenchymal stromal cells at niche

Cellular turnover varies immensely among the human body tissues. Skin and gut epithelia are replenished every 3–5 days. On the other hand, a neuron's lifespan is huge [26]. This turnover is regulated by stem cells in adult tissues. How these stem cells are *in vivo* and how they keep the homeostasis of tissue is a challenging subject. Despite this, it is known that the MSC and the stem cells live together in specific areas called niches.

A niche is an area of a tissue that provides a specific microenvironment, in which stem cells are present in an undifferentiated, quiescent and self-renewable state. The niche is composed of: (1) a population of stem cells; (2) a population of stromal cells, mainly MSC; (3) an extracellular matrix in which stem cells, stromal cells and molecular cues are embedded; (4) blood vessels support; and (5) neural inputs [27].

The niche is the place where humoral, neuronal, local (paracrine), positional (physical) and metabolic cues interact with each other to regulate stem cell fate [28]. MSC also lives in this environment and has a crucial role in the niche. The cross-talk between stem cells and MSC is very important to both cells. Cells of the niche, mainly MSC, interact with the stem cells to maintain them or promote their differentiation. And tissue homeostasis depends on this balance [27, 29].

The role of MSC in the niche has been studied in recent years. MSC may be the cell that sustains the niche and the cell that keeps the tissue stem cell in the quiescent state. MSC can secrete soluble factors, produce extracellular matrices due to its sensing of the extracellular signals and thus regulate stem cell fate [30].

MSC can be found in every vascularized tissue. Several studies have demonstrated a population of MSC in different tissues, mainly the ones highly vascularized. Following the minimal criteria defined by ISCT, several studies have demonstrated that MSC are the perivascular cells in tissues. Crisan et al. have isolated cells phenotypically positive for pericytes markers (CD146, NG2 and PDGF-Rβ2) from placenta, adipose tissue, pancreas and skeletal muscle and when cultured these cells shown MSC patterns [31]. Not only microvascular pericytes have been described to be the MSC origin cell but also adventitial perivascular cells [32].

Are the *in vitro* MSC the *in vivo* pericytes? Some authors state that MSCs are cell culture artifacts [33]. They disagreed that MSCs are pericyte because since our body is extremely vascularized thus the MSC population should be huge enough to guarantee efficient repair after injury. However our regeneration is not so efficient. In this sense, it has been shown that pericytes *in vitro* generate a cell similar to MSC, but *in vivo* all the pericyte functions may not release them to act as MSC [17]. In addition, if all MSC should be pericytes *in vivo*, then all MSC *in vitro* should be the same, and they are not. MSC from adipose tissue differs phenotypically (CD markers, secretion of molecules, etc.) from bone marrow-MSC, that differs from Wharton Jelly MSC, that differs from cord blood MSC etc. However, there is a hypothesis of an imprinting of tissue source on MSC properties that make tissue-MSC differs from each other [34]. All these opposite points of view show that the search for MSC *in vivo* continues.

Of note, all the knowledge on the MSC field achieved until now is obtained from cultured cells, expanded ex-vivo. In addition, in a plastic dish, MSC is not a pure population. The isolation methods and expansion in culture conditions did not exclude other cells from rising together. They are a heterogenous population in these conditions. Single cell RNA sequencing studies demonstrate that MSCs are heterogeneous and moreover MSC from different sites differs from each other [34–36].

Since most clinicals trials have been using ex-vivo expanded MSC and showing mild positive results *in vivo*, some researchers claim that the stimulation of the niche and their endogenous MSC should be a better option than administered exogenous MSC [30, 37, 38]. Thus, knowing how a niche works and how to properly stimulate it may result in better clinicals outcomes.

Hypoxic areas in the niches are common. At the bone marrow niche, the concentration of O2 is near 3%. Indeed, tissue O2 concentration may vary from 1 to 5% [39]. Several studies in rodents models as well as with human BM-MSC have demonstrated that a hypoxic condition increased osteogenic capabilities [40], increase the expression of pro-angiogenic factors [41], enhance MSC immunosuppression profile [42], maintain genomic stability [43], etc.

Since hypoxia has a huge effect on MSC metabolism, it is clear that energy metabolism can also be linked to MSC cross-talk to stem cells or its stemness. Several works have been studying the energy metabolic process at MSC. The homeostasis state of MSC can be regulated by metabolic signals leading to its stemness of MSC as described by Sun et al. [44]. They show that low levels of sodium lactate, upregulation of glycolysis, both induced by lysine demethylase 6B (KDM6B), can maintain MSC stemness. Indeed, energy metabolism is extremely important in the activation/differentiation of MSC [45]. At the pathological stage, glucose, fatty acid, and amino acid metabolism are altered at MSC. If those pathways could be restored, tissular homeostasis can also be restored [46].

Extracellular matrices (ECM) can also be regulated by MSC. Beyond the structural scaffold, ECM is an acellular 3D structure that is in close contact with the cells. ECM is composed of several proteins (mainly collagen and elastin), glycosaminoglycans and proteoglycans. ECM participates in cell adhesion and in signaling through mimicking several receptors. In addition, mechanical patterns of ECM can also interfere in cell response, such as stiffness [47, 48]. During injury, ECM can be remodeled. Stromal cells, including MSC, secrete more ECM to reconstruction, helping other cells to migrate to this injury site. We will exploit it below regarding MSC secretome.

5. Metabolically activated mesenchymal stromal cells

5.1 The MSC secretome

The MSC secretome is composed of a soluble fraction of bioactive molecules (cytokines, chemokines and growth factors) and particles (extracellular vesicles and exosomes, responsible for the delivery of microRNAs and proteins) with several regulatory effects such as (1) anti-inflammatory; (2) pro-angiogenic; (3) stimulation of endogenous progenitor cells; (4) anti-apoptotic; (5) anti-fibrotic; and (6) anti-oxidant [49]. In addition, there is secretion of extracellular vesicles (exosomes, microvesicles and apoptotic bodies). Inside these vesicles, there are a pool of active molecules (enzymes, receptors, cytokines, chemokines, miRNA, DNA) that can perform the same function of its mother cells (See **Figure 1**) [50, 51].

The whole MSC secretome, which is composed of proteins, nucleic acids, lipids, carbohydrates and extracellular vesicles can also be obtained from MSC-derived conditioned medium (MSC-CM). The soluble component of the secretome and their extracellular vesicles may be then separated with the use of specific methodologies as centrifugation, filtration and chromatography [2]. MSC-CM and extracellular vesicles are enriched with various regulatory components, including transforming factor-β (TGF- β), hepatic growth factor (HGF), indoleamine 2,3-dioxygenase-1 (IDO-1), prostaglandin E2 (PGE2), interleukin (IL)-10, IL-1 receptor agonist (IL-1Ra) and others. Thus, the exposure of different cells to MSC-CM or extracellular vesicles induces different responses depending on the secreted factor available [52].

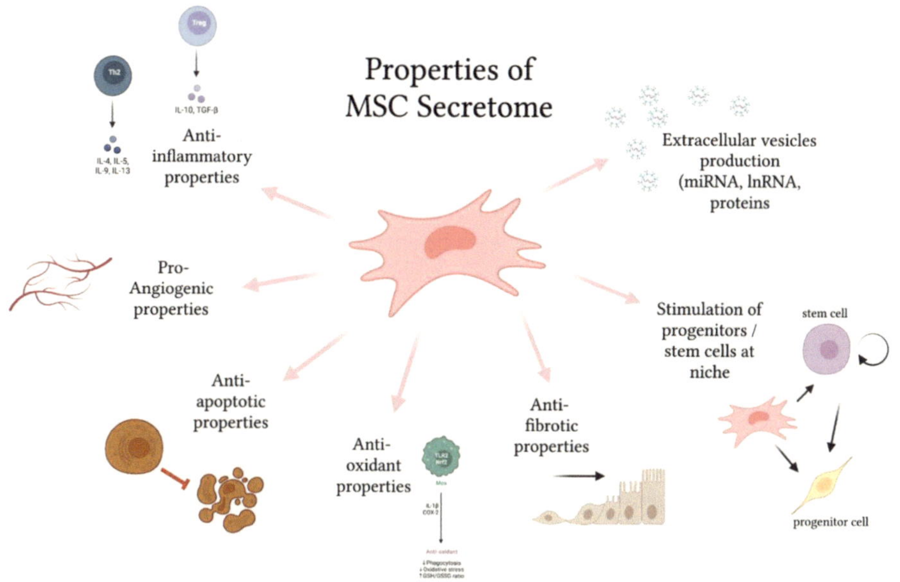

Figure 1.
The MSC secretome. Different molecules are secreted by MSCs. Some of them may induce modulation of immune response, by the expression of cytokines and chemokines to act as anti-inflammatory. It mainly occurs by induction of lymphocytes T helper 2 (TH2) and T regulatory cells (Treg). Molecules secreted by MSC can also be pro-angiogenic, anti-apoptotic, antioxidant and anti-fibrotic. Furthermore, MSCs secretome promotes the stimulation of other stem cells at niche. Extracellular vesicles, also considered part of MSC secretome, contain several molecules and RNA (miRNA, lnRNA) and proteins that may act over other cells through interaction with surface receptors or entering the contact with neighboring cells. IL, interleukin; TGF- β, transforming factor-β; COX-2, Cyclooxygenase-2; GSH, glutathione; GSSG, glutathione disulfide; miRNA, MicroRNA; lnRNA, long noncoding RNA; Th2, T helper 2; Treg, regulatory T cells; TLR2, toll-like receptor 2; Nrf2, nuclear factor erythroid 2-related factor 2; and Mox, oxidized phospholipids-activated macrophages phenotype. Created with BioRender.com.

As appointed by Filidou et al., the anti-inflammatory, anti-fibrotic and tissue regeneration properties of MSC-CM promote *in vivo* and *in vitro* beneficial effects on different disease models that, in general, damages the tissues. Specifically, the use of AdMSC-CM resulted in reduced expression of inflammatory chemokines and cytokines in human pulmonary subepithelial myofibroblasts in response to exposure to IL-1α and Tumor Necrosis Factor-α (TNF-α) and also to TGF-β-induced fibrotic responses in these cells. In addition, the authors showed a reduction in chemotaxis (CCL and CXCL), inflammatory (IL-1α) and fibrotic (collagen Type III) molecules mRNA and protein expression by CM derived from human AT-MSC [53]. Furthermore, an experimental model of preeclampsia induced by bacterial lipopolysaccharide (LPS) showed that human placenta-derived MSC-CM reduced expression levels of TNF-α and IL-6 in the mice placenta, while also reducing the expression of the anti-angiogenic factor sFlt-1 [54].

5.2 MSCs extracellular vesicles

The use of MSC of extracellular vesicles (MSC-EVs) has attracted attention for its ability to promote beneficial effects even when MSC itself is not present [55]. MSC biological characteristics may compromise its use as a therapeutic agent. MSCs proliferation decreases over culture passages, studies report concerns about increased tumorigenicity and the uncertainty of MSCs fate after venous injection calls attention to weak points of such therapeutic strategy [56].

MSC-EVs are classified according to their size, which ranges from apoptotic bodies (> 1000 nm), to microvesicles (100–1000 nm) and exosomes (30–200) [57]. Up to date, 45 MSC-EVs clinical trials are registered in Clinicaltrials.gov [clinicaltrials.gov "(mesenchymal stromal cells OR mesenchymal stem cells) AND (extracellular vesicle OR exosome OR microvesicle)", October 2022] of which 5 studies are currently at phase 3, including therapeutic approaches to rhinitis pigmentosa (NCT05413148), SARS-CoV-2 infection and acute respiratory distress syndrome (NCT05216562, NCT05354141), diabetes mellitus type 1 (NCT02138331) and stroke (NCT01716481).

The MSC-EVs content vary depending on the derived cell, microenvironment and physiological conditions, thus can be modulated by preconditioning methods, but are known to contain molecules such as messenger RNA, microRNAs, others regulatory RNAs (i.e., lncRNAs, microRNAs, ceRNAs, and circRNAs), enzymes, receptors, cytokines, chemokines and growth factors. Once released to the extracellular environment from the donor cell, MSC-EVs can be internalized by another cell via endocytosis or trigger responses through receptor-ligand interaction acting as a paracrine and endocrine agent. Furthermore, these MSC-derived EVs are capable of homing to injured tissue, having immunosuppressive effects or others similar to those promoted by transplanted MSCs [57].

MSC-EVs can be used in almost all therapy conditions that native MSCs are used or predicted to be; for instance, the MSC-derived exosomes were utilized in wound healing and was verified the promotion of collagen synthesis and proliferation and migration of fibroblasts and keratinocytes, important cells in the mechanisms of wound regeneration. Furthermore, it was detected that these effects are, greatly in part, promoted by microRNA in the exosomes. The therapeutic effects of microRNA derived from MSC-exosomes was widely reported in several studies showing benefits in the treatment of chronic skin ulcers, bone repair, promoting the immunomodulation in favor of inflammation resolution, improving angiogenesis, neurogenesis, macrophage polarization and limiting cardiac fibroblast proliferation, and improving tissue function after ischemia-reperfusion injury [55].

Using AT-MSC-derived exosomes Heo and Kim [58] reported a reduction in the gene expression of pro-inflammatory molecules as TNF-α, IL-6 and IL-8 which were induced by LPS in the THP-1 cell line, while the expression levels of anti-inflammatory CD163, ARG1, CD206, TGF-β1 and IL-10 were shown to be increased in the LPS + exosomes group. The treatment of human umbilical vein endothelial cells (HUVECs) with AT-MSC-derived exosomes increased the proliferation of HUVECs and gene expression level of pro-angiogenic genes like angiopoietin1 and flk1, while reducing the expression of those with detrimental vascular function as vasohibin-1 and thrombospondin-1. Remarkably, the expression of miR-132 and miR-146a were found increased in exosome-treated HUVECs, and these microRNAs bound to the anti-angiogenic genes thrombospondin-1 and vasohibin-1, respectively [58].

Furthermore, a study aiming to elucidate the role of MSC-EVs in mitochondrial damage showed a reversion of mitochondrial DNA deletion to the treated group that was not observed in injured renal tubular cells. Utilizing an *in vivo* model of acute kidney injury, the authors observed the same effect through up-regulation of mitochondrial factor A pathway activity. These findings suggest that MSC-EVs therapeutic effects can also be related to improvement of mitochondrial function in diverse diseases in addition to its role as anti-inflammatory, antioxidant and anti-apoptotic as observed in many other injury models [59].

Finally, use of MSC-EV are promisor therapies that comprehends the major effects attributed to MSC secretome, promoting desired improvements in regeneration and immunomodulation as that offered by paracrine effects credited to MSCs.

5.3 Activation signaling and pre-conditioning

The paracrine effect of MSC is highly dependent on the microenvironment around MSCs. The MSCs have some sensors receptors (i.e., TLRs, AhRs, TNFRs, and IFNRs) which act as an "antenna" that captures external signals that drive a special cellular effect. In contrast, in the absence of stimuli the MSCs show little to no expression of molecules responsible for their function, for instance, the immunomodulatory profile, such as the expression of human leukocyte antigen (HLA)-I and intercellular adhesion molecule-1 (ICAM-1).

The production of molecules from MSC secretome can be stimulated by the presence of inflammatory components that induce an immunomodulatory phenotype on MSCs [60]. The MSCs preconditioning with inflammatory factors such as IL-1β and interferon gamma (IFN-γ) result in augmented production of modulatory components by MSCs which can influence and regulate other cell types, such as macrophages, to acquire a regulatory phenotype [61]. Hence, exposure of MSCs to an inflammatory environment, containing for example IFN-γ and TNF-α cytokines, induces MSCs to start the production of specific molecules which will play a role as immunoregulators [62].

TNF-α is one of the first secreted cytokines during an inflammatory event. TNF-α binds to two distinct receptors, TNFR1 and TNFR2. While TNFR1 is expressed ubiquitously, few cellular populations express TNFR2, including immune cells and MSCs. In MSCs, TNFα/TNFR2 interaction promotes the expression or secretion of pro-angiogenic and cytoprotective mediators. Beldi et al. investigated the role of TNFR2 in MSCs and found that in comparison to TNFR2+ wild type MSCs, MSCs lacking TNFR2 were less immunosuppressive to CD4 and CD8 T cells when reducing cellular proliferation and cytokines production in T cells. Furthermore, while TNF-α stimuli did not result in increased expression of early HLA-I, MSC exposure to IFN-γ increased expression of HLA-I, an indicator of MSC activation [63].

Regarding the MSCs-EV, preconditioning may also be expected to happen. In fact, cultures of PBMCs in presence of MSC-derived exosomes preconditioned with TNF-α and IFN-γ, resulted in cytokines shifting: 34 inflammatory cytokines and chemokines were found to be downregulated and several anti-inflammatory, as IL-10, were upregulated. Moreover, preconditioning of MSC-exosomes with atorvastatin enhanced angiogenesis when compared to non-pretreated MSCs in myocardial infarction injury; and also TNF-α preconditioning of adipose tissue MSCs promoted higher osteoblast differentiation upon exosome treatment [64, 65].

Although showing interesting results during preclinical i*n vivo/in vitro* studies, the preconditioning of MSCs is still performed with human and non-human recombinant factors with lack of consistency at human clinical trials. To overcome this, the use of fresh human derived products can be an effective resource when we take in mind the use of preconditioning on the clinical scale. Thus, platelets or platelet-rich plasma have been proposed as a beneficial enhancer to therapeutic properties of MSCs. These platelets or platelet-rich plasma medium stimulates proliferation of MSCs and offer protection against oxidative stress, mainly due to the release of growth factors that exerts beneficial effects on MSCs. Further, the transfer of platelets mitochondria to MSCs stimulates wound-healing activity. And, the incubation of MSCs with full functional platelets, but not with dysfunctional mitochondria platelets, resulted in increased expression of pro-angiogenic genes [66].

Moreover, other similar approaches aiming to control extrinsic factors in MSCs modulation are available. Considering these aspects, some MSCs variability to its activity is found in response to (i) source or location, that is, Bone marrow-derived MSCs or Adipose tissue-derived MSCs, (ii) passage number in culture, and (iii) oxygen concentration and presence of different compounds in the environment, such as pharmacological agents. These extrinsic factors are useful methods of preconditioning MSCs and can be used to improve its therapeutic potential regulating the secretory MSCs profile. These effects can be reached using the hypoxic environment of cell culture, inflammatory cytokines, pharmacological compounds, and 3D cell culture models [60].

Finally, an interesting cell culture method of 3-dimensional culture can be used as a preconditioning factor as well. In this culture method, the physiological conditions seen as in the *in vivo* cell environment are replicated, as the spheroid 3D culture promotes a physiological-like environment, like those found in MSCs niche. In this spheroid culture, internal cells receive lower oxygen levels than MSCs in the surface of the 3D structure, creating a hypoxic environment. These spheroid cultured MSCs presented an increase in cytoprotective factors and enhanced proliferation, with increased immunomodulatory factors expression, along with elevated angiogenic, anti-fibrotic and anti-apoptotic activity [67].

6. Mesenchymal stromal cell dysfunction

Our knowledge on MSC is focused on how healthy MSC responds to an injury, by secreting several molecules, trying to rebuild the tissue homeostasis. At cellular therapy, healthy exogenous MSCs are administered to patients and in response to the injury, this exerts its regenerative role and helps heal the damage.

However, there are several conditions that can damp MSC capabilities of healing *in situ*. Autologous transplantations of MSC have mild results in clinical trials despite animal models generated great results [68]. Allogeneic transplantation of MSCs seems to have better outcomes. Moreover, the functional decline of MSCs has been associated with a pathophysiological driver of several diseases and aging [69].

Regenerative properties of endogenous MSC can be decreased *in vivo* and *in vitro*. Aging, metabolic changes due to pathologies and epigenetics changes can interfere at MSC *in vivo*. *In vitro*, MSC can be altered by cell culture passages (senescence), by storage at cryogenic conditions, by culture conditions (such as serum deprivation), by cell contact loss, by normoxia, etc. [69–71]. We will further exploit some of the MSC disabilities below. See **Figure 2**.

6.1 Aging: epigenetic and PMT at MSC disruption

Aging is a settled multifactorial process. Lopez-Otin has described 9 hallmarks that represent common denominators of aging: (1) genomic instability, (2) telomere attrition, (3) epigenetic alterations, (4) loss of proteostasis, (5) deregulated nutrient-sensing, (6) mitochondrial dysfunction, (7) cellular senescence, (8) stem cell exhaustion, and (9) altered intercellular communication [72]. Herein, we will focus on some of these hallmarks and its impact on MSC.

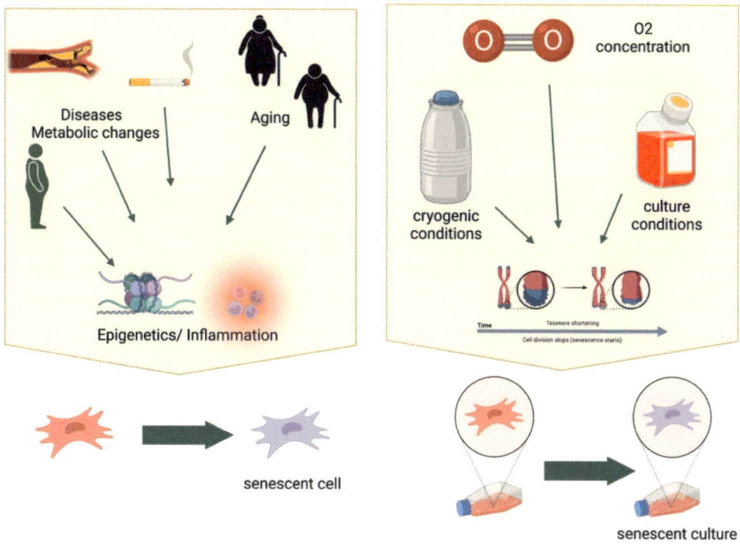

Figure 2.
Senescence In vivo and in vitro: Conditions that lead to MSC senescence. Several factors are correlated with MSC senescence. In vivo, pathologies, metabolic diseases and aging may interfere at epigenetic levels and/or generate chronic inflammation responses that can cause MSC senescence. In vitro, other factors contribute to a senescence culture such as O_2 concentration, culture and cryogenic conditions etc. mainly due to telomerase disruption. Created with BioRender.com.

Aging and age-related diseases have been associated with the higher number of senescent cells in the tissue [73]. In 1995, Dmiri et al. have described the quantification of the amount of beta-galactosidase as a biomarker of senescence. He demonstrated that a higher amount of beta-galactosidase is present at senescent cultured fibroblasts in human cells [74]. Since then, several works have been demonstrating that beta-galactosidase is not a reliable marker, so the search for a biomarker for senescence is still ongoing. It has been demonstrated that p16Ink4a-positive senescent cells accumulate with age in multiple tissues [75]. DNA damage response (DDR) is induced in healthy MSCs leading to the activation of the two main signaling pathways p19*ARF* and p16*INK4A* [73].

Evidence suggests that MSC senescence is a dynamic process driven by epigenetic and genetic changes. Moreover, aging can be impacted by both environmental and inherent factors. Genetics factors are associated with long term mutations in DNA that lead to failure of the replicative state of the cell. Environmental factors that do not change DNA, also affect cell cycle. To date, epigenetics refers to the study of heritable phenotypic alterations linked to differential gene expression when the same DNA sequence is maintained [76]. Epigenetic dysregulation is associated with (1) DNA-based mechanisms: DNA methylation and histone modifications (2) RNA-based mechanisms: noncoding RNAs and RNA modifications [69].

These genetics and epigenetic modifications can interfere directly with MSC by inducing the arrest of cell cycle, by producing a defective ECM niche production and by disrupting the MSC differentiation leading to tissue aberrations evidentiated at aging and disease [69, 76]. Several articles described differential methylation patterns at MSC isolated from young X olders patients. Moreover, these differential methylation patterns were also observed at long term cultures *in vitro* of MSC [77, 78].

Single-cell sequencing analysis of young X elderly BM-MSC have shown that young MSC have higher expression of genes related to tissue regeneration. Moreover, at young BM-MSC there is a cluster of cells that have a lower expression of genes of proliferation, that characterize them as quiescent cells, so stem cells. And these clusters of cells were not observed at elderly BM-MSC [79].

Not only epigenetic modification in DNA but also modifications in protein has huge importance in the differentiation processes. Protein post-transductional modifications (PTM) are protein modifications caused by adding groups of phosphates, acetyl, methyl, etc. in one or multiple amino acids and/or caused by proteolytic cleavage by ubiquitin [80]. These modifications can determine its activity state, localization, turnover, and interactions with other proteins. At MSC, PMT has been associated with differentiation to osteogenic lineage [81] Osteogenic differentiation of BM-MSC has been linked to *O*-GlcNAc cycling to the Runx2-dependent regulation of the early ALP marker [82].

In addition, aging decreases the number of stem cells in the niche, but not only it, aging also affects MSC and stem cell response due to metabolic and epigenetics changes [83]. Muscle stem cells (satellite cells) in aging tend to be converted to a fibroblast lineage instead of myogenic lineage [84, 85]. Several authors have been demonstrating that niche ECM stiffness leads to the aging process, dampening regeneration of the tissue and its homeostasis and moreover, leading to stem cell aging. In central nervous systems (CNS), ECM niche stiffness of oligodendrocyte progenitor cells (OPCs) have been related to aging processes mainly through the mechanoresponsive ion channel Piezo1 [83, 86].

Immunophenotypic profile of MSC can also be affected by senescence. Laschober et al. described that CD295 (leptin receptor or LEPR) have been found to increase during MSC senescence and it correlates with reduced proliferation capacities of MSC [87].

6.2 MSC senescence in culture conditions

In culture conditions, long term cultures are not welcome to be used in therapy due to its altered therapeutic profile. These cells became large and flattened ("sunny side up egg" morphology), less proliferative and less responsive. Senescence in culture characterized by the arrest of cell cycle. It is a known issue, as described by Hayclifk in fibroblast cultures [88]. Four types of senescence have been distinguished: replicative senescence (RS), oncogene-induced senescence (OIS), stress-induced premature senescence (SIPS), and developmental senescence [29, 89, 90].

Stress conditions at culture, such as adaptations to 2D culture, O2 concentration, confluency condition, the amount of nutrients even though the exposure to light lead to modifications that cause its arrest in the G0 phase of cell cycle of MSC [91, 92].

Cryopreservation is also a concern regarding MSC stability. Dimethyl sulfoxide (DMSO) has been the gold standard agent for cryobiology. However, the use of DMSO has been associated with *in vitro* toxicity. Since it has been associated with DNA methylation processes, DMSO affects many cellular processes and dysregulation of gene expression [93]. Mol et al. described that fresh culture MSC have a trend to have better outcomes for acute graft versus host disease (GvHD) and tissue injury in hemorrhagic cystitis than freeze-thawed MSC. Fresh MSCs have higher mRNA expression of IDO after 24 h IFNg priming, showing higher immunomodulatory properties than cryopreserved MSC [94].

6.3 Inflammation and senescence of MSC

Cycle arrest occurs due to a persistent DNA damage response (DDR) caused by either intrinsic (oxidative damage, telomere attrition, hyperproliferation) or external insults (ultraviolet, γ-irradiation, chemotherapeutic drugs) [95]. The more DNA damage, the more cell death, senescence and tissue dysfunction contributing to aging. Growing evidence has been describing that inflammation can also lead to DNA damage [96].

DNA damage induces the expression of type I interferons and other inflammatory factors [97]. The connection between DNA damage and inflammation is through the cytoplasmic DNA sensing pathway. Micronuclei formations (formed due to DNA damage during mitosis) can stimulate the cell senescence throughout cyclic GMP–AMP synthase (cGAS), a DNA sensor that stimulates STING (stimulator of interferon genes). To prevent undesired inflammation, besides cGAS-STING pathway, there are also the deoxyribonucleases (DNases) in the cytoplasm, digesting excessive DNA, serving as a negative regulator of cytoplasmic DNA. There are two major DNases in the cytoplasm: DNase2α (encoded by *DNaseII*) and TREX1 (originally designated DNaseIII). Intriguingly, both DNases are downregulated in senescent cells, contributing to aberrant cytoplasmic DNA sensing and inflammation [98].

Inflammaging, a term to define a chronic, low-grade sterile inflammation frequently observed during aging [99]. It is a macrophage centered process, involves several tissues and organs, including the gut microbiota, and is characterized by a complex balance between pro- and anti-inflammatory responses [100]. In elderly, the chronic inflammation observed is due to cells in tissue expressing pro-inflammatory cytokines, such as IL-1α, IL-6, TNF, and NF-κB activity and other inflammatory factors [101]. Chronic inflammation during aging and its negative outcome is supported by clinical data in kidney [102], liver [103], lung [104] etc.

Since MSC are perivascular cells and that they have a close connection with circulant factors in blood, it is possible to consider MSC with a central role in inflammaging, together with macrophages [105]. Rejuvenation strategies, such as culturing MSC with serum from older rats and parabiosis, showed a lower proliferation rate and survival of MSC exposed to serum from elderly subjects [106]. Thus, there are circulant molecules/cytokines that can impair MSC functions in aged individuals. Higher amounts of circulant beta-catenin and SMAD3 have been associated with senescence MSC profile [29]. More basic research must be done in this area.

Senescent cells are functional cells. Senescent cells were shown to secrete a range of inflammatory factors, which was termed the 'senescence-associated secretory phenotype' (SASP) [107]. The SASP mediates many of the cell-extrinsic functions of senescent cells. The SASP has its physiologic role: (1)by maintaining the SASP profile of the senescent cell (maintaining cell cycle arrest and SASP expression), (2) by eliciting immune response to generate a senescent cell clearance and (3) by secreting ECM and angiogenic factors leading to tissue regeneration [90, 108]. However, SASP has also deleterious effects by promoting inflammation (leading to inflammaging) and, potentially, tumor progression in neighboring cells. The correlation of SASP and inflammaging is beginning to be investigated using models to detect and eliminate the senescent cell (the INK-ATTAC model) [90, 108]. SAPS at MSC a is also related to higher secretion of extracellular microvesicles in aged subjects, as well its higher amount of microRNA content [109, 110] .

Interestingly, as it was described early in this chapter, MSCs have potent anti-inflammatory functions, whereas senescent MSCs play a pro-inflammatory role due to SASP, which has been considered a major cause of aged MSCs' detrimental effects [111]. In accordance with this, HMGB1 secreted by senescent fibroblasts is recognized by TLR4, followed by increase in SASP secretion [112]. These findings establish the critical role played by innate immune sensing mechanisms in regulating senescence [91].

6.4 Diseases and MSC

At MSC therapy, attention must be done regarding the pathological state of the patients at the harvest of MSC, since aging and pathological diseases can interfere at this isolated MSC. Moreover, when treating the patient, the pathogenic milieu where exogenous MSC is administered requires attention, because it may interfere with the MSC mechanism of action.

Obesity can impact BM-MSC. Ulum et al. described BM-MSC from patients with high body mass index (BMI) are more senescent, have disrupted differentiation to osteogenic and adipogenic cells, and highly expressed endoplasmic reticulum genes related to stress [113]. Diabetes can regulate AT-MSC as described by Abu-Shahba et al. They isolated AT-MSC from diabetic and non-diabetic patients and demonstrated that IL-1b is highly expressed in AT-MSC from diabetic patients [114].

7. Conclusion and new perspectives

The knowledge of MSC still requires much more research to elucidate its regenerative properties. More than 30 years of research and yet there is a lot to understand. The search for a better performance in MSCs cultures, the secretome profile, how to stimulate MSC to secrete higher amounts of such molecules using preconditioning techniques or niche stimulation, how MSC acts *in vivo*: a lot of questions with some clues, but far from the right answer.

Conflict of interest

The authors declare no conflict of interest.

Author details

Patricia Semedo-Kuriki[1]*, Gabriel Pereira[2], Danilo Cândido de Almeida[2] and Niels Olsen Saraiva Camara[1]

1 São Paulo University (USP), São Paulo, Brazil

2 Federal University of São Paulo (UNIFESP), São Paulo, Brazil

*Address all correspondence to: patricia.semedo@gmail.com

References

[1] Wright A, Arthaud-Day ML, Weiss ML. Therapeutic use of mesenchymal stromal cells: The need for inclusive characterization guidelines to accommodate all tissue sources and species. Frontiers in Cell and Development Biology. 2021;**9**:632717

[2] Harrell CR, Fellabaum C, Jovicic N, Djonov V, Arsenijevic N, Volarevic V. Molecular mechanisms responsible for therapeutic potential of mesenchymal stem cell-derived Secretome. Cells. 16 May 2019;**8**(5):467. DOI: 10.3390/cells8050467

[3] Krampera M, Le Blanc K. Mesenchymal stromal cells: Putative microenvironmental modulators become cell therapy. Cell Stem Cell. 2021;**28**:1708-1725

[4] Ebrahimi A, Ahmadi H, Pourfraidon Ghasrodashti Z, Tanide N, Shahriarirad R, Erfani A, et al. Therapeutic effects of stem cells in different body systems, a novel method that is yet to gain trust: A comprehensive review. Bosnian Journal of Basic Medical Sciences. 2021;**21**:672-701

[5] Kuntin D, Genever P. Mesenchymal stem cells from biology to therapy. Emerging Topics in Life Sciences. 2021;**5**:539-548

[6] Meacham CE, DeVilbiss AW, Morrison SJ. Metabolic regulation of somatic stem cells in vivo. Nature Reviews. Molecular Cell Biology. 2022;**23**:428-443

[7] Renesme L, Pierro M, Cobey KD, Mital R, Nangle K, Shorr R, et al. Definition and characteristics of mesenchymal stromal cells in preclinical and clinical studies: A scoping review. Stem Cells Translational Medicine. 2022;**11**:44-54

[8] Bianco P, Robey PG, Simmons PJ. Mesenchymal stem cells: Revisiting history, concepts, and assays. Cell Stem Cell. 2008;**2**:313-319

[9] Friedenstein AJ, Chailakhjan RK, Lalykina KS. The development of fibroblast colonies in monolayer cultures of Guinea-pig bone marrow and spleen cells. Cell and Tissue Kinetics. 1970;**3**:393-403

[10] Caplan AI. Mesenchymal stem cells. Journal of Orthopaedic Research. 1991;**9**:641-650

[11] Pittenger MF, Mackay AM, Beck SC, Jaiswal RK, Douglas R, Mosca JD, et al. Multilineage potential of adult human mesenchymal stem cells. Science. 1999;**284**:143-147

[12] Sipp D, Robey PG, Turner L. Clear up this stem-cell mess. Nature. 2018;**561**:455-457

[13] Galipeau J, Weiss DJ, Dominici M. Response to nature commentary "clear up this stem-cell mess.". Cytotherapy. Jan 2019;**21**(1):1-2

[14] Dominici M, Le Blanc K, Mueller I, Slaper-Cortenbach I, Marini F, Krause D, et al. Minimal criteria for defining multipotent mesenchymal stromal cells. The International Society for Cellular Therapy position statement. Cytotherapy. 2006;**8**:315-317

[15] Viswanathan S, Shi Y, Galipeau J, Krampera M, Leblanc K, Martin I, et al. Mesenchymal stem versus stromal cells: International society for cell & gene therapy (ISCT®) mesenchymal stromal cell committee position statement on nomenclature. Cytotherapy. 2019;**21**:1019-1024

[16] Amable PR, Teixeira MVT, Carias RBV, Granjeiro JM, Borojevic R. Protein synthesis and secretion in human mesenchymal cells derived from bone marrow, adipose tissue and Wharton's jelly. Stem Cell Research & Therapy. 2014;**5**:53

[17] Pittenger MF, Discher DE, Péault BM, Phinney DG, Hare JM, Caplan AI. Mesenchymal stem cell perspective: Cell biology to clinical progress. NPJ Regenerative Medicine. 2019;**4**:22

[18] Galipeau J, Krampera M, Barrett J, Dazzi F, Deans RJ, DeBruijn J, et al. International Society for cellular therapy perspective on immune functional assays for mesenchymal stromal cells as potency release criterion for advanced phase clinical trials. Cytotherapy. 2016;**18**:151-159

[19] García-Bernal D, García-Arranz M, Yáñez RM, Hervás-Salcedo R, Cortés A, Fernández-García M, et al. The current status of mesenchymal stromal cells: Controversies, unresolved issues and some promising solutions to improve their therapeutic efficacy. Frontiers in Cell and Development Biology. 2021;**9**:650664

[20] Zhou T, Yuan Z, Weng J, Pei D, Du X, He C, et al. Challenges and advances in clinical applications of mesenchymal stromal cells. Journal of Hematology & Oncology. 2021;**14**:24

[21] Spees JL, Lee RH, Gregory CA. Mechanisms of mesenchymal stem/stromal cell function. Stem Cell Research & Therapy. 2016;7:125

[22] Heo JS, Kim S, Yang CE, Choi Y, Song SY, Kim HO. Human adipose mesenchymal stem cell-derived exosomes: A key player in wound healing. Tissue Engineering and Regenerative Medicine. 2021;**18**:537-548

[23] Xia X, Chan KF, Wong GTY, Wang P, Liu L, Yeung BPM, et al. Mesenchymal stem cells promote healing of nonsteroidal anti-inflammatory drug-related peptic ulcer through paracrine actions in pigs. Science Translational Medicine. 30 Oct 2019;**11**(516):eaat7455. DOI: 10.1126/scitranslmed.aat7455

[24] Chen J, Luo L, Tian R, Yu C. A review and update for registered clinical studies of stem cells for non-tumorous and non-hematological diseases. Regenerative Therapy. 2021;**18**:355-362

[25] Yong KW, Choi JR, Mohammadi M, Mitha AP, Sanati-Nezhad A, Sen A. Mesenchymal stem cell therapy for ischemic tissues. Stem Cells International. 2018;**2018**:8179075

[26] Sender R, Milo R. The distribution of cellular turnover in the human body. Nature Medicine. 2021;**27**:45-48

[27] Mannino G, Russo C, Maugeri G, Musumeci G, Vicario N, Tibullo D, et al. Adult stem cell niches for tissue homeostasis. Journal of Cellular Physiology. 2022;**237**:239-257

[28] Ferraro F, Celso CL, Scadden D. Adult stem cells and their niches. Advances in Experimental Medicine and Biology. 2010;**695**:155-168

[29] Weng Z, Wang Y, Ouchi T, Liu H, Qiao X, Wu C, et al. Mesenchymal stem/stromal cell senescence: Hallmarks, mechanisms, and combating strategies. Stem Cells Translational Medicine. 2022;**11**:356-371

[30] Sagaradze GD, Basalova NA, Efimenko AY, Tkachuk VA. Mesenchymal stromal cells as critical contributors to tissue regeneration. Frontiers in Cell and Development Biology. 2020;**8**:576176

[31] Crisan M, Yap S, Casteilla L, Chen C-W, Corselli M, Park TS, et al. A

perivascular origin for mesenchymal stem cells in multiple human organs. Cell Stem Cell. 2008;**3**:301-313

[32] Craig DJ, James AW, Wang Y, Tavian M, Crisan M, Péault BM. Blood vessel resident human stem cells in health and disease. Stem Cells Translational Medicine. 2022;**11**:35-43

[33] Murray IR, West CC, Hardy WR, James AW, Park TS, Nguyen A, et al. Natural history of mesenchymal stem cells, from vessel walls to culture vessels. Cellular and Molecular Life Sciences. 2014;**71**:1353-1374

[34] Ménard C, Dulong J, Roulois D, Hébraud B, Verdière L, Pangault C, et al. Integrated transcriptomic, phenotypic, and functional study reveals tissue-specific immune properties of mesenchymal stromal cells. Stem Cells. 2020;**38**:146-159

[35] Medrano-Trochez C, Chatterjee P, Pradhan P, Stevens HY, Ogle ME, Botchwey EA, et al. Single-cell RNA-seq of out-of-thaw mesenchymal stromal cells shows tissue-of-origin differences and inter-donor cell-cycle variations. Stem Cell Research & Therapy. 2021;**12**:565

[36] Barrett AN, Fong C-Y, Subramanian A, Liu W, Feng Y, Choolani M, et al. Human Wharton's jelly mesenchymal stem cells show unique gene expression compared with bone marrow mesenchymal stem cells using single-cell RNA-sequencing. Stem Cells and Development. 2019;**28**:196-211

[37] Liu Y, Ma T. Metabolic regulation of mesenchymal stem cell in expansion and therapeutic application. Biotechnology Progress. 2015;**31**:468-481

[38] Zhou Y, Tsai T-L, Li W-J. Strategies to retain properties of bone marrow-derived mesenchymal stem cells ex vivo. Annals of the New York Academy of Sciences. 2017;**1409**:3-17

[39] Ivanovic Z. Hypoxia or in situ normoxia: The stem cell paradigm. Journal of Cellular Physiology. 2009;**219**:271-275

[40] Lennon DP, Edmison JM, Caplan AI. Cultivation of rat marrow-derived mesenchymal stem cells in reduced oxygen tension: Effects on in vitro and in vivo osteochondrogenesis. Journal of Cellular Physiology. 2001;**187**:345-355

[41] Chacko SM, Ahmed S, Selvendiran K, Kuppusamy ML, Khan M, Kuppusamy P. Hypoxic preconditioning induces the expression of prosurvival and proangiogenic markers in mesenchymal stem cells. American Journal of Physiology. Cell Physiology. 2010;**299**:C1562-C1570

[42] Liu Y, Yuan X, Muñoz N, Logan TM, Ma T. Commitment to aerobic glycolysis sustains immunosuppression of human mesenchymal stem cells. Stem Cells Translational Medicine. 2019;**8**:93-106

[43] Li T-S, Marbán E. Physiological levels of reactive oxygen species are required to maintain genomic stability in stem cells. Stem Cells. 2010;**28**:1178-1185

[44] Sun H, Zhang X, Dai J, Pan Z, Wu Y, Yu D, et al. Sodium lactate promotes stemness of human mesenchymal stem cells through KDM6B mediated glycolytic metabolism. Biochemical and Biophysical Research Communications. 2020;**532**:433-439

[45] Ning K, Liu S, Yang B, Wang R, Man G, Wang D-E, et al. Update on the effects of energy metabolism in bone marrow mesenchymal stem cells differentiation. Molecular Metabolism. 2022;**58**:101450

[46] van Gastel N, Carmeliet G. Metabolic regulation of skeletal cell fate and function in physiology and disease. Nature Metabolism. 2021;**3**:11-20

[47] Hynes RO. The extracellular matrix: Not just pretty fibrils. Science. 2009;**326**:1216-1219

[48] Diller RB, Tabor AJ. The role of the extracellular matrix (ECM) in wound healing: A review. Biomimetics (Basel). 1 Jul 2022;**7**(3):87. DOI: 10.3390/biomimetics7030087

[49] González-González A, García-Sánchez D, Dotta M, Rodríguez-Rey JC, Pérez-Campo FM. Mesenchymal stem cells secretome: The cornerstone of cell-free regenerative medicine. World Journal of Stem Cells. 2020;**12**:1529-1552

[50] Baez-Jurado E, Hidalgo-Lanussa O, Barrera-Bailón B, Sahebkar A, Ashraf GM, Echeverria V, et al. Secretome of mesenchymal stem cells and its potential protective effects on brain pathologies. Molecular Neurobiology. 2019;**56**:6902-6927

[51] Song N, Scholtemeijer M, Shah K. Mesenchymal stem cell immunomodulation: Mechanisms and therapeutic potential. Trends in Pharmacological Sciences. 2020;**41**:653-664

[52] Bogatcheva NV, Coleman ME. Conditioned medium of mesenchymal stromal cells: A new class of therapeutics. Biochemistry. 2019;**84**:1375-1389

[53] Filidou E, Kandilogiannakis L, Tarapatzi G, Spathakis M, Steiropoulos P, Mikroulis D, et al. Anti-inflammatory and anti-fibrotic effect of immortalized mesenchymal-stem-cell-derived conditioned medium on human lung Myofibroblasts and epithelial cells. International Journal of Molecular Sciences. 20 Apr 2022;**23**(9):4570. DOI: 10.3390/ijms23094570

[54] Nuzzo AM, Moretti L, Mele P, Todros T, Eva C, Rolfo A. Effect of placenta-derived mesenchymal stromal cells conditioned media on an LPS-induced mouse model of preeclampsia. International Journal of Molecular Sciences. 31 Jan 2022;**23**(3):1674. DOI: 10.3390/ijms23031674

[55] Hade MD, Suire CN, Suo Z. Mesenchymal stem cell-derived exosomes: Applications in regenerative medicine. Cells. 1 Aug 2021;**10**(8):1959. DOI: 10.3390/cells10081959

[56] Kou M, Huang L, Yang J, Chiang Z, Chen S, Liu J, et al. Mesenchymal stem cell-derived extracellular vesicles for immunomodulation and regeneration: A next generation therapeutic tool? Cell Death & Disease. 2022;**13**:580

[57] Harrell CR, Jovicic N, Djonov V, Arsenijevic N, Volarevic V. Mesenchymal stem cell-derived exosomes and other extracellular vesicles as new remedies in the therapy of inflammatory diseases. Cells. 11 Dec 2019;**8**(12):1605. DOI: 10.3390/cells8121605

[58] Heo JS, Kim S. Human adipose mesenchymal stem cells modulate inflammation and angiogenesis through exosomes. Scientific Reports. 2022;**12**:2776

[59] Zhao M, Liu S, Wang C, Wang Y, Wan M, Liu F, et al. Mesenchymal stem cell-derived extracellular vesicles attenuate mitochondrial damage and inflammation by stabilizing mitochondrial DNA. ACS Nano. 2021;**15**:1519-1538

[60] Ferreira JR, Teixeira GQ, Santos SG, Barbosa MA,

Almeida-Porada G, Gonçalves RM. Mesenchymal stromal cell Secretome: Influencing therapeutic potential by cellular pre-conditioning. Frontiers in Immunology. 2018;**9**:2837

[61] Philipp D, Suhr L, Wahlers T, Choi Y-H, Paunel-Görgülü A. Preconditioning of bone marrow-derived mesenchymal stem cells highly strengthens their potential to promote IL-6-dependent M2b polarization. Stem Cell Research & Therapy. 2018;**9**:286

[62] López-García L, Castro-Manrreza ME. TNF-α and IFN-γ participate in improving the Immunoregulatory capacity of mesenchymal stem/stromal cells: Importance of cell-cell contact and extracellular vesicles. International Journal of Molecular Sciences. 2 Sep 2021;**22**(17):9531. DOI: 10.3390/ijms22179531

[63] Beldi G, Khosravi M, Abdelgawad ME, Salomon BL, Uzan G, Haouas H, et al. TNFα/TNFR2 signaling pathway: An active immune checkpoint for mesenchymal stem cell immunoregulatory function. Stem Cell Research & Therapy. 2020;**11**:281

[64] Huang P, Wang L, Li Q, Tian X, Xu J, Xu J, et al. Atorvastatin enhances the therapeutic efficacy of mesenchymal stem cells-derived exosomes in acute myocardial infarction via up-regulating long non-coding RNA H19. Cardiovascular Research. 2020;**116**:353-367

[65] Lu Z, Chen Y, Dunstan C, Roohani-Esfahani S, Zreiqat H. Priming adipose stem cells with tumor necrosis factor-alpha preconditioning potentiates their exosome efficacy for bone regeneration. Tissue Engineering. Part A. 2017;**23**:1212-1220

[66] Levoux J, Prola A, Lafuste P, Gervais M, Chevallier N, Koumaiha Z, et al. Platelets facilitate the wound-healing capability of mesenchymal stem cells by mitochondrial transfer and metabolic reprogramming. Cell Metabolism. 2021;**33**:283-299.e9

[67] Kouroupis D, Correa D. Increased mesenchymal stem cell functionalization in three-dimensional manufacturing settings for enhanced therapeutic applications. Frontiers in Bioengineering and Biotechnology. 2021;**9**:621748

[68] Matta A, Nader V, Lebrin M, Gross F, Prats A-C, Cussac D, et al. Pre-conditioning methods and novel approaches with mesenchymal stem cells therapy in cardiovascular disease. Cells. 12 May 2022;**11**(10):1620. DOI: 10.3390/cells11101620

[69] Sui B-D, Zheng C-X, Li M, Jin Y, Hu C-H. Epigenetic regulation of mesenchymal stem cell homeostasis. Trends in Cell Biology. 2020;**30**:97-116

[70] Yang Y-HK, Ogando CR, Wang See C, Chang T-Y, Barabino GA. Changes in phenotype and differentiation potential of human mesenchymal stem cells aging in vitro. Stem Cell Research & Therapy. 2018;**9**:131

[71] Čater M, Majdič G. In vitro culturing of adult stem cells: The importance of serum and atmospheric oxygen. Advances in Experimental Medicine and Biology. 2022;**1376**:101-118

[72] López-Otín C, Blasco MA, Partridge L, Serrano M, Kroemer G. The hallmarks of aging. Cell. 2013;**153**:1194-1217

[73] van Deursen JM. The role of senescent cells in ageing. Nature. 2014;**509**:439-446

[74] Dimri GP, Lee X, Basile G, Acosta M, Scott G, Roskelley C, et al. A biomarker that identifies senescent human cells in culture and in aging skin in vivo. Proceedings of the National Academy of Sciences of the United States of America. 1995;**92**:9363-9367

[75] Krishnamurthy J, Ramsey MR, Ligon KL, Torrice C, Koh A, Bonner-Weir S, et al. p16INK4a induces an age-dependent decline in islet regenerative potential. Nature. 2006;**443**:453-457

[76] Avgustinova A, Benitah SA. Epigenetic control of adult stem cell function. Nature Reviews. Molecular Cell Biology. 2016;**17**:643-658

[77] Bork S, Pfister S, Witt H, Horn P, Korn B, Ho AD, et al. DNA methylation pattern changes upon long-term culture and aging of human mesenchymal stromal cells. Aging Cell. 2010;**9**:54-63

[78] Yu K-R, Kang K-S. Aging-related genes in mesenchymal stem cells: A mini-review. Gerontology. 2013;**59**:557-563

[79] Khong SML, Lee M, Kosaric N, Khong DM, Dong Y, Hopfner U, et al. Single-cell transcriptomics of human mesenchymal stem cells reveal age-related cellular subpopulation depletion and impaired regenerative function. Stem Cells. 2019;**37**:240-246

[80] Mann M, Jensen ON. Proteomic analysis of post-translational modifications. Nature Biotechnology. 2003;**21**:255-261

[81] Ying H, Pan R, Chen Y. Epigenetic control of mesenchymal stromal cell fate decision. In: Post-Translational Modifications in Cellular Functions and Diseases. London, IntechOpen; 2021

[82] Nagel AK, Ball LE. O-GlcNAc modification of the runt-related transcription factor 2 (Runx2) links osteogenesis and nutrient metabolism in bone marrow mesenchymal stem cells. Molecular & Cellular Proteomics. 2014;**13**:3381-3395

[83] Brunet A, Goodell MA, Rando TA. Ageing and rejuvenation of tissue stem cells and their niches. Nature Reviews. Molecular Cell Biology. 2022;**24**:45-62. DOI: 10.1038/s41580-022-00510-w

[84] Brack AS, Conboy MJ, Roy S, Lee M, Kuo CJ, Keller C, et al. Increased Wnt signaling during aging alters muscle stem cell fate and increases fibrosis. Science. 2007;**317**:807-810

[85] Stearns-Reider KM, D'Amore A, Beezhold K, Rothrauff B, Cavalli L, Wagner WR, et al. Aging of the skeletal muscle extracellular matrix drives a stem cell fibrogenic conversion. Aging Cell. 2017;**16**:518-528

[86] Segel M, Neumann B, Hill MFE, Weber IP, Viscomi C, Zhao C, et al. Niche stiffness underlies the ageing of central nervous system progenitor cells. Nature. 2019;**573**:130-134

[87] Laschober GT, Brunauer R, Jamnig A, Fehrer C, Greiderer B, Lepperdinger G. Leptin receptor/CD295 is upregulated on primary human mesenchymal stem cells of advancing biological age and distinctly marks the subpopulation of dying cells. Experimental Gerontology. 2009;**44**:57-62

[88] Hayflick L, Moorhead PS. The serial cultivation of human diploid cell strains. Experimental Cell Research. 1961;**25**:585-621

[89] Herranz N, Gil J. Mechanisms and functions of cellular senescence. The Journal of Clinical Investigation. 2018;**128**:1238-1246

[90] McHugh D, Gil J. Senescence and aging: Causes, consequences, and therapeutic avenues. The Journal of Cell Biology. 2018;**217**:65-77

[91] Kumari R, Jat P. Mechanisms of cellular senescence: Cell cycle arrest and senescence associated secretory phenotype. Frontiers in Cell and Development Biology. 2021;**9**:645593

[92] Cristofalo VJ, Lorenzini A, Allen RG, Torres C, Tresini M. Replicative senescence: A critical review. Mechanisms of Ageing and Development. 2004;**125**:827-848

[93] Erol OD, Pervin B, Seker ME, Aerts-Kaya F. Effects of storage media, supplements and cryopreservation methods on quality of stem cells. World Journal of Stem Cells. 2021;**13**:1197-1214

[94] Moll G, Alm JJ, Davies LC, von Bahr L, Heldring N, Stenbeck-Funke L, et al. Do cryopreserved mesenchymal stromal cells display impaired immunomodulatory and therapeutic properties? Stem Cells. 2014;**32**:2430-2442

[95] d'Adda di Fagagna F. Living on a break: Cellular senescence as a DNA-damage response. Nature Reviews. Cancer. 2008;**8**:512-522

[96] Zhao Y, Simon M, Seluanov A, Gorbunova V. DNA damage and repair in age-related inflammation. Nature Reviews. Immunology. 2022. DOI: 10.1038/s41577-022-00751-y

[97] Brzostek-Racine S, Gordon C, Van Scoy S, Reich NC. The DNA damage response induces IFN. Journal of Immunology. 2011;**187**:5336-5345

[98] Takahashi A, Loo TM, Okada R, Kamachi F, Watanabe Y, Wakita M, et al. Downregulation of cytoplasmic DNases is implicated in cytoplasmic DNA accumulation and SASP in senescent cells. Nature Communications. 2018;**9**:1249

[99] Franceschi C, Campisi J. Chronic inflammation (inflammaging) and its potential contribution to age-associated diseases. The Journals of Gerontology. Series A, Biological Sciences and Medical Sciences. 2014;**69**(Suppl. 1):S4-S9

[100] Franceschi C, Garagnani P, Vitale G, Capri M, Salvioli S. Inflammaging and "Garb-aging.". Trends in Endocrinology and Metabolism. 2017;**28**:199-212

[101] Tilstra JS, Robinson AR, Wang J, Gregg SQ, Clauson CL, Reay DP, et al. NF-κB inhibition delays DNA damage-induced senescence and aging in mice. The Journal of Clinical Investigation. 2012;**122**:2601-2612

[102] Akchurin OM, Kaskel F. Update on inflammation in chronic kidney disease. Blood Purification. 2015;**39**:84-92

[103] Brunt EM, Kleiner DE, Wilson LA, Unalp A, Behling CE, Lavine JE, et al. Portal chronic inflammation in nonalcoholic fatty liver disease (NAFLD): A histologic marker of advanced NAFLD-Clinicopathologic correlations from the nonalcoholic steatohepatitis clinical research network. Hepatology. 2009;**49**:809-820

[104] Balestro E, Calabrese F, Turato G, Lunardi F, Bazzan E, Marulli G, et al. Immune inflammation and disease progression in idiopathic pulmonary fibrosis. PLoS One. 2016;**11**:e0154516

[105] Lee B-C, Yu K-R. Impact of mesenchymal stem cell senescence on inflammaging. BMB Reports. 2020;**53**:65-73

[106] Zhang D-Y, Wang H-J, Tan Y-Z. Wnt/β-catenin signaling induces the

aging of mesenchymal stem cells through the DNA damage response and the p53/p21 pathway. PLoS One. 2011;**6**:e21397

[107] Coppé J-P, Desprez P-Y, Krtolica A, Campisi J. The senescence-associated secretory phenotype: The dark side of tumor suppression. Annual Review of Pathology. 2010;**5**:99-118

[108] Gil J. Cellular senescence causes ageing. Nature Reviews. Molecular Cell Biology. 2019;**20**:388

[109] Fafián-Labora J, Lesende-Rodriguez I, Fernández-Pernas P, Sangiao-Alvarellos S, Monserrat L, Arntz OJ, et al. Effect of age on pro-inflammatory miRNAs contained in mesenchymal stem cell-derived extracellular vesicles. Scientific Reports. 2017;**7**:43923

[110] Boulestreau J, Maumus M, Rozier P, Jorgensen C, Noël D. Mesenchymal stem cell derived extracellular vesicles in aging. Frontiers in Cell and Development Biology. 2020;**8**:107

[111] Biran A, Zada L, Abou Karam P, Vadai E, Roitman L, Ovadya Y, et al. Quantitative identification of senescent cells in aging and disease. Aging Cell. 2017;**16**:661-671

[112] Davalos AR, Kawahara M, Malhotra GK, Schaum N, Huang J, Ved U, et al. p53-dependent release of Alarmin HMGB1 is a central mediator of senescent phenotypes. The Journal of Cell Biology. 2013;**201**:613-629

[113] Ulum B, Teker HT, Sarikaya A, Balta G, Kuskonmaz B, Uckan-Cetinkaya D, et al. Bone marrow mesenchymal stem cell donors with a high body mass index display elevated endoplasmic reticulum stress and are functionally impaired. Journal of Cellular Physiology. 2018;**233**:8429-8436

[114] Abu-Shahba N, Mahmoud M, El-Erian AM, Husseiny MI, Nour-Eldeen G, Helwa I, et al. Impact of type 2 diabetes mellitus on the immunoregulatory characteristics of adipose tissue-derived mesenchymal stem cells. The International Journal of Biochemistry & Cell Biology. 2021;**140**:106072

Section 2

Preparation, Preservation and Translational Research

Chapter 3

Stem Cell-Derived Exosomes as New Horizon for Cell-Free Therapeutic Development: Current Status and Prospects

Devashree Vakil, Riddhesh Doshi, Flyn Mckinnirey and Kuldip Sidhu

Abstract

Exosomes have come a long way since they were first described in 1981 by Trams et al. as small lipid bilayer-enclosed vesicles of endocytic origin. Their ability to alter cell bioactivity combined with their advancing popularity as disease biomarkers and therapeutic delivery systems has compelled major Government institutions and regulatory authorities to invest further in this ever-growing field of research. Being relatively new, exosome research is besieged by challenges including but not limited to inefficient separation methods and preservation techniques, difficulties in characterization, and lack of standardized protocols. However, as excitement and research on exosomes increase, their relevance and capacity to elicit a distinct biological response is reinforced. Therefore, it is pertinent to further explore their potential as cell-free therapeutics. This review focuses on current difficulties and subsequent strategies to refine existing methodologies for efficient clinical translation of exosomes in a streamlined and cost-effective manner. The chapter is briefly divided into subsections, each relevant for sequential therapeutic development such as their classification, isolation, scaling up, storage, characterizations, regulatory requirements, therapeutic developments, and perspectives. Apart from literature search, we have endeavored to bring in our own experience in this field including some recent clinical developments.

Keywords: mesenchymal stem cells, exosomes, exosome characterization, signalosomes, assays

1. Introduction

This chapter reiterates the central dogma that mesenchymal stem cells (MSCs) ameliorate disease not just by virtue of their differentiation and self-renewal abilities, but in a paracrine manner, by secreting anti-inflammatory, immunomodulatory, and regenerative factors. Among these paracrine mediators, nanosized extracellular vesicles "exosomes" have generated supreme interest, owing to reports of their standalone therapeutic effect. Stem cell exosomes can circumvent the safety risks associated with the administration of cell therapy.

2. History and evolution of exosomes

A ground-breaking study by Chargaff and West in 1946 [1] for the first time detailed the phenomenon of plasma membrane fragments being shed off viable cells and forming "high particle weight lipoproteins." It was not until two decades later, when the vesicular particles isolated from body fluids were given some attention. Initially disregarded as artifacts of the separation technique [2], these were later believed to be associated with viruses [3]. Wolf and Prince then identified the usefulness of these serum-isolated extracellular vesicles and termed them "phospholipid-rich platelet dust" that could essentially be separated out by ultracentrifugation [4]. It was only in 1975, when particles of 30 to 60 nm diameter, containing an electron-dense core enveloped by a membrane, were recognized as microvesicles, and were firmly established as "breakdown products of normal cellular components," thus freeing them from any association with viruses' [5]. In 1981, Trams and co-workers coined the term "exosomes" for microvesicles harvested from tissue culture supernatant [6]. The exact physiological function of exosomes remained unknown, but reports of specific plasma membrane domains within sparked interest. In 1983, the phenomenon of formation and release of cellular vesicles by exocytosis was outlined by the works of Stahl and Johnstone, respectively [7, 8]. These reports individually identified exosomes as 50-nm spheres displaying receptors on their external surface and originating from a "non-lysosomal endocytic compartment." By mid to late 80s, the term exosome had caught on [9], and "exosome secretion pathway" was acknowledged for the existence of a novel intracellular trafficking pathway *via* shedding of cell membrane [10, 11]. In 1991, Johnstone identified cellular stress as the primary factor to aid internalization and shedding of archaic components of the plasma membrane in the form of exosomes, the mechanism of which was not yet known [12]. However, it was Johnstone's pivotal paper in 2005 [13], which underlined the biological significance of exosomes, establishing for the first time a fate for them that was beyond cellular waste. Following that, the field saw an exponential increase in exosome research including scale-up processes, use of direct modifications, and genetic engineering, paving the way for regenerative therapy. **Figure 1**, adapted from a landmark article in the field, depicts a timeline on exosome [14].

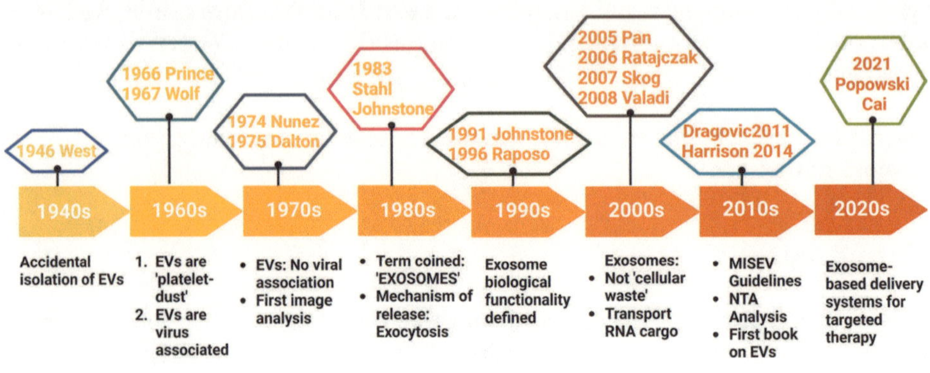

Figure 1.
Evolution and timeline of research conducted on exosomes. A brief timeline of when EVs were discovered, and coining of the term "exosomes."

3. Classification of exosomes

An inherent size overlap between distinct EV subtypes poses a challenge for characterization solely based on size [15]; however, the presence of markers associated with EV origin can further assist classification. Based on size, either small or large EV categories comprise of five major populations: exomeres, exosomes, migrasomes, apoptotic bodies, and large oncosomes [16]. The largest among these are apoptotic bodies, with a diameter > 800 nm (can go up to 5 μM), and consist of plasma membrane and cytoplasmic components of post-apoptotic (dying) cells. Smaller than these are microvesicles, or ectosomes ranging in size from 100 nm to 1 μm. These originate from an irregular blebbing of the plasma membrane. The smallest EVs, exosomes generate from multi-vesicular endosomes, and contain proteins, lipids, and nucleic acids [17, 18]. Their size is much under debate, with smallest exosomes at 30 nm and largest anywhere between 150 and 200 nm. Since exosomes are secreted only upon environmental and physiological cues like cellular stress, their selectively integrated cargo carries specific instructions for modulating target cells. Hence, this EV subclass is often referred to as "signalosomes." Interestingly, microvesicles and exosomes are both released by non-apoptotic cells (**Table 1**).

4. Exosome morphology

Dehydration during sample preparation in conventional electron microscopic techniques forces exosomes to reveal a cup-shaped structure, and they appear as flattened spheres [19, 20]. Cryo-electron microscopy helps exosomes remain fully hydrated and enables exosomes to retain a proper spherical morphology, thus it is a superior technique [21].

Type of EV	Mechanism of Biogenesis	Size	TYPE
Exomeres	Currently unknown	<50 nm	SMALL EVs
Exosomes	Inward budding of plasma membrane (PM) - endocytosis	30–150 nm	
Ectosomes or Microvesicles	Direct shedding into extracellular matrix (ECM) via outward budding of PM	100 nm – 1 μm	LARGE EVs
Migrasomes	Generated during cell migration, secreted into ECM	500–3000 nm	
Apoptotic Bodies	Outward blebbing of apoptotic cell membrane	800 nm– 5 μm	
Large Oncosomes	Released from membrane blebs of amoeboid tumor cells	1–10 μm	

Exosomes are the only EV class that originate through multivesicular body formation via endocytosis. Endosome-derived EVs are generally referred to as 'exosomes' throughout this chapter and may be interchangeably used with the term 'EVs'.

Table 1.
EV classification.

5. Biogenesis, release, and uptake of exosomes

Exosome biogenesis and their secretion involve a complex molecular pathway and exchange of material which is tightly regulated by each source cell [22, 23]. Cells secrete exosomes at different rates, depending on their type, metabolism, and other factors. The first step in this multistep pathway is invagination of the plasma membrane *via* endocytosis [24], forming endosomes, which mature to late endosome, also known as Multivesicular Bodies (MVBs), containing a population of ILVs (Intraluminal Endosomal Vesicles) [24]. The final fate of MVBs is to either i) undergo degradation *via* lysosomes or ii) fuse with plasma membrane and release the ILVs as exosomes into extracellular space [25–27]. Most crucial step is the process of channeling and deposition of a specific subset of proteins (including tetraspanins and some endosomal proteins), lipids (ceramide), and other macromolecules into the ILVs, for which the Endosomal Sorting Complex Required for Transport (ESCRT) pathway is recruited [28–30]. The ESCRT pathway is an intricate, Adenosine Tri Phosphate (ATP)-dependent process involving the use of four complexes—ESCRT-0, ESCRT-I, ESCRT-II, and ESCRT-III along with associated proteins Tsg101 and Alix among others [31–33]. Alternatively, the sorting of exosomal content and biogenesis may occur *via* an ESCRT-independent pathway [34, 35] that surpasses ceramide-mediated membrane budding [15]. Trafficking of exosomes to plasma membrane and their subsequent release involve binding to tether proteins mediated by Rab GTPases [36], followed by the fusion complex SNARE (soluble N-ethylmaleimide-sensitive factor attachment protein receptors) that brings membranes into close proximity [37], while sphingomyelinase mediates release [29]. The possible outcomes after exosome release are as follows:

a. capture by neighboring cells

b. reabsorption by the parent cell

c. relocation and uptake by a remote cell

d. entry into circulation via body fluids

Target cells uptake exosomes through (i) endocytosis mediated by clathrin, claveolin, or lipid rafts; (ii) direct plasma membrane fusion; or (iii) receptor-ligand interaction on the cells surface [38], as shown in **Figure 2**. Once internalized by the target cell, exosomes will fuse with an endocytic vesicle, releasing RNA and proteins in the cytosol. All cell types including stem cells secrete exosomes, found in various body fluids such as saliva, tears, plasma, serum, cerebrospinal fluid, bronchial fluid, synovial fluid, amniotic fluid, breast milk, urine, semen, lymph, bile, gastric acids [39–49]. However, this endosomal pathway for exosome biogenesis is the one factor that distinguishes exosomes from other extracellular vesicles (EVs) [50, 51]. Contrastingly, both Apoptotic bodies and Microvesicles are formed *via* outward blebbing of plasma membrane [15, 52–54].

6. Exosome composition

Stem cell exosomes partly replicate the content of their cells of origin [26, 55]. This discovery, coupled with the revelation that they represent a very specific subcellular compartment, the components of which are selectively sequestered, led to the hypothesis that exosomes are more than just cell debris. Stem cell exosomes comprise

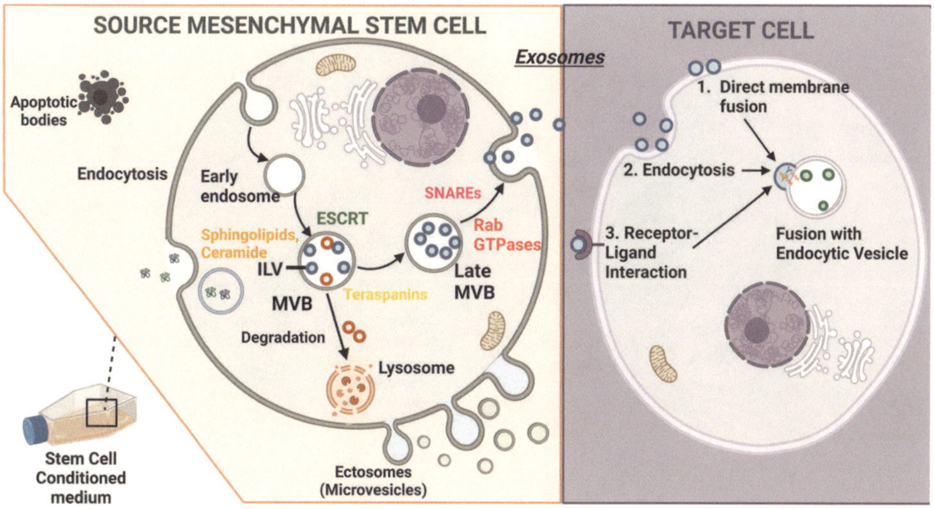

Figure 2.
MSC exosome biogenesis, release, and uptake. Pathway for stem cell exosome biogenesis, release, and three mechanisms for subsequent uptake by recipient cells (1). Direct fusion with target cell (2). Endocytosis (3). Binding to cell surface receptor on target cell and internalization.

a specific milieu of cytoplasmic and membrane proteins including receptors, enzymes, transcription factors, extracellular matrix proteins, lipids, and nucleic acids all of which target molecular pathways and are biologically active in recipient cells [56, 57]. Proteomic databases have now made available the protein composition of exosomes [58, 59]. Irrespective of their origin, all exosomes share few proteins, for instance a specific subset of endosomal, plasma membrane, and cytosolic proteins including cell adhesion molecules (CAMs), integrins, tetraspanins (CD9, CD63, and CD81), heat shock proteins (Hsp60, Hsp70, and Hsp90), biogenesis-related proteins (ALIX and TSG101), and Major Histocompatibility Complex—MHC-I/II proteins). Other transfer and fusion proteins such as Flotillins, Annexins, Heat Shock Proteins, Rab2, Rab7 may be up- or down-regulated depending on the tissue of origin [17, 60], which also include MVB Biogenesis proteins, prostaglandins, platelet-derived growth factor, latherin, transmembrane proteins, lysosome-associated membrane protein-2B, and other phospholipases [61–63]. The lipid bilayer of exosomes (**Figure 3**) with a characteristic thickness of 5 nm [64] is enriched in cholesterol, sphingomyelin and other sphingolipids, ceramide, phosphoglycerides like phosphatidyl serine, and diacylglycerol, which are usually conserved and specific to the parent cell [65]. Since lipids are a key component forming and protecting the exosome structure, lipid content is conserved, and variations are only observed among different cell types [66]. Lastly, another essential cargo that is conserved is nucleic acids, which are relatively distinct from the cytosolic pool of the parent cell. These include single- and double-stranded deoxyribonucleic acids (ssDNA and dsDNA), mitochondrial (mtDNA), and coding and non-coding ribonucleic acid (RNA) such as mRNA and microRNAs [67]. Cholesterol and sphingomyelin along with GPI-anchored proteins and Flotillin are also enriched in "lipid-rafts," implying a role for exosomes in transport [68]. Exosomes are a source of pro-inflammatory cytokines—Interleukins such as IL-1β, IL-6, and IL-8, monocyte chemoattractant protein 1 (MCP-1), Tumor Necrosis Factor-alpha (TNF-α)

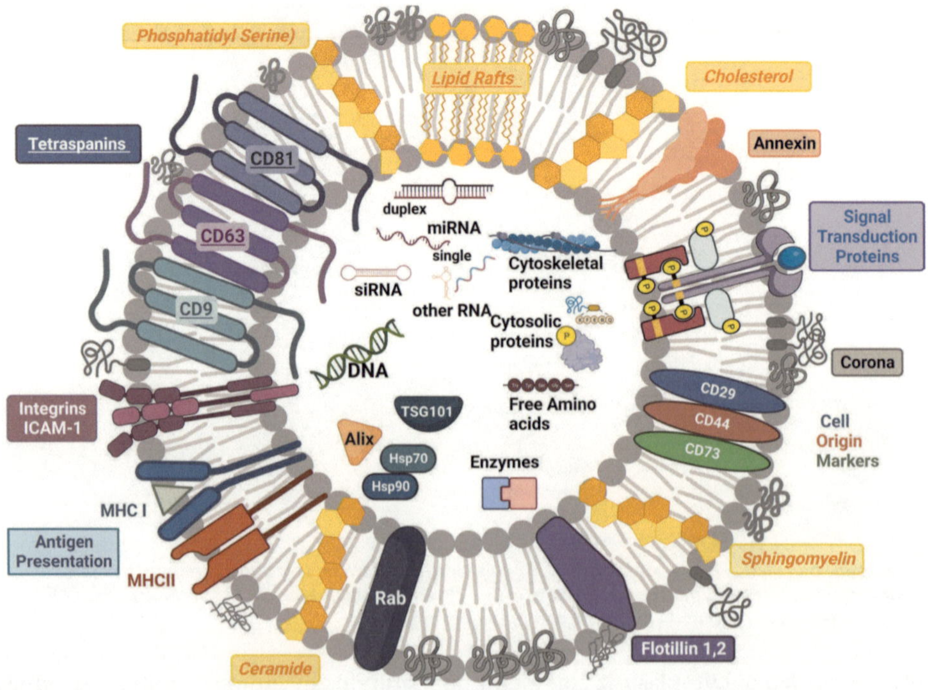

Figure 3.
Mesenchymal stem cell exosome structure – A representation. Exosome structure comprises of a cargo-bearing cytosol enveloped by a bilayer membrane carrying receptors, transmembrane proteins, integrins, lipid rafts, lipid molecules, and a range of surface markers, along with associated protein corona.

[69, 70]. **Figure 3** is a visualization of the identity and function of well-established components that make up the stem cell exosome structure, which consists of a lipid bilayer backbone, along with protein corona/glycocalyx attached to the surface (based on initial findings) [71].

7. Exosome function

The process of exosome biogenesis aids packaging and transfer of information in the form of lipids, nucleic acids, and proteins from the parent cell into the recipient [70]. Consequently, the precise biological functions of exosomes are dependent on their composition, and the reason stem cell exosomes receive tremendous attention is their biological fingerprint and functionality mirroring that of their parental cells [72]. Stem cell exosomes have generated a lot of scientific interests owing to their primary role as message and cellular cargo transporters in intercellular communication as well as involvement in processes such as coagulation, antigen presentation, immune modulation and inflammation, regeneration, cell differentiation, waste management, proliferation and apoptosis, tumor growth, and metastasis [73–85]. Owing to their specific lipid content, they can alter lipid composition, specifically cholesterol and sphingomyelin in target cells, thereby regulating target cell homeostasis [86]. Exosomes, owing to their unique composition, provide innovative opportunities as biomarkers for diagnosis and in treatments. Currently, many clinical trials have been registered for exosome-related studies [87].

8. Isolation methods

Stem cell exosomes are multi-component vesicles with heterogeneity among populations, which necessitates the development of an effective isolation method. Many isolation techniques have been developed, which enable high yield, variable purity, and quality of isolated exosomes. The relative merits and demerits of each of these procedures are summarized in **Table 2**.

Isolation Method	Basis of separation	Potential advantages	Potential limitations
Differential Ultracentrifugation	Density and size	High purity Affordability over time Ease of use Minimal technical expertise Reduced contamination risk Large yields	Low recovery rate Poor reproducibility Time consuming (>4 hrs) Labour-intensive Expensive set-up Aggregation of exosomes Presence of impurities (other EVs) Unsuitable for clinical use
Density Gradient Centrifugation	Size, mass and density	Standard protocol High purity High capacity	Narrow loading zone Low Yield Contamination due to size overlap Time consuming (>4 hrs)
Size Exclusion Chromatography (SEC)	Size	High yield Intact proteomic identity Preserves biophysical properties Improved distribution of exosomes injected in vivo Lower aggregation Intact vesicle structure Good reproducibility Enables quantitative detection	Long run time Requires specialized equipment Does not separate protein aggregates Not applicable for high throughput No scale-up
Ultrafiltration	Size (For Example, TFF)	High purity Applications in protein characterization studies Rapid No specific equipment High exosomal RNA yield [88] Easy scale-up	Low recovery Trapping of exosomes on membrane filter Shear-induced exosomal damage Reduced lifetime of membrane due to clogging No scale-up
Asymmetric Flow Field-Flow Fractionation	Size	Rapid isolation High purity No centrifugal force No shear stress	Equipment cost
Immunoaffinity Capture-based techniques	Molecular recognition	High affinity High specificity of isolation High yield (10–15 times) [89]. Easy to use and rapid No specialized equipment Cost effective at large scale	Low stability of antibodies High reagent cost Cumbersome Low yield Not applicable to all cell types Potential for false readouts

Isolation Method	Basis of separation	Potential advantages	Potential limitations
Polymer based Precipitation	Polymer-based precipitation	High recovery High yield Few steps and rapid Ease of use particles within size range clinical application Easy scale-up	Low purity of exosomes No method for removing the polymer Impaired downstream analysis
Microfluidic separation	Density, size, affinity	High yield Portable technology Cost effective Rapid High purity of exosomes	Low sample capacity, Complex set-up [90–92]

Table lists and summarizes advantages and limitations of techniques regularly employed for isolation and enrichment of exosomes.

Table 2.
The merits and demerits of stem cell exosome isolation methods.

8.1 Differential ultracentrifugation

Based on small size and low density, stem cell exosomes can easily be separated by ultracentrifugation that works on stepwise speed increments (300 x g to 2000 x g to 10,000 x g to 100,000 x g) or alternating between high and low speed (100,000 x g to 200,000 x g), ensuring that different sized particles are separated at different times. Despite its many shortcomings, ultracentrifugation is considered the gold standard for exosome isolation and accounts for 56% of all exosome isolation techniques used in research [93]. Since this process does not separate different sized EVs, the final product is more heterogeneous and may not be suitable for governing bodies [94].

8.2 Density gradient ultracentrifugation

There are two types of preconstructed density gradient media—moving-zone gradients and isopycnic gradients. Moving-zone gradients allow EVs of distinct sizes, but same density to separate out together. Once the sample is layered from top into a tube containing progressively increasing density from top to bottom, it is subjected to multiple rounds of ultracentrifugation allowing the exosomes to individually move toward the bottom, based upon their sedimentation rate. Moving gradient ultracentrifugation employs sucrose, deuterium oxide, or iodixanol-based gradients [95, 96]. When isopycnic density gradients (like caesium chloride) are employed, separation occurs because of differences in sedimentation rates/densities between exosomes and other impurities [97]. Exosomes concentrate at the density region of 1.10 and 1.21 g/cm^3, and pure exosome pellets are obtained by ultracentrifugation at 110,00 x g to 120,000 x g [98–100]. As inferred in **Table 1**, this may not be ideal, especially since the density separations of microvesicles as well as apoptotic bodies are remarkably close to exosomes.

8.3 Size exclusion chromatography (SEC)

SepharoseCL-2B, Sepharose CL-4B, Sephacryl S-100 columns, or SEC matrices exploit the property of smaller vesicles having longer diffusion paths in the paths between porous gels, leading to their retention in the columns. Since exosomes have large hydrodynamic radii, these are excluded from entering the pores, resulting in longer retention times within the column [101–103]. Widely used in many areas of biology, size exclusion chromatography (SEC) can be used to isolate EVs based on the molecular size, for example, QEV (izon), EC SEC columns (Stem cell). This method is used in collaboration with filtration or multiple columns and is reported to alter the characteristics of the EVs to a lesser extent (than other methods like Precipitation) [104].

8.4 Ultrafiltration

One of the most popular size-based separation methods is where particles in suspension are separated on basis of their size/molecular weight. Exosomes are isolated using membrane filters of specific molecular weight, using molecular weight cut-offs (MWCO). A great example is TFF (Tangential Flow Filtration), a method that filters EV sizes 100 kDa and above, using tangential flow of the fluid across the filter surface, avoiding filter cakes and clogging of the pores. This is a preferred method compared to UC. Most commercial large-scale EV purification use either 100 kDa or 300 kDa molecular weight cut-off filters. Because TFF has a history of use in biopharma, it can be easily translated into large-scale downstream processing. Companies such as Pall and Sartorius have developed TFF systems that can be modified to work with their 3D bioreactor systems making EV scale-up easier.

8.5 Asymmetric-flow field-flow fractionation (AF4)

AF4 employs a porous rectangular channel, which is subjected to parabolic flow around its axis, and sample retention and diffusivity are controlled by a cross-flow [93].

8.6 Immunoaffinity-capture

Precise isolation, based on antigen molecules highly concentrated on exosomal surface, targeted by specific fluorescently labeled antibodies immobilized on a polystyrene substrate is purified either using a microplate (ELISA) or using submicron-size magnetic beads [89]. Coupling with mass spectrometry significantly enhances the capacity and is referred to as "Mass Spectrometric immunoassay."

8.7 Polymer-based precipitation

Hydrophilic polymers like polyethylene glycol (PEG) are employed to tie up water molecules and thereby force less soluble exosomes from stem cell-secreted culture superfluates. The resulting exosome-precipitate is separated by low-speed centrifugation or ultrafiltration [105, 106].

To meet the demands of the ever-growing field of exosome therapeutics, there is a desperate need for a robust, highly reproducible and high-throughput isolation method that is still under development.

9. Exosome (EXO) characterization

9.1 Characterization based on physicochemical properties

Depending on their physiological origin, exosomes differ in the composition (quantity and quality) of their bioactive cargo capable of modulating and reprogramming recipient cells. While this property has conferred therapeutic prowess to exosomes, it also proves an impediment to the accurate assessment of their potency and efficacy. Consequently, a crucial question for clinical development of exosome therapeutics is to unravel the precise molecular composition and specific characteristics of exosomes. Unlike isolation, there is no available gold standard technique for either quantification or potency assessment of exosomes, owing mainly to inconsistencies in isolation methods and batch-to-batch variations. There are multiple guidelines available from ISEV (International Society for Extracellular Vesicles) to accurately characterize exosomes.

Table 3 represents a comprehensive list of the different physicochemical properties such as size, shape, surface charge, density, and porosity assessed by various Exosome Characterization methods.

Physicochemical Property	Exosome Parameter	Unit	Techniques Used
Size and Concentration	1. Particle Number 2. Particle Size 3. Particle/protein ratio	1. Particles/mL 2. nm 3. particles/mg	1. Nanoparticle Tracking Analysis (NTA), Tunable Resistive Pulse Sensing (TRPS), Quantitative Electron Microscopy (qEM) 2. NTA, Dynamic Light Scattering (DLS), TRPS 3. TRPS, calculated
Source Cells	1) Particle number	Cell number	Cell Counters, Flow Cytometer
Morphology	Structure and Size, Vesicular Diameter		Atomic Force Microscopy, Cryo-Transmission Electron Microscopy (TEM), TEM (side effects – dehydration of Exos)
Composition	1. Protein Content 2. Lipid Content 3. RNA Content 4. Cytokine Content	1. mg/mL 2. %, Absolute levels/ protein content 3. Yield (pg/uL) 4. Size Distribution 5. Relative measurement based on qualitative differences	1. micro-BCA/Bradford Assay, ELISA, MS 2. LC-ESI-MS/MS 3. RNA Sequencing, qPCR, Bioanalyzer 4. Cytokine Array, Luminex/ ProCartaPlex Assay
Identity	Transmembrane Proteins (Tetraspanins)	CD9/63/81, CD44, CD29, CD37, CD53, CD82	ELISA(concentration of marker), EXO Flow Cytometry using bead capture, NANO FLOW CYTOMETRY
	Transmembrane Integrins	MFG-E8, α3, α4	
	Antigen Presentation	MHC Class I, MHC Class II	
	Source Cell Markers	MSC: CD105, CD73, CD90, CD44 (Positive Surface	

Physicochemical Property	Exosome Parameter	Unit	Techniques Used
		Markers) CD34, CD45, HLA-DR (Negative Surface Markers)	
	Membrane Trafficking	Annexins: I, II, IV, V, VI, VII, XI; Rab 2, Rab 5c, Rab 10, Rab 7; Clathrin	
	ESCRT Proteins	Alix, Tsg 101	
	Heat Shock Proteins	Hsc70, HSP70, HSP90	
	Cytoskeletal Proteins	Tubulins: α1, α2, α6, β5, β3 Actin, Cofilin 1, Moesin	
	Enzymes	GAPDH, Pyruvate Kinase	
Identity	Signal transduction	Syntenin-1	ELISA(concentration of marker), Flow Cytometry (% expression)
	Lipid Rafts (Cytosolic recovered in Exos)	Flotillin-1, Flotillin-2	
	Lipoproteins	ApoA1	
	Miscellaneous	Lactadherin, Lamp2	
	Secretory Pathway	Negative Markers: Calnexin, GRP94	
Quality	Intact membrane	Calcein Staining, Cell Trace Violet Staining	Flow Cytometry, Fluorescent Microscopy

Table summarizes different physicochemical properties of stem cell exosomes that can be characterized by different methods as listed.

Table 3.
Exosome characterization methods.

9.1.1 Nano flow cytometry (nFCM)

nFCM is a high-resolution flow cytometric approach that allows generic quantitative and qualitative analysis of individual EVs as well as sorting of EV subsets (including exosomes), based on antibody and/or fluorescence staining. Nano Flow requires elaborate staining protocols that efficiently eliminate confounding variables such as high background noise caused by buffers, unbound fluorophore-conjugated antibodies, unincorporated dyes, protein aggregates, and other submicrometer-sized particles that can interfere with the EV measurements [107–109].

This is perhaps the only technique that enables analysis of exosome particles in low abundance (for instance, disease-related exosomes purified from clinical samples).

nFCM is the future of quantitative and standardized measurement of therapeutically significant nanoparticles, especially EVs. In our study, we optimized conditions for antibody-labeled, precipitation-enriched exosomes to be analyzed by nFCM based on exosome-specific markers.

9.1.2 Cytokine Array

We used a cytokine array kit to determine the expression of 36 different cytokine-related proteins in our EV/EXO preparations, using the Human Cytokine Array Kit

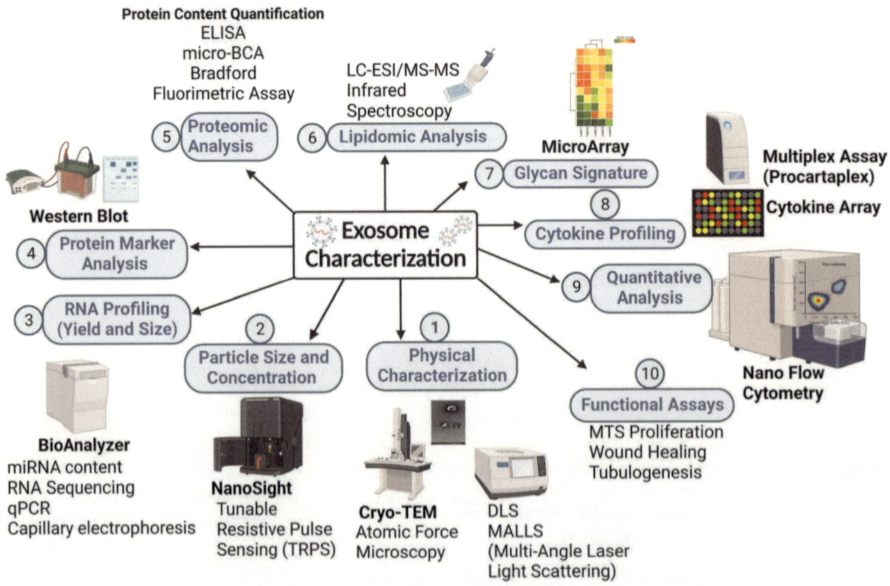

Figure 4.
Stem cell exosome quality control: Methods for multi-parameter characterization. Exosomes can be characterized based on multiple parameters that include qualitative and quantitative analyses such as their morphology, physical characteristics, size, concentration, cargo content, intact-ness, origin as well as their function.

(Proteome Profiler; R&D Systems) according to the manufacturer's instructions. This is essentially a membrane-based sandwich immunoassay, where a biotinylated antibody-stained exosome sample is incubated with an array membrane that is spotted with capture antibodies to the Exosomal target proteins and visualized using chemiluminescence. This process is semi-quantitative in that the signal produced is proportional to the amount of bound analyte. In our study, we discovered at least 106 different cytokines and growth factors bound to the Exosome membrane.

Figure 4 enlists the methods used to characterize Exosomes based on their physical properties, concentration, cargo, and function.

9.2 Characterization based on exosome potency

Exo potency is assessed by a matrix of functional assays that are performed under a tightly regulated validation process. Some of the *in vitro* potency assays that exploit the immunomodulatory properties of exosomes including immune cell signaling, wound closure, and angiogenesis are as listed below:

9.2.1 In vitro *potency assays*

9.2.1.1 Angiogenic evaluation of exosomes

Tube formation assay/vascular-like Network Formation Assay

To assay the angiogenic potential of MSC Exos, human endothelial cells are treated with exosome solution in variable ratios (1:10, 1:00, 1:1000) for 12 h in a 96 well plate, on Matrigel or Geltrex basement membrane matrices that enable vascular network formation. This assay works around the principle that MSC Exos significantly enhance formation of tube-like structures, thereby promoting angiogenesis in human endothelial cells. Validity of the assay can be reinstated by using non-supplemented cell growth media as a negative control and complete media as positive control. Imaging of capillary network is usually acquired with regular light microscopy; total tube length is measured using the Angiogenesis Analyzer plugin of ImageJ software and plotted for different Exo concentrations [110]. MSC Exos are internalized by many cells, including human Endothelial cells, and this assay proves the potency of Exos to promote angiogenesis *in vitro*.

9.2.1.2 Vascular endothelial growth factor (VEGF) immunoassay

VEGF plays a crucial role in angiogenesis and immunomodulation. This assay investigates the proteolytic stability of VEGF in Exo preparations, by mixing them in 1:1 ratio with Trypsin and analyzing *via* a membrane-based immunoassay [111]. This assay runs on the principle that the presence of MSC Exos protect secreted factors like VEGF from protease-mediated degradation. Functional testing of the angiogenesis assay is done by blocking VEGF with an anti-VEGF antibody along with protease treatment of Exos, which will inhibit tube formation in vascular endothelial cells. This is manifested in significantly reduced vascular network formation.

9.2.1.3 Immunomodulatory assessment of exosomes

9.2.1.3.1 T cell proliferation assay

MSC Exos inhibit T cell proliferation *in vitro*, leading to impaired T cell function, and 300,000 CFSE-labeled PBMCs from donors are stimulated with 5ug/mL PHA (Phytohemagglutinin) in a CD3-coated flat bottom 96 well-plate to induce mitogenesis at 96 h (Day4), after which Exos are introduced in varying doses. Flow cytometry determines the percentage of proliferating T cells by measuring percentage of viable CD3 positive cells, which also depict lower CFSE staining compared to non-PHA stimulated cells. No Exos Negative control is used to express maximum T cell proliferation. Maximum proliferation is carried out using Flow Cytometry. Production of CD3-stimulated T cells significantly decreases when treated with exosomes *in vitro*. Exos significantly inhibit T cell proliferation in a dose-dependent manner.

9.2.1.3.2 IL-10 release assay

Another *in vitro* potency assay for MSC exosomes is based on the release of IL-10 from PBMCs following incubation with exosomes. Post-incubation with Exos at 37C for 16-18 h, PBMCs are stimulated with LPS (Lipopolysaccharide) for 5 h. The Supernatant is assayed for IL-10 using ELISA and raw absorbance values are converted to concentration. Higher IL-10 secretion indicates high exosome potency of exosomes [112].

9.2.1.3.3 Macrophage polarization assay

This assay exploits the phenomenon that macrophages maintain a proinflammatory (classically activated M1) phenotype during active infections, and then switch back to a normal, anti-inflammatory M2 phenotype. Here, macrophages are incubated for 3 hours in the presence of PKH7-stained Exos and the principle of this assay is that the majority of the macrophages (>70%) should efficiently internalize Exos in their cytoplasm. This will significantly increase cell proliferation of the Exo-recipient macrophages, which is measured by Flow Cytometry using BrDU incorporation, as compared to Exo-untreated cells. This phenotype modulation of Macrophages from M1 to M2 can be quantified by assessing relative intensities of pro-inflammatory markers (like Ly6C, CD11b, CD40, and CD86) which are downregulated, while anti-inflammatory markers such as CD36, CD51, CD206 are upregulated [113].

Another form of potency assessment is to check the capacity of MSC Exos to suppress mRNA induction of Tumor Necrosis Factor Alpha (TNF-α) in M1 macrophages generated by LPS and IFN-γ (Interferon gamma) stimulation, in the presence or absence of MSC Exos for 24 h. After 24 hours, TNF-α mRNA levels are quantified by RT-qPCR. The functional end point of this assay is half minimal effective concentration or 50% decrease in levels of TNF-α, relative to control (EC_{50}) during the 24 hours [114].

Exosomes play a role in macrophage phenotype modulation by triggering their proliferation and polarization to decrease inflammation.

9.2.1.3.4 Exosome uptake by peripheral blood Mononucleocytes (PBMCs)

Exosome uptake by cells gives us an inkling about their ability to alter signaling in the recipient immune cells and subsequently their potency toward immunomodulation. In this simplistic assay, PBMCs are incubated for different time intervals (maximum being 48 hours) with pre-stained Exos in a ratio of Exo: PBMC = 5:1. The percentage and intensity of Exo uptake is quantified using either Flow Cytometry or fixing and visualizing with laser scanning confocal microscopy.

9.2.1.4 Wound healing scratch assay

This assay works on the simple principle that Exosomes activate human fibroblast cells to proliferate and migrate to the site of injury. Varying concentrations of Exosomes are added to wells containing Fibroblast cells, with manual scratch where cells lift off. After 24 hours, gap closure is measured using Image J Analysis software and visualization of gap closure under light microscope.

9.2.1.5 Multidrug resistance protein 1 (MRP1) assay

This assay investigates the ability of MSC Exos to downregulate the expression of MRP1 in a dose-dependent manner.

All *in vitro* assays should be performed in triplicates, with at least two independent experiments. It would be interesting to compare potency of freshly isolated stem cells Exos with frozen Exos and lyophilized Exos stored at different time intervals (3 months, 6 months) to obtain important insights into their stability over an extended period. It would also be interesting to see the differences in potency of Exos produced under hypoxic conditions as compared to normoxia-produced Exos.

9.2.2 In vivo *potency assays*

Biological *in vivo* potency assays need to be disease-specific, fit-for-purpose and should employ relevant functional end points. These cover multiple aspects of applicability, from administration route to dosing, and provide information about therapeutic effects and toxicity. Moreover, only an *in vivo* system will be able to precisely map exosome distribution and localization post-administration, circulation time, half-life, and target organs.

Most *in vivo* assays are extensions of their *in vitro* counterparts, including *in vivo* angiogenic assay (myocardial infarction model), *in vivo* macrophage modulation (M1-M2) in skeletal muscle injury model, *in vivo* wound healing assays.

10. Exosome preservation and storage

Stem cell-derived exosomes are being developed for variety of applications from primary diagnostics and drug delivery (engineered exosomes). Traditional storage method, cryopreservation has been found to cause decrease in exosome concentration and quality over extended period and is cost intensive. Herein, we discuss current developments in preservation of exosomes through techniques such as cryopreservation, freeze drying, and spray drying [115–117].

Cryopreservation involves freezing exosomes at temperatures much below the optimum required temperature for enzymatic and biochemical activity of constituent biomolecules. Common challenge is formation of ice crystals, which affects the functional and morphological properties of exosomes [118]. To overcome this, various cryoprotectants (CPAs) are used which controls the kinetics of ice formation and concentration of solute pockets, prevent aggregation of particles, and maintain osmotic balance internally and externally [119, 120]. Penetrating CPAs are permeable to the exosome membrane due to their low molecular weight (e.g., dimethyl sulfoxide, ethylene glycol, and glycerol), while non-penetrating CPAs form a glassy matrix or coating externally due to their high molecular weight (e.g., trehalose, sucrose, mannitol, and other sugars) [119, 121, 122]. Studies have shown that cocktails of these cryoprotectants aide in extending the shelf life of the exosomes at low temperatures [116]. Chung et al. [123] in their patent (US2020/0230174A1) used carboxylated Poly-lysine as cryoprotectant to avoid using DMSO. Critical parameter that needs to be optimized is the concentration of CPAs as higher concentration can have toxicity effects, while lower concentrations can cause cryoinjury [124].

Freeze-drying or lyophilization involves freezing exosome solutions followed by sublimation of ice to vapor phase under vacuum forming a dry powder/precipitate thereby maintaining biological properties [125]. Challenges with this method involve uncontrolled ice formation and stresses caused due to drying that affects the exosome stability and membrane integrity. Lyoprotectants such as trehalose, sucrose, buffers, or cocktails can be used to extend the product life [126, 127]. Driscoll et al. [128] compared sucrose, trehalose, and mannitol as lyoprotectants and found trehalose to be most economical and effective lyoprotectant in conjugation with manifold-based lyophilization. Kim et al. and Lim et al. in their patents present a technique to lyophilize exosomes in the presence of one or more lyoprotectants [129, 130]. Since lyophilized products involve storage temperatures between 4°C and room temperature as compared to cryopreservation at freezing conditions, this technique can increase dry state stability, reduce cold chain transportation, and can be used by dissolution in water.

Spray drying is a continuous drying process where wet solution is first atomized, followed by contacting with dry warm air, which leads to instant vaporization of moisture producing dry powder [131]. This technique reduces the risks associated with freezing and dehydration stresses. As this is a continuous operation, spray drying can be economical as well as scalable for large-scale manufacturing. Some critical parameters that affect the shelf life and quality of dry powder are initial feed concentration rate, atomization pressure, and outlet temperature. Behfar et al. [132] have patented a system of exosomes encapsulated in alginate by spray drying. A cyclone separator/electrostatic precipitator can be used for better retention of dry powder. However, experiments are required to be carried out to evaluate if the final product meets required critical quality attributes.

Other than preservation techniques, storage conditions such as pH, temperature, and number of freeze thaw cycles also affect the relative particle concentration, protein content, particle diameter, shelf life, and cellular uptake of exosomes [133, 134].

In summary, trehalose, a FDA approved non-reducing disaccharide sugar, presents as the most suitable option for cryopreservation and lyophilization by stabilizing the lipid membrane preventing exosome aggregation [135].

11. Exosome engineering

11.1 Targeted delivery of pay loads

Native exosomes present therapeutic properties and cargo similar to parent cell type thereby limiting the cargo quality and delivery of exosomes to targeted tissues or cells. Engineered exosomes with targeted delivery overcome these limitations by enhancing exosome efficacy as well as avoiding possible adverse reactions. Some of the methods to modify or load cargo onto native exosomes are as follows: (a) surface modification or passive cargo loading where the exosomes are modified chemically or physically directly to deliver desired cargo and (b) genetic engineering or active cargo loading where the parent cell type is genetically modified to secrete exosomes with specific cargo [136–140]. Some drawbacks to these modifications are compromised exosomal structure and possible induction of immunogenic effects as compared to native exosomes, Challenges linked to engineered exosomes are scalability, production cost, trained professionals, and downstream processing. Hence, there is a need to conduct further studies on these shortcomings at the same time maintaining exosome therapeutic value in conjugation with delivery systems.

11.2 Hydrogel and exosome engineering for tissue regeneration and enhanced secretion

Hydrogels are 3D network of physically or chemically crosslinked polymeric materials with a high affinity for water. Moreover, hydrogels have tissue-mimicking properties such as porous hydrophilic structure, biocompatibility, tunable stiffness, response to external stimuli (pH, temperature, enzymes, cells, etc.), and controlled degradability; thus, hydrogels present a suitable choice for targeted drug delivery system and tissue engineering.

Due to their porous structure and biocompatibility, exosomes are readily encapsulated in hydrogel matrix, offering a sustained and controlled bulk exosome delivery system for therapeutics. For instance, ADSC exosomes encapsulated in GelMA promote tendon regeneration [141], UMSC exosomes laden low-stiffness HA-MA hydrogels stimulate nervous tissue regeneration [142], UMSC-exosomes encapsulated in a hydrogel wound dressing encourage diabetic wound healing [143], and MSC-Exosomes in a sprayable fibrin heart patch regulate myocardial infarction [144]. Hydrogels can be printed with exosomes providing off-the-shelf solution to customized tissue constructs [145–147].

As hydrogels represent a native ECM such as environment and control over their functional chemistries, they can affect exosome secretion activity of hMSCs. Some examples are hydrogel composition and stiffness enhancing MSC secretome and exosome secretion profile [148–150]. Chen et al. [151] have developed a 3D bioprinting method, which augmented exosomes secretion compared to plastic cell culture.

Thus, hydrogels present an exciting drug delivery system and improve exosome secretion profile for native and engineered exosomes.

12. Exosomes: clinical and regulatory guidance

Thirty-seven clinical trials in Phase 1–4 are currently listed at clinicaltrilas.gov using the term extracellular vesicle (EV) or exosomes. The most abundant are in lung and respiratory diseases followed by graft-versus-host disease (www.clincaltrilas.gov). The rise in respiratory and lung diseases is a direct result of the current pandemic (Coronavirus-19), which has a fast track to authorization due to its disease class. Most jurisdictions look to the Food and Drug Administration (FDA) in the United States for guidance. The current guidance for industry is the same as the current guideline for mesenchymal stem cells, which requires a demonstration of safety and efficacy across multiple clinical trials and shows product purity and potency. Consumer alerts have been issued by authorities due to unregulated stem cells and exosomes. The International Society for Extracellular Vesicles (ISEV) and the European Network on Microvesicles and Exosomes in Health and Disease (ME-HaD) have formulated certain guidelines to foster their clinical use [152]. However, three is no harmonized regulatory framework around exosomes at the international level yet but some isolated approvals provided on case-to-case basis. The current status of ongoing trails in MSC derived exosomes can be seen in **Table 4** [87, 153].

Title	Condition	Locations
Effect of UMSCs Derived Exosomes on Dry Eye in Patients With cGVHD	Dry Eye	Zhongshan Ophthalmic Center, Guangzhou, Guangdong, China
Effect of Microvesicles and Exosomes Therapy on β-cell Mass in Type I Diabetes Mellitus (T1DM)	Diabetes Mellitus Type 1	Sahel Teaching Hospital, Sahel, Cairo, Egypt
Evaluation of Adipose Derived Stem Cells Exo.in Treatment of Periodontitis	Periodontitis	Beni-Suef University, Banī Suwayf, Egypt
Exosome of Mesenchymal Stem Cells for Multiple Organ Dysfunction Syndrome After Surgical repair of Acute Type A aortic Dissection	Multiple Organ Failure	Fujian Medical University, Fujian, China
MSC-Exos Promote Healing of MHs	Macular Holes	Tianjin Medical University Hospital, Tianjin, Chin
MSC EVs in Dystrophic Epidermolysis Bullosa	Dystrophic Epidermolysis Bullosa	Aegle therapeutics, Miami, Florida, USA

Title	Condition	Locations
The Use of Exosomes In Craniofacial Neuralgia	Neuralgia	Neurological Associates of West LA, Santa Monica, California, USA
Focused Ultrasound and Exosomes to Treat Depression, Anxiety, and Dementias	Neurodegenerative Diseases	Neurodegenerative Diseases
The Safety and the Efficacy Evaluation of Allogenic Adipose MSC-Exos in Patients With Alzheimer's Disease	Alzheimer Disease	Ruijin Hospital Shanghai Jiao Tong University School of Medicine, Shanghai, China
Allogenic Mesenchymal Stem Cell Derived Exosome in Patients With Acute Ischemic Stroke	Cerebrovascular Disorders	Saeed Oraei Yazdani, Tehran, Iran
iExosomes in Treating Participants With Metastatic Pancreas Cancer With KrasG12D Mutation	Stage IV Pancreatic Cancer, Pancreatic Ductal Adenocarcinoma, Metastatic Pancreatic Adenocarcinoma	M D Anderson Cancer Center, Houston, Texas, United States
A Pilot Clinical Study on Inhalation of Mesenchymal Stem Cells Exosomes Treating Severe Novel Coronavirus Pneumonia	COVID-19	Ruijin Hospital Shanghai Jiao Tong University School of Medicine, Shanghai, China
A Tolerance Clinical Study on Aerosol Inhalation of Mesenchymal Stem Cells Exosomes In Healthy Volunteers	Healthy	Ruijin Hospital Shanghai Jiao Tong University School of Medicine, Shanghai, China
Evaluation of Safety and Efficiency of Method of Exosome Inhalation in SARS-CoV-2 Associated Pneumonia.	COVID-19	Medical Centre Dynasty, Samara, Russian Federation
Organicell Flow for Patients With COVID-19	COVID-19	Landmark Hospital, Naples, Florida, United States
Safety and Efficiency of Method of Exosome Inhalation in COVID-19 Associated Pneumonia (COVID-19EXO2)	COVID-19	Medical Centre Dynasty, Samara, Russian Federation

Table 4.
Current clinical trials using MSC exosomes.

13. Exosomes: scale-up production

13.1 Upstream processing

In order to meet demand of aforementioned EVs in clinical trials and success thereof, upscale technologies must be employed [154, 155]. One of the major challenges to cellular and non-cellular biologics is upstream scaling of manufacture to bioreactor culturing. Generally, manufacture relies on the development from T flasks to multi-layer stacks; however, this method of culturing is restrictive in terms of surface-to-volume ratio. From 2D culture, the system can be scaled into roller bottles or spinner flasks. From there, lab-scale technologies can be employed which can then be transferred or scaled into high-capacity bioreactor systems. Manufacturing current Good Manufacturing Practice (cGMP) EVs to a commercial scale which are not only (i) reproduceable but (ii) cost effective remains a somewhat arduous task. Upscaling manufacture is a necessary step on route to commercialization; however, large investment is required to ensure a smooth translation from bench to bedside and gain successful regulatory acceptance. With upscaling comes less batch testing, less lot release criteria, less labor, less facility time, less consumables and reagents costs, and perhaps most importantly less impact of variation. However, there are also higher risk considerations including higher costs of failure, larger equipment costs and depreciation, more upfront research and product development, and undesired product changes [156]. Although much can be learnt from stainless steel bioreactors that are currently in use in other fields (Monoclonal antibody and vaccine production) at a scale of 20,000 Liters (L), cell and EV therapy scale-up solutions need to generate high volumes in single-use sterile bioreactor systems (SUBs). SUBs allow for less qualification and validation due to presterilization and minimal contact. Capacity of SUBs currently stands at 6000 L (Wuxi Biologics, China); however, downstream systems for purification and sensors are not necessarily compatible with the SUBs and product requirements.

13.1.1 Single-use technology

With an ever-changing landscape in bioreactor technology, it is important the technology of choice has longevity. A change of method nearing Phase 3 or commercialization could be devastating. System suitability relies on many factors including cell type, downstream processing, and carrier suitability. For EVs to reflect the expression pattern from the parent cells, it is a challenge with most historical characterization being completed on cells from 2D culture systems and most cell types being extremely sensitive to hydrodynamic conditions [157]. Hydrodynamic conditions within a bioreactor will significantly impact the biological performance of the cells and/or secreted molecules like EVs. This makes the choice of SUB of upmost importance [158]. Bioreactors enable the user to control gas, temperature, pH, and feed addition; however, these factors are reliant on which type of reactor is employed [159]. The main types of reactors used for EV manufacturing include those in which employ microcarriers or macrocarriers that are utilized in the following: 1. continuous stirred tank bioreactors, 2. hollow fiber reactors, 3. packed bed bioreactors, and 4. wave reactors [155, 160, 161]. By far the most widely used bioreactor type is the stirred tank reactor due to its high flexibility and low-operating costs, **Tables 1** and **2** outline a range of SUBs currently available and some of their limitations and advantages for EV production The Xcellerex (GE Healthcare), Allegro (Pall), and BIOSTAT (Sartorius) offer manufacturing platforms of SUBs with designs that closely match traditional reactors (**Table 5**) [162].

13.1.2 Substrate technology

Microcarriers and macrocarriers are generally used in SUB systems as the 3D surface the cells can attach and grow. Microcarriers come in many forms and can be characterized based on matrix, coating or size. This includes glass, diethylaminoethyl (DEAE)-dextran, acrylamide, polystyrene, collagen, and alginate [163]. To increase cell attachment, they are either coated (collagen) or non-coated (charged). Regulatory bodies require the culture system to be animal origin free for human use to avoid xenogeneic reactions that limits the choice. In the case of xeno-free microcarriers,

Type	Commercial examples	Advantages	Limitations
Wave (Rocking bed)	GE healthcare, Finesse (ThermoFisher),	Versatile	Not easily scaled, low cell recovery (beaching)
Stirred tank	Xcellerex (GE healthcare/Cytiva), Finesse (ThermoFisher), Mobius 2000 (Millipore), BIOSTAT (Sartorius), BioBLU (Eppendorf)	Functional at large volumes>50 L, most abundant	Stirred tank reactors have high shear force which can affect cell characteristics
Perfusion/ hollow fiber bioreactor	Fibercell (FiberCell systems), Quantum ell expansion	Minimal shear stress, isolation of EVs and cells easy	Very low throughput
Fixed bed	iCellis (Pall)	Scalable from lab, small footprint with large surface area	Difficult to remove cells, easy for EVs

Table 5.
Examples of SUBs commercially available.

some examples include Hillex and Star Plus (Sartorius, USA) as well as other dissolvable carriers like Synthemax (Corning, USA). One advantage of not being required to harvest the adherent cells as with EV production is the challenge of enzymatic detachment. Using microcarriers in stirred tank or wave reactors can cause issues including aggregation, engulfment, and beaching where the microcarriers and cells are stuck together or in the case of the wave bag reactor are caught in the corners of the bag. For EV manufacture using adherent cells, it is therefore preferred to use macrocarriers, generally made as disks (fibra-cel, Eppendorf) or strips from polytetrafluoroethylene (PTFE).

13.2 Downstream processing

Selective purification is necessary to isolate exosomes from other EV subsets as well as from the heterogeneous "soup" that is the stem cell secretome. This secretome contains a myriad of analytes, including cytokines, chemokines, enzymes, growth factors, Extracellular Matrix (ECM) proteins, and factors involved in ECM remodeling, different types of Extracellular Vesicles including Exosomes, microvesicles, apoptotic bodies, and others. The downstream processing of a mixed biological like the secretome requires careful precision. From regulatory standpoint purity, potency, safety, and efficacy of derived exosomes are of upmost importance for clinical relevance [164]. A pure product without contaminants from culture media and cells is critically important. Methods for EV isolation are broad and rely on alternative characteristics of the EVs for purification; however, challenges remain on which on scale vs. purity. Many methods are starting to become available to researchers and manufacturing organizations, for example, immunocapture of CD81-, CD63-, and CD9-positive molecules and microfluidics and the methods described below include Ultracentrifugation (UC), Precipitation, Size exclusion chromatography (SEC), and Tangential Filtration Flow (TFF) are most commonly used in larger-scale downstream processing. Currently, there are no well-defined methods for exosome isolation in high-efficiency and high-throughput, and the recommendation is a combination of the below methods.

14. Perspectives

Secretion of extracellular vesicle (EVs) is a universal phenomenon and hence must have biological implications. They have come a long way since first discovered in 70's by Peter Wolf [4]. Fast forward to today, there are extensive publications and clinical trial data to promote their role in diagnostics and therapeutics, including in drug delivery. However, the jury is still out there on EVs to fulfill these roles and gain prime time.

Among the two broader categories of EVs: ectosomes and exosomes, the latter are inward budding and hence endosomal origin with a size range of ~40 to 160 nm in diameter. Endosomes undergo a process of systematic invagination of the plasma membrane with the formation of multivesicular bodies that intersect with other intracellular vesicles (phagosomes) and organelles, contributing to diversity of exosome composition before they are released back to intercellular space. These subset of EVs as exosomes have attracted attention for their critical role in cell-to-cell communication and hence therapeutic values. In particular, the mesenchymal stem cells (MSCs)-derived exosomes seem to have strong anti-inflammatory and immunomodulatory roles and hence have been studied extensively.

The role of exosomes in intracellular and intercellular communication is quite apparent from many critical studies thus far. The developmental pathways for exosomes are highly regulated that are being unraveled at the molecular level; however, how they impact cells is still elusive. Nevertheless, recent studies including preclinical and clinical appraisals have demonstrated their role in mitigating symptoms of various diseases as delineated in this chapter. *Given the cellular therapy still facing huge challenges regarding cell differentiation, maturation, and integration, opportunity with exosomes to develop non-cellular active pharmaceutical ingredients (API)-based allogeneic therapy may be more appeasing to the regulatory authorities.*

This chapter has endeavored to address questions systematically by first reviewing the developments in the field and then putting perspectives on feasibility of their roles and the technical and regulatory hurdles involved. The scientific rigor behind exosomal research puts a demand on regulatory bodies to develop appropriate framework for promoting their developmental pathways toward human medicine.

15. Challenges

The current drug developmental paradigm critically requires defining the active ingredient in the product to its finest detail to eliminate heterogeneity and batch-to-batch variations before defining its pharmacology and pharmacokinetic profiling for final approval after clinical trials. This is followed up by rigor of their scale-up production under GMP from established master cell bank (MCB) with quality controlled (QC) protocols, cryopreservation if any, and shelf life for potency. This is a very arduous, time-consuming, and expensive journey in drug development and takes about 5–10 years with a projected cost of 3–4 billion USD.

15.1 Regulatory challenges

Purification of exosomes by various methods described in this chapter gives rise to a heterogeneous EV population consisting of 40–160 nm diameter vesicles, putatively classified into many subcategories based on size, Exo A, B, C; content, Exo 1, Exo 2, Exo 3; function, Exo α, Exo β, Exo γ; and source, Exo I, Exo II, Exo III [28]. However, as per societal guidelines, all EVs qualify as one based on their characterization. The major question is whether such heterogeneity matters, particularly when blood and blood-derived products are approved for therapeutic use. In fact, the pleotropic effects of exosomes are beneficial in some cases that include blood transfusion. Therefore, the current classification of exosomes API (active pharmaceutical ingredient) needs due diligence and redefining by the regulatory authorities so that it is at par with blood and blood products.

There are no harmonized protocols available as yet that endow regulatory approvals on exosome production for clinical development. Approvals that have been granted are on a case basis depending upon clinical needs, but no main stream approvals yet.

15.2 Technical challenges

Characterization: The current societal guidelines for characterizing exosomes include the following:

For size and number, Nanoparticle Tracking Analysis (NTA) or Dynamic Light Scattering (DLS) are both complementary techniques that offer different insights. DLS will generally measure a wider size range than NTA, but NTA offers greater resolution than DLS. There are inherent difficulties in both these techniques with wide variations within each sample and aggregation of exosomes affecting measurements. Stabilizing exosomes with some cryoprotectants such as trehalose can offer some respite. Alternatively, quantification using the Exo-Flow-ONE staining kit may be more accurate.

Exosome-specific surface markers such as CD9, 63, 81, 107, Alix, are generally assessed by using Western blot; however, the procedure is cumbersome and not efficient for a large number of samples. Additionally, nano-flowcytometry that is more efficient is not widely used because of the cost involved and lack of harmonized published protocols. Nevertheless, some recent publications including our own endeavour showed success with this technique [162–166]. Alternately super-resolution microscopy topology could be employed though it is neither cost effective nor efficient.

Proteomics and micro-RNA profiling are generally outsourced for bulk analyses that are more cost effective but relevance of such comprehensive data for drug development is far from clear, though may be relevant for diagnostic purposes.

Exosome potency assays include inhibiting T cell proliferation, macrophage phagocytosis, fibroblast activity, vascular-like network formation assay on Matrigel [167] and scratch wound healing assays [168] that are very useful and can be optimized quickly in the lab. Emerging *in vivo* potency assays such as EV-mediated wound healing [166, 169] (assess the biological response of exosomes in a disease model [170]. Critically understanding the mechanism of how exosomes elicit a response will assist in regulatory approval.

Scale-up production involves establishing a stem cell biobank made up of MSCs extracted from biologically available sources such as umbilical cord, bone, tooth, fat; however, these MSCs have finite life and reach senescence after 6–7 passages. To circumvent this problem, immortalized cell lines were used in the past that caused regulatory and safety issues. Current approach is therefore to reprogram pluripotent stem cells like iPSC (induced pluripotent stem cell) from human blood or skin tissue and differentiate these into MSCs so that there is an unending source of materials in the biobank. These iMSC derived from iPSCs continue to proliferate beyond passage 10–12 and do not senesce easily, likewise eMSC derived from hESC is other source of long-life stock from which MSCs can be derived easily. Producing these sources of iPSC and MSCs under GMP is very expensive proposition. We have recently moved from planar culture to 3D culture using bioreactors for improved exosome production. We observed an improvement in both quality and quantity of EVs produced. Incorporating such devices within aseptic environment is leading the way for transition to clinical trials and then clinics.

While cryopreservation of cell lines is a standardized procedure, storing exosomes is still an evolving field. Notionally, these can be stored in saline at -80°C with good keeping quality for years; however, there are inherent problem with exosomes aggregation and/or lysis happening during storage. Lyophilization is a quick and efficient way to store exosomes at room temperature for clinical purposes. Cryoprotectants like trehalose help in protecting against aggregation as discussed in this chapter.

Targeted delivery of exosome payloads by genetic engineering increases efficiency and efficacy. Considerable progress has been made toward understanding the logistics for exosome delivery. The controlled release of exosomes at the site of injury is by using various biodegradable gels, and extracellular matrices are on trials as in our labs. Particularly in this regard is 3D printing of exosomes onto bandages as therapeutics is of great relevance in this field.

16. Conclusions and perspectives

In conclusion, exosomes field has emerged as a critical area of therapeutic development as a third pillar of medicine with proof of principle and good science behind it. However, bringing it to fruition requires liaising with the regulatory authorities to harmonize framework around quality control protocols, which will further facilitate clinical trials and more importantly bring focus and excitement for continued funding in this field.

Conflict of interest

The authors declare no conflict of interest.

Author details

Devashree Vakil, Riddhesh Doshi, Flyn Mckinnirey and Kuldip Sidhu*
CK Cell Technologies Pty Ltd., Sydney, NSW, Australia

*Address all correspondence to: k.sidhu@ckcelltechnologies.com

References

[1] Chargaff E, West R. The biological significance of the thromboplastic protein of blood. The Journal of Biological Chemistry. 1946;**166**:189-197

[2] Prince AM, Adams WR. Virus-like particles in human plasma and serum: Role of platelet lysosomes. Journal of the National Cancer Institute. 1966;**37**(2): 153-166

[3] Levine PH et al. Relationship between clinical status of leukemic patients and virus-like particles in their plasma. Cancer. 1967;**20**(10):1563-1577

[4] Levine PH, Wolf P. The nature and significance of platelet products in human plasma. British Journal of Haematology. 1967;**13**:269-288

[5] Dalton AJ. Microvesicles and vesicles of multivesicular bodies versus "virus-like" particles. Journal of the National Cancer Institute. 1975;**54**(5):1137-1148

[6] Trams EG, Lauter CJ, Salem N Jr, Heine U. Exfoliation of membrane ecto-enzymes in the form of micro-vesicles. Biochimica et Biophysica Acta. 1981; **645**(1):63-70

[7] Pan B-T, Johnstone RM. Fate of the transferrin receptor during maturation of sheep reticulocytes in vitro: Selective externalization of the receptor. Cell. 1983;**33**(3):967-978

[8] Harding C, Heuser J, Stahl P. Receptor-mediated endocytosis of transferrin and recycling of the transferrin receptor in rat reticulocytes. Journal of Cell Biology. 1983;**97**(2): 329-339

[9] Johnstone RM, Bianchini A, Teng K. 1989 reticulocyte maturation and exosome release: Transferrin receptor containing exosomes shows multiple plasma membrane functions. Blood. 1989;**74**(5):1844-1851

[10] Harding C, Heuser J, Stahl P. Endocytosis and intracellular processing of transferrin and colloidal gold-transferrin in rat reticulocytes: Demonstration of a pathway for receptor shedding. European Journal of Cell Biology. 1984;**35**:256-263

[11] Pan B-T, Teng K, Wu C, Adam M, Johnstone RM. Electron microscopic evidence for externalization of the transferrin receptor in vesicular form in sheep reticulocytes. The Journal of Cell Biology. 1985;**101**:942-948

[12] Johnstone RM, Mathew A, Mason AB, Teng K. Exosome formation during maturation of mammalian and avian reticulocytes: Evidence that exosome release is a major route for externalization of obsolete membrane proteins. Journal of Cellular Physiology. 1991;**147**(1):27-36

[13] Johnstone RM. Revisiting the road to the discovery of exosomes. Blood Cells, Molecules & Diseases. 2005;**34**(3): 214-219

[14] Couch Y, Buzàs EI, Di Vizio D, Gho YS, Harrison P, Hill AF, et al. A brief history of nearly EV-erything - the rise and rise of extracellular vesicles. Journal of Extracellular Vesicles. 2021 Dec;**10**(14):e12144. DOI: 10.1002/jev2.12144

[15] Anand S, Samuel M, Kumar S, Mathivanan S. Ticket to a bubble ride: Cargo sorting into exosomes and extracellular vesicles. Biochim Biophys Acta Proteins Proteom. 2019;**1867**(12): 140203

[16] Kang T, Atukorala I, Mathivanan S. Biogenesis of extracellular vesicles. New Frontiers: Extracellular Vesicles Subcellular Biochemistry. 2021;**97**:19-43

[17] Février B, Raposo G. Exosomes: Endosomal-derived vesicles shipping extracellular messages. Current Opinion in Cell Biology. 2004;**16**(4):415-421

[18] Mathivanan S et al. ExoCarta 2012: Database of exosomal proteins, RNA and lipids. Nucleic Acids Research. 2011;**40**(D1):D1241-D1244

[19] Lobb RJ, Becker M, Wen SW, Wong CSF, Wiegmans AP, Leimgruber A, et al. Optimized exosome isolation protocol for cell culture supernatant and human plasma. Journal of Extracellular Vesicles. 2015;**4**:27031

[20] Chuo STY, Chien JCY, Lai CPK. Imaging extracellular vesicles: Current and emerging methods. Journal of Biomedical Science. 2018;**25**:91

[21] Yuana Y, Koning RI, Kuil ME, Rensen PC, Koster AJ, Bertina RM, et al. Cryo-electron microscopy of extracellular vesicles in fresh plasma. Journal of Extracellular Vesicles. 2013;**31**:2

[22] Fei X, Li Z, Yang D, Kong X, Lu X, Shen Y, et al. Neddylation of Coro1a determines the fate of multivesicular bodies and biogenesis of extracellular vesicles. Journal of Extracellular Vesicles. 2021;**10**:e12153

[23] Hao Y, Song H, Zhou Z, Chen X, Li H, Zhang Y, et al. Promotion or inhibition of extracellular vesicle release: Emerging therapeutic opportunities. Journal of Controlled Release. 2021;**340**: 136-148

[24] Jeppesen DK, Fenix AM, Franklin JL, Higginbotham JN, Zhang Q, Zimmerman LJ, et al. Reassessment of exosome composition. Cell. 2019;**177**: 428-445

[25] Huotari J, Helenius A. Endosome maturation. The EMBO Journal. 2011;**30**: 3481-3500

[26] Thery C, Zitvogel L, Amigorena S. Exosomes: Composition, biogenesis and function. Nature Reviews. Immunology. 2002;**2**:569-579

[27] Kowal J, Tkach M, Thery C. Biogenesis and secretion of exosomes. Current Opinion in Cell Biology. 2014; **29**:116-125

[28] Kalluri R, LeBleu VS. The biology, function, and biomedical applications of exosomes. Science. 2020;**367**(6478): eaau6977. DOI: 10.1126/science.aau697

[29] Hessvik NP, Llorente A. Current knowledge on exosome biogenesis and release. Cellular and Molecular Life Sciences. 2018;**75**:193-208

[30] Hurley JH, Hanson PI. Membrane budding and scission by the Escrt machinery: It's all in the neck. Nature Reviews. Molecular Cell Biology. 2010; **11**:556-566

[31] Henne WM, Buchkovich NJ, Emr SD. The Escrt pathway. Developmental Cell. 2011;**21**:77-91

[32] Colombo M, Moita C, van Niel G, Kowal J, Vigneron J, Benaroch P, et al. Analysis of ESCRT functions in exosome biogenesis, composition and secretion highlights the heterogeneity of extracellular vesicles. Journal of Cell Science. 2013;**126**(Pt 24):5553-5565

[33] Frankel EB, Audhya A. ESCRT-dependent cargo sorting at multivesicular endosomes. Seminars in Cell and Developmental Biology. 2018; **74**:4-10

[34] Juan T, Fürthauer M. Biogenesis and function of ESCRT-dependent extracellular vesicles. Seminars in Cell and Developmental Biology. 2018;**74**: 66-77

[35] Stuffers S, Sem Wegner C, Stenmark H, Brech A. Multivesicular endosome biogenesis in the absence of ESCRTs. Traffic. 2009;**10**(7): 925-937

[36] van Niel G et al. The tetraspanin CD63 regulates ESCRT-independent and dependnent endosomal sorting during melanogenesis. Developmental Cell. 2011;**21**:708-721

[37] Zhen Y, Stenmark H. Cellular functions of Rab GTPases at a glance. Journal of Cell Science. 2015;**128**: 3171-3176

[38] Trajkovic K, Hsu C, Chiantia S, Rajendran L, Wenzel D, Wieland F, et al. Ceramide triggers budding of exosome vesicles into multivesicular endosomes. Science. 2008;**319**:1244-1247

[39] Liu J, Jiang F, Jiang Y, Wang Y, Li Z, Shi X, et al. Roles of exosomes in ocular diseases. International Journal of Nanomedicine. 2020;**15**:10519-10538

[40] Goto T, Fujiya M, Konishi H, Sasajima J, Fujibayashi S, Hayashi A, et al. An elevated expression of serum exosomal microRNA-191, −21, −451a of pancreatic neoplasm is considered to be efficient diagnostic marker. BMC Cancer. 2018;**18**:116

[41] Serum Grigor Eva AE, Tamkovich SN, Eremina AV, Tupikin AE, Kabilov MR, Chernykh VV, et al. Exosomes in tears of healthy individuals: Isolation, identification, and characterization. Biochemistry (Moscow), Supplement Series B: Biomedical Chemistry. 2016;**10**:165-172

[42] Akers JC, Ramakrishnan V, Kim R, Skog J, Nakano I, Pingle S, et al. miR-21 in the extracellular vesicles (EVs) of cerebrospinal fluid (CSF): A platform for glioblastoma biomarker development. PLoS One. 2013;**8**:e78115

[43] Pisitkun T, Shen R-F, Knepper MA. Identification and proteomic profiling of exosomes in human urine. Proceedings of the National Academy of Sciences of the United States of America. 2004;**101**: 13368-13373

[44] Caby M-P, Lankar D, Vincendeau-Scherrer C, Raposo G, Bonnerot C. Exosomal-like vesicles are present in human blood plasma. International Immunology. 2005;**17**:879-887

[45] Vojtech L, Woo S, Hughes S, Levy C, Ballweber L, Sauteraud RP, et al. Exosomes in human semen carry a distinctive repertoire of small non-coding RNAs with potential regulatory functions. Nucleic Acids Research. 2014; **42**:7290-7304

[46] Zlotogorski-Hurvitz A, Dayan D, Chaushu G, Korvala J, Salo T, Sormunen R, et al. Human saliva-derived exosomes: Comparing methods of isolation. The Journal of Histochemistry and Cytochemistry. 2015;**63**:181-189

[47] Yuan Z, Bedi B, Sadikot RT. Bronchoalveolar lavage exosomes in lipopolysaccharide-induced septic Lung injury. Journal of Visualized Experiments. 2018;**135**:e57737

[48] Dixon CL, Sheller-Miller S, Saade GR, Fortunato SJ, Lai A, Palma C, et al. Amniotic fluid exosome proteomic profile exhibits unique pathways of term and preterm labor. Endocrinology. 2018; **159**:2229-2240

[49] Li Z, Wang Y, Xiao K, Xiang S, Li Z, Weng X. Emerging role of exosomes in

the joint diseases. Cellular Physiology and Biochemistry. 2018;**47**:2008-2017

[50] Milasan A, Tessandier N, Tan S, Brisson A, Boilard E, Martel C. Extracellular vesicles are present in mouse lymph and their level differs in atherosclerosis. Journal of Extracellular Vesicles. 2016;**5**:76

[51] Yoon SB, Chang JH. Extracellular vesicles in bile: A game changer in the diagnosis of indeterminate biliary stenoses? HepatoBiliary Surgery and Nutrition. 2017;**6**:408-410

[52] Yoshida Y, Yamamoto H, Morita R, Oikawa R, Matsuo Y, Maehata T, et al. Detection of DNA methylation of gastric juice-derived exosomes in gastric cancer. Integrative Molecular Medicine. 2014;**1**: 17-21

[53] Tan SS, Yin Y, Lee T, Lai RC, Yeo RW, Zhang B, et al. Therapeutic MSC exosomes are derived from lipid raft microdomains in the plasma membrane. Journal of Extracellular Vesicles. 2013;**2**. Article ID: 22614. DOI: 10.3402/jev.v2i0.22614

[54] Del Conde I, Shrimpton CN, Thiagarajan P, Lopez JA. Tissue-factor-bearing microvesicles arise from lipid rafts and fuse with activated platelets to initiate coagulation. Blood. 2005;**106**: 1604-1611

[55] Raposo G, Stoorvogel W. Extracellular vesicles: Exosomes, microvesicles, and friends. The Journal of Cell Biology. 2013;**200**: 373-383

[56] Colombo M, Raposo G, Théry C. Biogenesis, secretion, and intercellular interactions of exosomes and other extracellular vesicles. Annual Review of Cell and Developmental Biology. 2014; **30**:255-289

[57] Valadi H et al. Exosome mediated transfer of mRNAs and microRNAs is a novel mechanism of genetic exchange between cells. Nature Cell Biology. 2007; **9**(6):654-659

[58] Shabbir A, Cox A, Rodriguez-Menocal L, Salgado M, Van Badiavas E. Mesenchymal stem cell exosomes induce proliferation and migration of normal and chronic wound fibroblasts and enhance angiogenesis in vitro. Stem Cells and Development. 2015;**24**:1635-1647

[59] Théry C et al. Proteomic analysis of dendritic cell-derived exosomes: A secreted subcellular compartment distinct from apoptotic vesicles. Journal of Immunology. 2001;**166**(12):7309-7318

[60] Poliakov A, Spilman M, Dokland T, Amling CL, Mobley JA. Structural heterogeneity and protein composition of exosome-like vesicles (prostasomes) in human semen. The Prostate. 2009; **69**(2):159-167

[61] van Niel G, Porto-Carreiro I, Simoes S, Raposo G. Exosomes: A common pathway for a specialized function. Journal of Biochemistry. 2006; **140**(1):13-21

[62] Kalra H et al. Vesiclepedia: A compendium for extracellular vesicles with continuous community annotation. PLoS Biology. 2012;**10**: e1001450

[63] Kim DK et al. EVpedia: An integrated database of high-throughput data for systemic analyses of extracellular vesicles. Journal of Extracellular Vesicles. 2013;**2**:20384

[64] Conde-Vancells J, Rodriguez-Suarez E, Embade N, Gil D, Matthiesen R, Valle M, et al. Characterization and comprehensive proteome profiling of exosomes secreted by hepatocytes.

Journal of Proteome Research. 2008;7: 5157-5166

[65] Subra C, Grand D, Laulagnier K, Stella A, Lambeau G, Paillasse M, et al. Exosomes account for vesicle-mediated transcellular transport of activatable phospholipases and prostaglandins. Journal of Lipid Research. 2010;**51**: 2105-2120

[66] Doyle LM, Wang MZ. Overview of extracellular vesicles, their origin, composition, purpose, and methods for exosome isolation and analysis. Cell. 2019;**8**(7):727

[67] Minciacchi VR, Freeman MR, Di Vizio D. Extracellular vesicles in cancer: Exosomes, microvesicles and the emerging role of large oncosomes. Seminars in Cell & Developmental Biology. 2015;**40**:41-51

[68] Jenjaroenpun P, Kremenska Y, Nair VM, Kremenskoy M, Joseph B, Kurochkin IV. Characterization of RNA in exosomes secreted by human breast cancer cell lines using next-generation sequencing. PeerJ. 2013;**5**(1): e201E

[69] Ikonen E. Roles of lipid rafts in membrane transport. Current Opinion in Cell Biology. 2001;**13**:470-477

[70] Barnes BJ, Somerville CC. Modulating cytokine production via select packaging and secretion from extracellular vesicles. Frontiers in Immunology. 2020;**29**(11): 1040. DOI: 10.3389/fimmu.2020. 01040

[71] Jan AT, Rahman S, Badierah R, Lee EJ, Mattar EH, Redwan EM, et al. Expedition into exosome biology: A perspective of Progress from discovery to therapeutic development. Cancers (Basel). 2021;**13**(5):1157

[72] Jung HH et al. Cytokine profiling in serum-derived exosomes isolated by different methods. Scientific Reports. 2020;**10**(1):1-11

[73] Su D, Tsai HI, Xu Z, Yan F, Wu Y, Xiao Y, et al. Exosomal PD-L1 functions as an immunosuppressant to promote wound healing. Journal of Extracellular Vesicles. 2020;**9**:1709262

[74] Yin Y, Chen H, Wang Y, Zhang L, Wang X. Roles of extracellular vesicles in the aging microenvironment and age-related diseases. Journal of Extracellular Vesicles. 2021;**10**:e12154

[75] Zhu D, Duo Y, Suo M, Zhao Y, Xia L, Zheng Z, et al. Tumor-Exocytosed exosome/aggregation-induced emission Luminogen hybrid Nanovesicles facilitate efficient tumor penetration and photodynamic therapy. Angewandte Chemie, International Edition. 2020;**59**: 13836-13843

[76] Zarnowski R, Noll A, Chevrette MG, Sanchez H, Jones R, Anhalt H, et al. Coordination of fungal biofilm development by extracellular vesicle cargo. Nature Communications. 2021;**12**:6235

[77] Merino-González C, Zuñiga FA, Escudero C, Ormazabal V, Reyes C, Nova-Lamperti E, et al. Mesenchymal stem cell-derived extracellular vesicles promote angiogenesis: Potencial clinical application. Frontiers in Physiology. 2016;**7**:24

[78] Eirin A, Zhu X-Y, Puranik AS, Tang H, McGurren KA, van Wijnen AJ, et al. Mesenchymal stem cell–derived extracellular vesicles attenuate kidney inflammation. Kidney International. 2017;**92**:114-124

[79] Waldenström A et al. Role of exosomes in myocardial remodeling. Circulation Research. 2014;**114**:315-324

[80] van der Pol E, Boing AN, Harrison P, Sturk A, Nieuwland R. Classification, functions, and clinical relevance of extracellular vesicles. Pharmacological Reviews. 2012;**64**:676-705

[81] Jaiswal R, Johnson MS, Pokharel D, Krishnan SR, Bebawy M. Microparticles shed from multidrug resistant breast cancer cells provide a parallel survival pathway through immune evasion. BMC Cancer. 2017;**17**:104

[82] Schneider E, Winzer R, Rissiek A, Ricklefs I, Meyer-Schwesinger C, Ricklefs FL, et al. CD73-mediated adenosine production by CD8 T cell-derived extracellular vesicles constitutes an intrinsic mechanism of immune suppression. Nature Communications. 2021;**12**:5911

[83] Rak J, Guha A. Extracellular vesicles–vehicles that spread cancer genes. BioEssays. 2012;**34**:489-497

[84] Camussi G, Deregibus MC, Bruno S, Grange C, Fonsato V, Tetta C. Exosome/microvesicle-mediated epigenetic reprogramming of cells. American Journal of Cancer Research. 2011;**1**:98-110

[85] Bellingham SA, Guo BB, Coleman BM, Hill AF. Exosomes: Vehicles for the transfer of toxic proteins associated with neurodegenerative diseases? Frontiers in Physiology. 2012;**3**:124

[86] Mashouri L, Yousefi H, Aref AR, Ahadi AM, Molaei F, Alahari SK. Exosomes: Composition, biogenesis, and mechanisms in cancer metastasis and drug resistance. Molecular Cancer. 2019; **18**(1):75

[87] Available from: www.Clinicaltrials.gov

[88] Gurunathan S, Kang MH, Jeyaraj M, Qasim M, Kim JH. Review of the isolation, characterization, biological function, and multifarious therapeutic approaches of exosomes. Cell. 2019;**8**(4): 307

[89] Zeringer E, Barta T, Li M, Vlassov AV. Strategies for isolation of exosomes. Cold Spring Harbor Protocols. 2015;**2015**(4):319-323. DOI: 10.1101/pdb.top074476[90] Iliescu FS, Vrtačnik D, Neuzil P, Iliescu C. Microfluidic technology for clinical applications of exosomes. Micromachines (Basel). 2019; **10**(6):392

[91] Lin S, Yu Z, Chen D, Wang Z, Miao J, Li Q, et al. Progress in microfluidics-based exosome separation and detection Technologies for Diagnostic Applications. Small. 2020;**16**: 1903916. DOI: 10.1002/smll.201903916

[92] Chan KC, Ko JM, Lung HL, Sedlacek R, Zhang ZF, Luo DZ, et al. Catalytic activity of matrix metalloproteinase-19 is essential for tumor suppressor and anti-angiogenic activities in nasopharyngeal carcinoma. International Journal of Cancer. 2011; **129**:1826-1837

[93] Zarovni N, Corrado A, Guazzi P, Zocco D, Lari E, Radano G, et al. Integrated isolation and quantitative analysis of exosome shuttled proteins and nucleic acids using immunocapture approaches. Methods. 2015;**87**:46-58

[94] Mateescu B, Kowal EJK, van Balkom BWM, Bartel S, Bhattacharyya SN, Buzás EI, et al. Obstacles and opportunities in the functional analysis of extracellular vesicle RNA - an ISEV position paper. Journal of Extracellular Vesicles. 2017; **6**(1):1286095

[95] Cantin R, Diou J, Belanger D, Tremblay AM, Gilbert C. Discrimination between exosomes and HIV-1:

Purification of both vesicles from cell-free supernatants. Journal of Immunological Methods. 2008;**338**:21-30

[96] Théry C et al. Isolation and characterization of exosomes from cell culture supernatants. Current Protocols in Cell Biology. 2006;**3**:22

[97] Miranda KC, Bond DT, Levin JZ, Adiconis X, Sivachenko A, Russ C, et al. Massively parallel sequencing of human urinary exosome/microvesicle RNA reveals a predominance of non-coding RNA. PLoS One. 2014;**9**:e96094

[98] Livshits MA, Khomyakova E, Evtushenko EG, Lazarev VN, Kulemin NA, Semina SE, et al. Isolation of exosomes by differential centrifugation: Theoretical analysis of a commonly used protocol. Scientific Reports. 2015;**5**:17319. DOI: 10.1038/srep17319

[99] Li W, Shao B, Liu C, Wang H, Zheng W, Kong W, et al. Noninvasive diagnosis and molecular phenotyping of breast cancer through microbead-assisted flow cytometry detection of tumor-derived extracellular vesicles. Small Methods. 2018;**2**:1800122. DOI: 10.1002/smtd.201800122

[100] Suthar J, Parsons ES, Hoogenboom BW, Williams GR, Guldin S. Acoustic Immunosensing of exosomes using a quartz crystal microbalance with dissipation monitoring. Analytical Chemistry. 2020;**92**:4082-4093. DOI: 10.1021/acs.analchem.9b05736

[101] Feng Y, Huang W, Wani M, Yu X, Ashraf M. Ischemic preconditioning potentiates the protective effect of stem cells through secretion of exosomes by targeting Mecp2 via miR-22. PLoS One. 2014;**9**:e88685

[102] Lai RC, Arslan F, Lee MM, Sze NSK, Choo A, Chen TS, et al. Exosome secreted by MSC reduces myocardial ischemia/reperfusion injury. Stem Cell Research. 2010;**4**:214-222

[103] Kang D, Oh S, Ahn SM, Lee BH, Moon MH. Proteomic analysis of exosomes from human neural stem cells by flow field-flow fractionation, nanoflow liquid chromatography-tandem mass spectrometry. Journal of Proteome Research. 2008;**7**:3475-3480

[104] Gamez-Valero A, Monguio-Tortajada M, Carreras-Planella L, Franquesa M, Beyer K, Borras FE. Size-exclusion chromatography-based isolation minimally alters extracellular vesicles' characteristics compared to precipitating agents. Scientific Reports. 2016;**6**:33641

[105] Batrakova EV, Kim MS. Using exosomes, naturally equipped nanocarriers, for drug delivery. Journal of Controlled Release. 2015;**219**:396-405

[106] Nordin Z et al. Ultrafiltration with size-exclusion liquid chromatography for high yield isolation of extracellular vesicles preserving intact biophysical and functional properties. Nanomedicine. 2015;**11**:879-883

[107] van der Vlist EJ, Hoen ENM NT, Stoorvogel W, GJA A, MHM W. Fluorescent labeling of nano-sized vesicles released by cells and subsequent quantitative and qualitative analysis by high-resolution flow cytometry. Nature Protocols. 2012;**7**:1311-1326

[108] Kormelink TG, Arkesteijn GJA, Nauwelaers FA, van den Engh G, Hoen ENM NT, MHM W. Prerequisites for the analysis and sorting of extracellular vesicle subpopulations by high-resolution flow cytometry. Cytometry Part A. 2016;**89**:135-147

[109] Enmn HT, van der Vlist EJ, Aalberts M, HCH M, Bosch BJ, Bartelink W, et al. Quantitative and qualitative flow cytometric analysis of nanosized cell-derived membrane vesicles. Nanomedicine and Nanotechnology. 2012;**8**:712-720

[110] Carpentier G, Berndt S, Ferratge S, Rasband W, Cuendet M, Uzan G, et al. Angiogenesis analyzer for imageJ. A comparative morphometric analysis of "Endothelial Tube Formation Assay" and "Fibrin Bead Assay". Science Reports. 14 Jul 2020;**10**(1): 11568. DOI: 10.1038/s41598-020-67289-8

[111] Wang X, Wang H, Cao J, Ye C. Exosomes from adipose-derived stem cells promotes VEGF-C-dependent Lymphangiogenesis by regulating miRNA-132/TGF-β pathway. Cellular Physiology and Biochemistry. 2018;**49**: 160-171

[112] Jiao J, Milwid JM, Yarmush ML, Parekkadan B. A mesenchymal stem cell potency assay. Methods in Molecular Biology. 2011;**677**:221-231

[113] Lo Sicco C, Reverberi D, Balbi C, Ulivi V, Principi E, Pascucci L, et al. Mesenchymal stem cell-derived extracellular vesicles as mediators of anti-inflammatory effects: Endorsement of macrophage polarization. Stem Cells Translational Medicine. 2017;**6**: 1018-1028

[114] Willis GR, Kourembanas S, Mitsialis SA. Toward exosome-based therapeutics: Isolation, heterogeneity, and fit-for-purpose potency. Frontiers in Cardiovascular Medicine. 2017;**9**(4):63

[115] Ziaee A, Albadarin AB, Padrela L, Femmer T, O'Reilly E, Walker G. Spray drying of pharmaceuticals and biopharmaceuticals: Critical parameters and experimental process optimization approaches. European Journal of Pharmaceutical Sciences. 2019;**15**(127): 300-318

[116] Bahr MM, Amer MS, Abo-El-Sooud K, Abdallah AN, El-Tookhy OS. Preservation techniques of stem cells extracellular vesicles: A gate for manufacturing of clinical grade therapeutic extracellular vesicles and long-term clinical trials. International Journal of Veterinary Science and Medicine. 2020;**8**(1):1-8

[117] Charoenviriyakul C, Takahashi Y, Nishikawa M, Takakura Y. Preservation of exosomes at room temperature using lyophilization. International Journal of Pharmaceutics. 2018;**553**(1–2):1-7

[118] Gelibter S, Marostica G, Mandelli A, Siciliani S, Podini P, Finardi A, et al. The impact of storage on extracellular vesicles: A systematic study. Journal of Extracellular Vesicles. 2022;**11**(2): e12162

[119] Bosch S, de Beaurepaire L, Allard M, Mosser M, Heichette C, Chrétien D, et al. Trehalose prevents aggregation of exosomes and cryodamage. Scientific Reports. 2016; **8**(6):36162

[120] Jeyaram A, Jay SM. Preservation and storage stability of extracellular vesicles for therapeutic applications. The AAPS Journal. 2017;**20**(1):1

[121] Kusuma GD, Barabadi M, Tan JL, Morton DAV, Frith JE, Lim R. To protect and to preserve: Novel preservation strategies for extracellular vesicles. Frontiers in Pharmacology. 2018; **29**(9):1199. DOI: 10.3389/fphar. 2018.01199

[122] Budgude P, Kale V, Vaidya A. Cryopreservation of mesenchymal

stromal cell-derived extracellular vesicles using trehalose maintains their ability to expand hematopoietic stem cells in vitro. Cryobiology. 2021;**1**(98): 152-163

[123] Chung C, Namin S, Ganey T, Sicard R. Exosome composition and method of manufacture – PATENT. 2020. Available from: https://patents.google.com/patent/US20200230174A1/en

[124] Best BP. Cryoprotectant toxicity: Facts, issues, and questions. Rejuvenation Research. 2015;**18**(5): 422-436. DOI: 10.1089/rej.2014.1656 Epub 2015 Sep 22

[125] Nireesha GR, Divya L, Sowmya C, Venkateshan N, Babu NM, Lavakumar V. Lyophilization/freeze drying - an review. International Journal of Novel Trends in Pharmaceutical Sciences. 2013;**3**(4):87-98

[126] Bari E, Perteghella S, Catenacci L, Sorlini M, Croce S, Mantelli M, et al. Freeze-dried and GMP-compliant pharmaceuticals containing exosomes for acellular mesenchymal stromal cell immunomodulant therapy. Nanomedicine. 2019;**14**(6):753-765

[127] Frank J, Richter M, de Rossi C, Lehr CM, Fuhrmann K, Fuhrmann G. Extracellular vesicles protect glucuronidase model enzymes during freeze-drying. Scientific Reports. 2018; **8**(1):12377

[128] Driscoll J, Yan IK, Patel T. Development of a lyophilized off-the-shelf mesenchymal stem cell-derived acellular therapeutic. Pharmaceutics. 2022;**14**(4):849

[129] Kim KI, Cho BS, Yi YW. 2021 (54) Method for lyophilizing exosome. EP3777533A1 WO 2020/027466. Available from: https://patents.google.com/patent/EP3777533A1/en

[130] Lim SK. Method for lyophilising an exosome WO2018070939A1. 2018. Available from: https://patents.google.com/patent/WO2018070939A1/en

[131] Kikuchi H, Yamauchi H, Hirota S. A spray-drying method for mass production of liposomes. Chemical & Pharmaceutical Bulletin. 1991;**1**(39): 1522-1527

[132] Behfar A, Terzic A, Cook W, Crespo R. Exosome delivery technology US20160324794A12016. Available from: https://patents.google.com/patent/US20160324794/en

[133] Cheng Y, Zeng Q, Han Q, Xia W. Effect of pH, temperature and freezing thawing on quantity changes and cellular uptake of exosomes. Protein & Cell. 2019;**10**(4):295-299

[134] Sivanantham A, Jin Y. Impact of storage conditions on EV integrity/surface markers and cargos. Life. 2022; **12**(5):697

[135] El Baradie KBY, Nouh M, O'Brien F III, Liu Y, Fulzele S, Eroglu A, et al. Freeze-dried extracellular vesicles from adipose-derived stem cells prevent hypoxia-induced muscle cell injury. Frontiers in Cell and Development Biology. 2020;**20**(8):181

[136] Lino MM, Simões S, Tomatis F, Albino I, Barrera A, Vivien D, et al. Engineered extracellular vesicles as brain therapeutics. Journal of Controlled Release. 10 Oct 2021;**338**:472-485. DOI: 10.1016/j.jconrel.2021.08.037. Epub 2021 Aug 21

[137] Mishra A, Singh P, Qayoom I, Prasad A, Kumar A. Current strategies in tailoring methods for engineered

exosomes and future avenues in biomedical applications. Journal of Materials Chemistry B. 2021;**9**(32): 6281-6309

[138] Yu WW, Wan QQ, Wei Y, Li YT, Li QH, Ye T, et al. Engineered extracellular vesicles: Regulating the crosstalk between the skeleton and immune system. Journal of Engineering and Regenerative Medicine. 2022;**3**(3): 270-282

[139] Harishkumar M, Radha M, Yuichi N, Muthukalianan GK, Kaoru O, Shiomori K, et al. Designer exosomes: Smart Nano-communication tools for translational medicine. Bioengineering (Basel). 2021;**8**(11):158. DOI: 10.3390/bioengineering8110158

[140] Song Y, Kim Y, Ha S, Sheller-Miller S, Yoo J, Choi C, et al. The emerging role of exosomes as novel therapeutics: Biology, technologies, clinical applications, and the next. American Journal of Reproductive Immunology. 2021;**85**(2):e13329

[141] Liu H, Zhang M, Shi M, Zhang T, Lu W, Yang S, et al. Gelatin Methacrylate Loaded with Adipose-Derived Mesenchymal Stem Cell Exosomes Promotes High-Quality Healing of Injured Tendon by Activating both SMAD1/5/9 and SMAD2/3. Rochester, NY: Social Science Research Network; 2021

[142] Liu Z, Tong H, Li J, Wang L, Fan X, Song H, et al. Low-stiffness hydrogels promote peripheral nerve regeneration through the rapid release of exosomes. Frontiers in Bioengineering and Biotechnology. 2022;**23**(10):922570

[143] Ma S, Hu H, Wu J, Li X, Ma X, Zhao Z, et al. Functional extracellular matrix hydrogel modified with MSC-derived small extracellular vesicles for chronic wound healing. Cell Proliferation. 2022;**55**(4):e13196. DOI: 10.1111/cpr.13196

[144] Yao J, Huang K, Zhu D, Chen T, Jiang Y, Zhang J, et al. A minimally invasive exosome spray repairs heart after myocardial infarction. ACS Nano. 2021;**15**(7):11099-11111. DOI: 10.1021/acsnano.1c00628

[145] Born LJ, McLoughlin ST, Dutta D, Mahadik B, Jia X, Fisher JP, et al. Sustained released of bioactive mesenchymal stromal cell-derived extracellular vesicles from 3D-printed gelatin methacrylate hydrogels. Journal of Biomedical Materials Research. Part A. 2022;**110**(6): 1190-1198

[146] Hu Y, Wu B, Xiong Y, Tao R, Panayi AC, Chen L, et al. Cryogenic 3D printed hydrogel scaffolds loading exosomes accelerate diabetic wound healing. Chemical Engineering Journal. 2021;**426**:130634. DOI: 10.1016/j.cej.2021.130634

[147] Feng M, Hu S, Qin W, Tang Y, Guo R, Han L. Bioprinting of a blue light-cross-linked biodegradable hydrogel encapsulating amniotic mesenchymal stem cells for intrauterine adhesion prevention. ACS Omega. 2021; **6**(36):23067-23075

[148] Abdeen AA, Weiss JB, Lee J, Kilian KA. Matrix composition and mechanics direct proangiogenic signaling from mesenchymal stem cells. Tissue Engineering. Part A. 2014;**20** (19–20):2737-2745

[149] Abdeen AA, Lee J, Li Y, Kilian KA. Cytoskeletal priming of mesenchymal stem cells to a medicinal phenotype. Regenerative Engineering and Translational Medicine. 2017;**3**(1):5-14

[150] Lenzini S, Debnath K, Joshi JC, Wong SW, Srivastava K, Geng X, et al. Cell–matrix interactions regulate functional extracellular vesicle secretion from mesenchymal stromal cells. ACS Nano. 2021;**15**(11):17439-17452

[151] Chen J, Zhou D, Nie Z, Lu L, Lin Z, Zhou D, et al. A scalable coaxial bioprinting technology for mesenchymal stem cell microfiber fabrication and high extracellular vesicle yield. Biofabrication. 2022;**14**(1):015012

[152] Lötvall J, Hill AF, Hochberg F, Buzás EI, di Vizio D, Gardiner C, et al. Minimal experimental requirements for definition of extracellular vesicles and their functions: A position statement from the International Society for Extracellular Vesicles. Journal of Extracellular Vesicles. 22 Dec 2014;**3**: 26913. DOI: 10.3402/jev.v3.26913

[153] Lee B-C, Kang I, Yu K-R, Schallmoser K. Clinical medicine therapeutic features and updated clinical trials of mesenchymal stem cell (MSC)-derived exosomes. Journal of Clinical Medicine. 2021;**10**:711

[154] Tavassoli H, Alhosseini SN, Tay A, Chan PPY, Kah S, Oh W, et al. Large-scale production of stem cells utilizing microcarriers: A biomaterials engineering perspective from academic research to commercialized products. Biomaterials. 2018;**181**:333-346. DOI: 10.1016/j.biomaterials.2018.07.016

[155] Simaria AS, Hassan S, Varadaraju H, Rowley J, Warren K, Vanek P, et al. Allogeneic cell therapy bioprocess economics and optimization: Single-use cell expansion technologies. Biotechnology and Bioengineering. 2014; **111**(1):69-83

[156] Mizukami A, Swiech K. Mesenchymal stromal cells: From discovery to manufacturing and commercialization. Stem Cells International. 2018;**11**(2018):4083921. DOI: 10.1155/2018/4083921

[157] Neubauer P, Stojkovic DS, Faria Bellani C, Ajeian J, Duffy L, Miotto M, et al. Scale-up Technologies for the Manufacture of adherent cells. Frontiers in Nutrition. 2020;**7**:575146

[158] Stolberg S, Mccloskey KE. Can Shear Stress Direct Stem Cell Fate? 2009. Available from: www.interscience.wiley.com

[159] Pigeau GM, Csaszar E, Dulgar-Tulloch A. Commercial scale manufacturing of allogeneic cell therapy. Commercial scale manufacturing of allogeneic cell therapy. Frontiers in Medicine. 2018;**5**:233

[160] dos Santos FF, Andrade PZ, da Silva CL, Cabral JMS. Bioreactor design for clinical-grade expansion of stem cells. Biotechnology Journal. 2013;**8**(6): 644-654

[161] Sharma R, Harrison STL, Tai SL. Advances in bioreactor Systems for the Production of biologicals in mammalian cells. CBEN. 2022;**9**:42-62. DOI: 10.1002/cben.202100022

[162] Nogueira DES, Cabral JMS, Rodrigues CAV. Single-use bioreactors for human pluripotent and adult stem cells: Towards regenerative medicine applications. Bioengineering. 2021;**8**(5):1-24. DOI: 10.3390/bioengineering8050068

[163] Koh B, Sulaiman N, Fauzi MB, Law JX, Ng MH, Idrus RBH, et al. Three dimensional microcarrier system in mesenchymal stem cell culture: A systematic review. Cell & Bioscience. 2020;**10**(1):75. DOI: 10.1186/s13578-020-00438-8

[164] Théry C, Witwer KW, Aikawa E, Jose Alcaraz M, Anderson JD, Andriantsitohaina R, et al. Minimal information for studies of extracellular vesicles 2018 (MISEV2018): A position statement of the International Society for Extracellular Vesicles and update of the MISEV2014 guidelines. Journal of Extracellular Vesicles. 23 Nov 2018; 7(1):1535750. DOI: 10.1080/20013078.2018.1535750

[165] Cao G, Chen B, Zhang X, Chen H. Human adipose-derived mesenchymal stem cells-derived Exosomal microRNA-19b promotes the healing of skin wounds through modulation of the CCL1/TGF-β signaling Axis. Clinical, Cosmetic and Investigational Dermatology. 2020;**13**: 957-971

[166] Silva AM, Lázaro-Ibáñez E, Gunnarsson A, Dhande A, Daaboul G, Peacock B, et al. Quantification of protein cargo loading into engineered extracellular vesicles at single-vesicle and single-molecule resolution. Journal of Extracellular Vesicles. 2021;**10**(10):e12130

[167] Wolf M, Poupardin RW, Ebner-Peking P, Andrade AC, Blöchl C, Obermayer A, et al. A functional corona around extracellular vesicles enhances angiogenesis, skin regeneration and immunomodulation. Journal of Extracellular Vesicles. 2022;**11**(4):e12207

[168] Reinisch A, Hofmann NA, Obenauf AC, Kashofer K, Rohde E, Schallmoser K, et al. Humanized large-scale expanded endothelial colony–forming cells function in vitro and in vivo. Blood. 2009;**113**(26):6716-6725

[169] Jain R, Vakil D, Cunningham C, Sidhu K. Human mesenchymal stem cells conditioned media promotes the wound healing process - an in vitro study. Journal of Stem Cell Therapy and Transplantation. 2019;**3**(1):28-30

[170] Ebner-Peking P, Krisch L, Wolf M, Hochmann S, Hoog A, Vári B, et al. Self-assembly of differentiated progenitor cells facilitates spheroid human skin organoid formation and planar skin regeneration. Theranostics. 2021;**11**(17): 8430-8447

Chapter 4

Evaluation and Characterization of Human Bone Marrow Mesenchymal Stromal Cells Cryopreserved in Animal Component-Free, Chemically Defined, Serum-Free Conditions

Suresh Kannan, Swaroop Bhagwat, Pawan Kumar Gupta and Udaykumar Kolkundkar

Abstract

Mesenchymal stromal cells (MSCs) have the potential to treat various disease indications and are the future of cell therapy-based regenerative medicine. Typically, MSCs cryopreserved in serum-containing freezing formulation are supplied at the clinical site, which necessities that this formulation is removed before the administration. This is a cumbersome process, and there is an immediate need for identifying serum-free, xeno-free cryopreservation medium that can be readily used. Here, we analysed two commercially available serum-free, xeno-free, defined freezing media *viz.*, CryoStor 5 (CS5) and CryoStor 10 (CS10) on their effect on human bone marrow MSCs at different freezing cell densities (5, 10, 12.5, 15 and 25 million cells per ml) over a period of 6 months and compared them to the in-house PlasmaLyte A (PLA)-based cryopreservation media. We found that the MSCs cryopreserved in CS5 and CS10 showed similar characteristics as compared with the in-house freezing media for the various parameters analysed including post-thaw recovery, viability, phenotypic marker expression, CFU-F ability and trilineage differentiation potential of the MSCs. Our results show that human MSC could be successfully cryopreserved using serum-free and xeno-free cryopreservation media and can be delivered to the bedside without any manipulations.

Keywords: bone marrow mesenchymal stromal cells, phenotypic characterization, cryopreservation, stability, apoptosis, serum-free and xeno-free cryopreservation media

IntechOpen

1. Introduction

The potential use of Bone marrow mesenchymal stromal cells (BM-MSCs) to treating various disease indications is in steady increase [1] and demands huge availability of clinical quality BM-MSCs to meet the growing demands. Typically, MSCs frozen in cryopreservation solution were supplied to the clinical site, and the method of freezing along with the composition of freezing media itself plays an important role in determining the characteristics of MSCs before infusion [2, 3]. One essential requirement of cGMP grade formulation reagents in cryopreservation is that they are free from animal serum proteins and toxic chemicals, as the xenogeneic compounds possess the risk of transmission of animal viral, prion and zoonose contamination [4].

Presently, the common practice of cryopreserving MSCs is in PlasmaLyte A supplemented with albumin and a cryoprotectant of DMSO and/or dextran [5, 6]. Conventionally, 10% DMSO is used because they readily penetrate cell membranes and thus confer protection to the intracellular components [7, 8]. But it is proven that freezing media containing a higher concentration of DMSO is toxic to patients [8, 9]. Efforts to reduce toxicity include the removal of DMSO prior to transfusion or decreasing the amounts used in the freezing process. But these post-thaw manipulations consume lots of time, decrease the viability of the cellular product and further require aseptic zone for processing and concentration of cells at site [8]. Several groups have investigated MSCs cryopreservation using low concentrations of DMSO and or many alterative to DMSO like polyvinylpyrrolidone (PVP), methylcellulose, polyethene glycol (PEG) trehalose and polymer mimics without or with DMSO from 2.5 to 7.5% [10–13]. But it is necessary that these in-house cryopreservation formulations are prepared from USP grade or cGMP grade reagents to meet the safety requirements and may pose many regulatory challenges while filing an NDA and obtaining regulatory clearances. One of the possible alternatives is to screen the commercially available GMP compliance USP grade, serum-free cryopreservation media for MSCs. Several groups have evaluated the commercial MSC cryopreservation for a short period of time from 1 week to 1 month, but it is necessary to study the real-time stability testing for at least 6 months. In this study, we evaluated the commercially available cryopreservation solutions – CryoStor (CS) 5 and CS10 from BioLife Solutions, USA for their ability to cryopreserve BMMSCs and compared them to our in-house developed PlasmaLyte A-based cryopreservation formulation. We analysed post-thaw viability, total cell recovery, trilineage differentiation, phenotypic marker expression, CFU-F potential and apoptotic cell percentage at a testing interval of 1 week, 1, 2, 3 and 6 months time point and discussed the data here.

2. Materials and methods

2.1 Isolation and culture of human bone marrow-derived MSCs

The BM was collected from healthy donors of age group between 19 and 40 years, after obtaining approval from the Institutional Ethics Committee of the Kasturba Hospital, Manipal and signed informed consent. The mononuclear cells (MNC) were isolated following the density gradient centrifugation method using Lymphoprep (Axis-Shield PoC) as described previously [9]. The isolated MNCs were seeded at a density of 50 million cells per T-75 flask in Dulbecco's modified Eagles medium-Knock Out (DMEM-KO) supplemented with 10% FBS (Hyclone, Waltham, MA), 2 mM L-glutaMAX (Invitrogen, Carlsbad, CA) and 1X penstrep (Invitrogen,

Carlsbad, CA). The cells were passaged when they reached 80–90% confluency using trypsin (0.25%)/ Ethylenediamine tetra acetic acid (EDTA; 1 mM) (Gibco, USA). MSCs from three donors were pooled at passage 2 (P2) in equal proportion and cultured at a seeding density of 1000 cells per cm^2 in bFGF (2 ng/ml) enriched KO-FBS complete medium till P5.

2.2 Cryopreservation

Cryopreservation of BM-MSCs involves freezing of cells in 1 ml of formulation medium containing 10% (v/v) Dimethyl sulfoxide (DMSO) as a cryoprotectant with 5% (v/v) human serum albumin and quantified (QS) with PlasmaLyte A (Baxter Inc) in 5 ml Daikyo Crystal Zenith (CZ) vials from West Pharma. The commercial formulation of Cryostor CS10 and CS5 (BioLife Solutions Inc) were used for cryopreservation of BM-MSCs in CZ vials for comparison. MSCs after trypsinisation were centrifuged to remove the trypsin and the pellet was dislodged by gentle tapping to the pellet, 2–8°C refrigerated cryopreservation media was slowly added. The cells were resuspended at five different concentrations (5, 10, 12.5, 15 and 25 million cells per ml) in these three formulations. These cells were aliquoted @ 1 ml per CZ vials and cryopreserved by slow freezing @ 1°C/min till 80°C by keeping them in Cryomed controlled rate freezer (CRF). During this time an aliquot of the cells was taken to measure the pre-freeze 7AAD/viability. The frozen vials were then shifted to canister racks and were stored in the vapour phase of liquid nitrogen (LN_2).

2.3 Thawing

Each sample was thawed according to the internal Stempeutics SOP. Briefly, after a minimum of 1-week storage in liquid nitrogen, the vials were retrieved and thawed immediately by placing them in a water batch at 37°C till the ice crystals were just disappearing. The thawed cells were transferred into 15 ml tube and the sample was diluted at 1:9 with complete media. The cells were centrifuged at 1200 rpm for 10 minutes and the pellet was then resuspended in KO-FBS complete media.

2.4 Testing frequency, cell count and manual viability assessments

The samples from three different cryopreservation mediums at five different freezing concentrations were analysed at time points of 1 week, 1 month, 2 months, 3 months and 6 months. The total cell recovery (TCR) and viability analysis by manual method and by flow cytometry method were carried out immediately after the thawing procedure. The total cell count and manual viability were determined by the trypan blue (Fluka) exclusion method. The number of viable (non-stained) and non-viable (stained) cells were enumerated microscopically. TCR is calculated by adding the total number of stained and unstained cells. The viability percentage was calculated by dividing the total number of viable cells by TCR and multiplied by 100. For estimation of viability by flow cytometry, the cells were stained with 7AAD and were analysed following the protocol as described previously [14].

2.5 Immunophenotyping

We analysed a set of two positive (CD90-PE, CD73-PE) and two negative cell surface markers (CD14-FITC and CD19-FITC) by flow cytometry. All the antibodies used for these studies were purchased from BD Pharmingen, San Diego. The cryopreserved cells were thawed and resuspended in wash buffer containing phosphate buffer saline (PBS) supplemented with 1% (v/v) FBS and 1% (w/v) sodium azide for analysis. The cells were incubated with saturating concentrations of fluorescein isothiocyanate (FITC) or phycoerythrin-(PE) conjugated antibodies at 4°C for 30 minutes in dark. After that, the cells were washed with wash buffer three times and re-suspended in 0.5 ml of wash buffer. The labelled cells were analysed in EasyCyte (Guava Technology) flow cytometer after setting the instrument parameters with respective isotype-matched controls. For every sample, 10,000 events were captured and the data was analysed using Guava Express Pro software (Guava Technologies). Fluorescence intensity of 25% or above its isotype control is considered an antigenic event and was used for calculation.

2.6 Differentiation potential

The differentiation potential of frozen MSCs was analysed by their ability to differentiate into osteogenic, adipogenic and chondrogenic lineages. Osteogenic differentiation was induced by culturing P5 BM-MSCs in the KO-FBS supplemented with 10^{-8} M dexamethasone, 30 μg/ml ascorbic acid and 10 mM β-glycerophosphate (all Sigma-Aldrich). For adipogenic differentiation, cells were cultured in the KO-FBS supplemented with 1 μM dexamethasone, 0.5 mM isobutylmethylxanthine (IBMX), 1 μg/ml insulin and 100 μM indomethacin (all Sigma-Aldrich). The chondrogenic differentiation was induced using STEMPRO (Invitrogen) chondrogenesis differentiation medium. After 21 days of differentiation, the cells were fixed and stained with Von Kossa, Oil Red O and Safranine O, respectively, for osteo, adipo and chondro differentiation cultures. The images were captured using Nikon Eclipse 90i microscope (Nikon Corporation, Japan, www.nikon.com) and Image-Pro Express software (Media Cybernetics, Inc., Silver Spring, MD, www.mediacy.com).

2.7 CFU-F assay

For CFU-F assay, 100 MSCs from 5 different freezing concentrations of three cryopreservation media at six-month time points were plated in KO-FBS (n = 2 of each condition) on a 100 mm^2 cell culture dish. After 14 days in culture, the plates were stained with crystal violet and the number of colonies was counted.

2.8 Apoptosis analysis

Apoptosis analysis was carried out using Tali apoptosis kit following the manufacturer's instructions (Tali Apoptosis kit – Annexin V Alexa Fluor 488 and propidium iodide, Cat # A10788. Life technologies). Briefly, 1 million cells/ml were resuspended in annexin binding buffer (ABB) and 5 ul of Annexin V were added per 100 μl of sample. The samples were incubated for 20 minutes in dark, centrifuged and resuspended in 100 μl of ABB. After adding 1 μl of PI and incubation for 5 min, the samples were read in Tali® Image-Based Cytometer.

2.9 Statistical analysis

All values are expressed as mean ± SEM (standard error of mean). Data were analysed by using Graphpad Prism (version 5, Graphpad Software Inc., La Jolla,

CA, USA). Two-way ANOVA (Analysis of variances) was performed in order to compare means between groups and Bonferroni post-tests were carried out to find out the significance of the variables tested. P value <0.05 was considered significant.

3. Results

3.1 Post-thaw viability – 7 Amino-actinomycin D (7AAD)

BM-MSCs were scaled-up to P5 in multiple numbers of ten cell stacks and the cultures were frozen in various cell concentrations ranging from 5×10^6 to 25×10^6 per ml in 5 ml CZ vials using 3 different cryopreservation media (PLA, CS10 and CS5). Pre-freeze viability of BM-MSCs in PLA, CS10 and CS5 formulation media was >98.6 ± 1.2% (mean ± standard deviation). Upon, thawing, BM-MSCs viability of PLA, CS10 and CS5 of 5 different cell concentrations (5, 10, 12.5, 15 and 25 million cells per ml) were measured by flow cytometry using 7AAD at different time points viz., 1 week, 1, 2, 3 and 6 months. All the samples have shown >92% (**Figure 1**) viability by 7AAD in all the time points. No significant difference in percentage of viable cells was observed after six-month storage when compared to one-week storage. Two-way analysis of variance (ANOVA) was performed to assay differences over different time points and among different cryopreservation media. The results indicate that CS5 is equally good to that of CS10 and PLA and there are no significant differences in comparison with either time points or different cryopreservation media.

3.2 Total cell recovery and viability

The Post-thaw cell recovery and viability of five different concentrations at 5 different time points (1st (first) week, 1st month, 2nd month, 3rd month and 6th month) in 3 different cryopreservation media were assayed using trypan blue dye

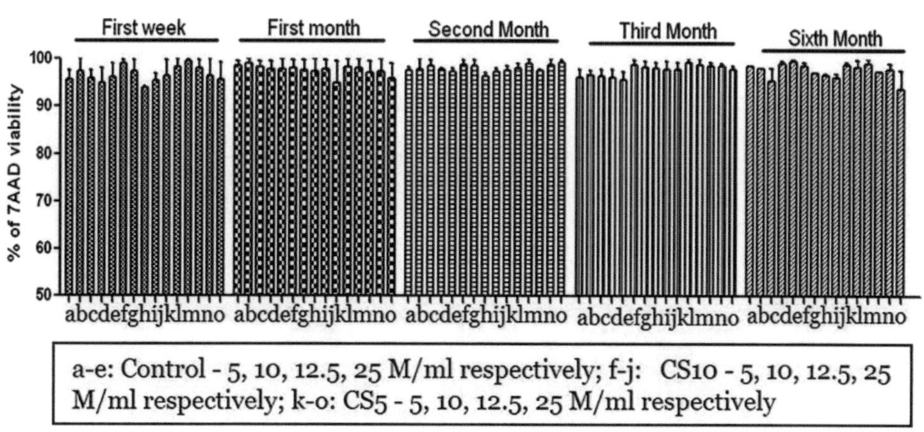

Figure 1.
Post-thaw viability by 7AAD of BM-MSCs following cryopreservation with PlasmaLyte a based cryopreservation solution (control), CryoStor (CS) -10, and CS5 variants. Cell viability assessed at 5 different time points (1st week, 1st month, 2nd month, 3rd month and 6th month) at 5 different freezing concentration (5, 10, 12.5, 15 and 25 million cells per ml) shows no significant differences among three cryopreservation media.

exclusion assay. The total viable and non-viable cell count was taken as total cell recovery and shown in **Figure 2A**. The results demonstrated that no cell loss was observed upon thawing the samples at all time points in 3 cryopreservation media. There are **Figure 2B** also no significant differences in total cell recovery percentage in the lowest and highest freezing densities (5 and 25 million cells/ml respectively) at 3rd and 6th month of recovery after freezing (). Simultaneously viability by dye exclusion method (DEM) was analysed in all of the samples and the results were shown in **Figure 3**. Viability by DEM was between 85 to 95% in all test samples (**Figure 2C**).

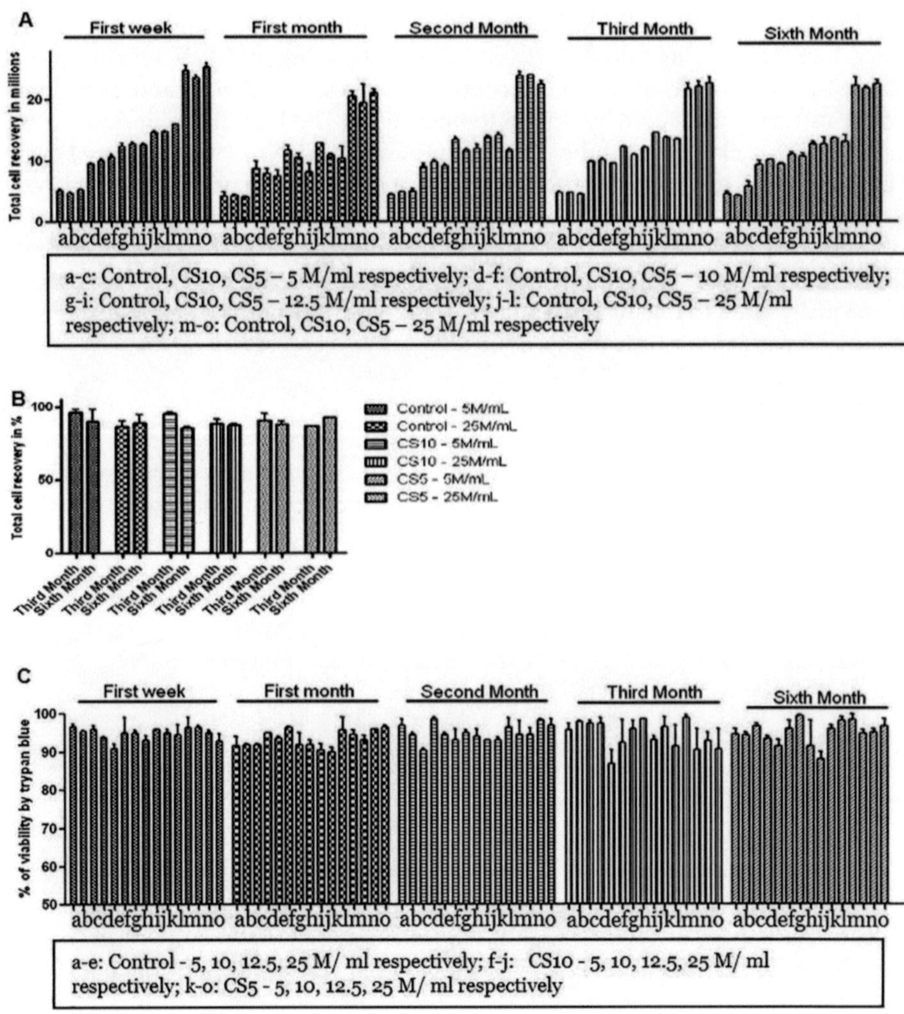

Figure 2.
Post-thaw Total cell recovery (TCR) of BM-MSCs cryopreserved in CS10, CS5 compared to control. (A) TCR in millions as assessed by summing total viable and non-viable cell counts after trypan blue staining. (B) TCR in percentage between the lowest and highest freezing densities at 3rd and 6th month. (C) Viability percentage as calculated by trypan blue exclusion method. There is no significant differences among the three cryopreservation media.

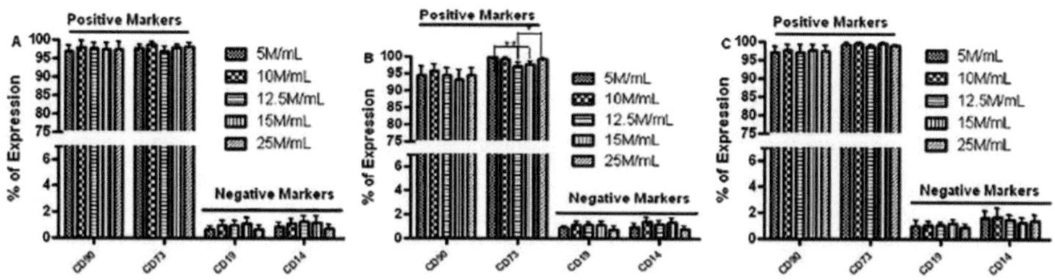

Figure 3.
Immuno-phenotype of BM-MSCs following cryopreservation with PlasmaLyte A-based cryopreservation solution (control) (A), CryoStor (CS) 10 (B), and CS5 (C) at six month time point. Two positive markers of CD90 and CD73 and two negative markers of CD19 and CD14 were analysed. No significant differences among three cryopreservation media at 5 different freezing concentrations except for CD 73 expression variation among CS10 group.

3.3 Immunophenotyping

The surface marker expression was evaluated and analysed by using flow cytometry for BM-MSCs cryopreserved in control, CS10 and CS5 of 5 different cell concentrations. The freeze–thaw BM-MSCs in all three cryopreservation media across all time points were positive for CD90 and CD73 and negative for CD19 and CD14. BM-MSCs showed similar expression of CD markers in CS5 compared with that of CS10 and control (**Figure 3**). There are no significant differences in positive and negative marker expression observed among different seeding densities in any of those three groups except for CD 73 expression variation in CS10 group. There is a significant difference (n = 5,*$p < 0.05$) between 12.5 and 25 million per ml and this significance is even stronger (n = 5, **$p < 0.01$) between 5 and 12.5 million per ml. Howsoever, these expression levels were more than 95% for positive markers and < 2.5% for negative markers in all three cryopreservation media at different cell densities.

3.4 Differentiation potential

We investigated the in-vitro functional tri-lineage differentiation potential of P5 BM-MSCs at the 6th month time point after cryopreservation in PLA-based cryopreservation medium (control), CS10 and CS5 at 5 different cell freezing concentrations. The differentiation towards adipocytes was evident by the formation of fatty vacuole deposits and was observed by Oil Red O staining (**Figure 4A**). The differentiation towards osteoblasts was observed by Von Kossa staining (**Figure 4B**) and that of chondrocytes was observed by safranine O staining (**Figure 4C**) after 21 days of differentiation induction. The cells cryopreserved at different cell densities in different cryopreservation media stained for all three lineages, showing that the trilineage differentiation ability is maintained in all these conditions. The results of CS5 and CS10 were comparable with that of the control cryopreservation medium for all 5 different cell freezing concentrations.

3.5 CFU-F assay

Clonogenic potential of cryopreserved MSCs at 5 different cell freezing concentrations (5, 10, 12.5, 15 and 25 million cells per ml) in each of the 3 different

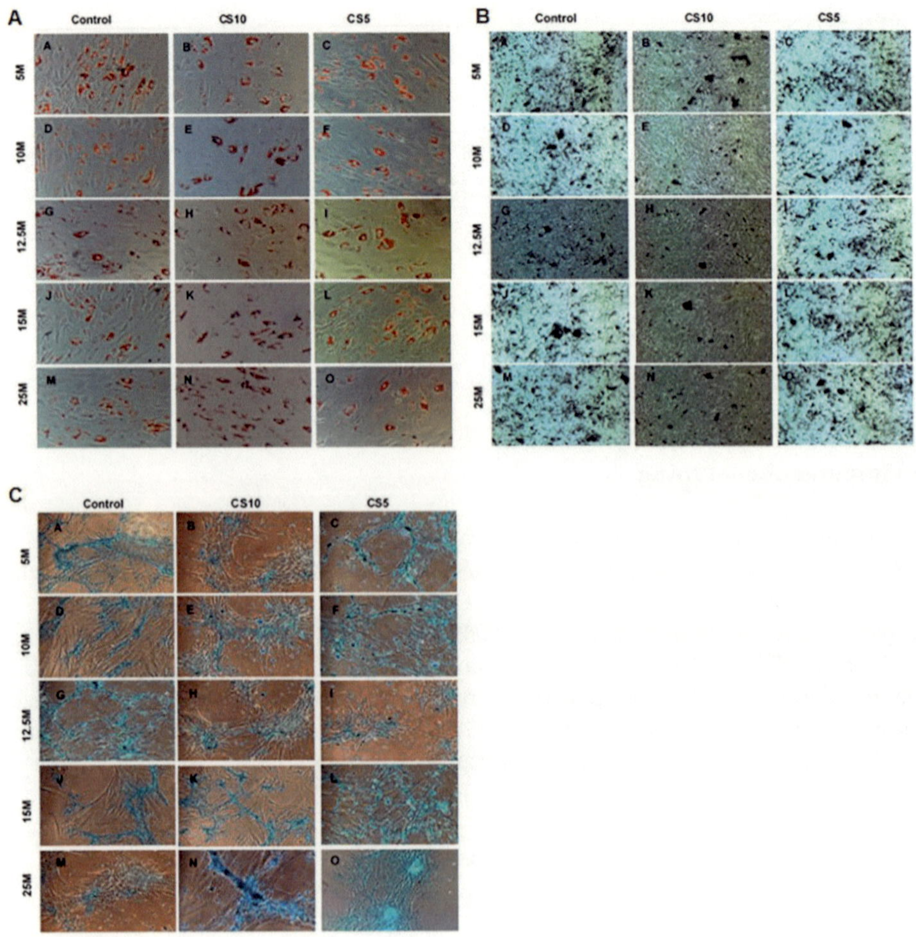

Figure 4.
Trilineage differentiation of BM-MSC in 3 different cryopreservation medium at 6th month time point. BM-MSC cryopreserved in PlasmaLyte a based cryopreservation solution (control), CryoStor (CS) 10 and CS5 were differentiated into (A) adipocytes (B) osteocytes and (C) chondrocytes. Adipocytes stained with oil red O, osteocytes with Von kossa staining and chondrocytes by safronin O stain. All pictures captured with magnification of 10X.

cryopreservation media (control, CS10 and CS5) at 6-month time points were assessed by counting the number of colonies with more than 50 cells. There is no significant change in the number of colonies formed by cells cryopreserved at different concentrations in all the three different cryopreservation media, except for two different cell concentrations in CS5 storage (**Figure 5**). The number of colonies formed by 5 and 10 million per ml freezing concentration ($n = 2$, $P < 0.001$) was significantly lower compared to other concentrations in CS5.

3.6 Apoptosis analysis

The comparative post-thaw apoptosis assay was performed at 6th month time point of BM-MSCs cryopreserved in control, CS10 and CS5 of 5 different freezing concentrations. The percentage of apoptotic cells increased with the decrease in freezing concentration, with the lowest concentration of 5 million/ml having the

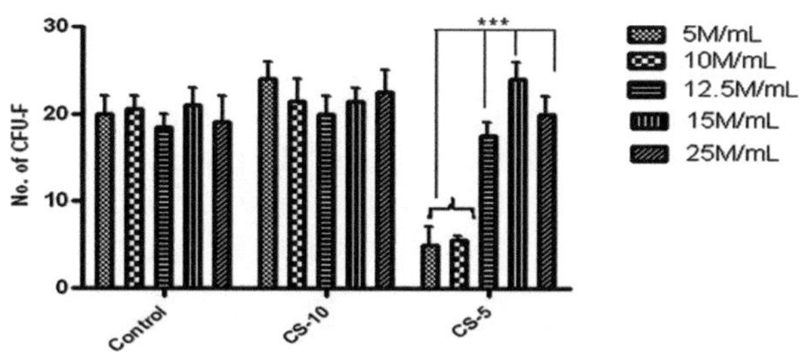

Figure 5.
*CFU-F assay of BM-MSC in 3 different cryopreservation medium at 6th month time point. The number of colonies formed in 5 different freezing concentration at 6th month time point show no statistically difference between control and CS10. There is a significant difference (***P < 0.001) observed in CS5 at lower freezing density of 5 and 10 million cells per mL compared to other 3 freezing concentration (12.5, 15 and 25 million cells per mL).Abbreviation:CFU — colony forming unit fibroblast.*

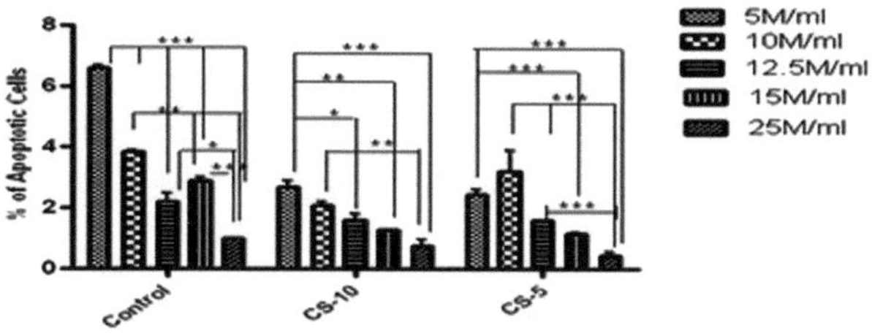

Figure 6.
*Apoptosis assay. Percentage of apoptotic cells in 3 cryopreservation media (control, CS5, CS10) in 5 different freezing concentration at 6th month time point reveals significant increase in percentage of apoptosis from 5 million cells per ml to 25 million cells per ml in all three cryopreservation media. (*P < 0.05,**p < 0.01***P < 0.001).*

highest value and vice versa in all the three freezing media (**Figure 6**). The differences are significant (n = 2,*P < 0.05, **P < 0.01, ***P < 0.001) except between 25 and 15 million cells per ml of CS10 and CS5. Overall, CS10 and CS5 have a lesser percentage of apoptotic cells compared to control in all concentrations tested.

4. Discussion

The demand for MSCs in clinical application is growing and necessitates availability of good quality cryopreserved MSCs with minimal pre-transplantation manipulations at the clinical trial site or final usage of the product. Optimal cryopreservation protocols including freezing density were not yet tested or published globally and no consensus has reached for the standard freezing density of MSCs. Moreover, published efforts were guided by HSCs cryopreservation protocols [15, 16]. Cryopreservation technology for MSCs is still evolving, as MSCs seem to lose viability very rapidly and are mostly attributed due to the rapid development of apoptotic processes and cryopreservation-induced delayed-onset cell death (CIDOCD) [17]. Most commonly, MSCS were cryopreserved in formulations containing 10% DMSO that needs to be removed from the final cellular product before infusion for human use. The removal of DMSO has its own challenge and requires the sample to be handled in a GMP compliance area and any improper removal may cause several ailments like headache, nausea, vomiting, sedation, high blood pressure, bradycardia or anaphylactic shock to the patients [18].

The total substitute of DMSO is not advisable, as it affects the viability and decreases the shelf life of MSCs during cryopreservation. Though many in-house cryopreservation formulations have been reported [19, 20], it is important that all their constituents meet the bio-safety standards. The use of approved and commercially available freezing formulations will ease the complications and facilitates hassle-free filing to NDA (New Drug Application), auditing, screening and testing etc.

In the present study, we have critically evaluated 5% and 10% DMSO containing commercial cryopreservation media (CS5 & CS10, respectively) for post-thaw viability, CFU-F, phenotypic markers expression, differentiation potential and percentage apoptotic cells of cryopreserved BMMSCs and compared to in-house formulations (PlamsLyte A with 10% DMSO and 5% HSA) at different time points up to 6 months. PlamsLyte A is an isotonic solution with the physiochemical properties closely resembling human plasma and is widely used as perioperative fluid [21]. PlasmaLyte A is commonly used in cryopreservation of MSCs and in our earlier studies it is known to preserve the viability and efficiency of BM-MSCs for clinical trials [22].

We have demonstrated comparable results for all the characteristics analysed in all these formulations at different time points. Our results showed that the post-thaw viability analysed by two methods viz., trypan blue dye exclusion and 7AAD were > 85% in all conditions tested up to 6 months time period. Few studies reported a reduction in post-thaw viability in time with the use of DMSO-based cryoprotectant [10, 23] but they used different freezing concentrations and DMSO percentages. The viability of >85% of what we obtained should not be a problem as there are reports stating that even with 70% viability, the cells were demonstrated to have enhanced immunosuppression within 6 months of time period [24]. We also demonstrated that the total cell recovery of BM-MSCs is >85% in all these conditions and it probably seems that higher cell freezing densities yields lower cell recovery compared to lower freezing density, but the differences in not significant in the conditions tested. It may be probably that the lower concentration cells tend to thaw faster and dilute DMSO faster during post-thawing procedures maintaining cell integrity and lower osmotic cell shock to the cells.

The phenotypic marker expression of MSCs cryopreserved in all the three different cryopreservation media was comparable with >95% expression for CD90 and CD73 and < 3% for CD19 and CD14 markers at all tested variables. There was no significant difference in expression of CD markers in all three freezing media with respect to different freezing densities at different time points except for CD 73 markers in different CS10 concentrations. It should not be a matter of concern as the expression in all of them is above 95% as stated by ISCT guidelines. Though contradictory studies showed the stable expression [25] or decreased expression [23] of phenotypic markers over different time points, we have not observed such difference in any of the time points with respect to freezing media or different freezing densities.

The trilineage differentiation ability of MSCs is also not compromised in any of these three formulations. There are not many studies, which compare the differentiation characteristics of MSCs at different freezing densities and various time points. Nevertheless, a study by Naaldijk *et al*. 2012 found a slight reduction in osteogenesis capacity of MSCs with higher DMSO concentrations [2]. In our study, we did not do any quantification but we observe that all of these conditions retain the trilineage differentiation capabilities.

Additionally, we evaluated the CFU-F ability and apoptotic cell percentage in cryopreserved BM MSC in different cryopreservation formulations at different cell densities at 6-month time point. These extended assays were done to comply with the stability testing of new drug substances and products of ICH guidelines 21 CFR 31.2.23(a) (7) (ii) which requires a minimum testing period of 6 months to confirm the functionality of the product. The CFU-F ability is one of the markers of stemness and proliferation capacity of MSCs. We observed a similar competence in the number of CFU-Fs formed at different cell concentrations in all three cryopreservation media, except for the two lowest cell densities in CS5 formulation. There are no studies that report the effect of freezing cell concentrations on CFU-F, however, a study reports no difference in number of CFU-Fs between 5% and 10% DMSO concentration in freezing media [26]. With regard to percentage of apoptotic cells at 6 months time point, we found that the lower freezing densities are prone to higher apoptotic rates compared to higher freezing densities in all the formulations. Higher level of intrinsic proteolytic activity may be higher in lower freezing concentration, as the DMSO availability per cell is high, compared to higher freezing cell density, where the DMSO availability per cell is low. Usually, post-thaw activation of caspase-3 demonstrated the proteases activity and subsequently increase intrinsic proteolytic activity following cryopreservation [20, 27]. Hence higher freezing density of BM-MSCs has a lesser apoptotic percentage compared to a lower freezing density.

5. Conclusion

In this study, we demonstrated the possibility of using reduced 5% DMSO containing cryopreservation media (CS5) for cryopreserving BM-MSCs without any impact on viability, phenotypic characteristics and functional properties of MSCs. This was the first study to provide the characterization and comparison of human BM-MSCs cryopreserved in different freezing densities ranging from as low as 5 M cells per ml to higher freezing densities as 25 M cells per ml in two commercially available variants of CS (CS10 and CS5) and comparing it to in-house formulation.

Based on our presented data, we can conclude that chemically defined reduced DMSO-based formulation of CS5 addresses challenges and minimizes the post-cryopreservation manipulation of MSCs for clinical use. However, these data needs to be backed up by safety and efficacy studies, with long-term stability program up to 1 year with more intrinsic molecular, proteomics analysis and immune-suppressive ability of cryopreserved MSCs before employing the CS5 for the cryopreservation of MSCs for therapeutic applications.

Acknowledgements

This work was fully funded by Stempeutics Research Pvt. Ltd., India.

Author contribution statement

SK: Design of studies, data analysis, interpretation and manuscript writing. SB: Design of studies, perform experiment, data collection and manuscript writing. PKG: Design of studies and manuscript correction. UK: Design of studies and data analysis. Correction and final approval of the manuscript.

Conflict of interest

The authors declare no conflict of interest.

Ethics statement

The authors confirm that the ethical policies of the journal, as noted on the journal's author guidelines page, have been adhered to and the appropriate ethical review committee approval has been received. Obtaining bone marrow from consenting healthy donors was approved by the Institutional Ethics Committee (IEC) at the Manipal Hospital, Bangalore, India.

Author details

Suresh Kannan, Swaroop Bhagwat, Pawan Kumar Gupta and Udaykumar Kolkundkar
Stempeutics Research Pvt. Ltd, Bengaluru, India

*Address all correspondence to: uday.kumar@stempeutics.com

References

[1] Rodriguez-Fuentes DE, Fernandez-Garza LE, Samia-Meza JA, Barrera-Barrera SA, Caplan AI, Barrera-Saldana HA. Mesenchymal stem cells current clinical applications: A systematic review. Archives of Medical Research. 2021;**52**(1):93-101

[2] Naaldijk Y, Staude M, Fedorova V, Stolzing A. Effect of different freezing rates during cryopreservation of rat mesenchymal stem cells using combinations of hydroxyethyl starch and dimethylsulfoxide. BMC Biotechnology. 2012;**12**:49

[3] Fu X, Xu B, Jiang J, Du X, Yu X, Yan Y, et al. Effects of cryopreservation and long-term culture on biological characteristics and proteomic profiles of human umbilical cord-derived mesenchymal stem cells. Clinical Proteomics. 2020;**17**:15

[4] Bhat S, Viswanathan P, Chandanala S, Prasanna SJ, Seetharam RN. Expansion and characterization of bone marrow derived human mesenchymal stromal cells in serum-free conditions. Scientific Reports. 2021;**11**(1):3403

[5] Pal R, Hanwate M, Totey SM. Effect of holding time, temperature and different parenteral solutions on viability and functionality of adult bone marrow-derived mesenchymal stem cells before transplantation. Journal of Tissue Engineering and Regenerative Medicine. 2008;**2**(7):436-444

[6] Petrenko Y, Chudickova M, Vackova I, Groh T, Kosnarova E, Cejkova J, et al. Clinically relevant solution for the hypothermic storage and transportation of human multipotent mesenchymal stromal cells. Stem Cells International. 2019;**2019**:5909524

[7] Antebi B, Asher AM, Rodriguez LA 2nd, Moore RK, Mohammadipoor A, Cancio LC. Cryopreserved mesenchymal stem cells regain functional potency following a 24-h acclimation period. Journal of Translational Medicine. 2019;**17**(1):297

[8] Awan M, Buriak I, Fleck R, Fuller B, Goltsev A, Kerby J, et al. Dimethyl sulfoxide: A central player since the dawn of cryobiology, is efficacy balanced by toxicity? Regenerative Medicine. 2020;**15**(3):1463-1491

[9] Pal R, Hanwate M, Jan M, Totey S. Phenotypic and functional comparison of optimum culture conditions for upscaling of bone marrow-derived mesenchymal stem cells. Journal of Tissue Engineering and Regenerative Medicine. 2009;**3**(3):163-174

[10] Liu Y, Xu X, Ma X, Martin-Rendon E, Watt S, Cui Z. Cryopreservation of human bone marrow-derived mesenchymal stem cells with reduced dimethylsulfoxide and well-defined freezing solutions. Biotechnology Progress. 2010;**26**(6):1635-1643

[11] Ntai A, La Spada A, De Blasio P, Biunno I. Trehalose to cryopreserve human pluripotent stem cells. Stem Cell Research. 2018;**31**:102-112

[12] Biggs CI, Bailey TL, Ben G, Stubbs C, Fayter A, Gibson MI. Polymer mimics of biomacromolecular antifreezes. Nature Communications. 2017;**8**(1):1546

[13] Thirumala S, Gimble JM, Devireddy RV. Evaluation of methylcellulose and dimethyl sulfoxide as the cryoprotectants in a serum-free freezing media for cryopreservation of adipose-derived adult stem

cells. Stem Cells and Development. 2010;**19**(4):513-522

[14] Gottipamula S, Sharma A, Krishnamurthy S, Majumdar AS, Seetharam RN. Human platelet lysate is an alternative to fetal bovine serum for large-scale expansion of bone marrow-derived mesenchymal stromal cells. Biotechnology Letters. 2012;**34**(7):1367-1374

[15] Meneghel J, Kilbride P, Morris GJ. Cryopreservation as a key element in the successful delivery of cell-based therapies-a review. Frontiers in Medicine (Lausanne). 2020;7:592242

[16] Xie J, Ekpo MD, Xiao J, Zhao H, Bai X, Liang Y, et al. Principles and protocols for post-cryopreservation quality evaluation of stem cells in novel biomedicine. Frontiers in Pharmacology. 2022;**13**:907943

[17] Baust JM, Corwin W, Snyder KK, Van Buskirk R, Baust JG. Cryopreservation: Evolution of molecular based strategies. Advances in Experimental Medicine and Biology. 2016;**951**:13-29

[18] Kollerup Madsen B, Hilscher M, Zetner D, Rosenberg J. Adverse reactions of dimethyl sulfoxide in humans: A systematic review. F1000Res. 2018;7:1746

[19] Erol OD, Pervin B, Seker ME, Aerts-Kaya F. Effects of storage media, supplements and cryopreservation methods on quality of stem cells. World Journal of Stem Cells. 2021;**13**(9):1197-1214

[20] Baust JG, Gao D, Baust JM. Cryopreservation: An emerging paradigm change. Organogenesis. 2009;**5**(3):90-96

[21] Weinberg L, Collins N, Van Mourik K, Tan C, Bellomo R. Plasma-Lyte 148: A clinical review. World Journal of Critical Care Medicine. 2016;**5**(4):235-250

[22] Gupta PK, Dutta S, Kala S, Nekkanti M, Desai SC, Mahapatra SS, et al. Phase IV postmarketing surveillance study shows continued efficacy and safety of Stempeucel in patients with critical limb ischemia due to Buerger's disease. Stem Cells Translational Medicine. 2021;**10**(12):1602-1613

[23] De Rosa A, De Francesco F, Tirino V, Ferraro GA, Desiderio V, Paino F, et al. A new method for cryopreserving adipose-derived stem cells: An attractive and suitable large-scale and long-term cell banking technology. Tissue Engineering. Part C, Methods. 2009;**15**(4):659-667

[24] Samuelsson H, Ringden O, Lonnies H, Le Blanc K. Optimizing in vitro conditions for immunomodulation and expansion of mesenchymal stromal cells. Cytotherapy. 2009;**11**(2):129-136

[25] Cooper K, Viswanathan C. Establishment of a mesenchymal stem cell bank. Stem Cells International. 2011;**2011**:905621

[26] Ock SA, Rho GJ. Effect of dimethyl sulfoxide (DMSO) on cryopreservation of porcine mesenchymal stem cells (pMSCs). Cell Transplantation. 2011;**20**(8):1231-1239

[27] Bissoyi A, Pramanik K. Role of the apoptosis pathway in cryopreservation-induced cell death in mesenchymal stem cells derived from umbilical cord blood. Biopreservation and Biobanking. 2014;**12**(4):246-254

Chapter 5

Nanotechnology-Based Stem Cell Therapy: Current Status and Perspectives

Ponpandian Samuel, Shenbagamoorthy Sundarraj and D.N.P. Sudarmani

Abstract

The nanoparticles or nanobots are equivalent to the size of biological molecules of the human body and this is claimed to be the massive advantage of nanotechnology. Currently, top-down and bottom-up fabrication methods are being adopted to synthesize nanomaterials. Hence, the products developed from nanotechnology can be used for assessment of several biological parameters under *in vitro* and *in vivo* conditions. Effective production of nanoparticles, accompanied by the advent of novel characterization studies, enables us to manipulate the arrangement of atoms distributed on the surface of the nanomaterials to make it functionally more effective than before. In addition to the support imparted by nanotechnology, it also plays a primary role in the field of diagnostics. Another important outcome of nanotechnology is nanomedicine, which deals with the site-specific delivery of drugs with the aid of fabricated nanosystems. The advent of technology in recent years has enabled researchers to build novel forms of drug delivery systems like liposomes, dendrimers, nanoparticles and nanocrystals, which in turn ensure the précised delivery of drugs to suitable targets. Several need-based and value-added applications of nanotechnology are enlisted in the chapter.

Keywords: nanoparticles, nanobots, liposomes, nanosystems, nanocrystals, dendrimers

1. Introduction

Nanotechnology and stem cell therapies are two diverse fields and recent prominent areas of research in the direction of improvement to solve challenges during the treatment. Stem cell therapy, also known as regenerative medicine, promotes the repair mechanism of dysfunctional, incapacitated or wounded tissue by stem cells or their derivatives. Stem cell therapy is a treatment that has been done with stem cells. Stem cell therapy holds promise for treating a broad spectrum of diseases, such as cancer, heart disease, diabetes and neurodegenerative diseases. Researchers are still studying various sources for stem cells, which are applied for stem-cell treatment [1]. It is fascinating that the integration of two disciplines, nanotechnology and stem cell sciences, divulges new ways to identify the role of molecular apparatus in the mechanism of the differentiation of stem cells regulation and elucidate more about the stem cell-based treatment strategies for insight into the human disease, prevention and theranostics.

Nanotechnology-based approaches in stem cell research have been established by utilizing biocompatible, biodegradable, solubility, stability, specificity, multimodality and efficacy for undergoing attachment to cognate receptors. Researchers have already shown that the following nanoparticle has been developed for the applications in stem cells differentiation and regeneration therapy, such as superparamagnetic iron oxide nanoparticles (SPIONs)-(ferucarbotran) NPs [2, 3], auto-assembled peptide [4], magnetic NPs [5], polyelectrolyte NPs [6], cerium oxide NPs [7], graphene oxide NPs [8], poly-ε-caprolactone [9], ZnO NPs [10, 11], SiO_2-NPs [12], iron oxide NP [13], collagen nanofiber [14], retinoic acid loaded with polymeric nanoparticles [15], tri-$CaPSO_4$ (tricalcium phosphate) [16], carbon nanofiber [17, 18], graphene-oxide nanoparticles (GO-NPs) [19], AuNPs [20, 21], PANPs [22], Au@BSA@PLL [23], USPIO [24], PFCE-NPs [25] and tri-Ca-silicate [26].

In the latest time, the application of nanotechnology in stem cell research has engaged better advances, which is attractive to an emerging interdisciplinary field. Stem cell nanotechnology is developing towards stem cell isolation, lineage and differentiation, stem cell imaging, active tracking, regenerative medicine and tissue engineering of stem cells (**Figure 1**). Nevertheless, stem cell nanotechnology also faces many challenges similar to any emerging interdisciplinary field. The mechanism of interaction between nanomaterials and stem cells still needs to be elucidated well as nanomaterials and nanostructures are modified to enhance the function of stem cells, and the action of metabolizing nanomaterials inside stem cells is arduous. The fabrication of multifunctional or homogenous nanostructures developed by existing knowledge and principles has been a great challenge in synthesizing, modifying and characterizing the quality and stability of nanomaterials and the mechanism of interacting with the stem. However, stem cell nanotechnology shows great fascinating

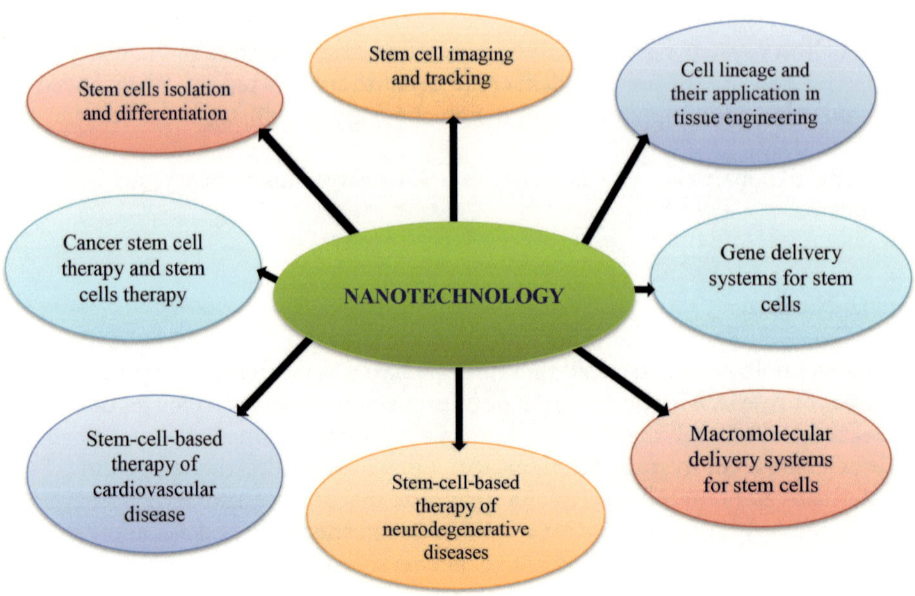

Figure 1.
Applications of nanotechnology in stem cell.

scenarios, stem cells are emerging for the application of the drug and macromolecular delivery for degenerative diseases [27].

2. Application of nanotechnology in stem cells isolation and differentiation

A crucial point in stem cell-based therapy is the segregation of appropriate cell types. The magnetic cell isolation technique was used to isolate specific types of cells. Magnetic nanoparticles can be used to label the stem cells for identification from a pool of different cell types by magnetic-activated cell sorting (MACS) [28]. This process involves combining MNPs with monoclonal antibodies (MAB) directed against unusual cell surface antigens, which causes the magnetic field of the cells expressing these antigens to be retained. It has been demonstrated that MNPs and anti-CD34 antibodies work together to efficiently label and distinguish peripheral blood progenitor cells from the blood. An uninterrupted quadruple magnetic flow sorter consisting of a flowing carrier and a quadrupole magnet with 1.42 T maximum field loudness and optimal field strength was able to separate these cells from mononuclear cells suspension of whole blood when MNP-conjugated anti-CD34 proteins were used to label CD34-cells. The CD34 cells collected had a purity of 60–96%, a retrieval rate of 18–60%, an improvement rate of 12–169 and a throughput of (1.7–9.3) 104 cells/s [29]. The optimized cells could be employed for cell transplantation-based regenerative medicine.

For SC proliferation and differentiation, scaffold-dependent nanomaterials and associated polymers have been used. Different scaffolds have been investigated with a focus on nanotubes, nanoparticles and nanofibers to control the differentiation of SC. Carbon nanotubes (CNTs) and titanium dioxide (TiO_2) are viable possibilities for scaffold creation, such as bone replacement therapy, due to their outstanding mechanical properties [30]. The impact of biological molecules and intricate interactions with scaffold substances improves SC development. Due to their exceptional electrical, mechanical and refractive indices and extensive surface topographical features, many nanomaterial-based scaffolds have been used in tissue engineering applications; nevertheless, this sector is focused on graphene and graphene oxide (GO) as non-toxic scaffolds [31, 32].

The researchers identified several peptide sequences that can firmly attach to NSCs. The new peptide (HGEVPRFHAVHL, HGE) was combined with quantum dots, Zhao et al. [30] discovered that the 48/34 kDa proteins on the membranes of NSCs produced from monkey ESCs but not human ESCs were particularly identified by this HE-quantum dot combination. According to this work, ESC-conjugated particular peptides may be used to examine the lineages they have committed to, and they may also be a mechanism for separating differentiated cells from ESC-differentiated cell populations. iPSCs often need to be cultivated on the feeder layer cells to preserve their pluripotency. Graphene (G) and graphene oxide (GO) have lately been established as cell culture substrates due to their biocompatibility at low concentrations and 2D structure with a large surface area. G and GO can support the culture of mouse iPSCs by allowing stem cells to differentiate. The cell proliferation and differentiation properties are induced by graphene materials (**Figure 2**). While iPSCs cultured on GO surfaces exhibit faster rates of adherence and proliferation than those on glass surfaces, iPSCs cultured on G surfaces show a similar effect on cell adhesion and proliferation [34]. Another benefit of GO is that it keeps the iPSCs in the undifferentiated stage while speeding up the differentiation [33].

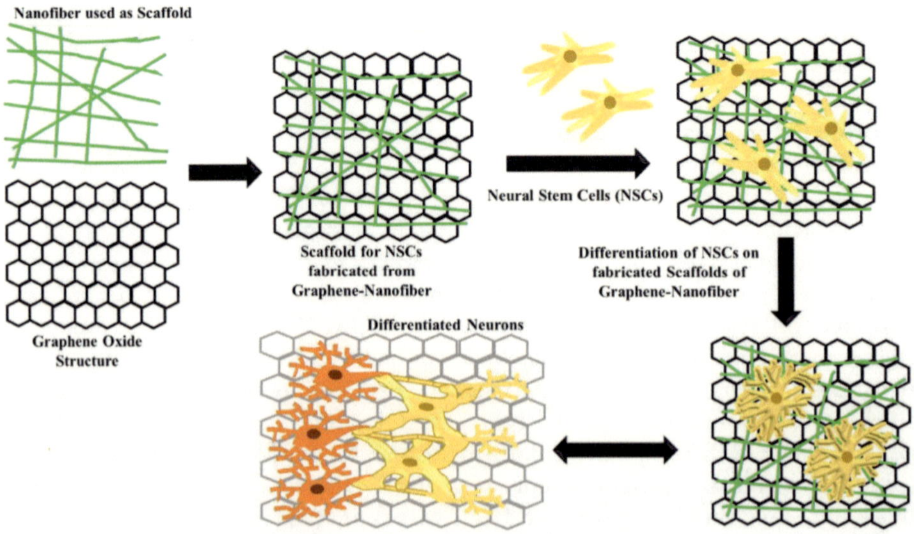

Figure 2.
Schematic representative of scaffold structure fabricated from graphene–nanofiber for differentiation of neural stem cells. The illustration was created by Asil et al. [33], and published in Appl. Sci. 2020, 10(14), 4852; https://doi.org/10.3390/app10144852. Licensed under CC by 4.0.

3. Application of nanotechnology in stem cell-based regeneration

One of the main concerns in cell therapy is whether SCs can be guided to specific areas to repair damaged brain regions [35]. In earlier studies, cells could only be observed after the animals were sacrificed. The mechanism of stem cell homing to specific tissues was identified using novel labelling techniques or materials without compromising SC proliferation, differentiation or migration, which is vital for tissue engineering and regenerative medicine. Proper labelling enables the practical detection of transplanted cells and the tracking of cells at the defect site to ascertain their role in tissue regeneration. For example, hMSCs were labeled with different nanoparticles, including quantum dots, fluorescence-labeled silica nanoparticles, gold nanoparticles and super-paramagnetic iron oxide nanoparticles, to follow these cells during live imaging and ascertain whether SCs are taking part in repair processes. Magnetic nanoparticles (dMNPs) coated with polyamidoamine dendrimer are used to incorporate the pluripotent transcription factors Oct4, Sox2, Lin 28 and Nanog to manufacture lentiviruses that generate iPSCs [36]. The generated lentivirus was ten times more potent than viruses produced by the liposome method. After generating iPSCs, these cells were labeled with fluorescent magnetic nanoparticles. The fluorescence signals were observed using fluorescence microscopy, and the magnetic nanoparticles were located using magnetic resonance imaging. Successful cellular uptake and long-term retention of these nanoparticles in cells are advantageous for monitoring and labelling these cells after implantation. Even though these different nanoparticles can penetrate and mark cells effectively and efficiently, cytotoxicity has been raised as a concern about nanoparticle application. The cytotoxic effect of a substance is influenced by its size, shape, content, surface charge and hydrophobicity. These NP characteristics result in a rise in cytosolic reactive oxygen species, chromosomal aberrations and cell death. Therefore, it is crucial to focus on enhancing this nanoparticle's cellular absorption for monitoring and labelling while reducing their cytotoxicity and interference with cellular differentiation in the context

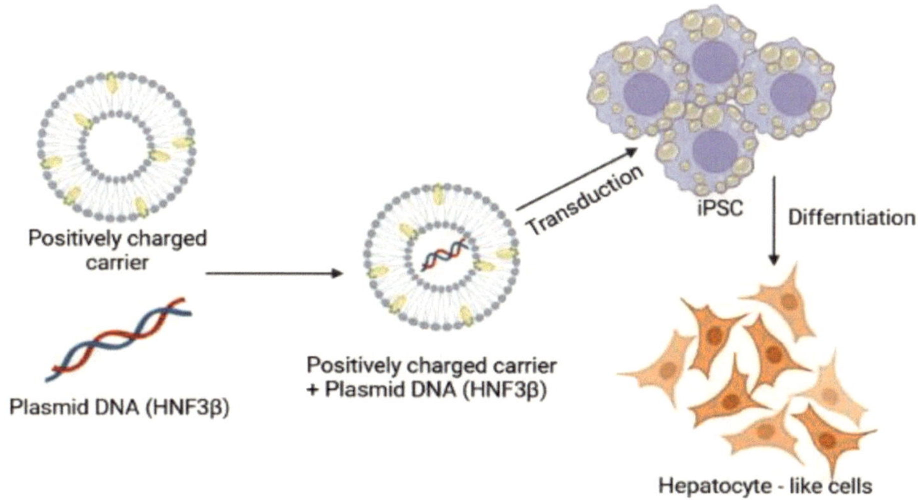

Figure 3.
Nanoparticle mediated transduction of induced pluripotent stem cells differentiate into specialized cells.

of biocompatibility. Future research on NP applications should concentrate on advancements in tissue-specific cell labelling, imaging and tracking (**Figure 3**).

Metallic NP-induced processes stimulate the proliferation and differentiation of stem cells through several mechanisms, including the alteration of signaling pathways, the production of reactive oxygen species and the tinkering of numerous transcription factors. Metallic nanoparticles have a potential impact on stem cell differentiation and proliferation both *in vivo* and *in vitro*. The superparamagnetic properties of the IONPs (iron oxide nanoparticles), often referred to as superparamagnetic iron oxide (SPIO) NPs, enable them to travel to the injured region, making them a potentially valuable tool for the treatment of degenerative illnesses. Human MSCs (hMSCs) can multiply when treated with SPIO- (Ferucarbotran) nanoparticles, which work by reducing intracellular H_2O_2. They can also speed up the cell cycle by upregulating proteins like cyclin D1, cyclin B, and cyclin-dependent kinase 4. Consequently, SPIO-NPs can be exploited as a secure supply of nanomaterials to promote the proliferation of stem cells [37].

4. Application of nanotechnology in stem cell imaging

For imaging and stem cell tracing, nanoparticles such as gold nanorods, MNPs and quantum dots have been utilized due to their distinct characteristic features. In cellular imaging, immunoassays, DNA hybridization and optical barcoding, QDs have been employed successfully. A new practical platform introduced for bioanalytical sciences and biomedical engineering is provided by quantum dots. Mesenchymal stem cells (MSCs) can internalize the quantum dots coupled with an antibody against the mortalin protein to produce i-QD composites, which label the MSCs. The normal adipocyte, osteocyte and chondrocyte development that the i-QD tagged MSCs underwent *in vitro* and *in vivo* strongly suggests that i-QDs can be used for *in vivo* imaging diagnostics and tracing of stem cells in the distribution of mouse body [38]. QDs can be created nanoprobes with unique functions that can be utilized for molecular imaging, gene or medication administration or molecule tracing. These biomolecules can be added to QDs, such as liposomes, PEG, peptides or antibodies. MNPs were used for molecular imaging, and stem cell tracing in addition to quantum dots. Superparamagnetic iron oxide nanoparticles (SPIONs) were found to be multifunctional MRI-based contrast agents and the same can be used for labelling and tracking transplanted stem cells [39, 40]. Dextran-coated iron oxide nanoparticles were covalently bounded to fluorescent molecules to define HSC labelling and the engraftment process. Fluorophores were conjugated to the dextran coat for fluorescence-activated cell sorting and purification, which removed false signals from nanoparticle contaminants that were not sequestered [27].

With no significant toxicity *in vitro* or *in vivo*, a short-term specified incubation technique was devised, effectively labeled both cycling and quiescent HSCs. Immunodeficient mice were given purified primary human cord blood cells that were CD34-positive and lineage-depleted, allowing tagged human HSCs to be found in the recipient mice's bones. The cell populations that had snatched up the nanoparticles were precisely quantified, and their destiny after transplantation was monitored using flow cytometry. The presence of MNPs-labeled human stem cells in the bone marrow was confirmed by flow cytometry analysis. There has been substantial research in stem cell treatment for various central nervous system (CNS) illnesses [41]. The Endorem-labeled GFP MSCs were transplanted to rats intravenously into the femoral vein or intra-cerebrally into the hemisphere oblique to the lesion the cells were grafted. A 4.7-T Bruker spectrometer was used to check on rats with grafted stem cells once per week from three to seven weeks after transplantation. On MR scans, the lesion appeared as a hyper-intense signal. Its intensity matched GFP labelling or Prussian blue staining. One week following a transverse spinal cord lesion, MSCs tagged with Endorem were also delivered intravenously into the femoral vein. The lesion cavity appeared as inhomogeneous tissue with a significant hyper-intensive signal on MR images of longitudinal spinal cord slices from animals without spinal grafts but with a lesion. Dark hypo-intense patches were seen as lesions in transplanted animals.

Histological analysis revealed that transplanted mice had a substantial iron positive while lesioned control animals had only a few iron-containing cells. The lesion in grafted animals was significantly smaller than in control rats, indicating that the grafted MSCs had a beneficial impact on lesion repair. There are numerous successful uses of MR tracking in various organs, including the heart, liver, kidney and pancreatic islets. Fluorescent MNPs (FMNPs) can combine with the BRCAA1 antibody to create BRCAA1 antibody-labeled FMMNP probes, BRCAA1 protein showed signs of overexpression in ES CCE cells [27].

5. Application of nanotechnology in stem cell tracking

Conventional methods to track implanted stem cell fate predicted *in vitro* cell labelling for cell transplantation, subsequent follow-up of cell engraftment and existence concluded by the analysis of histological sections of sacrificed animals or tissue biopsies, by this invasive technique did not permit long-term and continuous experimentation. Recent advances in stem cell therapy need more precise and non-invasive methods for qualitatively and quantitatively tracking transplanted cells inside the host to facilitate the understanding of the prognosis of neurodegenerative treatment and eventually improve patients' health.

The traditional techniques have been developed by the improvement of specific contrast agents, such as endogenous biomolecules with intrinsic fluorescence, exogenous fluorescent proteins or non-fluorescent organic dyes, which have been used for fundamentally two labelling modalities. Which are direct labelling and indirect labelling. (a) direct labelling is the cell incubated with specific intracellular probes; (b) indirect labelling is the cell tracked through the expression of the indicator by a reporter gene inserted in the genome of the cells. Direct methods are simple to apply and less expensive, although potential limitations include fast signal decay due to cell proliferation and subsequent insufficient marker distribution between daughter cells. Alternatively, an indirect technique is much more stable but needs genetic manipulation of cells and has not been suitable for clinical applications. Generally, the active contrast agents frequently present disadvantages like photo-bleaching over time, interference derived from tissue auto-fluorescence, chemical and metabolic degradation *in vivo* and even low transfection efficiency in primary cells and thus are not considered suitable for *in vivo* imaging. Several engineered nanoparticles with unique magnetic and optical properties have been established and employed in biomedicine due to their capability to offer real-time monitoring of tracking intracellular processes at a biomolecular level [42]. The transplanted stem cells labeled by these nanoparticles can be detected by multiple imaging methods, such as magnetic resonance imaging (MRI), nuclear imaging, single-photon emission computed tomography imaging (SPECT), positron emission tomography-computed tomography (PETCT) and photoacoustic imaging [43–46].

Stem cells are tracked by the functional modification of nanoparticles such as Gd-based nanoparticles are the most extensively used T1-contrast agent for labelling and tracking stem cells [47]. The GD-based nanoparticles composed of spherical europium-doped gadolinium oxysulfide (Gd_2O_2S: Eu^{3+}) have been fabricated and observed by MRI, X-ray imaging and photoluminescence imaging. The number of MSCs labeled by Gd_2O_2S: Eu^{3+} have feasible cell tracking in animal models [48].

Au-based nanoparticles are potential contrast agents of photoacoustic imaging developing as a modern method for tracking cells *in vivo*. More significantly, MSCs can be directly labeled by Au-based nanoparticles, so their differentiation after transplantation *in vivo* has been noticed using photoacoustic imaging *in vivo* [49, 50]. Huang et al. [51] synthesized Au-based nanoparticles (AA@ICG@PLL) with dual-modal imaging (CT and near-infrared fluorescence) for labelling and tracking MSCs of mice. AA@ICG@PLL exhibited excellent cellular uptake by MSCs and biocompatibility due to the modification of indocyanine green (ICG) and poly-L-lysine (PLL).

Hsieh et al. [52] described a QD-based NP for labelling human MSCs, in which CdSe was used as the core, and the shell was encapsulated by ZnS. Chen et al. [53] stated an AgS_2 QD-based NP for tracking human MSCs transplanted in the mouse by employing fluorescence imaging. Li et al. [54] developed QDs-based nanoparticles (RGD-β-CDQDs) to label and track human MSCs, which were fabricated of QDs, β-cyclodextrin (β-CD) and Cys-Lys-Lys-ArgGly-Asp (CKKRGD) peptide. The QDs altered by b-CD had greater cellular uptake and eased the differentiation of MSCs, due to the small molecule dexamethasone and siRNA carried by b-CD. Further significantly, the labeled MSCs have been identified for one month.

Super-paramagnetic iron oxide nanoparticles (SPIO NPs) are synthesized for labelling MSCs. The labeled MSCs have been tracked in an animal model by MRI [55–58]. Furthermore, these labeled MSCs still upheld differentiation potential. Lee et al. [59] synthesized SPIO NPs with the modification of poly lactic-co-glycolic acid (PLGA) and then utilized fluorescent dye Cy5.5 to functionalize the synthesized nanoparticles for labelling and tracking MSCs to explore the interactions between PLGA-SPIO NPs and MSCs.

Ma et al. [60] stated up-conversion-based nanoparticles, which were fabricated by $NaYF_4$:Yb^{3+}, Er^{3+} NPs, poly (acrylic acid) (PAA) and poly (allylamine hydrochloride)

(PAH), as a fluorescence maker for tracking bone marrow MSCs *in vitro*. Kang et al. [61] developed a UC-based NP with NIR-controllable properties to label MSCs. In a remote-controllable way, stem cell differentiation was regulated. Moreover, Ren et al. [62] produced conversion-based nanoparticles $NaYF_4$:Yb/Erusedligand free labelling and tracking mouse bone MSCs.

Huang et al. [63] synthesized mesoporous silica nanoparticles altered by fluorescein isothiocyanate, and the labeled MSCs have been identified by imaging to track their viability *in vivo*. Due to clathrin-mediated endocytosis, the nanoparticles have been internalized into MSCs and showed greater cellular uptake. Additionally, Chen and Jokerst [64] used silica nanoparticles to label MSCs and then track the MSCs by ultrasound imaging. The results exhibited that silica nanoparticles have expressively increased the ultrasound signal of MSCs *in vivo*. Yao et al. [65] described unique core-shell nanoparticles in which the core is composed of cobalt protoporphyrin IX (CoPP)-loaded mesoporous silica nanoparticles, and the shell is a ^{125}I-conjugated/ spermine-modified dextran polymer, to label and guide the transplantation of MSCs by PA imaging and SPCT nuclear imaging. Chen et al. [66] developed three sizes of silicon carbide nanoparticles to label MSCs and showed dual-modality imaging of photoluminescence and photoacoustic imaging. Cyanine dye-doped silica nanoparticles have been used to label hMSC without affecting stemness surface marker expression, proliferation, viability and differentiation capability into osteocytes [67].

Lim et al. [68] fabricated bicyclononyne (BCN)-conjugated glycol chitosan nanoparticles (BCN-NPs) as dual-modal stem cell imaging probes for the cellular imaging system. Yin et al. [69] described an organic semiconducting polymer nanoparticle (OSPNC) as a contrast agent for tracking MSCs. The developed cationic nanoparticles revealed the intensive tissue imaging due to the meaningfully higher signal-to-noise (SNR) and improved the cellular uptake for human MSCs because of their biocompatibility, appropriate size and optimized surface property.

6. Nano patterns drive the fate of stem cells into a specific cell lineage

The instructional and tissue-specific niches of stem cells play a variety of activities, including migration, adhesion and proliferation. Extracellular matrix (ECM) components of nanoscale feature-sized fibrillary collagens, elastin and glycosaminoglycans are particularly prone to affect SCs. The topography and part structure of the ECM can compel stem cells to develop into particular cell lineages.

The crucial step in stem cell-based therapies is to direct SCs with accurate fabrication in a defined direction. Nanotechnologists have created several synthetic nanoplatforms that mimic the topological characteristics of the natural SCs niche to stimulate stem cell activation. The attachment of SCs surface proteins to topography is a fundamental aspect of the mechanism via which stem cells interpret and respond to nanotopographical signals. Focused adhesion, a form of integrin-mediated cell attachment to ECM components, is essential for stem cell regulation. Gene and protein levels will vary with mechanical stimulation and regulating focal adhesions will affect the stem cell differentiation pathway. Integrin-mediated adhesion signaling and other factors that impact the state of the SCs include cytoskeleton (CSK) stress, SC structure and nuclear dynamics. Arginine-glycine-aspartate is a crucial peptide episode in ECM proteins that regulates cell adherence (RGD). Recent studies have concentrated on the effect of RGD-containing nanopatterns on stem cell activity.

Cao et al. [70] planned the synthesis of several charged or neutral oligopeptide motifs connected with RGD using quartz substrates as a model and were employed for surface modification. They showed that, in the presence of RGD, positively charged oligopeptide patterns hinder osteogenic development, but negatively charged and neutral oligopeptide patterns may promote it.

Wang et al. [71] investigated the effects of RGD nanospacings ranging from 37 to 124 nm on the conduct of MSC. RGD nanopatterns were developed on PEG hydrogels. Cells were exposed to these nanopatterns at the highest serum level for eight days. They differentiated SCs into adipogenic and osteogenic lineages with large and small nanospacings. Stem cell activity is influenced by the symmetry, size, and regularity of surface nano-topographic features, which have been shown to have a substantial impact. Park et al. [32] showed that MSC activity, which includes differentiation, development and spreading, is significantly dependent on the diameter (d) of self-assembled layers (SAL) of TiO_2 nanotubes. They showed that osteogenic differentiation of MSC can be significantly reduced by increasing the diameter of the tube to 50 nm or higher after separating SCs into osteogenic cells through a tube having a 15 nm diameter. Researchers have looked into how different-pitch nanogrooves affect the ability of SCs to self-renew, differentiate and proliferate.

Currently, an important area of research involves the merger of SC nanotechnology (SC-NTech) and tissue engineering ideas. Nano-engineered 3-D scaffolds are frequently used to make it possible for SCs to differentiate into specific cell lineages. These three-dimensional scaffolds might be biodegradable, allowing cells to produce their own ECM as the synthetic scaffold degrades. For example, using nanofibrous scaffolds in bone tissue engineering intensely increased the differentiation of SCs into osteogenic cells compared to controls.

7. Application of nanotechnology in stem-cell-based tissue engineering

The principle of tissue engineering combined with stem cells enables the development of a stem cell-based therapeutic strategy for human diseases. Stem cell and progenitor cell steering differentiation is presently one hotspot, the differentiation of stem cells that conjugate 3D materials is deliberated as the most perspective tissue engineering. Recently, the developments of several micro/nanofabrication technologies have been used to stimulate stem cells to develop into 3D biodegradable scaffolds. Nanostructured scaffolds are fabricated to initiate stem cells to turn into specific cell types compromising the tissues and organs in the body. Inside these scaffolds, cells secrete their matrix, and as the scaffold degrades, they form a 3D tissue structure that mimics the body's natural tissues. Gelain et al. [72] described that they had established a 3D cell culture system using an exclusive peptide nanofiber scaffold with mouse adult neural stem cells. They prepared 18 different peptides, which directly integrate various functional motifs to stimulate cell adhesion, differentiation and bone marrow homing and engraftment activities. These functionalized peptides are self-assembled into nanofiber scaffolds where cells have been completely entrenched by the scaffold in 3D. Without the addition of neurotrophic factors and soluble growth factors, two of these scaffolds functionalized with bone marrow homing motifs significantly enhanced the survival of the neural stem cells and also encouraged differentiation towards cells expressing neuronal and glial markers.

Carbon nanotube patterns have been used to improve the growth and alignment of MSCs. The MSCs revealed in CNT growth patterns, and the cell culture results showed that the CNT designs have no harmful consequence on the MSCs [73]. The outcomes demonstrated that CNT patterns have enormous potential as a new platform for basic research and applications expanding stem cells.

Stem cell differentiation is diligently related to their microenvironment. The regulation of stem cells is contingent on their dealings with a highly specialized microenvironment or niche. Secreted factors, stem cell-neighboring cell interactions, extracellular matrix (ECM) and mechanical properties collectively made the stem cell microenvironment. The stem cell niche secretes suitable chemicals to direct the differentiation and development of stem cells. Mineral components are essential to stem cell localization; matrix components are vital to the restraint of stem cells, and bone-forming osteoblasts are also important to the maintenance and proliferation of stem cells, the calcium-sensing receptor located on the surface of HSCs, and other cells are critical to stem cells finding their niche.

Nanotechnology has been employed to create artificial *in vivo* conditions like stem cell microenvironments to discover the fundamental mechanisms of the conversion into differentiated cells. A better solution is presently under exploration: growing the stem cells on a so-called 'lab-on-a-chip'. They synthesize a silicon chip with a thousand nanoreservoir cavities, which surface contains about a thousand reservoir cavities, with each reservoir only about 500 nm across. A reservoir that holds liquid chemicals similar to the stem cells has been exposed to the niche. Each reservoir is covered with a lipid bilayer model resembling a cell membrane. These reservoir bilayers also hold the same voltage-gated channels found in cells. A small charge of electricity has been applied to any individual reservoir to open the channels and allow the chemicals to spill out, delivering them to develop any particular stem cell. The nanoreservoir chip technology also allows the opportunity of growing cells layer by layer, making compound tissues, which are otherwise challenging to produce.

Substrate topography impacts a wide range of stem cell behaviors in a manner discrete from surface chemistry. One physical difference in the topography of divergent basement membranes is the size of pores and ridges. *In vivo* cell never see flat surfaces: on the nanoscale, no basement membrane or extracellular matrix is flat. The great majority of features in the extracellular environment are in the submicron to the nanoscale range, confirming that an individual cell interacts with numerous topographic features. Nanofibrous structures have favorably modulated osteoblast, osteoclast and fibroblast activities towards the implant or scaffold materials. Nanofibrous matrices are presented as scaffolds that have improved structural similarity to target tissues than their bulk counterparts because leading mechanisms in tissues are nanoscale structures, and cells seem to adhere and proliferate enhanced on nanoscale structures than on bulk materials. The synthesis of natural polymer-based nanofibers is advantageous because of their proven biocompatibility and biodegradation. Strategic aspects of natural polymers include less immune reaction, nontoxicity, hydrophilicity, enhanced cell adhesion and proliferation. The electrospinning method was adapted to fabricate natural polymer nanofibers. Chitosan and alginate, abundant natural polymers have been widely used in tissue engineering, but none had been fabricated into nanostructured matrices until recent years. Uzieliene et al. [74] described that they effectively used chitosan and alginate-based nanofibrous matrices to mimic the extracellular matrix of articular cartilage that mainly contains type II collagen and proteoglycans (glycosaminoglycan, GAG). A nanopit template was created with a conglomeration surface less than 100 nm in diameter. The flat culture surface and nutrient medium of nanopit align ordered the stem cell has been not differentiated. The stem cell could grow to the calcified ossature cell in the nutrient medium concurrent with well-ordered and unordered aligned nanopit. The surface of the transplanted tissue is the nanoengineering surface that has induced the stem cell to propagate into the ossature. Surface character plays a significant role in stem cell proliferation.

8. Nanoparticle-mediated gene delivery systems for stem cells

Recent research has previously revealed the therapeutic uses of embryonic stem cells (ESCs), and the generation of progenitor cells with *in vivo* reconstitution properties has also been described for the treatment of severe hereditary, excruciating and degenerative illnesses [75]. A fundamental barrier to the therapeutic uses of these pluripotent cells is the lack of non-invasive and live cell imaging of grafted cells to manage biodistribution (*in-vivo* tracking). Additionally, reproducible methods for the effective intracellular distribution of biomolecules such as RNA, DNA, peptides and proteins are required to control ES cell development should be developed.

Fluorescent multi-walled nanotubes of carbon (dMNTs-C) functionalized with polyamidoamine are very successful at penetrating the CCE embryonic stem cell line in mice [76]. As they are easier to use and can be produced in large quantities than viral vectors, which are riskier for therapeutic use, dendrimers could be a viable non-viral transmission vector. It has been found that dendrimer-modified polyamidoamine (PAMAM) MNPs significantly boost the efficacy of gene delivery [77, 78]. The dMNTs will be a modern method of gene transfer for ESCs and will be used in ES research. Nanoparticles such as MNPs [79] and QDs can penetrate human MSC cells and sustain themselves in ES cells for a long time. Previous studies have shown that SiO_2-coated CdTe nanoparticles can bind to and support inside of induced-differentiated neurons, hematopoietic cells and endothelial cells while exhibiting minimal cytotoxicity at the applied dose. It is simple to show that these transplanted stem cells with MNPs formed teratomas made up of tissues from all three germ layers [41]. Recently, a biological delivery technique that uses nanoneedles and atomic force microscopy (AFM) to transport genes into living cells was created [80].

El-Kharrag et al. [81] examined polymer-based nanoparticles (NPs) for the delivery of mRNA and nucleases to human granulocyte colony-stimulating factor (GCSF)-mobilized CD34+ cells, which might also be employed for *in vivo* administration. The effectiveness of NP-mediated *ex vivo* administration was closely associated with the charge of the nanoparticles and exhibited minimal toxicity. When compared directly to electroporation, NP-mediated gene editing allowed for a 3-fold decrease in reagent usage while maintaining comparable efficiency. Furthermore, employing nanoparticles showed increasing human HSC engraftment capacity in the NSG mice xenograft model. Finally, successfully stored mRNA- and nuclease-loaded nanoparticles were lyophilized, preserving their transfection capacity following rehydration.

9. Nanoparticles as macromolecular delivery systems for stem cells

Stem cells are unique cells found in the body and are rightly called internal repair systems. The unique properties of stem cells are the ability to proliferate extensively and differentiate into specific cells that are used in therapeutic procedures against dreadful diseases. The biggest challenge ahead of using this cell is to find an effective way to maintain the division and differentiation of stem cells under tightly regulated patterns. It is found that several macromolecules, such as DNA, RNA, proteins and peptides, regulate these pathways effectively. These macromolecules can be introduced into the stem cells at the right time to make it possible. Conventional methods will not work out because of the complexities associated with the cells and macromolecules. Although physical methods such as electroporation and nucleofection could bring out promising results simultaneously, it causes irreversible damage to the cells under some circumstances. Research could also bring in viral vectors concurrently, causing drawbacks, such as toxicity and mutagenesis, and has not been forward to accomplish the transfer. Nanoparticles are found to be very effective after surface modification. They considered several parameters such as size, shape and design of nanoparticles made by Zhu et al. [82] to deliver the plasmid into mouse embryonic fibroblast cells to reprogram it into pluripotent cells. A plasmid carrying OSKM (arginine terminated polyamidoamine) nanoparticle was used to carry out the shipment of macromolecule into the target. In another experiment, Sohn et al. [83] used acid-senitive polyketal-based nanoparticles to activate pluripotency in bone marrow mononuclear cells. The outcome that was fertile in polyketal-based nanoparticles would produce multiple reprogrammed cells. Besides, mesoporous silica nanoparticles were analyzed by Chen et al. [84] for their efficacy against induced pluripotent stem cells. The outcomes of mesoporous silica nanoparticles lead the way by limited cytotoxicity against induced pluripotent stem cells. Positively charged (cationic) nanoparticles were chosen to deliver hepatocyte nuclear 3b factor. It was found that it increased the mRNA concentration in stem cells with liver-specific genes and activated those cells to differentiate into cells resembling hepatocytes with similar functions.

10. Application of nanotechnology in stem-cell-based therapy of neurodegenerative diseases

Neurodegenerative diseases (ND) can be defined as the gradual degeneration of neurons, which are considered the fundamental unit of the nervous system (**Figure 4**). Neural degeneration could affect the patient, their family members and society. Henceforward, there is no prominent treatment procedure to handle the disease even though the symptoms of ND can be slowed down [85]. A team of doctors, neuroscientists and bioengineers are required to standardize procedures to treat the disease effectively [86]. The neurogenesis cascade of complex mechanism leads to the synthesis of neurons, which makes up the CNS. Stems cells derived from different sources can be subjected to a sequence of processes such as proliferation and differentiation. The differentiated stem cells can be used as ideal drug candidates for cell-based therapy [87].

The key objectives of cell-based therapy are to protect the neurons and to enhance the differentiation and regeneration potential. In recent years, applications of stem cells in cell therapy to treat neurodegenerative diseases have gained more attention among researchers [88]. The neuroprotective effect of stem cells has been scientifically proven [89]. Transplantation of stem cells is positive regulation in Parkinson's disease (PD), spinal muscular dystrophy (SMD) and amyotrophic lateral sclerosis (ALS) [90]. Nanomaterials are considered by the biomedical domain, an effective tool to carry value-added drugs and to deliver those chemicals to the specified target. Since nanoparticles are special and unique properties [91], the nanotechnology domain can be coupled with cell therapy to extend better treatment to people

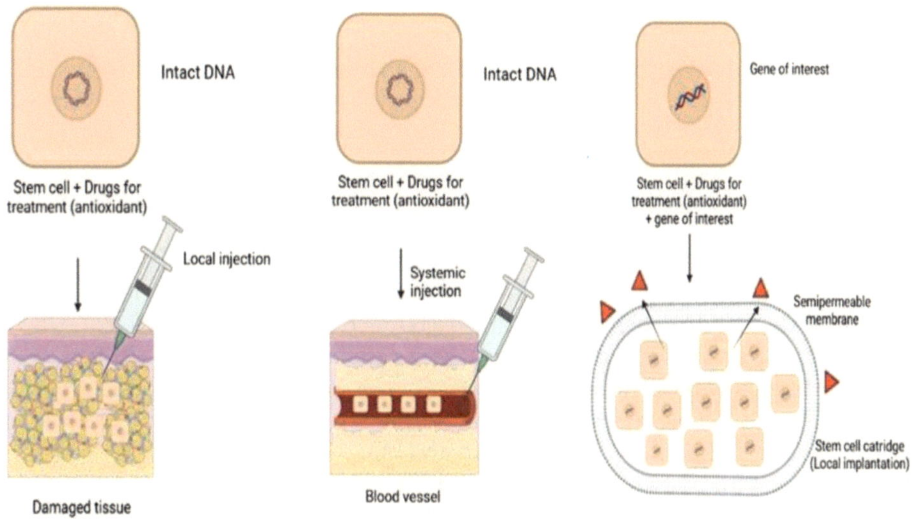

Figure 4.
Current novel methods to utilize stem cells in cell therapy in treatment of neurodegenerative disease.

experiencing severe forms of neurodegenerative diseases. The microenvironment of the CNS is called the stem cell niche. When the stem cell niche is administered with pro-neurogenic factors (proteins), it can kindle the proliferation and differentiation of endogenous and exogenous neural stem cells [92]. After successful internalization of the nanoparticles, it can stimulate neurogenesis, which is a much-expected point in the treatment of neurodegenerative diseases. But still, researchers are needed to find out the mystery behind the relationship between stem cells and nanoparticles. If those mysteries might be identified then it may be easy to use the stem cell—nanoparticle complex to treat the neurodegenerative disease effectively.

A high translational potential exists for neural stem cells (NSCs) in transplantation therapy for neural repair. A vital objective of regenerative neurology is to increase the therapeutic potential of these cells through genetic engineering. Major non-viral vectors for the safe bioengineering of NSCs include magnetic nanoparticles (MNPs), which have imperative advantages over viral vectors in terms of safety, scalability and use.

11. Nanomedicine in cancer stem cell therapy

Conventional therapies such as chemotherapy and radiation therapy remove the tumor but not the cancer stem cell (CSC). Permanent removal of cancer stem cells could result in long-lasting remission of disease, remarkable reduction in metastasis and seems to boost the immune status of the patients. Again, cancer stem cell (CSC) therapies are controlled by nanotechnology and carry them with therapeutic payloads (TPL) [93]. Nanoparticles are engineered in such a way that to attack the cells with over-expressed receptor proteins called CD44. Hyaluronic acid situated on the surface of the B16F10 cells could lead to the bonding with CD44. The study confirms that nanoparticles are remarkably useful for the shipment of CSC suppression antitumour drugs [94].

A novel therapeutic procedure such as nucleus-targeted drug delivery (NTDD) can help researchers to reverse the drug resistance in CSC. Silica nanoparticles are engineered in such a way as to attack the nucleus of the CSC. Surface modulation coupled with thermal sensitive exposure could help to reach the nucleus efficiently. Nucleus-targeted drug delivery facilitates the apoptosis of CSC, which in turn is caused by chemotherapy and thermotherapy [95].

12. Nanoparticles as macromolecular delivery systems for glioblastoma

Macromolecular drug delivery has taken a bounce in the last 20 years [96]. Due to the robust development in the biotechnology domain, enough novel methods have been developed based on macromolecules such as DNA, RNA, siRNA, proteins and peptides. The U.S Food and Drug Administration (FDA) has classified macromolecular drugs into vaccines, blood and blood components and allergen extracts are used for diagnosis and treatment. In the modern drug delivery system (DDS), an important property called "active targeting" is exploited to make the DDS deliver the drugs selectively to the target without affecting the healthy neighboring cells [97]. Glioblastoma is a deadly form of malignant tumor of the central nervous system (CNS). Current treatment relies on giving radiation therapy followed by a chemotherapeutics regime using a DNA alkylating agent. Life expectancy is also less even after undergoing a series of treatment procedures. The cancer progression leads to the impact of GBM (Glioblastoma) after reaching into the deeper areas of the brain. Hence standard and alternative methods are required to extend drug delivery effectively. In this method, one such effective tool is the solid-lipid nanoparticle developed by Kuo et al. [98]. These nanoparticles are conjugated with metallotransferrin antibodies. Further, the transcytosis property of the nanoparticle across human brain-microvascular endothelial cells was examined and found to be very effective, and at the same time, it inhibits the growth of U87MG cells *in vitro*.

13. Conclusion

Stem cell nanotechnology begins new avenues for the manufacture, study and potential application of SCs in regenerative medicine. For imaging and labelling, drug or gene administration, tissue engineering scaffolds and stem cell proliferation monitoring, nanomaterials such as fluorescent CNTs, QDs, fluorescent MNPs and fluorescent CNTs, among others, have been used. Differentiation-engineered nanostructures have been employed, and it is anticipated that they would speed up the detection and monitoring of microenvironmental signals. Despite numerous challenges, stem cell nanotechnology offers new opportunities that will considerably improve the identification and tracking of SC-fate and will develop novel stem cell therapies. As a result, stem cell-based therapies would be furnished as an alternative and effective remedy for genetic disorder.

DOI: http://dx.doi.org/10.5772/intechopen.109275

Author details

Ponpandian Samuel[1*], Shenbagamoorthy Sundarraj[2*] and D.N.P. Sudarmani[3]

1 Department of Biotechnology, Ayya Nadar Janaki Ammal College (Autonomous), Sivakasi, Tamil Nadu, India

2 Postgraduate and Research Department of Zoology, V.O.Chidambaram College, Tuticorin, Tamil Nadu, India

3 Department of Zoology, Ayya Nadar Janaki Ammal College (Autonomous), Sivakasi, Tamil Nadu, India

*Address all correspondence to: samuel_sf432@anjaconline.org and shenvsun@gmail.com

References

[1] Dong Y, Wu X, Chen X, Zhou P, et al. Nanotechnology shaping stem cell therapy: Recent advances, application, challenges, and future outlook. Biomedicine & Pharmacotherapy. 2021;**137**:111236

[2] Huang D-M, Hsiao J-K, Chen Y-C, Chien L-Y, et al. The promotion of human mesenchymal stem cell proliferation by superparamagnetic iron oxide nanoparticles. Biomaterials. 2009;**30**:3645-3651

[3] Goodfellow FT, Simchick GA, Mortensen LJ, Stice SL, Zhao Q. Tracking and quantification of magnetically labelled stem cells using magnetic resonance imaging. Advanced Functional Materials. 2016;**26**:3899-3915

[4] Prakash S, Khan A, Paul A. Nanoscaffold based stem cell regeneration therapy: Recent advancement and future potential. Expert Opinion on Biological Therapy. 2010;**10**:1649-1661

[5] Ishii M, Shibata R, Numaguchi Y, Kito T, et al. Enhanced angiogenesis by transplantation of mesenchymal stem cell sheet created by a novel magnetic tissue engineering method. Arteriosclerosis, Thrombosis, and Vascular Biology. 2011;**31**:2210-2215

[6] Maia J, Santos T, Aday S, Agasse F, et al. Controlling the neuronal differentiation of stem cells by the intracellular delivery of retinoic acid-loaded nanoparticles. ACS Nano. 2011;**5**:97-106

[7] Das S, Singh S, Dowding JM, Oommen S, et al. The induction of angiogenesis by cerium oxide nanoparticles through the modulation of oxygen in intracellular environments. Biomaterials. 2012;**33**:7746-7755

[8] Chen G-Y, Pang DW-P, Hwang S-M, Tuan H-Y, Hu Y-C. A graphene-based platform for induced pluripotent stem cell culture and differentiation. Biomaterials. 2012;**33**:418-427

[9] Li P, Fu L, Liao Z, Peng Y, et al. Chitosan hydrogel/3D-printed poly(ε-caprolactone) hybrid scaffold containing synovial mesenchymal stem cells for cartilage regeneration based on tetrahedral framework nucleic acid recruitment. Biomaterials. 2021;**278**:121131

[10] Augustine R, Dominic EA, Reju I, Kaimal B, et al. Investigation of angiogenesis and its mechanism using zinc oxide nanoparticle-loaded electrospun tissue engineering scaffolds. RSC Advances. 2014;**4**:51528-51536

[11] Ahtzaz S, Nasir M, Shahzadi L, Amir W, et al. A study on the effect of zinc oxide and zinc peroxide nanoparticles to enhance angiogenesis-pro-angiogenic grafts for tissue regeneration applications. Materials and Design. 2017;**132**:409-418

[12] Gallina C, Capelôa T, Saviozzi S, Accomasso L, et al. Human mesenchymal stem cells labelled with dye-loaded amorphous silica nanoparticles: Long-term biosafety, stemness preservation and traceability in the beating heart. Journal of Nanobiotechnology. 2015;**13**:77

[13] Skelton RJP, Khoja S, Almeida S, Rapacchi S, et al. Magnetic resonance imaging of iron oxide-labelled human embryonic stem cell-derived cardiac progenitors. Stem Cells Translational Medicine. 2016;**5**:67-74

[14] Bagher Z, Azami M, Ebrahimi-Barough S, Mirzadeh H, et al. Differentiation of Wharton's jelly-derived mesenchymal stem cells into motor neuron-like cells on three-dimensional collagen-grafted nanofibers. Molecular Neurobiology. 2016;**53**:2397-2408

[15] Ferreira R, Fonseca MC, Santos T, Sargento-Freitas J, et al. Retinoic acid-loaded polymeric nanoparticles enhance vascular regulation of neural stem cell survival and differentiation after ischaemia. Nanoscale. 2016;**8**:8126-8137

[16] Nuschke A, Rodrigues M, Rivera J, Yates C, et al. Epidermal growth factor tethered to β -tricalcium phosphate bone scaffolds via a high-affinity binding peptide enhances survival of human mesenchymal stem cells/multipotent stromal cells in an immune-competent parafascial implantation assay in mice. Stem Cells Translational Medicine. 2016;**5**:1580-1586

[17] Naskar D, Ghosh AK, Mandal M, Das P, et al. Dual growth factor loaded nonmulberry silk fibroin/carbon nanofiber composite 3D scaffolds for in vitro and in vivo bone regeneration. Biomaterials. 2017;**136**:67-85

[18] Lalwani G, D'agati M, Gopalan A, Patel SC, et al. Three-dimensional carbon nanotube scaffolds for long-term maintenance and expansion of human mesenchymal stem cells. Journal of Biomedial Materials Research Part A. 2017;**105**:1927-1939

[19] Zhou X, Nowicki M, Cui H, Zhu W, et al. 3D bioprinted graphene oxide-incorporated matrix for promoting chondrogenic differentiation of human bone marrow mesenchymal stem cells. Carbon N. Y. 2017;**116**:615-624

[20] Kim T, Lee N, Arifin DR, Shats I, et al. In vivo micro-CT imaging of human mesenchymal stem cells labelled with gold-poly-L-lysine nanocomplexes. Advanced Functional Materials. 2017;**27**:1604213

[21] Chen Y-W, Hsieh S-C, Yang Y-C, Hsu S, et al. Functional engineered mesenchymal stem cells with fibronectin-gold composite coated catheters for vascular tissue regeneration. Nanomedicine: Nanotechnology, Biology, and Medicine. 2018;**14**:699-711

[22] Qin X, Chen H, Yang H, Wu H, et al. Photoacoustic imaging of embryonic stem cell-derived cardiomyocytes in living hearts with ultrasensitive semiconducting polymer nanoparticles. Advanced Functional Materials. 2018;**28**:1704939

[23] Ning X, Bao H, Liu X, Fu H, et al. Long-term in vivo CT tracking of mesenchymal stem cells labelled with Au@BSA@PLL nanotracers. Nanoscale. 2019;**11**:20932-20941

[24] Mathiasen AB, Qayyum AA, Jørgensen E, Helqvist S, et al. In vivo MRI tracking of mesenchymal stromal cells labelled with ultrasmall paramagnetic iron oxide particles after intramyocardial transplantation in patients with chronic ischemic heart disease. Stem Cells International. 2019;**2019**:1-10

[25] Constantinides C, McNeill E, Carnicer R, Al Haj Zen A, et al. Improved cellular uptake of perfluorocarbon nanoparticles for in vivo murine cardiac 19F MRS/MRI and temporal tracking of progenitor cells. Nanomedicine: Nanotechnology, Biology, and Medicine. 2019;**18**:391-401

[26] Wu H, Liu T, Xu Z, Qian J, et al. Enhanced bacteriostatic activity, osteogenesis and osseointegration of silicon nitride/polyetherketoneketone composites with femtosecond laser

induced micro/nano structural surface. Applied Materials Today. 2020;**18**:100523

[27] Wang Z, Ruan J, Cui D. Advances and Prospect of nanotechnology in stem cells. Nanoscale Research Letters. 2009;**4**:593

[28] Rodrigues GMC, Fernandes TG, Rodrigues CAV, Cabral JMS, Diogo MM. Purification of human induced pluripotent stem cell-derived neural precursors using magnetic activated cell sorting. In: Turksen K, editor. Stem Cells and Good Manufacturing Practices: Methods, Protocols, and Regulations. New York, NY: Springer New York; 2015. pp. 137-145

[29] Jing Y, Moore LR, Williams PS, Chalmers JJ, et al. Blood progenitor cell separation from clinical leukapheresis product by magnetic nanoparticle binding and magnetophoresis. Biotechnology and Bioengineering. 2007;**96**:1139-1154

[30] Zhao C, Tan A, Pastorin G, Ho HK. Nanomaterial scaffolds for stem cell proliferation and differentiation in tissue engineering. Biotechnology Advances. 2013;**31**:654-668

[31] Mao HY, Laurent S, Chen W, Akhavan O, et al. Graphene: Promises, facts, opportunities, and challenges in nanomedicine. Chemical Reviews. 2013;**113**:3407-3424

[32] Park D, Park J, Lee J, Shim CJ, et al. Fabrication and characterization of graphene oxide-coated plate for efficient culture of stem cells. Journal of Tissue Engineering and Regenerative Medicine. 2021;**18**:775-785

[33] Carradori D, Eyer J, Saulnier P, Préat V, des Rieux, A. The therapeutic contribution of nanomedicine to treat neurodegenerative diseases via neural stem cell differentiation. Biomaterials. 2017;**123**:77-91

[34] Masoudi Asil S, Ahlawat J, Guillama Barroso G, Narayan M. Application of nanotechnology in stem-cell-based therapy of neurodegenerative diseases. Applied Sciences. 2020;**10**:4852

[35] Halim A, Qu K-Y, Zhang X-F, Huang N-P. Recent advances in the application of two-dimensional nanomaterials for neural tissue engineering and regeneration. ACS Biomaterials Science & Engineering. 2021;7:3503-3529

[36] Ruan S, Wang Z, Ji J, et al. Efficient preparation and labelling of human induced pluripotent stem cells by nanotechnology. International Journal of Nanomedicine. 2011;**2011**:425

[37] Jasmin. In vitro labeling mesenchymal stem cells with superparamagnetic iron oxide nanoparticles: Efficacy and cytotoxicity. Methods in Molecular Biology. 2020;**2118**:235-250

[38] Ohyabu Y, Kaul Z, Yoshioka T, Inoue K, et al. Stable and nondisruptive in vitro/in vivo labelling of mesenchymal stem cells by internalizing quantum dots. Human Gene Therapy. 2009;**20**:217-224

[39] Li L, Jiang W, Luo K, Song H, et al. Superparamagnetic Iron oxide nanoparticles as MRI contrast agents for non-invasive stem cell labelling and tracking. Theranostics. 2013;**3**:595-615

[40] An L, Tao Q, Wu Y, Wang N, et al. Synthesis of SPIO nanoparticles and the subsequent applications in stem cell labelling for Parkinson's disease. Nanoscale Research Letters. 2021;**16**:107

[41] Maxwell DJ, Bonde J, Hess DA, Hohm SA, et al. Fluorophore-conjugated Iron oxide nanoparticle labelling and analysis of engrafting human hematopoietic stem cells. Stem Cells. 2008;**26**:517-524

[42] Accomasso L, Gallina C, Turinetto V, Giachino C. Stem cell tracking with nanoparticles for regenerative medicine purposes: An overview. Stem Cells International. 2016;**2016**:1-23

[43] Sutton EJ, Henning TD, Pichler BJ, Bremer C, Daldrup-Link HE. Cell tracking with optical imaging. European Radiology. 2008;**18**:2021-2032

[44] Andreas K, Georgieva R, Ladwig M, Mueller S, et al. Highly efficient magnetic stem cell labelling with citrate-coated superparamagnetic iron oxide nanoparticles for MRI tracking. Biomaterials. 2012;**33**:4515-4525

[45] Cheng S-H, Yu D, Tsai H-M, Morshed RA, et al. Dynamic in vivo SPECT imaging of neural stem cells functionalized with radiolabelled nanoparticles for tracking of glioblastoma. Journal of Nuclear Medicine. 2016;**57**:279-284

[46] Patrick PS, Kolluri KK, Zaw Thin M, Edwards A, et al. Lung delivery of MSCs expressing anti-cancer protein TRAIL visualised with 89Zr-oxine PET-CT. Stem Cell Research & Therapy. 2020;**11**:256

[47] Tseng C-L, Shih I-L, Stobinski L, Lin F-H. Gadolinium hexanedione nanoparticles for stem cell labelling and tracking via magnetic resonance imaging. Biomaterials. 2010;**31**:5427-5435

[48] Santelli J, Lechevallier S, Baaziz H, Vincent M, et al. Multimodal gadolinium oxysulfide nanoparticles: a versatile contrast agent for mesenchymal stem cell labelling. Nanoscale. 2018;**10**:16775-16786

[49] Donnelly EM, Kubelick KP, Dumani DS, Emelianov SY. Photoacoustic image-guided delivery of plasmonic-nanoparticle-labelled mesenchymal stem cells to the spinal cord. Nano Letters. 2018;**18**:6625-6632

[50] Nafiujjaman M, Kim T. Gold nanoparticles as a computed tomography marker for stem cell tracking. Methods in Molecular Biology. 2020;**2126**:155-166

[51] Huang J, Huang J, Ning X, Luo W, et al. CT/NIRF dual-modal imaging tracking and therapeutic efficacy of transplanted mesenchymal stem cells labelled with Au nanoparticles in silica-induced pulmonary fibrosis. Journal of Materials Chemistry B. 2020;**8**:1713-1727

[52] Hsieh S-C, Wang F-F, Lin C-S, Chen Y-J, et al. The inhibition of osteogenesis with human bone marrow mesenchymal stem cells by CdSe/ZnS quantum dot labels. Biomaterials. 2006;**27**:1656-1664

[53] Chen G, Tian F, Li C, Zhang Y, et al. In vivo real-time visualization of mesenchymal stem cells tropism for cutaneous regeneration using NIR-II fluorescence imaging. Biomaterials. 2015;**53**:265-273

[54] Li J, Lee WY, Wu T, Xu J, et al. Multifunctional quantum dot nanoparticles for effective differentiation and long-term tracking of human mesenchymal stem cells in vitro and in vivo. Advanced Healthcare Materials. 2016;**5**:1049-1057

[55] Jasmin, Souza GT, Andrade Louzada R, Rosado-de-Castro PH, et al. Tracking stem cells with superparamagnetic iron oxide nanoparticles: Perspectives and considerations. International Journal of Nanomedicine. 2017;**12**:779-793

[56] Azevedo-Pereira RL, Rangel B, Tovar-Moll F, Gasparetto EL, et al. Superparamagnetic iron oxide nanoparticles as a tool to track mouse neural stem cells in vivo. Molecular Biology Reports. 2019;**46**:191-198

[57] Shahror RA, Wu CC, Chiang YH, Chen KY. Tracking superparamagnetic

iron oxide-labeled mesenchymal stem cells using MRI after intranasal delivery in a traumatic brain injury murine model. Journal of Visualized Experiments. 21 Nov 2019;(153). DOI: 10.3791/60450

[58] Rawat S, Gupta S, Bhat M, Dinda AK, Mohanty S. Efficient labeling of human mesenchymal stem cells using iron oxide nanoparticles. Methods in Molecular Biology. 2020;**2150**:113-120

[59] Lee SH, Park DJ, Yun WS, Park J-E, et al. Endocytic trafficking of polymeric clustered superparamagnetic iron oxide nanoparticles in mesenchymal stem cells. Journal of Controlled Release. 2020;**326**:408-418

[60] Ma Y, Ji Y, You M, Wang S, et al. Labelling and long-term tracking of bone marrow mesenchymal stem cells in vitro using NaYF4:Yb3+,Er3+ upconversion nanoparticles. Acta Biomaterialia. 2016;**42**:199-208

[61] Kang H, Zhang K, Pan Q, Lin S, et al. Remote control of intracellular calcium using upconversion nanotransducers regulates stem cell differentiation In vivo. Advanced Functional Materials. 2018;**28**:1802642

[62] Ren N, Liang N, Yu X, Wang A, et al. Ligand-free upconversion nanoparticles for cell labelling and their effects on stem cell differentiation. Nanotechnology. 2020;**31**:145101

[63] Huang D, Hung Y, Ko B, Hsu S, et al. Highly efficient cellular labelling of mesoporous nanoparticles in human mesenchymal stem cells: Implication for stem cell tracking. The FASEB Journal. 2005;**19**:2014-2016

[64] Chen F, Jokerst JV. Stem cell tracking with nanoparticle-based ultrasound contrast agents. Methods in Molecular Biology. 2020;**2126**:141-153

[65] Yao M, Shi X, Zuo C, Ma M, et al. Engineering of SPECT/photoacoustic imaging/antioxidative stress triple-function nanoprobe for advanced mesenchymal stem cell therapy of cerebral ischemia. ACS Applied Materials & Interfaces. 2020;**12**:37885-37895

[66] Chen F, Zhao ER, Hu T, Shi Y, et al. Silicon carbide nanoparticles as a photoacoustic and photoluminescent dual-imaging contrast agent for long-term cell tracking. Nanoscale Advances. 2019;**1**:3514-3520

[67] Accomasso L, Rocchietti EC, Raimondo S, Catalano F, et al. Fluorescent silica nanoparticles improve optical imaging of stem cells allowing direct discrimination between live and early-stage apoptotic cells. Small. 2012;**8**:3192-3200

[68] Lim S, Yoon HY, Jang HJ, Song S, et al. Dual-modal imaging-guided precise tracking of bioorthogonally labelled mesenchymal stem cells in mouse brain stroke. ACS Nano. 2019;**13**:10991-11007

[69] Yin C, Wen G, Liu C, Yang B, et al. Organic semiconducting polymer nanoparticles for photoacoustic labelling and tracking of stem cells in the second near-infrared window. ACS Nano. 2018;**12**:12201-12211

[70] Cao F-Y, Yin W-N, Fan J-X, Tao L, et al. Evaluating the effects of charged oligopeptide motifs coupled with RGD on osteogenic differentiation of mesenchymal stem cells. ACS Applied Materials & Interfaces. 2015;**7**:6698-6705

[71] Wang X, Ye K, Li Z, Yan C, Ding J. Adhesion, proliferation, and differentiation of mesenchymal stem cells on RGD nanopatterns of varied nanospacings. Organogenesis. 2013;**9**:280-286

[72] Gelain F, Bottai D, Vescovi A, Zhang S. Designer self-assembling

peptide nanofiber scaffolds for adult mouse neural stem cell 3-dimensional cultures. PLoS One. 2006;**1**:e119

[73] Lee J-R, Ryu S, Kim S, Kim B-S. Behaviors of stem cells on carbon nanotube. Biomaterials Research. 2015;**19**:3

[74] Uzieliene I, Bironaite D, Bernotas P, Sobolev A, Bernotiene E. Mechanotransducive biomimetic systems for chondrogenic differentiation In vitro. International Journal of Molecular Sciences. 2021;**22**:9690

[75] Park I-H, Lerou PH, Zhao R, Huo H, Daley GQ. Generation of human-induced pluripotent stem cells. Nature Protocols. 2008;**3**:1180-1186

[76] Cui D, Zhang H, Wang Z, Asahi T, Osaka T. Effects of dendrimer-functionalized multi-walled carbon nanotubes on murine embryonic stem cells. ECS Transactions. 2008;**13**:111-116

[77] Lee JW, Kim B-K, Kim HJ, Han SC, et al. Convergent synthesis of symmetrical and unsymmetrical PAMAM dendrimers. Macromolecules. 2006;**39**:2418-2422

[78] Pan B, Cui D, Sheng Y, Ozkan C, et al. Dendrimer-modified magnetic nanoparticles enhance efficiency of gene delivery system. Cancer Research. 2007;**67**:8156-8163

[79] Syková E, Jendelová P. Magnetic resonance tracking of transplanted stem cells in rat brain and spinal cord. Neurodegenerative Diseases. 2006;**3**:62-67

[80] Han SW, Nakamura C, Obataya I, Nakamura N, Miyake J. A molecular delivery system by using AFM and nanoneedle. Biosensors & Bioelectronics. 2005;**20**:2120-2125

[81] El-Kharrag R, Berckmueller KE, Madhu R, Cui M, et al. Efficient polymer nanoparticle-mediated delivery of gene editing reagents into human hematopoietic stem and progenitor cells. Molecular Therapy. 2022;**30**:2186-2198

[82] Zhu K, Li J, Lai H, Yang C, Guo C, Wang C. Reprogramming fibroblasts to pluripotency using arginine-terminated polyamidoamine nanoparticles based non-viral gene delivery system. International Journal of Nanomedicine. 2014;**9**:5837-5847

[83] Sohn Y-D, Somasuntharam I, Che P-L, Jayswal R, et al. Induction of pluripotency in bone marrow mononuclear cells via polyketal nanoparticle-mediated delivery of mature microRNAs. Biomaterials. 2013;**34**:4235-4241

[84] Chen W, Tsai P-H, Hung Y, Chiou S-H, Mou C-Y. Nonviral cell labelling and differentiation agent for induced pluripotent stem cells based on mesoporous silica nanoparticles. ACS Nano. 2013;**7**:8423-8440

[85] Nguyen H, Zarriello S, Coats A, Nelson C, et al. Stem cell therapy for neurological disorders: A focus on aging. Neurobiology of Disease. 2019;**126**:85-104

[86] Nemeth CL, Fine AS, Fatemi A. Translational challenges in advancing regenerative therapy for treating neurological disorders using nanotechnology. Advanced Drug Delivery Reviews. 2019;**148**:60-67

[87] Lindvall O. Developing dopaminergic cell therapy for Parkinson's disease-give up or move forward? Movement Disorders. 2013;**28**:268-273

[88] Hoveizi E, Ebrahimi-Barough S, Tavakol S, Sanamiri K. In vitro differentiation of human iPS cells into neural like cells on a biomimetic Polyurea. Molecular Neurobiology. 2017;**54**:601-607

[89] Hunsberger JG, Rao M, Kurtzberg J, Bulte JWM, et al. Accelerating stem cell trials for Alzheimer's disease. Lancet Neurology. 2016;**15**:219-230

[90] Chang K-A, Lee J-H, Suh Y-H. Therapeutic potential of human adipose-derived stem cells in neurological disorders. Journal of Pharmacological Sciences. 2014;**126**:293-301

[91] Corradetti B, Ferrari M. Nanotechnology for mesenchymal stem cell therapies. Journal of Controlled Release. 2016;**240**:242-250

[92] Santos T, Maia J, Agasse F, Xapelli S, et al. Nanomedicine boosts neurogenesis: New strategies for brain repair. Integrative Biology. 2012;**4**:973

[93] Tabassum N, Verma V, Kumar M, Kumar A, Singh B. Nanomedicine in cancer stem cell therapy: From fringe to forefront. Cell and Tissue Research. 2018;**374**:427-438

[94] Li Y, Shi S, Ming Y, Wang L, et al. Specific cancer stem cell-therapy by albumin nanoparticles functionalized with CD44-mediated targeting. Journal of Nanobiotechnology. 2018;**16**:99

[95] Li H, Yan W, Suo X, Peng H, et al. Nucleus-targeted nano delivery system eradicates cancer stem cells by combined thermotherapy and hypoxia-activated chemotherapy. Biomaterials. 2019;**200**:1-14

[96] Pagels RF, Prud'homme RK. Polymeric nanoparticles and microparticles for the delivery of peptides, biologics, and soluble therapeutics. Journal of Controlled Release. 2015;**219**:519-535

[97] Ryu JS, Kuna M, Raucher D. Penetrating the cell membrane, thermal targeting and novel anticancer drugs: The development of thermally targeted, elastin-like polypeptide cancer therapeutics. Therapeutic Delivery. 2014;**5**:429-445

[98] Kuo Y-C, Chao I-W. Conjugation of melanotransferrin antibody on solid lipid nanoparticles for mediating brain cancer malignancy. Biotechnology Progress. 2016;**32**:480-490

Chapter 6

3D Culturing of Stem Cells: An Emerging Technique for Advancing Fundamental Research in Regenerative Medicine

Sonali Rawat, Yashvi Sharma, Misba Majood and Sujata Mohanty

Abstract

Regenerative medicine has been coming into spotlight ever since the realisation that conventional treatments are not enough, and the need for specific therapies has emerged. This, however, has paved way for cell-free therapy using extracellular vesicles. A two-dimensional (2D) cell culture model is widely recognised as the "gold standard" for researching cellular communications ex vivo. Although the 2D culture technique is straightforward and easy to use, it cannot replicate the in vivo ECM interactions & microenvironment. On the contrary, 3D culture culturing technology has emerged which include structures such as spheroids and organoids. Organoids are small replicas of in vivo tissues and organs, which faithfully recreate their structures and functions. These could be used as models to derive stem cells based EVs for manufacturing purposes. The linkages between infection and cancer growth, as well as mutation and carcinogenesis, may be modelled using this bioengineered platform. All in all, 3D culturing derived EVs serves as a novel platform for diagnostics, drug discovery & delivery, and therapy.

Keywords: organoid, 2D & 3D cell culture, extracellular matrix, carcinogenesis, bioengineered platform, drug testing

1. Introduction

Cell culture has become an indispensable tool for elucidating fundamental biophysical and biomolecular mechanisms that govern how cells construct into tissues and organs, how well these tissues function, and how that function is disrupted in disease. Cell culture is now used extensively in biomedical research, biomedical engineering, stem cell therapy, and commercial applications. In regenerative medicine, stem cells are the central players, however the shortcomings and risks associated with cellular therapy are higher as compared to non-cellular therapy thereby EVs are better. Although adherent, two-dimensional (2D) cell culture has long been the norm, recent research has shifted toward three-dimensional (3D) structures and more feasible biochemical and biomechanical microenvironments. Understanding the *in vivo* processes that leads to the formation and purpose of tissues and organs requires deciphering the mechanisms underlying these behaviours. Laboratory experiments should ideally be carried out using a user-defined three-dimensional (3D) model that closely mimics the cell's microenvironment [1–3]. However, challenges in developing such a model include the building of the tissue-tissue interface, control of the spatiotemporal distributions of oxygen, carbon dioxide, nutrients or waste, and further followed by the customization of other microenvironmental factors known to regulate activities *in vivo*. It is well understood that cells adapt to their surroundings by reacting to local signals and cues, which has implications for cell proliferation, differentiation, and function [4, 5]. Traditional culture methods for growing mammalian cells *in vitro* are far removed from the complexities that cells encounter in real-life tissues. One of the most noticeable physical differences is the shape and geometry that cells acquire when grown on a flat substrate, such as a cell culture plate or flask. When cells grow on two-dimensional (2D) surfaces, they flatten and remodel their internal cytoskeletons. Lacking the ability to form more natural tissue-like structures, existing *in vitro* 2D cell culture models are frequently a poor substitute when used to study cell growth and various associated aspects [6]. This has a substantial impact on cell performance, as well as, the outcomes of biological assays. Monolayers of cultured cells, for example, are assumed to be more sensitive to therapeutics. Moreover, due to the limited cell interactions, culturing cells on rigid surfaces may increase cell proliferation, but adversely impact cell differentiation. A more adequately engineered cell culture environment might enhance drug discovery predictive accuracy and aid in the interpretation of tissue morphogenesis [4–6].

Some significant aspects of cancer cells, for example, cannot be adequately modelled in 2D cultures. To overcome the limitations, novel 3D cell culture platforms that better mimic in vivo conditions are now being developed, which are sometimes referred to as spheroid or organoid culture. In many cases, these new platforms have shown to be more capable of stimulating *in vivo*-like cell fates for the processes under investigation. 3D research shows that increasing the dimensionality of the extracellular matrix (ECM) surrounding cells from 2D to 3D has a significant impact on cell proliferation, differentiation, mechano-responses, and cell survival [5, 7, 8].

For example, Extracellular vesicles (EVs) are membrane-enclosed structures that are released by almost all cell types. They transport biologically active molecules such as RNAs, lipids, and proteins from the delivering cell to the target cell, allowing for a novel mode of intercellular communication. The use of EVs as diagnostic tools is highly influenced not only by the molecular cargo but also by the quantity of EVs derived from various cell subpopulations in tissues and body fluids. Moreover, overall mechanisms and factors influencing EV release are still unknown [9, 10]. Organoids are obtained from animal or patient samples, cultured in 3D matrices like Matrigel under well-defined conditions, and retain the cellular heterogeneity found in in vivo epithelial tissues. As a result, they depict one of the most cutting-edge technologies for studying human diseases, allowing for the investigation of pathways and factors that influence EV release.

Although these findings suggest that 3D systems should be used anytime feasible, the system of choice is often governed by the specific process of interest, and there is currently no universal 3D platform; additionally, 2D cell culture approaches can still recapitulate in vivo behaviour for many bioactivities, and new advances in substrate configuration continue to offer new capabilities for this platform [11]. All in all, 3D platforms are probable to become a more appealing alternative to 2D cell culture as technology advances to enable a broader range of processes. Technological advances have opened up new avenues for cell culture and the formation of 3D tissue-like frameworks. This is primarily due to research activities between cell biology and biophysical sciences, which has introduced new materials and manufacturing techniques to produce technologies tailored to support 3D cell growth in vitro [12]. The culture of cells in 3D is rapidly progressing, as evidenced by the increasing number of publications in the scientific literature. The adoption, validation, and implementation of these novel strategies will guarantee the effectiveness of this technology. This will most likely take time as the scientific community recognises the limitations of traditional 2D cell culture and recognises the value of new methods to reliably culture cells in 3D.

2. Technologies for 2D and 3D cell culture

2.1 2D cell culture techniques

Traditional 2D cell culture relies on the cells adhering to a flat surface, typically a glass or polystyrene petri dish, to provide mechanical support. Culturing in 2D monolayers allows the cell access to a significant amount of growth factors and nutrients in the form of media, resulting in homogeneous proliferation and expansion [13, 14]. However, a majority of these 2D approaches do not allow for regulation of cell shape, which influences biophysical cues that affect cell biological properties *in vivo*. Micropatterned surfaces, such as cell-adhesive islands, microwells, and micropillars, have indeed been developed to control cell shape in 2D culture and aid in the investigation of the effects of cell shape on bioactive components [15]. This induced polarity may alternate cell functions such as expansion and migration for perceiving soluble components and other microenvironmental signals. One of the strategies to completely eradicate apical-basal directionality is by the sandwich culture procedure, which adds a layer of ECM atop the cells and provides a mimic of 3D ECM that can be used to mitigate the effect of cell polarisation in 2D cell culture (**Figure 1**). The sandwich culture method, which involves placing cells between two layers of ECM, polyacrylamide, collagen or any type of suitable ECM, has long been shown to produce cell cultures with morphology and function that more closely resemble the *in vivo* behaviour. This is especially important in drug discovery, in which scientists aim to understand pharmacokinetic profile in relation to the organs [15]. The sandwich method was discovered to reduce oxygen diffusion sufficiently, resulting in an 80% survival rate over 5 days, compared to a 32% success rate over 2 days in a mixing process population culture. Many researchers have been able to examine the consequences of pharmacokinetics, which is crucial to consider when modelling physiological and pathological events, thanks to the sandwich culture [16, 17]. Another strategy could including micropatterning, which is a designed 2D surface that allows imprinting and alteration to create a 2D microenvironment for cell culture that contains distinct physiochemical factors, topography, stiffness, and mechanical load. In a typical 2D cell culture, cells are subjected to a homogeneous surface free of defects that could interfere with their development [18, 19]. Moreover, cells cultured on patterned and un-patterned PLLA surfaces differentiated at a slower rate than cells cultured on tissue culture-treated polystyrene (PS), the experiment's control. In terms of lipid production, it was discovered that a later time points, shaped

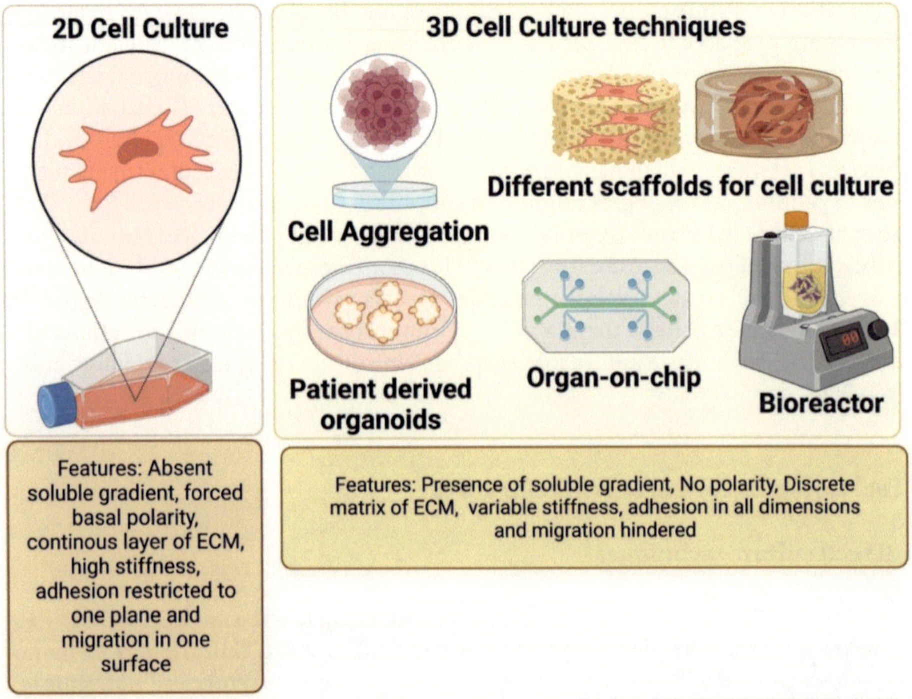

Figure 1.
The representative diagram showing different kinds of cell culture techniques and their impact on cells fate.

PLLA surfaces produced the most lipids, followed by PS, and then non-patterned PLLA [20, 21]. The findings revealed that the micro-patterning and surface type both influenced the rate of cell differentiation. Overall, since the early 1900s, two-dimensional (2D) cell culture has been the technique used to culture cells, which plays an important role in research but has many drawbacks due to 2D models imprecisely representing tissue cells *in vitro* [22–24].

2.2 3D cell culture techniques

2.2.1 Aggregate cultures and the formation of spheroids

A 3D culture model is supposed to provide a tissue-like microenvironment in which cells can proliferate, aggregate, and differentiate. This could have an application in predicting the effect of a drug on cells. For several reasons, cells cultured in 3D respond differently to drugs than cells cultured in 2D [25, 26]. Changes in physical and biological features between 2D and 3D cultures make 2D cultured cells more susceptible to drug effects than 3D cultured cells because 2D cultured cells cannot maintain constant morphology like 3D cells. Since 3D cultured cells have greater depth than 2D cells, the variation in shape between 2D and 3D cultured cells creates a change in local pH levels within the cells. Lower intracellular pH levels have been shown to reduce drug efficacy, giving back to drug resistance. Further, Microfluidics, microchips, embryoid bodies (EBs), collagen gels (GELs), and hanging-drop culture are all methods for spheroid cultures [26, 27]. Various studies have established the experiments in two different 3D culture methods based on the differentiation and proliferation of embryonic stem cells (ESCs). To promote adherence, cells were cultured as embryoid bodies (EBs) in either a collagen type I gel (GEL) or in non-tissue culture-treated dishes. GEL and unattached EB cultures produced cluster morphology that was similar, with defined boundaries and the occasional cavity. The existence of enlarged masses along the edges of the gel form presented a marked difference in the morphology of GEL cultures. Over the same 12-day period, free EBs had more continuous change in the genotypic expression profiles of cytoskeletal genes than GEL cultures [28, 29]. These findings suggest that, despite having similar morphologies, the gene expression of each kind of 3D culture is unique in its adaptation to its microenvironment. Similarly, hanging drop method is another technique to establish spheroids but because of the difficulty in maintenance measures, like changing of media, the traditional hanging-drop method does not allow for extended cell culture.

All such 3D scaffolds and associated cell-encapsulation techniques provide valuable tools for understanding how the ECM influences cell fate. During the last decade, major advances have been made in the techniques for encapsulating cells in 3D using tissue engineering scaffolds with customised biochemical and biophysical components. Majorly, biopolymers derived from animal tissues are especially popular because they contain similar biochemical components to those found in cells' native tissue and may encourage tissue regeneration. However, one of the most pressing issues is the inability to individually control the key elements required in modulating cell bioactivities, such as biochemical properties, matrix elasticity, and macro-porosity [26, 28, 29]. Therefore, a prefabricated scaffolds has the advantage of a configurable biochemical composition, matrix elasticity, and micro-architectures. These scaffolds can be made using polymer phase separation, 3D printing, lyophilizing, gas foaming, stereo-lithography and porogen leaching with soluble templates to form pores or channels. However, current methods for creating prefabricated scaffolds frequently involve procedures that create circumstances which are too severe for cells to survive, such as extreme pressure, non-physiological osmotic pressure, and the use of organic solvents [28, 29]. As a result, cell diffusion is primarily used to deliver cells into scaffolds, and this method is frequently associated with low cell penetration rates and poor scaffold cellularization. Hydrogels made up of various types of biopolymers have been broadly used as scaffolds in contrast to prefabricated scaffolds because of their ease of cell encapsulation. They have tissue-like water content as well as effortlessly controllable biochemical and mechanical characteristics. Most hydrogels, on the other hand, are composed of micron/nanometre-sized mesh that is frequently too small to facilitate post fabrication cellularization and lack the microtopography required for influencing cell shape and supporting cell mobility, proliferation, and matrix production [30–32]. The main disadvantage of hydrogels in tissue regeneration is that matrix degradation simultaneously changes biochemical elements and matrix elasticity, both of which require careful control. Furthermore, matching the rate of hydrogel degradation with the rate of tissue formation is extremely difficult, which is essential for maintaining the shape and structural stability of tissue engineering.

Making multiple layers of cell sheets is another method for engineering organs and tissues without relying on constructed scaffolds. A plethora of studies have successfully replicated cardiomyocyte pulsatile function and functional dopaminergic neurons in a 3D construct by stacking multiple cell sheets or on the single cell sheet by twitching the mechanical properties of scaffold [29, 31–33].

Apart from this, bioreactors are designed to study cell behaviour during the development of micro-tissues or organs and to generate more cells for clinical use or laboratory research. Large-scale bioreactors involve simple systems like spinner flasks and rotating wall bioreactors, which enable for semi-adherent cell growth, in addition to more complicated systems like gravimetric bioreactors. The impacts of fluid transport on a cell membrane scale have been investigated using bioreactors with cell-sized conduits. Micro-bioreactors have also shown promise in drug screening and controlling the cell microenvironment.

2.2.2 Organoid formation from diseased microenvironment and microfluidic 3D cell culture

The shift from 2D to 3D culture techniques is a significant step toward more biologically relevant tissue models. However, 3D culture techniques do not yet capture the multicellular intricacies of tissues, they lack vasculature, do not provide precise control over gradients, and exchange medium at discrete time points rather than continuously. Microfluidic techniques enable spatial control of fluids in micrometre-sized channels, which can be used to investigate the biological significance of 3D culture models (**Figure 1**) [29]. Early examples of spatial patterning of adhesion molecules and hydrogels, which are still employed in microfluidic 3D cell culture, are depicted in **Figure 1**. The three foremost drivers for using microfluidic methodologies in 3D cell culture today are as follows:

1. The ability to co-culture cells in a controlled manner.

2. Generating and controlling (signalling) gradients.

3. Perfusion/flow integration.

An even more functional aspect that can be introduced using microfabrication techniques is mechanobiological aspects such as active stretch and tension. Microfluidic devices can conquer the drawbacks of traditional cell/stem cell culture techniques and tissue engineering approaches by better simulating *in vivo* interplay between ECM and cells thereby enabling high-resolution in situ imaging [25]. The combination of the unique benefits of microfluidics and the breadth of possibilities offered by stem cell technologies can also provide alternatives for the management of neurodegenerative diseases such as Alzheimer's and Parkinson's, as well as other disorders or injuries of the central or peripheral nervous system. This method has even progressed so far as to suggest the development of devices known as "brain-on-a-chip" [34, 36]. In neuro-regeneration, for example, these systems enable the development of uniform populations of neuronal and glial cells. The potential to co-culture cells in a 3D arrangement, better controlled signalling, and the ability to combine diffusion and laminar flow are the most significant advantages of microfluidic cell culture. Microfluidics allows researchers to precisely control stem cell/cell numbers and growth conditions, as well as arrange or design cells in spatially controlled positions and track cell responses to various internal/external mechanical, chemical, and optical stimuli [35–37]. Furthermore, microfluidics techniques enable high-throughput studies of single cells in microenvironments closely mimicking biologically relevant conditions by creating gradients of mechanical forces and different chemical agents. These benefits are classified into four categories: (a) fabrication characteristics, (b) Biomaterial ingredients, (c) Biochemical properties, and (d) bio-physico-mechanical (**Table 1**) [37–39].

S. no.	Properties	Benefits
1	Fabrication characteristics	In tissue engineering and biological research, soft lithography is the most commonly used method for fabricating microfluidic devices. This technique entails replica moulding, embossing, and printing. In this regard, following two topics should be cited: large scale assays and sensor integration. Microfluidic large scale integration can be achieved by parallelizing assays with valves or droplet-based microfluidics and designing a single platform to achieve a series of successive steps. This strategy can be used to investigate the consequences of various biochemical factors on stem cell behaviour and fate, as well as to develop controlled gene/drug delivery systems and nanotoxicological assays [40, 41].
2	Biomaterial ingredients	The biomaterial properties of microfluidics-based tissue engineering devices are composed of various factors; microfabrication, numerous biomaterial methods to ECM and biological materials have been used. The extracellular matrix (ECM) of cells and stem cells is made up of various biological molecules such as proteins, proteoglycans, and soluble factors. Distinct natural or synthetic materials were used to make these microdevices, which can then be modified to mimic the ECM [43, 44].
3	Biochemical properties	Various biochemical molecules have been employed in microdevices to better mimic the physiological properties of the target tissue. In their *in vivo* microenvironment, stem cells are exposed to a variety of soluble signalling cues, including extracellular calcium ions, various growth factors, nutrients, and oxygen. The most studied conventional and microfluidics strategies for this purpose are automated culture systems, soluble gradients, and temporal exposure regimens. Microfluidic systems can also be used to achieve automated temporal control of the delivery of soluble factors. A gradient generating microfluidic device, for example, was used to create a continuous growth factor gradient containing a mixture of PDGF, EGF, and FGF2 [45].
4	Bio-physico-mechanical	Controlling physical and mechanical factors such as spatial confinement, biomolecular tensions, shear stress, surface rigidity, and topography is possible with microfluidics. Other benefits of microfluidics in stem cell research include control over fluidic flow, which affects dynamic cell culture situations, the availability of shear stress as found naturally in different organs, and the availability of different medium compositions [46–48].

Table 1.
Benefits of different aspects of 3D cell culture.

Overall, the fabrication of an artificial human organ (even going as far as a brain) is beginning to be considered possible due to the multidisciplinary overlap of biology and engineering combined with emerging new trends such as microfluidics, stem cells, and nanotechnology; this has been imagined as a "Organ-on-a-chip." Organs on a chip will provide much better mimicking of real human physiology and will be beneficial for tissue engineering, disease modelling, and drug screening; however, much more well-designed research in this field is still required [37–39, 49].

3. Impact of culture strategy on the secretion and content of extracellular vesicles

Extracellular vesicles are a novel modality in the scope of diagnosis, drug delivery & regenerative medicine. These are nanoscale vesicles which can be sub divided based upon their mode of synthesis, size and content. The sub category among them which has recently gained popularity is the small vesicles identified in the range of 30–150 nm & found to be synthesised via the endosomal route, also commonly known as exosomes. These vesicles are naturally found to be involved in mediating inter-cellular communication. They carry a diverse compass of functional molecules including DNA, RNA, miRNA, Proteins, Enzymes & many other are yet under discovery. The release & content of these vesicles are largely influenced by the cell microenvironment & the extracellular cues which are received in a mechanosensitive manner [9]. These vesicles happen to package the content from inside the cell & deliver them to the cell-in-need. These are primarily known to fuse with the cells & release their content in order to facilitate the functioning, however, at certain instances they might as well be phagocytosed. Recent gain of limelight have brought up diverse and divergent functions & applications of these small vesicles, for e.g. Tian et al., suggested the role of these vesicles in the detection of breast cancer. They suggested 8 EV proteins that could serve as a biomarker for diagnosing and differentiation between non metastatic and metastatic breast cancer [50]. Being membrane bound structures, these circulating EVs succeed in preserving their content & can be easily captured by any kind of normal cells. When released from the cancer cells, EVs have been evidenced to enhance the malignancy by causing malignant transformations in the recipient normal cells. Tumorigenicity and cancer spread are highly attributable to the intercellular communication in the tumour microenvironment & in the blood stream via the release of EVs. It was stated by Bebelman et al., that cancer cells are found to exhibit higher EV secretion as compared to the non-cancerous cells [51]. This could be attributed to the fact that cancer cells hold a diminish property of contact inhibition, therefore forming a tightly packed 3D layered & inculcated mass of cells which advances into a tumour. In order to study the biology of this deadly near-pandemic disease, it is essential that the conditions subjected to the tumour in vivo be replicated in in vitro set ups so that the exact nature & habits of a tumour could be elucidated. This has led to invoking the question regarding the culture conditions in which cancer cells & their fate are studied. The most popular choice of cell culture is based upon a 2D sub culturing method wherein the cells are allowed to form a sheet like structure and their propensities are studied [25, 27]. A 2D culture set up is popular due to its ease of handling and other properties as discussed in the previous section, however it does not necessarily confirm that the results being obtained from such a set up are actually a simulation of the in vivo scenario. On that account, 3D culture is the recent technique of choice of researchers as it is expected & anticipated to model the tumour microenvironment in vivo. Pertaining the same, the EVs which are released from the 3D kind of culture have also been found to mimic the in vivo secretions in a more identical manner. Upon comparison of EVs from 2D and 3D culture of the same cell types, it was also found that there were differences in both EV secretion and EV content.

3.1 Effect of 3D culture on EV secretion

Extracellular vesicles have been the modality of interest for diagnostic & therapeutic purposes. However, their less yield hinders their successful commercialization. Thereby in order to enhance the yield of EVs for commercial purposes, many strategies have been explored. One of these strategies is the culturing of cells in a 3D manner [46]. Kim et al., compared the secretion of mesenchymal stem cell derived EVs in a 2D monolayer culture format vs. a 3D culture format via spheroid formation using the hanging drop method & the poly-HEMA coating. From this study, they found that exosome secretion was significantly enhances upon culturing in a 3D format. This led them to sought the cause of increase in EV production, which was found to be the creation of hypoxic niche in a 3D culture format, along with the increased cell density and circular cellular morphology [52]. This finding has also been evidenced by many other studies, for e.g. Yan et al., cultured umbilical cord derived MSCs in a hollow fibre bioreactor & found that the EV secretion was increased by 7.5 folds as compared to the EV release in monolayer culturing [53]. Similar finding was also reported by Haraszti et al., wherein 3D culturing of MSCs resulted in 20 fold increase in exosomes concentration when combining 3D culturing of MSCs along with isolation via differential ultracentrifugation. They also established another technique wherein they combined 3D culturing with Tangential flow filtration which thereby enhanced the yield up to 27 folds. Patel et al., developed a culture system by combining a tubular perfusion bioreactor system & a 3D printed scaffold, wherein they found a 100-fold increase in EV production by endothelial cells [54]. Not just in primary cells, but these results are also observed in cancer cells based 3D Culturing [55]. Hwang et al. suggested that the EV release was increased upon 3D culturing in colorectal cancer [56].

3.2 Effect of 3D culture on EV content

The effect of 3D culturing of cells on EV content is exceptionally significant. 3D culturing has proven time and again that it is a better model to study the cell-ECM & cell-cell interactions as the dynamics of 2D and 3D culturing are poles apart. As a means of cell-cell communication, EVs regulate vibrant interplay mechanisms, and thereby 3D culturing leads to modifications in the cargo of EVs as the stimuli perceived by the cells is varied as compared to 2D culture. It was concluded in one of the studies that 3D culturing leads to an overall depression of protein expression while upregulation of miRNA cargo in EVs due to the downregulation of ARF6 pathway influencing the cell arrangement & secretion profile thereof [57]. Many studies have presented varied views on this matter. It was found that culturing of HeLa cell line in a 3D manner results in the secretion of EVs which were up to 96% similar in their RNA profile with the circulating EVs collected from the plasma of a cervical cancer patient, however the genetic profile of EVs i.e. DNA was unaltered [58]. It is also speculated that EV release in 3D culture systems is aided by the higher expression of transporters [59]. Due to the mechano-sensing based activities in cells upon culturing in the 3D microenvironment, the gene and protein expression of the same cells are differential. For e.g., Eguchi et al. observed that upon 3D culturing, the neuroendocrine adenocarcinoma cells formed large organoids in a steady growing pattern which further expressed numerous stem cell specific markers, neuroendocrine markers and intercellular adhesion molecules. While in case of 2D culturing, it was found that cells had a faster growth rate, while intercellular adhesion molecules were decreased and mesenchymal transition was increased. It was deduced thereby that the 3D culturing of cells leads to the formation of more realistic tumoroids in terms of morphology & gene expression [60]. Furthermore due to enhanced intercellular communication in 3D culturing, EVs which are involved in transcellular transport are more in number when compared to the 2D culture system. Tu et al., realised 3D culturing as a better model for tumour progression, as they estimated the miRNA content of exosomes and protein expression of GPC-1, and found that the trend observed in EVs derived from spheroids presented higher relevance to the progression of pancreatic cancer [61]. Not just in cancer, but 3D culturing has also been tested for primary cells like Mesenchymal Stem Cells (MSCs). It was found the culturing of MSCs in a 3D manner leads to multi-fold increase in the exosome concentration & enrichment of cargo such that they were more efficient in their uptake capabilities and improved the viability of the recipient cells [62]. Furthermore, it was deduced that the cargo content of MSCs-EVs was vividly distinct when derived from a 3D culture microenvironment. The results of a microarray suggested that expression of 193 miRs were varied wherein 68 miRs were up regulated and 125 miRs were downregulated [63]. Yu et al., also explored the EV dynamics in 3D vs. 2D culture of mesenchymal stem cells & observed that there was a 2.5 fold increase in exosome production upon 3D culturing along with 2.9 fold increase in the enrichment of proteins. Furthermore, they also suggested that exosomes derived from 3D culturing of MSCs had heightened expression of osteo-inductive genes and proteins, which could be attributed to the upregulation of YAP pathway [64]. 3D culture system derived exosomes were also proven to possess extended anti-inflammatory effects and were able to restore the homeostatic balance of Th17 and Treg cells in a model of periodontal inflammation. This was suggested to be happening as a result of enhanced expression of miR1246 in the 3D derived exosomes, thereby affecting the Nfat5 expression which plays a role in Th17 polarity [65]. However, there have also been studies that suggest that EVs isolated from MSCs cultured in a 3D manner did not sufficiently execute the properties which are a characteristic of their parent cells like immunomodulation & anti-fibrotic activity. It was observed that the level of IDO was significantly downregulated when the cells were cultured in 3D & there was a rise in pro-inflammatory capacity of macrophages upon culturing of EVs derived from 3D culture as compared to 2D culture. This could be due to the extensive networking and interactions between the cells itself during 3D culturing & the decline in cell volume thereby affecting the packaging of EVs so released [66].

3D culturing for EV derivation is a budding area of research, and so there is yet nothing conclusive about the possible effects of 3D culturing on exosome release & cargo profile. There have been many contrasting views that support or reject the hypothesis of culturing cells in a 3D manner. It can be accepted that 3D culture model might best be able to replicate the cancer biology due to its ability to replicate tumour like interactions & features in vitro, and concurrent release of in vivo like EVs. Such a culture model could aid the advancements in identification of cancer biomarkers which may be specifically analysed in a simulated manner. While 3D culturing is being preferred in carcinoma-based studies, culturing of primary cells in a 3D microenvironment is still a topic of debate. There have been divergent perspectives of researchers regarding the derivation of EVs from 3D culture of primary cells however, it still needs to be developed further to enhance its benefits, more than its shortcomings. EVs are a recent popular candidate for therapeutics, drug loading & delivery, and 3D culture has shown tremendous potential in enhancing their yield, therefore it could be an interesting application & strategy to exploit these modalities for commercial manufacturing of the EVs.

4. Future prospects of using organoid as drug screening and EVs as biomarker analysis

Research in cancer has relied heavily for a considerable amount of time on cancer cell lines as a model system. Recent research has made use of high-throughput screening of broad panels of cancer cell lines to detect patterns of drug sensitivity and to correlate drug sensitivity to genetic changes [67]. These high-throughput cell-line-based research paint a picture of a complex network of biological variables that affect sensitivity to most cancer medicines. It is possible, for example, that there is no direct connection between sensitivity to a particular medicine and individual genetic changes. Instead, the outcome of drug sensitivity may be determined by the complicated interactions that occur between several genetic changes, which are difficult to find. Therefore, despite the new information that has recently become available, it is still difficult to develop algorithms that can accurately predict the drug sensitivity of a patient's tumour based on the spectrum of genomic alterations that are present, in the context of the individual's specific genetic background [68]. Even though there is a great deal of information accessible on the biology of cancer, there are still a great deal of questions regarding this international health issue [69]. There is a clear and pressing requirement to keep researching and developing improved therapies for cancer patients. Incomplete or inaccurate modelling of cancer is one of the primary roadblocks in the way of the development of additional treatment regimens. This is because, at times, cancer models can only poorly recapitulate clinical conditions. Because of this, a significant number of medications that produce encouraging outcomes in cancer models fail when tested in humans. Therefore, despite the fact that animal models appear to provide useful insights into the fundamentals of cancer biology, it is vital to keep in mind that these models frequently fail to faithfully recreate the pathogenic processes that take place in patients [70]. As a result, the field of oncology requires the development of new methodologies and approaches to create fresh targeted medicines and to continue lowering the number of fatalities caused by cancer.

An important advance in scientific methodology over the course of time has led to the development of three-dimensional (3D) organoid culture as well as 3D printed scaffolds, both of which are able to simulate human biology as well as diseases more accurately. In 1946, Smith and Cochrae were the first people to use the term "organoid," which means "resembling an organ," to describe a case of cystic teratoma. This term was used to describe the growth of a cystic teratoma [71]. However, the term "organoid" now has a more restricted definition. This definition states that organoids are self-assembled in vitro 3D structures, which are primarily generated from primary tissues or stem cells such as adult stem cells, induced pluripotent stem cells (iPSCs), and embryonic stem cells (ESCs). The production of organoids is dependent on the self-assembly and differentiation of cells, as well as the signalling signals from the extracellular matrix (ECM) and the conditioned medium. This is true regardless of the circumstances. When the three-dimensional constructions are complete, they are able to replicate the intricate features of their real-life organ equivalents, as well as undergo genetic engineering, long-term expansion, and cryopreservation [72]. Organoids and 3D cultures have emerged because of several attempts to replicate the biology of human organs, such as stem cell development in 2D cultures with or without a 3D matrix, cell culture on a microfluidic device (organ-on-a-chip), and bio-printing of cells. Opportunities for medication discovery and human disease study have been expanded because of these modelling initiatives [73]. The term "organoids" refers to three-dimensional structures that can self-organise through the processes of self-renewal and tenogenic differentiation. These structures are formed from pluripotent stem cells that have been cultivated from organ-specific tissues. Organoids have a distinct organisation that places them in the category of micro physiological systems. This is because they are capable of both self-renewal and self-organisation, and, more importantly, that they display organ functionalities that are analogous to those of the tissue(s) from which they originated. Therefore, it is essential to establish cultures of functional tissues, but these cultures should be devoid of the mesenchymal, stromal, immune, and neuronal cells that interspace tissues in vivo. This will allow for the development and maintenance of optimal conditions for organoid design. In fact, this process is dependent on the construction of artificial extracellular matrices in order to allow organoid self-organisation into structures that are analogous to the architecture of the native tissue [69, 74]. To this day, organoids have been successfully created from the intestine, liver, pancreas, colon, and prostate of murine animals, as well as from the small intestine, colon, stomach, and prostate of human beings. The fact that these organoids can be grown for an extended period and, according to whole-genome sequencing, match the patient tissue from whence they originated suggests that their phenotypic and genetic traits are consistent [75].

Patient derived organoids (PDOs) have recently proven valuable in translational research because these models can be maintained for an extended period and cryopreserved. In addition, PDOs are genetically stable, which makes them a perfect choice for modelling diseases. In addition, PDO models are helpful because they enable the expansion of normal cells as well as tumour cells in parallel, which contributes to the formation of a living tumour organoid biobank. PDO models, on the other hand, solely represent the epithelial tissues of organs; they do not include the stroma, nerves, or vasculature that are seen in real organs, which is an essential distinction to make. When organoids are generated from different types of tissues, different types of growth components are required (**Figure 2**) [74].

The use of murine and human embryonic stem cell lines and induced pluripotent stem cell lines to generate organoids gets around the limited availability of high-quality human primary material. However, in order to perform directed differentiation, in-depth knowledge of the factors involved in germ layer formation and subsequent lineage specification is required. In contrast to the employment of ESCs, the utilisation of iPSC lines necessitates the performance of an additional step. Specifically, the expression of OCT4, KLF4, SOX2, and MYC29 is required in order to transform somatic cells into iPSCs. Following this step, embryonic stem cells (ESCs) and

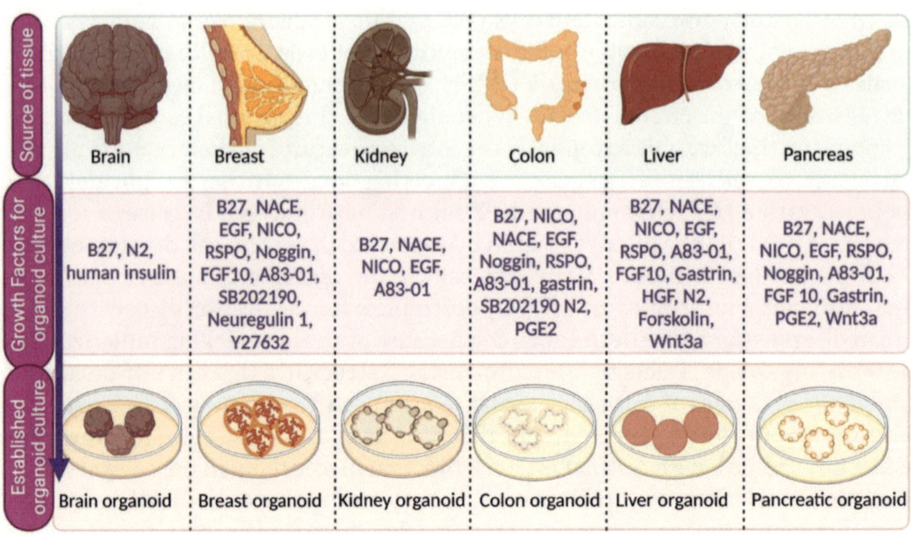

Figure 2.
Culture additives/growth factors used for generation of different organoid models.

induced pluripotent stem cells (iPSCs) are subjected to germ layer and tissue-specific patterning factors. Next, the cells are embedded in Matrigel in order to facilitate the development of 3D architecture. Finally, the cells are treated with differentiation factors in order to produce the desired organoids.

The first successful mesoderm-derived organoids were reported not too long ago. Renal organoids were formed by manipulating GSK3 and FGF signalling pathways in human iPSCs. This was done while the cells were in an intermediate mesodermal state. The architecture and segmentation of human foetal nephrons into ducts, tubules, and glomeruli is replicated in these organoids, which have the same name [76]. The human renal organoids provide a 3D model to study human renal development and disease under well-defined conditions, thus overcoming various limitations of previous models such as 2D monolayers, short-term 3D aggregates, and co-cultures with mouse fibroblasts [77, 78].

The existence of cell types other than those intended for lineage in ESCs and iPSCs is one of the most distinguishing characteristics of organoids created from primary tissue as opposed to those generated from ESCs and iPSCs. This is due to the fact that the factors that are used for the directed differentiation of ESCs and iPSCs are not completely efficient in driving all of the cells toward the lineage of choice. As a result, many ectodermal and endodermal organoids, such as those of the intestine, stomach, and kidney, have reported the limited presence of mesenchymal cell types [79, 80]. Despite these developments, some tissues remain resistant to organoid culture but have been successfully cultivated in 3D as whole-tissue explants or organotypic/mechanically supported cultures (for example, skin or ovary) [81]. Understanding the endogenous stem cell microenvironment and signalling pathways driving lineage specification in organoid cultures is crucial. Our limited knowledge in these areas for certain tissues prevents us from logically designing niche parameters for organoid formation. Identifying stem cells is not necessary for growing primary tissue units but understanding the stem cell niche is essential for long-term culture sustainability [82]. Small-molecule modulators of critical signalling pathways and organ-specific hormones could facilitate organoid growth from organs like the ovaries. A tight dependence on growth factor/signalling gradients for stem cell renewal and lineage specification may also complicate organoid formation from tissues. Microfluidics could be utilised to create in-vivo-like concentration gradients [83]. In vivo stem cell behaviour and differentiation are also highly impacted by local biomechanical factors, such as interactions with the extracellular matrix49. In order to create more robust organoid culture models for a larger spectrum of tissues, researchers are screening for substrates and ECM factors that influence cell behaviour in vitro [84].

In the field of cancer research, the improvement of culture techniques has been applied to the study of EVs, which models the environment and physiological conditions that are present in the area surrounding tumours. The role of EVs in tumour physiology is not limited to cell-to-cell communication; rather, they are also a promising source of biomarkers, a tool to deliver drugs, and a mechanism to induce antitumor activity [10, 85]. Extracellular vesicles, also known as EVs, were discovered in the 1980s and were initially thought to be carriers of waste products that were produced inside of cells [86]. After it was discovered that RNA could be found enclosed within the lipid membrane of EVs, approximately 20 years later, EVs began to be seen in a different light; specifically, as crucial mediators in the process of intercellular communication [87]. The nucleic acids contained in EV RNAs were found to be distinct from those found in the cell from which they originated, displaying distinct sequences and even concentration profiles. In the course of time, research has demonstrated that EVs are responsible for transporting nucleic acids, such as RNA and DNA, as well as a wide variety of biomolecules, which includes proteins and lipids, into and out of cells [88]. For instance, in cancer, the EV content of the tumour is tumour-like, and a class of EVs known as exosomes help the progression of the tumour by signalling to the tumour cells that they should establish the pre-metastatic niche. In another scenario, EVs that are released from cells that have been infected with a virus such as HIV can contain fragments of viral RNA as well as viral proteins; consequently, the function of EVs in HIV is unclear at the present time. In addition, extracellular vesicles released by breast cancer cells have been shown to contribute to the spread of the disease to the brain and to have triggered the breakdown of the barrier that separates the blood and the brain. In general, EVs can break through natural barriers such as the blood-brain barrier and others. These mechanisms can be exploited to deliver therapeutic agents to parts of the body that are difficult to reach. EVs have also been shown to play a role in reproductive biology, the differentiation of stem cells, angiogenesis, and a variety of other biological processes [89]. Various clinical trials are ongoing and have been completed where they have used EVs for therapy of various cancer types [90]. EVs are critically important for tumour communication with their intended target cells. Therefore, the study and modification of EVs have opened so many doors for diagnosis and therapy. It is well-established that elevated levels of circulating EVs are linked to the development of most cancers. Blood EV concentration has also been shown to correlate with tumour volume in several tumour types. These EVs have become the substrate for biomarker mining in a variety of cancers, including prostate cancer, due to the valuable information they transport about the tumour (**Table 2**) [91–93].

Three-dimensional (3D) culture enables cell growth in a physiological topology, and organoids and spheroids continue to release EVs, which are essential for tumour communication with targeted cells, and the released EVs are functional (**Figure 3**). The extracellular vesicles that are released by pancreatic cancer organoids have the

Id/reference	Disease	EV source	Stage	Goal	Status
NCT01294072	Colon cancer	Plant	Phase 1	To see if curcumin could be delivered using plant EVs	Active
NCT03608631	Pancreatic cancer	Mesenchymal stromal cells	Phase 1	To evaluate the side effects of mesenchymal stromal EVs on pancreatic cancer cells	Not recruiting
NCT01159288	Lung cancer	Dendritic cells	Phase 2	To determine if patients' conditions improve after treatment with EVs	Completed
Dai et al.	Colorectal cancer	Ascites	Phase 1	To determine the role of ascites EVs in immunotherapy	Completed
Morse et al.	Lung cancer	Dendritic cells	Phase 1	Role of dendritic EVs in immunotherapy	Completed
Besse et al.	Non-small cell lung cancer	Dendritic cells	Phase 2	To assess the role of dendritic EVs on NSCLC patients	Complete

Table 2.
Human clinical trials using EVs for therapeutic purposes.

ability to activate p38 MAPK and increase the expression of F-box protein 32 and UBR2 in myotubes. In the case of colorectal cancer stem cells, 3D cultures exhibit a higher level of EVs release in comparison to 2D conformations. The presence of APC mutations in colon cancer organoids that activated the WNT pathway resulted in an increase in the amount of EVs released in cultures based on Matrigel. This release was presumably likely helped along by the presence of collagen, which is a component of the extracellular matrix and is present in this sort of gel [61]. Collagen is a component of the gel. Additionally, a further potential hypothesis is that the greater expression of transporters in 3D cultures may be partially responsible for the release of EVs [94–96]. It was shown that tumoroids of colon cancer cells with improved stemness had significant levels of expression of the ATP-binding cassette transporter G1, which is a cholesterol lipid efflux pump. Similarly, inhibiting this transporter prevents the release of EVs and leads to an increase in the number of vesicles found inside the cell [97].

Research has been done to investigate the spontaneous effect of EVs derived from normal cells in order to use them as natural antitumor agents. For instance, extracellular vesicles produced from glia have been shown to have an anticancer effect in spheroids composed of glioma cells. This effect was demonstrated by a gradual reduction in the tumour potential to invade surrounding tissue. Another example is the EVs that are produced by mesenchymal stem cells (MSCs). These EVs could initiate angiogenesis and preserve vascular homeostasis in activated endothelial cells [98, 99]. On the other hand, most of the publications centre their attention on the prospect of loading EVs with anticancer medicines and biomolecules such as amino acids, lipoproteins, or nucleic acids. In a microfluidic system that contained a variety of cell types, an anticancer effect of EVs that were loaded with a particular miRNA (miR-497) was evaluated [100]. These kinds of devices are helpful when used in conjunction with an extracellular matrix because doing so makes it possible to investigate migration in response to a factor that is controlled via microfluidic channels. In this experiment, the non-small cell lung cancer cell (NSCLC) line A549 was

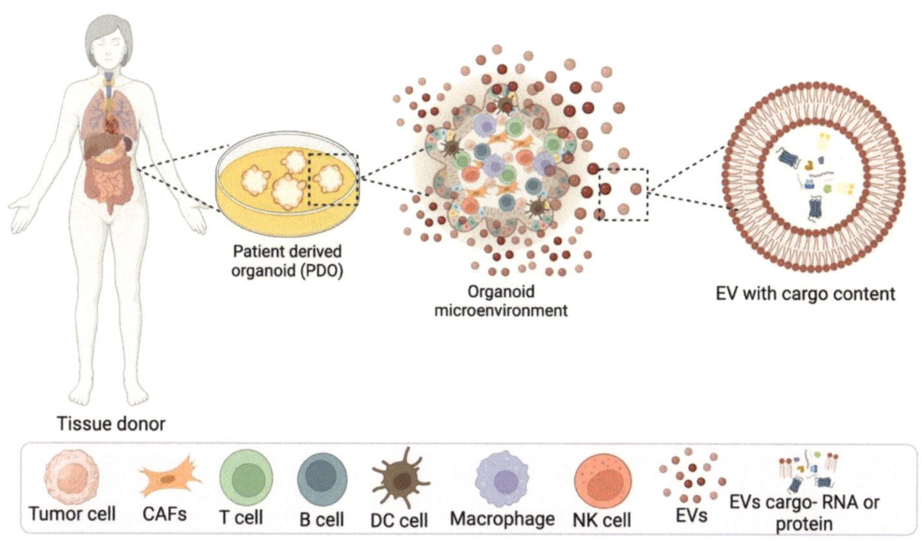

Figure 3.
The schematic diagram showing the establishment of patient derived organoid cell culture depicting the organoid microenvironment and respective EV distribution.

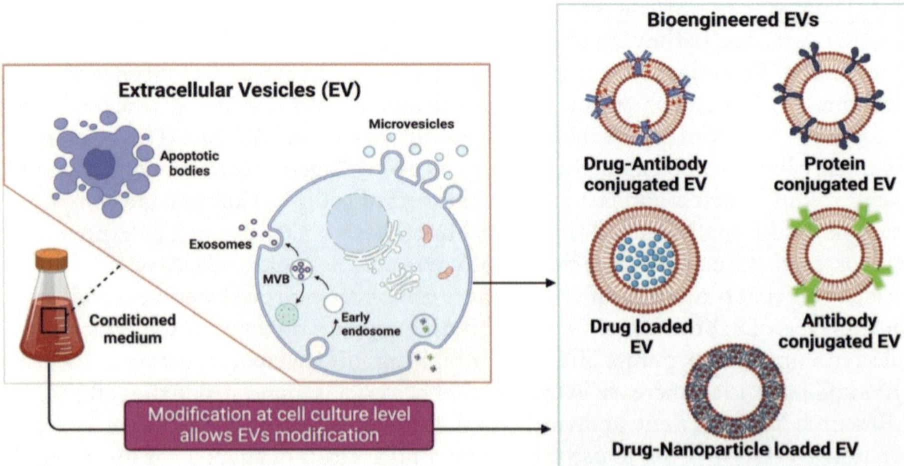

Figure 4.
The schematic diagram shows the isolation of EVs and modification of EVs for targeted delivery.

cultured alongside human umbilical vein endothelial cells (HUVEC). Both cell types were grown in a dish (HUVEC). When the experiment was run under these conditions, the production of tubes by endothelial cells was prevented, and the amount of tumour migration was significantly reduced in comparison to the control. Both types of cells were separated in the microfluidic devices using the matrigel component. This was done so that the limitations of cocultures that are related with cell separation after analysis could be avoided [101, 102]. This is a fascinating illustration of how 3D culture may be used to recreate the physiological intricacy of tumours (**Figure 3**).

One of the most fascinating uses of 3D cultures is the large-scale and standardizable production of EVs. This is one of the most exciting applications of 3D cultures because till yet there is no established biomanufacturing platform for EVs, which poses restriction for clinical translation (**Figure 4**). The utilisation of bioreactor flasks is a straightforward method that can be utilised because these flasks boost the creation of EVs that are discharged by tumour cells. Utilising cell cultures on microfluidic substrates is a more interesting application of this technology [103, 104]. These automated systems can manufacture therapeutic exosomes, which could also be modified, and harvest them in real-time from the cultures that are performed on the chip. As a device of this kind has been utilised in the process of isolating leukocytes from human blood. An alternative method that has been utilised is a 3D-printed scaffold-perfusion bioreactor system to investigate the impact that dynamic cultures have on the production of EVs from endothelial cells. Because of this method, the cells were able to keep up their level of functionality (i.e., pro-vascularization bioactivity or pro-angiogenic gene expression) [95, 101].

5. Conclusion

Three-dimensional (3D) cell culture models are more functionally important than two-dimensional (2D) cell cultures and include a broad range of structures such as embryoid bodies, spheroids, patches, and scaffolds. Whereas, organoids and

3D culture systems are becoming being recognised as a tool for advancing medical research without relying on animal models. Improvements to these methods may even result in new methods for creating 3D models. Understanding the limitations of the systems is critical for their improvement and determining the model's suitability for investigating EVs. Overall, cellular architecture influences the concentration and cargo profile of EVs. Many studies, for example, found that 3D in vitro systems secreted more EVs than their 2D culture counterparts. Moreover, the possibility of a necrotic core developing in multicellular cultures poses a unique challenge to isolating EVs from 3D in vitro systems. The necrotic core can generate EVs composed primarily of apoptotic bodies rather than small vesicles or large vesicles. To address the challenges, developing cell viability criteria and measures to normalise the outcomes compared against controls such as 2D cell monolayer cultures are required. With the start of human clinical trials for EV therapeutics, these challenges become even more important. Despite the widespread use of organoids in biology, the technology is still in its infancy for certain disorders. Most neurodevelopmental or neuropsychiatric disorders, such schizophrenia, Parkinson, or autism, are examined using animal models. Autism spectrum diseases or Parkinson's have clinical heterogeneity (epilepsy, sleep disruptions, motor difficulties), making organoid culture challenging to use. Researchers have been developing techniques to create more mature and complicated brain organoids. Organoids can be utilised to explore developmental brain injuries and disorders (DBD) Stem cells are linked to several disorders. Scientists do not know how stem cells develop abnormalities or which type of specialised cell to generate. The organoids method can answer questions about stem cells in diseases like emphysema, when lung stem cells fail to heal damage. Further, scientists have suggested utilising organoids to screen medications that can produce specialised cell types for hereditary illnesses like cystic fibrosis, where ciliated cells that remove mucus from the lung do not work properly. Generate organoids from cystic fibrosis patient tissues, then design a medication to make ciliated cells operate better in organoid cultivation. Since scientists can co-culture organoids with immune cells, the approach can be used to investigate autoimmune disease mechanisms and screen medications.

Acknowledgements

The authors would like to acknowledge the Biorender software for figures.

Conflict of interest

The authors declare that they have no competing interest.

Abbreviations

2D	two-dimensional
3D	three-dimensional
APC	adenomatous polyposis coli

ATP	adenosine triphosphate
DNA	deoxyribonucleic acid
EBs	embryoid bodies
ECM	extracellular matrix
EGF	epidermal growth factor
ESCs	embryonic stem cells
EVs	extracellular vesicles
FGF	fibroblast growth factor
GELs	collagen gels
GPC-1	Glypican 1
HIV	human immunodeficiency virus
HUVEC	human umbilical vein endothelial cell
IDO	indoleamine 2,3-dioxygenase
MAPK	mitogen-activated protein kinase
miR	micro RNA
MSCs	mesenchymal stem/stroma cells
Nfat5	nuclear factor of activated T cells
NSCLC	non-small cell lung cancer
PDGF	platelet derived growth factor
PDOs	patient derived organoids
PLLA	ploy (l-lactic acid)
PS	polystyrene
RNA	ribonucleic acid
UBR2	ubiquitin protein ligase E3 component N-recognin 2

Author details

Sonali Rawat, Yashvi Sharma, Misba Majood and Sujata Mohanty*
Stem Cell Facility, DBT-Centre of Excellence for Stem Cell Research, All India Institute of Medical Sciences, India

*Address all correspondence to: drmohantysujata@gmail.com

References

[1] Antoni D, Burckel H, Josset E, Noel G. Three-dimensional cell culture: A breakthrough in vivo. International Journal of Molecular Sciences. 2015;**16**(3):5517-5527

[2] Dhaliwal A. Three dimensional cell culture: A review. Materials and Methods. 2012;**2**:162. DOI: 10.13070/mm.en.2.162

[3] Sarkar C, Mondal M, Torequl Islam M, Martorell M, Docea AO, Maroyi A, et al. Potential therapeutic options for COVID-19: Current status, challenges, and future perspectives. Frontiers in Pharmacology. 2020;**11**:572870

[4] Chaiyawat P, Chokchaichamnankit D, Lirdprapamongkol K, Srisomsap C, Svasti J, Champattanachai V. Alteration of O-GlcNAcylation affects serine phosphorylation and regulates gene expression and activity of pyruvate kinase M2 in colorectal cancer cells. Oncology Reports. 2015;**34**(4):1933-1942

[5] Kolind K, Leong KW, Besenbacher F, Foss M. Guidance of stem cell fate on 2D patterned surfaces. Biomaterials. 2012;**33**:6626-6633. DOI: 10.1016/j.biomaterials.2012.05.070

[6] Mabry KM, Payne SZ, Anseth KS. Microarray analyses to quantify advantages of 2D and 3D hydrogel culture systems in maintaining the native valvular interstitial cell phenotype. Biomaterials. 2016;**74**:31-41

[7] Edmondson R, Broglie JJ, Adcock AF, Yang L. Three-dimensional cell culture systems and their applications in drug discovery and cell-based bio-sensors. Assay and Drug Development Technologies. 2014;**12**:207-218. DOI: 10.1089/adt.2014.573

[8] Dupont S, Morsut L, Aragona M, Enzo E, Giulitti S, Cordenonsi M, et al. Role of YAP/TAZ in mechanotransduction. Nature. 2011;**474**(7350):179-183

[9] Ludwig N, Whiteside TL, Reichert TE. Challenges in exosome isolation and analysis in health and disease. International Journal of Molecular Sciences. 2019;**20**(19):4684

[10] Möller A, Lobb RJ. The evolving translational potential of small extracellular vesicles in cancer. Nature Reviews. Cancer. 2020;**20**(12):697-709

[11] Ravi M, Paramesh V, Kaviya SR, Anuradha E, Solomon FP. 3D cell culture systems: Advantages and applications. Journal of Cellular Physiology. 2015;**230**(1):16-26

[12] Huh D, Hamilton GA, Ingber DE. From 3D cell culture to organs-on-chips. Trends in Cell Biology. 2011;**21**(12):745-754

[13] Fu J, Wang YK, Yang MT, Desai RA, Yu X, Liu Z, et al. Mechanical regulation of cell function with geometrically modulated elastomeric substrates. Nature Methods. 2010;**7**(9):733-736

[14] Ihalainen TO, Aires L, Herzog FA, Schwartlander R, Moeller J, Vogel V. Differential basal-to-apical accessibility of lamin A/C epitopes in the nuclear lamina regulated by changes in cytoskeletal tension. Nature Materials. 2015;**14**(12):1252-1261

[15] Xu X, Wang W, Kratz K, Fang L, Li Z, Kurtz A, et al. Controlling major cellular processes of human mesenchymal stem cells using microwell structures. Advanced Healthcare Materials. 2014;**3**(12):1991-2003

[16] LeCluyse EL, Audus KL, Hochman JH. Formation of extensive canalicular networks by rat hepatocytes cultured in collagen-sandwich configuration. American Journal of Physiology-Cell Physiology. 1994;**266**(6):C1764-C1774

[17] Dunn JC, Tompkins RG, Yarmush ML. Hepatocytes in collagen sandwich: Evidence for transcriptional and translational regulation. The Journal of Cell Biology. 1992;**116**(4):1043-1053

[18] Ezzell RM, Toner M, Hendricks K, Dunn JC, Tompkins RG, Yarmush ML. Effect of collagen gel configuration on the cytoskeleton in cultured rat hepatocytes. Experimental Cell Research. 1993;**208**(2):442-452

[19] Jones HM, Barton HA, Lai Y, Bi YA, Kimoto E, Kempshall S, et al. Mechanistic pharmacokinetic modeling for the prediction of transporter-mediated disposition in humans from sandwich culture human hepatocyte data. Drug Metabolism and Disposition. 2012;**40**(5):1007-1017

[20] Chaubey A, Ross KJ, Leadbetter RM, Burg KJ. Surface patterning: Tool to modulate stem cell differentiation in an adipose system. Journal of Biomedical Materials Research Part B: Applied Biomaterials. 2008;**84**(1):70-78

[21] Stevens MM, George JH. Exploring and engineering the cell surface interface. Science. 2005;**310**(5751):1135-1138

[22] Wan Y, Wang Y, Liu Z, Qu X, Han B, Bei J, et al. Adhesion and proliferation of OCT-1 osteoblast-like cells on micro-and nano-scale topography structured poly (L-lactide). Biomaterials. 2005;**26**(21):4453-4459

[23] Underhill GH, Bhatia SN. High-throughput analysis of signals regulating stem cell fate and function. Current Opinion in Chemical Biology. 2007;**11**(4):357-366

[24] Scadden DT. The stem-cell niche as an entity of action. Nature. 2006;**441**(7097):1075-1079

[25] Costa EC, Moreira AF, de Melo-Diogo D, Gaspar VM, Carvalho MP, Correia IJ. 3D tumor spheroids: An overview on the tools and techniques used for their analysis. Biotechnology Advances. 2016;**34**(8):1427-1441

[26] Cushing MC, Anseth KS. Hydrogel cell cultures. Science. 2007;**316**(5828):1133-1134

[27] Dhandayuthapani B, Yoshida Y, Maekawa T, Kumar DS. Polymeric scaffolds in tissue engineering application: A review. International Journal of Polymer Science. 2011;**2011**:19. Article 290602. DOI: 10.1155/2011/290602

[28] Ferreira LP, Gaspar VM, Mano JF. Design of spherically structured 3D in vitro tumor models—Advances and prospects. Acta Biomaterialia. 2018;**75**:11-34

[29] Godugu C, Patel AR, Desai U, Andey T, Sams A, Singh M. AlgiMatrix™ based 3D cell culture system as an in-vitro tumor model for anticancer studies. PLoS One. 2013;**8**(1):e53708

[30] Peppas NA, Bures P, Leobandung WS, Ichikawa H. Hydrogels in pharmaceutical formulations. European Journal of Pharmaceutics and Biopharmaceutics. 2000;**50**(1):27-46

[31] Talukdar S, Kundu SC. A non-mulberry silk fibroin protein based 3D in vitro tumor model for evaluation of anticancer drug activity.

Advanced Functional Materials. 2012;**22**(22):4778-4788

[32] Hoffman AS. Hydrogels for biomedical applications. Advanced Drug Delivery Reviews. 2012;**64**:18-23

[33] Bi YA, Kazolias D, Duignan DB. Use of cryopreserved human hepatocytes in sandwich culture to measure hepatobiliary transport. Drug Metabolism and Disposition. 2006;**34**(9):1658-1665

[34] Huh D, Hamilton GA, Ingber DE. From 3D cell culture to organs-on-chips. Trends in Cell Biology. 2011;**21**(12):745-754

[35] Choi SW, Yeh YC, Zhang Y, Sung HW, Xia Y. Uniform beads with controllable pore sizes for biomedical applications. Small. 2010;**6**(14):1492-1498

[36] Gu L, Mooney DJ. Biomaterials and emerging anticancer therapeutics: Engineering the microenvironment. Nature Reviews. Cancer. 2016;**16**(1):56-66

[37] Torisawa YS, Takagi A, Nashimoto Y, Yasukawa T, Shiku H, Matsue T. A multicellular spheroid array to realize spheroid formation, culture, and viability assay on a chip. Biomaterials. 2007;**28**(3):559-566

[38] Pineda ET, Nerem RM, Ahsan T. Differentiation patterns of embryonic stem cells in two-versus three-dimensional culture. Cells, Tissues, Organs. 2013;**197**(5):399-410

[39] Hsiao AY, Tung YC, Kuo CH, Mosadegh B, Bedenis R, Pienta KJ, et al. Micro-ring structures stabilize microdroplets to enable long term spheroid culture in 384 hanging drop array plates. Biomedical Microdevices. 2012;**14**(2):313-323

[40] Fennema E, Rivron N, Rouwkema J, van Blitterswijk C, De Boer J. Spheroid culture as a tool for creating 3D complex tissues. Trends in Biotechnology. 2013;**31**(2):108-115

[41] DeForest CA, Sims EA, Anseth KS. Peptide-functionalized click hydrogels with independently tunable mechanics and chemical functionality for 3D cell culture. Chemistry of Materials. 2010;**22**(16):4783-4790

[42] Nii M, Lai JH, Keeney M, Han LH, Behn A, Imanbayev G, et al. The effects of interactive mechanical and biochemical niche signaling on osteogenic differentiation of adipose-derived stem cells using combinatorial hydrogels. Acta Biomaterialia. 2013;**9**(3):5475-5483

[43] Nichol JW, Koshy ST, Bae H, Hwang CM, Yamanlar S, Khademhosseini A. Cell-laden microengineered gelatin methacrylate hydrogels. Biomaterials. 2010;**31**(21):5536-5544

[44] Lutolf MP, Hubbell JA. Synthetic biomaterials as instructive extracellular microenvironments for morphogenesis in tissue engineering. Nature Biotechnology. 2005;**23**(1):47-55

[45] Haraguchi Y, Shimizu T, Sasagawa T, Sekine H, Sakaguchi K, Kikuchi T, et al. Fabrication of functional three-dimensional tissues by stacking cell sheets in vitro. Nature Protocols. 2012;**7**(5):850-858

[46] Shimizu T, Yamato M, Kikuchi A, Okano T. Cell sheet engineering for myocardial tissue reconstruction. Biomaterials. 2003;**24**(13):2309-2316

[47] Yeatts AB, Choquette DT, Fisher JP. Bioreactors to influence stem cell fate: Augmentation of mesenchymal stem cell signaling pathways via dynamic culture systems. Biochimica et Biophysica

Acta (BBA)—General Subjects. 2013;**1830**(2):2470-2480

[48] Whitesides GM. The origins and the future of microfluidics. Nature. 2006;**442**(7101):368-373

[49] Bhatia SN, Ingber DE. Microfluidic organs-on-chips. Nature Biotechnology. 2014;**32**(8):760-772

[50] Tian F, Zhang S, Liu C, Han Z, Liu Y, Deng J, et al. Protein analysis of extracellular vesicles to monitor and predict therapeutic response in metastatic breast cancer. Nature Communications. 2021;**12**(1):1-3

[51] Bebelman MP, Janssen E, Pegtel DM, Crudden C. The forces driving cancer extracellular vesicle secretion. Neoplasia. 2021;**23**(1):149-157

[52] Kim M, Yun HW, Park DY, Choi BH, Min BH. Three-dimensional spheroid culture increases exosome secretion from mesenchymal stem cells. Tissue Engineering and Regenerative Medicine. 2018;**15**(4):427-436

[53] Yan L, Wu X. Exosomes produced from 3D cultures of umbilical cord mesenchymal stem cells in a hollow-fiber bioreactor show improved osteochondral regeneration activity. Cell Biology and Toxicology. 2020;**36**(2):165-178

[54] Patel DB, Luthers CR, Lerman MJ, Fisher JP, Jay SM. Enhanced extracellular vesicle production and ethanol-mediated vascularization bioactivity via a 3D-printed scaffold-perfusion bioreactor system. Acta Biomaterialia. 2019;**95**:236-244

[55] Haraszti RA, Miller R, Stoppato M, Sere YY, Coles A, Didiot MC, et al. Exosomes produced from 3D cultures of MSCs by tangential flow filtration show higher yield and improved activity. Molecular Therapy. 2018;**26**(12):2838-2847

[56] Hwang WL, Lan HY, Cheng WC, Huang SC, Yang MH. Tumor stem-like cell-derived exosomal RNAs prime neutrophils for facilitating tumorigenesis of colon cancer. Journal of Hematology & Oncology. 2019;**12**(1):1-7

[57] Rocha S, Carvalho J, Oliveira P, Voglstaetter M, Schvartz D, Thomsen AR, et al. 3D cellular architecture affects microRNA and protein cargo of extracellular vesicles. Advanced Science. 2019;**6**(4):1800948

[58] Thippabhotla S, Zhong C, He M. 3D cell culture stimulates the secretion of in vivo like extracellular vesicles. Scientific Reports. 2019;**9**(1):1-4

[59] Namba Y, Sogawa C, Okusha Y, Kawai H, Itagaki M, Ono K, et al. Depletion of lipid efflux pump ABCG1 triggers the intracellular accumulation of extracellular vesicles and reduces aggregation and tumorigenesis of metastatic cancer cells. Frontiers in Oncology. 2018;**8**:376

[60] Eguchi T, Sogawa C, Okusha Y, Uchibe K, Iinuma R, Ono K, et al. Organoids with cancer stem cell-like properties secrete exosomes and HSP90 in a 3D nanoenvironment. PLoS One. 2018;**13**(2):e0191109

[61] Tu J, Luo X, Liu H, Zhang J, He M. Cancer spheroids derived exosomes reveal more molecular features relevant to progressed cancer. Biochemistry and Biophysics Reports. 2021;**26**:101026

[62] Cao J, Wang B, Tang T, Lv L, Ding Z, Li Z, et al. Three-dimensional culture of MSCs produces exosomes with improved yield and enhanced therapeutic efficacy for cisplatin-induced acute kidney

injury. Stem Cell Research & Therapy. 2020;**11**(1):1-3

[63] Yang L, Zhai Y, Hao Y, Zhu Z, Cheng G. The regulatory functionality of exosomes derived from hUMSCs in 3D culture for Alzheimer's disease therapy. Small. 2020;**16**(3):1906273

[64] Yu W, Li S, Guan X, Zhang N, Xie X, Zhang K, et al. Higher yield and enhanced therapeutic effects of exosomes derived from MSCs in hydrogel-assisted 3D culture system for bone regeneration. Materials Science and Engineering C. 2022;**112646**

[65] Zhang Y, Chen J, Fu H, Kuang S, He F, Zhang M, et al. Exosomes derived from 3D-cultured MSCs improve therapeutic effects in periodontitis and experimental colitis and restore the Th17 cell/Treg balance in inflamed periodontium. International Journal of Oral Science. 2021;**13**(1):1-5

[66] Kusuma GD, Li A, Zhu D, McDonald H, Inocencio IM, Chambers D, et al. Effect of 2D and 3D culture microenvironments on mesenchymal stem cell-derived extracellular vesicles potencies. Frontiers in Cell and Development Biology. 14 Feb 2022;**147**

[67] Barretina J, Caponigro G, Stransky N, Venkatesan K, Margolin AA, Kim S, et al. The cancer cell line Encyclopedia enables predictive modelling of anticancer drug sensitivity. Nature. 2012;**483**(7391):603-607

[68] Tentler JJ, Tan AC, Weekes CD, Jimeno A, Leong S, Pitts TM, et al. Patient-derived tumour xenografts as models for oncology drug development. Nature Reviews. Clinical Oncology. 2012;**9**(6):338-350

[69] Dart A. Organoid diversity. Nature Reviews. Cancer. 2018;**18**(7):404-405

[70] Caponigro G, Sellers WR. Advances in the preclinical testing of cancer therapeutic hypotheses. Nature Reviews. Drug Discovery. 2011;**10**(3):179-187

[71] Kretzschmar K, Clevers H. Organoids: Modeling development and the stem cell niche in a dish. Developmental Cell. 2016;**38**(6):590-600

[72] Sato T, Vries RG, Snippert HJ, Van De Wetering M, Barker N, Stange DE, et al. Single Lgr5 stem cells build crypt-villus structures in vitro without a mesenchymal niche. Nature. 2009;**459**(7244):262-265

[73] Kim J, Koo BK, Knoblich JA. Human organoids: Model systems for human biology and medicine. Nature Reviews. Molecular Cell Biology. 2020;**21**(10):571-584

[74] Fatehullah SH, Tan N. Barker, organoids as an in vitro model of human development and disease. Nature Cell Biology. 2016;**2016**(18):246-254

[75] Lancaster MA, Knoblich JA. Organogenesis in a dish: Modeling development and disease using organoid technologies. Science. 2014;**345**(6194):1247125

[76] Takasato M, Er PX, Chiu HS, Maier B, Baillie GJ, Ferguson C, et al. Kidney organoids from human iPS cells contain multiple lineages and model human nephrogenesis. Nature. 2015;**526**(7574):564-568

[77] Valente MJ, Henrique R, Costa VL, Jerónimo C, Carvalho F, Bastos ML, et al. A rapid and simple procedure for the establishment of human normal and cancer renal primary cell cultures from surgical specimens. PLoS One. 2011;**6**(5):e19337

[78] Xia Y, Nivet E, Sancho-Martinez I, Gallegos T, Suzuki K, Okamura D, et al. Directed differentiation of human pluripotent cells to ureteric bud kidney progenitor-like cells. Nature Cell Biology. 2013;**15**(12):1507-1515

[79] Spence JR, Mayhew CN, Rankin SA, Kuhar MF, Vallance JE, Tolle K, et al. Directed differentiation of human pluripotent stem cells into intestinal tissue in vitro. Nature. 2011;**470**(7332):105-109

[80] Noguchi TA, Ninomiya N, Sekine M, Komazaki S, Wang PC, Asashima M, et al. Generation of stomach tissue from mouse embryonic stem cells. Nature Cell Biology. 2015;**17**(8):984-993

[81] Taguchi A, Kaku Y, Ohmori T, Sharmin S, Ogawa M, Sasaki H, et al. Redefining the in vivo origin of metanephric nephron progenitors enables generation of complex kidney structures from pluripotent stem cells. Cell Stem Cell. 2014;**14**(1):53-67

[82] Lancaster MA, Knoblich JA. Organogenesis in a dish: Modeling development and disease using organoid technologies. Science. 2014;**345**(6194):1247125

[83] Soen Y, Mori A, Palmer TD, Brown PO. Exploring the regulation of human neural precursor cell differentiation using arrays of signaling microenvironments. Molecular Systems Biology. 2006;**2**(1):37

[84] Flaim CJ, Chien S, Bhatia SN. An extracellular matrix microarray for probing cellular differentiation. Nature Methods. 2005;**2**(2):119-125

[85] Majood M, Rawat S, Mohanty S. Delineating the role of extracellular vesicles in cancer metastasis: A comprehensive review. Frontiers in Immunology. 19 Aug 2022;**13**:966661. DOI: 10.3389/fimmu.2022.966661. PMID: 36059497. PMCID: PMC9439583

[86] Li Y, Tang P, Cai S, Peng J, Hua G. Organoid based personalized medicine: From bench to bedside. Cell Regeneration. 2020;**9**(1):1-33

[87] Majood M, Rawat S, Mohanty S. Delineating the role of extracellular vesicles in cancer metastasis: A comprehensive review. Frontiers in Immunology. 2022;**0**:4697

[88] Bin Zha Q, Yao YF, Ren ZJ, Li XJ, Tang JH. Extracellular vesicles: An overview of biogenesis, function, and role in breast cancer. Tumor Biology. 2017;**2**:39

[89] Yáñez-Mó M, Siljander PR, Andreu Z, Bedina Zavec A, Borràs FE, Buzas EI, et al. Biological properties of extracellular vesicles and their physiological functions. Journal of Extracellular Vesicles. 2015;**4**(1):27066

[90] Abels ER, Breakefield XO. Introduction to extracellular vesicles: Biogenesis, RNA cargo selection, content, release, and uptake. Cellular and Molecular Neurobiology. 2016;**36**(3):301-312

[91] Abdollahi S. Extracellular vesicles from organoids and 3D culture systems. Biotechnology and Bioengineering. 2021;**118**(3):1029-1049

[92] Tominaga N, Kosaka N, Ono M, Katsuda T, Yoshioka Y, Tamura K, et al. Brain metastatic cancer cells release microRNA-181c-containing extracellular vesicles capable of destructing blood–brain barrier. Nature Communications. 2015;**6**(1):1-2

[93] Ferguson S, Weissleder R. Modeling EV kinetics for use in early cancer detection. Advanced Biosystems. 2020;**4**(12):1900305

[94] Pang B, Zhu Y, Ni J, Thompson J, Malouf D, Bucci J, et al. Extracellular vesicles: The next generation of biomarkers for liquid biopsy-based prostate cancer diagnosis. Theranostics. 2020;**10**(5):2309

[95] Szvicsek Z, Oszvald Á, Szabó L, Sándor GO, Kelemen A, Soós AÁ, et al. Extracellular vesicle release from intestinal organoids is modulated by Apc mutation and other colorectal cancer progression factors. Cellular and Molecular Life Sciences. 2019;**76**(12): 2463-2476

[96] Hwang WL, Lan HY, Cheng WC, Huang SC, Yang MH. Tumor stem-like cell-derived exosomal RNAs prime neutrophils for facilitating tumorigenesis of colon cancer. Journal of Hematology & Oncology. 2019;**12**(1):1-7

[97] Yang J, Zhang Z, Zhang Y, Ni X, Zhang G, Cui X, et al. ZIP4 promotes muscle wasting and cachexia in mice with orthotopic pancreatic tumors by stimulating RAB27B-regulated release of extracellular vesicles from cancer cells. Gastroenterology. 2019;**156**(3):722-734

[98] Namba Y, Sogawa C, Okusha Y, Kawai H, Itagaki M, Ono K, et al. Depletion of lipid efflux pump ABCG1 triggers the intracellular accumulation of extracellular vesicles and reduces aggregation and tumorigenesis of metastatic cancer cells. Frontiers in Oncology. 2018;**8**:376

[99] Yeon JH, Jeong HE, Seo H, Cho S, Kim K, Na D, et al. Cancer-derived exosomes trigger endothelial to mesenchymal transition followed by the induction of cancer-associated fibroblasts. Acta Biomaterialia. 2018;**76**:146-153

[100] Murgoci AN, Cizkova D, Majerova P, Petrovova E, Medvecky L, Fournier I, et al. Brain-cortex microglia-derived exosomes: Nanoparticles for glioma therapy. ChemPhysChem. 2018;**19**(10):1205-1214

[101] Jeong GS, Han S, Shin Y, Kwon GH, Kamm RD, Lee SH, et al. Sprouting angiogenesis under a chemical gradient regulated by interactions with an endothelial monolayer in a microfluidic platform. Analytical Chemistry. 2011;**83**(22):8454-8459

[102] Jeong K, Yu YJ, You JY, Rhee WJ, Kim JA. Exosome-mediated microRNA-497 delivery for anti-cancer therapy in a microfluidic 3D lung cancer model. Lab on a Chip. 2020;**20**(3):548-557

[103] Guerreiro EM, Vestad B, Steffensen LA, Aass HC, Saeed M, Øvstebø R, et al. Efficient extracellular vesicle isolation by combining cell media modifications, ultrafiltration, and size-exclusion chromatography. PLoS One. 2018;**13**(9):e0204276

[104] Bordanaba-Florit G, Madarieta I, Olalde B, Falcón-Pérez JM, Royo F. 3D cell cultures as prospective models to study extracellular vesicles in cancer. Cancers. 2021;**13**(2):307

Chapter 7

Female Germline Stem Cells: A Source for Applications in Reproductive and Regenerative Medicine

Hong-Thuy Bui, Nhat-Thinh Nguyen, Truc Phuong Lam Do, Anh My Le Ba and Nguyen Van Thuan

Abstract

One of the most significant findings in stem cell biology is the establishment of female germline stem cells (FGSCs) in the early 21st century. Besides the massive contribution of FGSCs to support ovarian function and fertility of females, the ability to create transgenic animals from FGSCs have high efficiency. Whether FGSCs can differentiate into mature oocytes for fertilization and complete embryonic development is a significant question for scientists. FGSCs were shown to produce oocytes, and the fertilized oocytes could generate offspring in mice and rats. This discovery has opened a new direction in human FGSCs research. Recently, cryopreservation of ovarian cortical tissue was already developed for women with cancer. Thus, isolation and expansion of FGSCs from this tissue before or after cryopreservation may be helpful for clinical fertility therapies. Scientists have suggested that the ability to produce transgenic animals using FGSCs would be a great tool for biological reproduction. Research on FGSCs opened a new direction in reproductive biotechnology to treat infertility and produce biological drugs supported in pre-menopausal syndrome in women. The applicability of FGSCs is enormous in the basic science of stem cell models for studying the development and maturation of oocytes, especially applications in treating human disease.

Keywords: ovarian stem cell, primordial germ cell, female germline stem cell, oocyte-like cell, regenerative medicine

1. Introduction

It is widely known that mammals begin their lives with a fixed number of oocyte-containing follicles, which do not increase but only degenerate after birth. This explains why fertility decreases with age, and the phenomenon of menopause in women is an indication that reserve resources of oocytes have been depleted over age. This differs from the male that can produce sperm throughout their life due to the presence of spermatogonia stem cells (SSCs), which could allow them to constantly proliferate and differentiate to maintain their persistent spermatogenesis, allowing male mammals to have a long-lasting reproductive age [1]. Since the 1950s, the dogma in reproductive biology has been widely accepted that primordial germ cells (PGCs)-derived oogonia in mammals cease proliferation and differentiate into primary oocytes shortly after birth, which will arrest in prophase of meiosis I until fertilization triggers the completion of meiosis [2]. In other words, shortly after birth, mammalian ovaries can differentiate to produce new oocytes to compensate for the consumption of ovulation. This explanation was well accepted due to the shorter gestational age of females compared to that of males. This theory of a fixed ovarian reserve had been the central principle in the field of reproduction.

The evolutionary and molecular processes of female reproductive aging have been highly debated. In the early 21st century, scientists at Harvard Medical University ignited the debate about the unexpected ability of mouse ovaries to regenerate immature oocytes after destruction [3]. Then, the report raised numerous questions by showing that these proliferative ovarian cells, termed female germline stem cells (FGSCs), could produce immature oocytes. Transplantation of these FGSCs into the ovaries of adult mice, was able to differentiate them into mature eggs that are able to ovulate, fertilize, and produce viable offspring [4]. This study opens a new direction in the study of stem cells in human ovaries. If we succeed in establishing the human FGSCs (hFGSCs), they will play a very significant role in reproductive medicine and the treatment of menopause symptoms in women. This would not only recover fertility in infertile women but also delay early menopause in women and treat pre-menopausal syndrome for women without hormone replacement therapy. Notably, the preservation of hFGSC can restore fertility and endocrine function for patients after cancer treatment. This is a huge challenge for scientists "Do hFGSCs exist in humans as they do in mice or not?". Recent studies from researchers supporting the existence of oogenesis in postnatal mammalian ovaries raised some questions, and opened a new avenue for the investigation of stem cells in human ovarian tissue. Stem cells are thought to have numerous uses in cell therapy. Chemotherapy, radiation, genetic induction, or hormonal stress can all result in ovarian failure [5]. Additionally, premature ovarian failure (POF), which affects 1% of young women, is a common cause of ovarian dysfunction before the age of 40 [6]. Furthermore, young women are also rendered sterile by some diseases causing oocyte loss, such as polycystic ovary syndrome [7]. However, very little progress has been made in solving this problem. With assisted reproductive technologies, the findings that FGSCs play a role in fertility provide the future application for clinical therapies [8]. Cryopreservation of ovarian cortical tissue has already been developed for female patients with cancer. Isolation and expansion of FGSCs from this tissue before or after cryopreservation may be helpful for new fertility applications [9]. Some progress has been made in addressing this issue through assisted reproductive technologies. It has been discovered that FGSCs play a role in fertility, offering the potential for future clinical applications. However, the question arises whether the egg cells are produced from hFGSCs or not. If hFGSCs exist in ovaries, why do women still have the phenomenon of menopause? The remaining challenge is to clearly elucidate the origin, roles, and capabilities of these cells, and to be able to use them for therapeutic applications. To this end, studies in mammalian models other than the mouse need to be done because several mechanisms of biological processes for oocytes in mice are different from those in humans. This review will present recent studies on the existence of germline stem cells (GSCs) in the mammalian ovary and summarize the current understanding of ovarian germline stem cells (OSCs) and FGSCs (**Figure 1**).

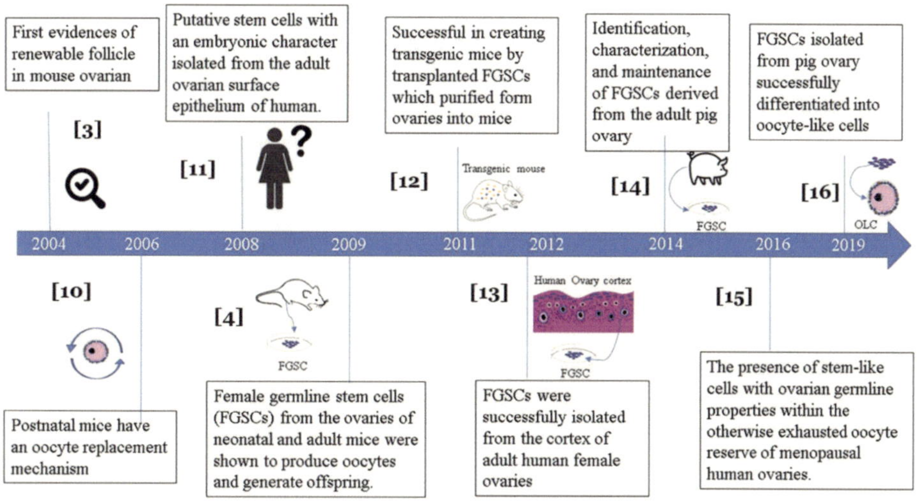

Figure 1.
The timeline of major discoveries in the history of female germline stem cells research.

2. The existence of germline stem cells in adult ovary

For several decades, it has been believed that females are born with a limited pool of oocytes and lose their capacity for oocyte renewal under perturbed conditions. The presence of mammalian FGSCshas been a highly debated area of reproductive technology since 2004, the Harvard University group suggested a source of cells for oogenesis during the reproductive period [3]. In this study, they discovered that the atresia follicle formation rate in mice was lower than the consumption rate of non-atretic follicles. Hence, they believed that there is a renewal of follicles in mice. Moreover, immunostaining showed the presence of proliferating germ cells by expression of germ cell markers. In their final set of experiments, wild-type mouse ovaries were grafted to transgenic GFP-positive mouse ovarian bursa to provide additional evidence for ongoing folliculogenesis in postnatal life. They found that GFP was positive after 3–4 weeks in the grafted ovarian tissue, leading to the conclusion that oogenesis continues in the postnatal ovary. This has sparked much debate about whether germline stem cells do exist in the postnatal ovary. In addition, it was reported that both immature and follicles were detected after transplanting the bone marrow-derived cells from the adult mice into the ovary of infertility mice [10]. Both reports opened many arguments on whether FGSCs existed in the postnatal ovary. Kerr's study in 2006, in which the number of ovarian follicles in mice were counted at different ages, revealed that the average number of follicles did not significantly decrease between 7 and 100 days after birth. Since postnatal mice mature sexually in around 8 weeks (56 days) and ovulation consumes a portion of the follicular pool, it is suggested that postnatal mice have an oocyte replacement mechanism [11]. The culture of cells attained from scrapings of the human ovarian surface epithelium (OSE) resulted in the formation of large oocyte-like cells (OLCs) expressing zona pellucida proteins [12], leading the authors to suggest that putative germ cells within the OSE of the postnatal ovary differentiate from mesenchymal progenitors in the ovarian tunica albuginea. In line with this possibility, small round (2–4 µm diameter) c-kit/stage-specific embryonic antigen (SSEA)-positive cells were isolated from human OSE cells. These cells expressed early PGC markers, including

OCT4 (POU5F1), NANOG, and SOX2 [13]. In 2009, the first successful result in isolating and purifying the FGSCs from the ovary of neonatal and adult mice was reported, by using a magnetic bead sorting technique against Vasa protein, a germline-specific marker [4]. In order to dissipate the doubts about the existence of putative FGSCs in postnatal mammalian ovaries, this group utilized another germline cell-specific protein, Fragilis, to isolate and purify putative FGSCs in postnatal mice. They successfully purified the cells using magnetic sorting techniques, which showed the same characteristics as FGSCs isolated by Vasa protein. Then, FGSCs were transplanted with different genes and subsequently implanted into the ovaries of infertile female mice to create transgenic animals after mating with normal male mice [14]. This research provided significant evidence to support the existence of germline stem cells in postnatal female mammals and opened a new direction in the study of stem cells in the human ovary. Suppose the FGSCs can be successfully isolated from humans, the benefit of this type of cells will play a key role in reproductive studying, medicine, and treatment for menopausal syndrome in women, and other relevant clinical applications. Infertile women can have the ability to give birth again. Besides, further investigation on FGSCs can open a new method for delaying early menopause in women and treating pre-menopause syndrome for women without using hormone replacement therapy.

In 2012, remarkably, FGSCs were able to be isolated from the cortex of adult human ovaries and differentiated into oocyte-like structure cells *in vitro*. Moreover, xenotransplantation of hFGSCs modified to express GFP into immunodeficient female mice resulted in the development of follicles harboring GFP-positive oocytes in the human ovarian cortex after 1–2 weeks [15]. Further, these isolated FGSCs formed into follicles containing oocytes when transplanted into the immunodeficiency mice and could be expanded for months and spontaneously generate 35–50 μm oocytes [16]. This discovery has opened a new direction in research on human FGSCs. Therefore, FGSCs can have an important role in the treatment of diseases caused by infertility females or in extending the period of menopause, as well as the application of stem cell therapy.

We have successfully established pig FGSC from ovarian tissue *in vitro* culture, and porcine PGCs-like Putative Stem Cells (PSCs) continue to maintain their germ stem cell identity *in vitro* and can differentiate into OLCs under appropriate culture conditions. Moreover, experimental evidence showed that PGCs-like PSCs are probably generated from Vasa-positive stem cells *in vitro*. Finally, we demonstrated the critical role of ovarian cell-derived regulatory factors and the proximal stem cell niche in the establishment of porcine PSCs [17]. In addition, many studies provided evidence available to support the existence and potential of putative germline stem cells in the adult mammalian ovary, such as bovine, monkey, and human [18–21]. A finding demonstrated the presence of stem-like cells with ovarian germline properties within the otherwise exhausted oocyte reserve of menopausal human ovaries. Using immunomagnetic enrichment based on membrane DDX4 expression followed by single-cell sorting under a dielectric field, large culture-derived OLCs expressing markers of mature and haploid oocytes were obtained from fertile women as well as menopausal women [22].

Besides, other groups failed to observe replenishment of the follicle pool by donor bone marrow-derived cells [23] or after chemical depletion [24]. They also failed to observe the generation of new follicles even after depletion with busulphan toxin, thus putting into question a need to explain the regeneration of follicle numbers, a finding further supported by mathematical modeling [25]. Lei and Spradling traced the numbers of follicles over time using tamoxifen-induced random labeling of cells. They argued that the follicle pool is highly stable with a half-life of 10–11 months, which would make the follicle pool at birth large enough to support ~500 ovulations required during the life of a mouse [26]. In a study to confirm the existence of FGSCs in postnatal mouse ovaries, transplantation of premeiotic female PGCs and companion pre-follicular cells into the ovaries of adult mice has been shown to be capable of supporting the formation of new follicles. However, the transplanted PGCs could only form follicles with their pre-follicular cells and vice versa [27]. Although the authors concluded that neo-oogenesis does not normally occur in the ovaries of adult mice, the results nonetheless provide an answer to the important question of whether adult ovaries can support neo-oogenesis from transplanted PGCs. Taken together, we suggest that germline stem cells themselves may not persist in postnatal and adult mammalian ovaries but that progenitor cells/small PSCs in the ovary may instead differentiate into germline stem cells under appropriate conditions [17]. Therefore, although experimental evidence supports the existence of cells with germline progenitor/stem cell characteristics in ovaries of various species, including humans, the existence of GSCs in postnatal ovaries remains ambiguous.

3. Location of female germline stem cells (FGSCs) in the ovary

In female mammalian species, during the embryonic stage, a subset of blastula cells can form PGCs by germ cell determination under some signal induction. Millions of germ cells are formed during embryogenesis. However, most of the germ cells degenerate after embryonic development. Most oogonia die in this period, while the remaining enter the first meiotic division. These latter cells, called primary oocytes, proceed to the first meiotic prophase after replicating their genomes. These primary oocytes are then arrested at this stage of development until the first menstrual cycle. Only a few numbers of oocytes periodically resume meiosis after puberty. Millions of germ cells are produced during embryonic development, but only hundreds of oocytes mature during a female's lifetime [28]. In a menstrual cycle, when the primary oocyte enters metaphase I, its nucleus (germinal vesicle) breaks down, and the metaphase spindle migrates to the periphery of the cells. In telophase, the chromosomes are evenly divided, but one of the two daughter cells retains almost all of its cytoplasmic components, while the other cell has almost no cytoplasm. The smaller and larger cells are called the first polar body and the secondary oocyte, respectively. Moreover, the same phenomenon takes place during the second division of meiosis. Nearly all the cytoplasm is retained by the mature egg (the ovum), and a second polar body receives little more than a haploid nucleus. Thus, the purpose of oogenic meiosis is to conserve the volume of cytoplasm in a single oocyte. Ovulation begins shortly thereafter, in which the follicle ruptures and the secondary oocyte is released into the uterine tube, yet the second meiotic division has not occurred yet. Meiosis of a secondary oocyte is completed only if a sperm succeeds in penetrating its barriers. Meiosis II then resumes, producing one haploid ovum that, at the instant of fertilization by a (haploid) sperm, becomes the first diploid cell of the new offspring (a zygote) [29, 30].

The general idea in reproductive biology is that FGSCsdifferentiate into primordial oocytes through fetal development and that oogenesis begins with a pool of primordial follicles, which is the case in the majority of animals. Several studies have reported the presence of FGSCs in the ovary. Parte and colleagues in 2011 reported the discovery of very small pluripotent stem-like cells deposited in the OSE of adult rabbits, sheep, monkeys, and menopausal humans [18]. Two different populations of putative stem cells (PSCs) of varying sizes were found in scraped OSE. While the larger 4–7 μm cells with cytoplasmic localization of Oct-4 and little expression of SSEA-4 were likely the tissue-committed progenitor stem cells, the smaller

1–3 μm very small embryonic-like PSCs were pluripotent in nature. To demonstrate characteristics of these cells derived from OSE, the PSCs underwent spontaneous differentiation, c-Kit, DAZL, GDF-9, VASA, and ZP4 germ cell markers were used to immunolocalize in oocyte-like structures. Mammalian ovaries include a unique population of extremely small embryonic-like PSCs and tissue-committed progenitor stem cells that have the ability to develop into oocyte-like structures *in vitro*, contradicting the conventional belief that OSE is a bipotent source of oocytes and granulosa cells.

In 2014, we indicated that the ovary contains a considerable number of undifferentiated cells with stem cell characteristics. These might remain in the adult ovary and cannot proliferate normally, but they can undergo proliferation and differentiate into OLCs under appropriate conditions. PSCs were found to comprise a heterogeneous population based on c-kit expression, cell size, and expressed stem and germ cell markers. Analysis of PSCs molecular progression during establishment showed that these cells undergo cytoplasmic-to-nuclear translocation of Oct4 in a manner reminiscent of gonadal PGCs. Flow cytometry analysis revealed abundant PSCs proliferation after isolation and culture for 1 week. Of these, 4.65% of the cells were positive for the germ cell marker Vasa, and some were also positive for additional germ and stem cell markers, such as Fragilis, Thy-1, SSEA4, and c-kit. At this time, two populations of PSCs were observed: one with a cell diameter of 5–7 μm and one with a cell diameter of 10–12 μm. The cells became identical in size after 2 weeks in culture, at 10–12 μm, with an increasing percentage of cells positive for germ and stem cell markers [17]. About 2.8% of all mouse testicular cells were c-kit positive [31] and had the capacity to become multipotent germline stem cells, whereas c-kit-negative cells go on to become SSCs [32]. We similarly observed two distinct subsets of cells (c-kit positive versus c-kit negative) within the PSCs population. This finding was strengthened by immunofluorescence analysis showing that, after 1 month in culture, most of the PSCs expressed high levels of the reprogramming factor Oct4. In contrast, only 22% of the PSCs expressed high levels of c-kit [17].

4. FGSCs aging and stem cell niche in the ovary

While much evidence support the existence of OSCs, it raises the question that ovarian have follicle reserve. Why do they not appear to contribute to postnatal follicle formation, and why does the phenomenon of menopause occur only in females? The researchers believe that the FGSCs aging directly determines ovarian aging.

The stem cell niche is the key to elucidating the entire mechanism of stem cell senescence; Schofield proposed the hypothesis in 1978 that the components surrounding stem cells act as a microenvironment that promotes their growth and protects them from external damage [33]. Other than FGSCs, stem cell niche can be found in almost any stem cell, such as intestinal, myocardial, neural, and hematopoietic stem cells. Stem cell niche can support the growth of stem cell, and disturbance of these niches can cause stem cell damage and eventually leads to certain diseases. Ovarian stem cell niche aid FGSCs to continually proliferate to differentiate into postnatal follicles and oocytes, by regulating to divide into new stem cells and differentiate into germ cells. FGSCs niche is extensively studied in Drosophila Melanogaster. Although the structure and the function of FGSCs niche in mammals have not been fully understood, the component of the niche is believed to be similar to that of drosophila, thus predicting that it may at least be composed of follicular membrane-stromal cells, granulosa cells, extracellular matrix, blood vessels, immune system-related cells, and cytokines. It was suggested that damage to the stem cell niche is a major cause of the ovarian recession and is more closely related to aging in the ovary than the stem cells themselves. Thus, factors that damage the stem cell niche may have a critical impact on ovarian regression.

Factors that lead to stem cell decay are nutritional and energy deficiencies in the stem cell niche. Insufficiency of energy is caused by mitochondrial depletion associated with aging women, thus only supplies limited amounts of ATP. As a result, the risk of birth defects and infertility is increased due to the reduced energy of FGSCs provided by stem cell niche. The immune system has a critical role in the maintenance of OSC niche to support FGSCs. The perivascular compartment of the stem cell niche forms a bridge to connect the niche and both cellular and humoral aspects of immunity. Cellular immunity that provides support in stem cell niche are monocytes, macrophages, and T cells. Cytokines and immunoglobulins are the humoral aspects of immunity that help stem cell niche maintenance. These are essential for the derivation of OSCs into new germline cells. Weakened immunity due to aging is causing difficulty in the maintenance of the stem cell niche and resulting in ovarian recession.

Dysfunctional gonads can be regenerated by transplanting niche cells of germline stem cells (mostly Sertoli cells or mesenchymal cells). Thus, it could be helpful as first-line therapy to permanently restore gonadal function in POF and cancer patients. Thus, it is clinically more important to reestablish the niche of the FGCSs than to inhibit the aging of the FGCSs themselves in order to delay ovarian aging. The study of the FGSC niche is relatively new, and there are still several issues that we still need to know about. To be able to use for the application of the FGSC niche in clinical practice, these should be addressed and clarified. Understanding the mechanism and application of stem cell niche in FGCS can be crucial in regenerative medicine. Although the result of transplanting FGSCs into infertile female ovaries remains controversial, this has the potential to regenerate.

Researchers have put out solutions to answer this question. One of the most widely accepted theories is that stem cell functions would decline with age, which would result in a loss of renewal capacity [34]. They explained that the aging of FGSC is related to the aging of the stem cell niche. The niche is a specialized microenvironment that gives stem cells specialized cues in the form of adhesion molecules, differentiation and self-renewal-regulating signals, spatial organization, and metabolic support to stem cells. As a result, the niche is crucial for controlling stem cells' fundamental processes and protecting them from cell damage and toxins. Changes in the niche may result in to decline in stem cell function. To demonstrate this hypothesis, Bukovsky observed that the niche of FGSCs formed during early embryonic development consists of nonspecific ovarian monocyte-derived cells (MDCs), T cells, and vascular endothelial cells, In contrast to the nests of adult ovarian germinal stem cells, which are made up of primary CD14 + MDCs, activated HLA-DR + MDCs, and T cells [12]. Furthermore, when the ovarian tissues of older mice were transplanted into young mice, the young mice's ovarian tissues were found to have fewer follicles and no mature follicles [35].

Another hypothesis proposed by researchers suggests that ovarian function may decline by systemic aging-related signals despite the presence of oogonial stem cells [36]. For example, progressive loss of ovarian estrogen (E2) production drives reproductive aging and menopause [37]. To demonstrate the hypothesis, they have shown that mouse OSCs express E2 receptor-α (Erα) by RT-PCR and western blot analysis. To test for potential interactions of E2-activated Erα with meiotic regulatory pathways in OSCs, chromatin immunoprecipitation (ChIP)-PCR assays were applied to assess the Stra8 promoter. Results showed that Erα occupied a consensus ER response element (ERE) in the Stra8 promoter. Moreover, E2 treatment increased the number of GFP-positive cells. Thus, OSCs have differentiated in response to E2. In reverse, Era-deficient shows a loss of Stra8 expression and oocyte numbers. This study will provide more information on how changes in ovarian estradiol production with aging in women are related to age-related ovarian dysfunction and reproductive aging.

Moreover, oxidative stress also has an essential role in the aging of FGSCs. ROS is a chemically reactive oxygen atom or group of atoms produced during cellular metabolism. The development of ovarian granulosa cells was inhibited by ROS, which also damaged mitochondria and lowered the production of the anti-oxidative enzyme. Additionally, it might diminish ovarian function and trigger an inflammatory response, impairing fertility [38]. Resveratrol (RES) is a naturally occurring substance with many pharmacological roles, including antioxidant, anti-inflammatory, immune-regulating, cell-protective, anti-tumor, and anti-apoptotic effects. RES therapy can be employed to enhance ovarian follicle function by lowering TNF-levels, which was validated by lowering LH levels and the ratio of LH/FSH, two markers of ovarian function. The scientists discovered that RES significantly increased body weight, ovarian index, follicle quantity, and decreased follicular atresia in POF mice.

5. Isolation, maintenance, and characterization of FGSCs

After debating the existence of germline stem cells in ovaries, proponents of their existence continue to question what types of FGSCs exist in ovaries and their characteristics. FGSCs' sizes vary considerably, ranging from 2 to 8 μm. Tilly's group focused on bigger-sized (5–8 μm) OSCs [16], whereas smaller (2–4 μm) pluripotent stem cells, very small embryonic-like stem cells (VSELs) were found in ovary surface epithelium (OSE) [39]. However, OSCs of both sizes express germ-line markers and differentiate into OLCs. Thus, they concluded that there are two different populations of stem cells, the small-sized, pluripotent VSELs, and the bigger OSCs. VSELs are pluripotent stem cells produced from epiblast that are identical to PGCs and persist in small numbers in adult gonads [39]. In contrast, OSCs are tissue-specific progenitors that are bigger and have different gene expressions from pluripotent VSELs. It has been reported that VSELs are the most primitive population of quiescent SCs found in adult tissues compared to OSCs, which quickly divide and produce germ cell nests before differentiating into oocytes. FSH receptor (FSHR) expression was observed on both very small embryonic-like stem cells (VSELs) and ovarian stem cells (OSCs) by immune-localization and immunophenotyping studies. FSH treatment increases germ cell clusters and stimulates stem cells to undergo proliferation and clonal expansion to form germ cell nests. This was further confirmed by the differential expression of OCT-4 in VSELs and NUMB in OSCs. Immunohistochemical expression of OCT-4, proliferation, and FSHR were noted on stem cells located in the OSE of ovarian sections of sheep. Therefore, the establishment of FGSCs is significant for many applications.

Using adult porcine ovaries to isolate, identify, and characterize FGSCs to elucidate their origin and then examine the capability of these cells to proliferate, grow, and differentiate. These cells were heterogeneous, depending on both c-kit expression and cell size, and also expressed stem cell and germline markers. Importantly, we clearly demonstrated that cells with characteristics of early PGCs are present in the adult porcine ovaries. Once FGSCs were established, they could be expanded *in vitro* for months without the loss of identifying markers and proliferative potential. Under appropriate conditions, FGSCs can be differentiated into OLCs. These have the potential to make new oocytes, support ovarian function and fertility, and may support therapeutic strategies in humans [17]. The methods used to isolate, maintain, and characterize FGSCs from each of the research teams studied to date are summarized in **Figure 2**.

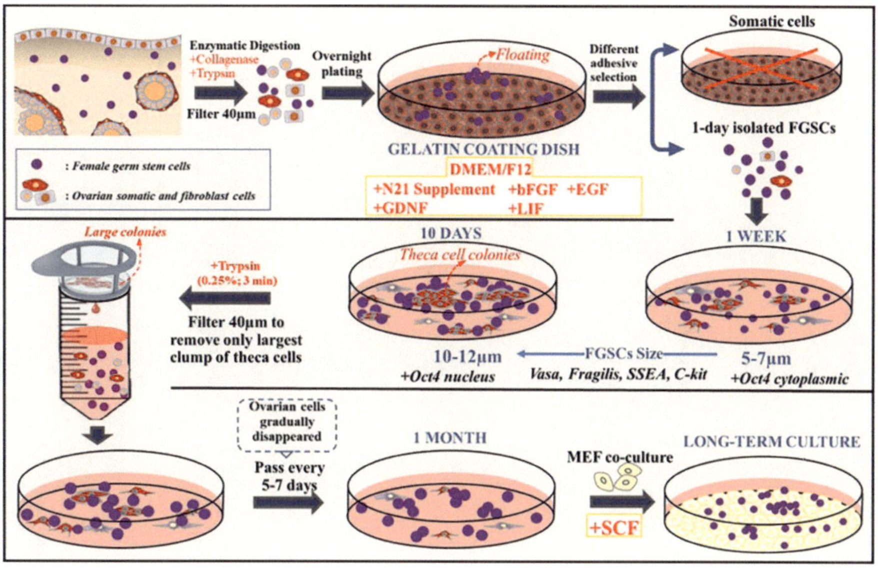

Figure 2.
The female germline stem cells (FGSCs) were successfully isolated from porcine ovarian tissue by enzymatic digestion and different adhesive selection, then cultured in vitro, in DMEM/F-12 and N21 free-serum supplemented. Two populations of FGSCs, one with a cell diameter of 5-7μm and the other with a cell diameter of 10-12μm, were observed with spherical morphology and expressed specific germline characteristics (Vasa, Stella, Oct4, c-kit). The cells became identical at 10-12 μm after ten days in culture, Oct4 expression was reduced in the cytoplasm and augmented in the nucleus, forming groups of cells clustered around theca cell colonies. The large clumps of ovarian cells will be removed during cell passages by the filter of 40μm, and FGSCs were maintained for one month. After one month, following the gradual death of the ovarian cells, the FGSCs were transferred onto mitomycin C-treated MEF feeder layers for long-term culture with the addition of stem cell factor.

The formation of germ cell nests has provided evidence in support of neo-oogenesis from the stem cells [40]. VSELs may endure radiotherapy and chemotherapy and preserve lifelong homeostasis. Due to the loss of function brought on by disturbing ecology, its impairment results in host aging, and the existence of overlapping pluripotency markers raises the possibility that it may also be linked to epithelial ovarian cancer. Many methods for isolation and culture of FGSC have been studied in recent years. Briefly, FGSCs can be isolated by enzymatic digestions (collagenase and trypsin) and purified with either MVH/Fragilis-magnetic bead sorting or/with Flow Cytometer-based SSEA-4, FRAGILIS, DDX4 (VASA) or long-term passage methods to eliminate ovarian somatic cells. Isolated FGSCs are able to be cultured in both absence or presence of feeder cell layers, MEF, STO, or granulose cells. The gelatin-coated surface is typically used in the culture of non-sorting cells. The concentration of FBS varies from 5 to 15% and is used in maintenance and differentiation medium. Besides that, growth factors GDNF, EGF, and bFGF are important for the expansion of FGSCs; LIF is used for FGSCs maintenance, while bone morphogenetic protein and retinoic acid can significantly induce germ cell differentiation [41, 42]. In addition, SCF was also suggested to increase the number of Fragilis- and MVH-positive cells and enhance colony formation efficiency [17, 43]. Other supplements such as antibiotic and antimycotic, pyruvate, non-essential amino acids, antioxidants, insulin, transferrin, and selenium are added to increase the viability and proliferation of ovarian cells. Cultured cells normally express general pluripotent, germ cell, and oocyte markers. While Oct4 is always strongly expressed in cultured OSCs, others, such as SOX2 and Nanog, are still in conflict. VASA,

Species	Isolation	Condition	Result	References
Mouse	MACS- based MVH	MEM α, 10% FBS, bFGF, EGF, GDNF, LIF Co-culture: STO	FGSCs/ Offspring.	[4]
	FACS-based DDX4	MEM α, 10% FBS, N2 supplement, bFGF, EGF, GDNF, LIF Co-culture: MEF	Oogonial stem cells	[15]
	FACS- and MACS-based DDX4	MEM α, 10% FBS, N2 supplement, bFGF, EGF, GDNF, LIF Co-culture: MEF	Oogonial stem cells	[16]
	Preplate culture	DMEM, 15% FBS, LIF, Co-culture: MEF	Oogonial stem cells	[45]
	FACS-based DDX4	MEM α, 10% FBS, bFGF, EGF, GDNF, LIF Co-culture: STO	FGSCs/ Offspring	[14]
	MACS-based Fragilis and MVH	MEM α, 10% FBS, LIF, SCF Co-culture: MEF	Putative stem cells	[43]
	MACS- based MVH	MEM α, 15% FBS, N2 supplement, LIF, GDNF, bFGF Co-culture: STO	FGSCs	[46]
	Whole ovarian cells culture	DMEM/F12, 5% FBS, LIF	Ovarian stem cells	[47]
	Repetitive different adhesion selection	MEM α, 10% FBS, N2 supplement, LIF, GDNF, bFGF, EGF	FGSCs	[37]
Rat	MACS-based Fragilis	MEM α, 10% FBS, bFGF, LIF, EGF, GDNF Co-culture: STO	FGSCs/Transgenic offspring	[48]
Sheep	OSE cells scraping	DMEM/F12, 10% FBS, FSH	Ovarian stem cells	[40]
Porcine	OSE cells scraping	DMEM/F12, 20% knock-out serum, LIF, EGF	FGSCs	[49]
	Differential adhesion selection	DMEM/F12, B27, bFGF, EGF, GDNF, LIF, SCF Co-culture: MEF	FGSCs	[17]
	Differential adhesive selection	DMEM/F12, N21 supplement, bFGF, EGF, GDNF, LIF	FGSCs	[50]
Bovine	Ovarian tissue culture	MEM α, bFGF, FSH	Ovarian stem cells	[20]
Monkey	Colony pickup	DMEM/F12, 10% FBS, LIF Co-culture: MEF	FGSCs	[19]
Human	FACS-based DDX4	MEM α, 10% FBS, N2 supplement, bFGF, EGF, GDNF, LIF	Oogonial stem cells	[15]
	FACS- and MACS-based DDX4	MEM α, 10% FBS, N2 supplement, bFGF, EGF, GDNF, LIF	Oogonial stem cells	[16]

bFGF: basic fibroblast growth factor; EGF: epidermal growth factor; FBS: fetal bovine serum; FGSCs: female germline stem cells; GDNF: glial cell line-derived neurotrophic factor; LIF: leukemia inhibitory factor; MEF: mouse embryonic fibroblast; OSE: Ovarian surface epithelium; SCF: stem cell factor; STO: SIM embryonic fibroblasts.

Table 1.
Summary of isolation and culture of FGSCs in mammalian.

PRDM1, FRAGILIS, and DAZL are considered germline-associated genes. Oocyte-specific markers, including ZP1–3, GDF9, NOBOX, and SCP3, are rarely detected or weakly expressed [15, 44]. A summary of these methods is shown in **Table 1**.

6. FGSCs share characteristics with epiblast-derived PGCs

We have successfully established FGSCs from porcine ovaries and demonstrated that these FGSCs derived from PGCs have been retained and become inactivated. These PGCs were thought to exist only during the fetal period, and all transformed into oocytes before the individual was born [17]. In addition, a study has also shown similar patterns of gene expression profiles between FGSCs and PGCs [51].

In the current model, PGCs are derived from a small number of epiblast cells and are identified before differentiation into different germ layers begins. PGCs then undergo a complex migration process, passing through the abdominal cavity, along the developing hindgut, and finally into the dorsal mesentery, where gonads develop. Despite the apparent importance of these early developmental processes for future fertility, little is understood. It was suggested that PGCs temporarily reside in an "Allantoic Core Domain" (ACD), which they propose has similar functions to the Spemann organizer, consisting of a stem cell pool that extends the body axis in a posterior direction—contributing not only to the germ cell lineage but also the three germ layers. This creates a solid interface between the future umbilical cord and the developing embryo [52]. The stem cells in the ACD express Oct4, Blimp1, Stella, and Fragilis—markers thought to be specific for PGCs but appear to contribute also to other tissues [53]. These observations and the fact that hematopoietic stem cells also migrate from the proximal epiblast to the embryonic aorta-gonad-mesonephros during the same development period. Implies that it is theoretically possible that there may be "intermixing" or that there is a common precursor pool for PGCs and a subpopulation of bone marrow stem cells. Cell lineage tree analysis based on somatic mutations accumulated in microsatellites has shown that oocytes form a completely distinct cluster from other cell populations, suggesting no mixing of germline progenitor pools with other cell types but the bone marrow stem cells [54]. It is conceivable that very rare subpopulations of these cells would be missed in this type of analysis. Their result also shows that aging and unilateral ovariectomy increase the number of mitotic divisions of oocytes. This may be explained by the recruitment of oocytes in the order in which they first differentiate during development. However, it is also consistent with the idea that oocytes are continuously produced from circulating stem cells. Many researchers have suggested that if GSCs exist, they are most likely derived from normal developmental precursors of oocytes, that is, PGCs or oogonia (which have not yet differentiated into oocytes and can undergo mitosis) [55, 56]. The close relationship of PGCs to pluripotent cells is evidenced by the fact that they can be returned to a pluripotent phenotype called embryonic germ cells *in vitro* without genetic manipulation after isolation from the embryo [57, 58].

In 2014, we investigated the developmental origin of porcine PSCs. In normal development, c-kit, SSEA1, and SSEA4 are expressed by the majority of pre-gonadal PGCs and are progressively downregulated when PGCs enter meiosis in the embryonic ovary [59]. In contrast, Vasa protein is detectable only when PGCs enter the gonadal ridges and remains elevated in human fetal and postnatal oocytes [60]. VASA (DDX4)-negative VSEL stem cells (2–4 μm) isolated from the human OSE express genes typical of ESCs, such as NANOG and SOX2, thereby indicating their undifferentiated status. After culture for 3 weeks under differentiation conditions, VASA-negative cells are transformed into OLCs expressing VASA and ZP2, a marker for oocytes. In the present study, small Vasa-positive porcine PSCs (5–7 μm in diameter) began to reduce their expression of Nanog, Sox2, and Rex1 after 1 week in culture, indicating their transformation to a differentiating status. Previous investigations showed that Vasa-positive VSEL stem cells isolated from adult organs express several characteristic markers of early PGCs, including fetal-type alkaline phosphatase, Oct4, SSEA-1, CXCR4, Stella, Fragilis, Nobox, and Hdac6. Since the porcine PSCs described herein similarly express a number of typical, early PGC markers, these findings might indicate a close association of PSCs with Vasa-positive VSELs and epiblast-derived PGCs [17].

7. Self-renewal capacity of FGSC

The studies have focused on the development of FGSCs into oocytes both *in vivo* and *in vitro*. Transplantation of the GFP-FGSCs back into ovaries leads to the generation of fertilization-competent eggs that produce embryos and offspring [4]. Furthermore, GFP_FGSCs have generated GFP-positive OLCs enclosed in host somatic cells, as characterized by morphology and expression of oocyte-specific markers after injection into adult human ovarian cortical tissue and transplantation into an immune-deficient mouse [61].

Maintaining and extending FGSCs *in vitro* is a crucial step to obtaining fully active germ cells. The effects of various supplements on FGSCs proliferation have been evaluated to optimize conditions for *in vitro* culture of FGSCs. Ovarian tissue plays an essential role in maintaining the properties of FGSCs *in vitro* culture [62]. Follicle-stimulating hormone (FSH) and basic fibroblast growth factor (bFGF) were considered to induce the proliferation of FGSCs and retain their potential for spontaneous differentiation into oocyte-like structures in extended cultures [39]. Preliminary studies by our group have successfully established porcine FGSCs and maintained them for more than 6 months without loss of proliferative potential. Expression of identified germline markers was also maintained. The estimated cell doubling time was 48–72 hours. Subsequently, long-term culture increased the number of differentiated cells among FGSCs, but many FGSCs that were positive for both BrdU and Oct4 or Vasa retained high proliferative potential [17].

GSK3 inhibitors are involved in this process by promoting the proliferation of FGSCs by activating both β-CATENIN and E-CADHERIN [63]. FGSCs that exhibit pluripotency are highly capable of self-renewal, which involves a number of genes and signaling pathways. CADHERIN-22 (CDH22), a member of the cadherin superfamily, functions in FGSC maintenance and self-renewal through its interaction with JAK–STAT and β-CATENIN. The knockdown of CDH22 strongly affected FGSC proliferation by its inhibition and triggering apoptosis, decreased phosphorylation levels of p-JAK2 and p-STAT3, and led to the downregulation of β-catenin [64, 65]. CDH22 also interacts with PI3K to phosphorylate AKT3 and increase the expression of N-myc and cyclin family of FGSCs to promote self-renewal [65].

A study in 2012 evaluated the effects of leukemia inhibitory factor (LIF) and other growth factors, epidermal growth factor (EGF), beg, and glia-derived neurotrophic factor (GDNF), on the proliferation and colony formation of FGSCs. Results showed that these growth factors promote FGSCs proliferation through activation and upregulation of β-CATENIN and E-CADHERIN, and LIF has a significant positive effect on cell colony number [63]. Activation of the GDNF signaling pathway is mediated by GFRα1 (GDNF receptor), which is related to circGFRα1 by leading to the expression of GFRα1 [66]. In addition, YAP1, an effector of the Hippo signaling pathway, regulates FGSCs proliferation and differentiation *in vitro* and ovarian function [67]. The Hedgehog (Hh) signaling pathway plays a vital role in the fertility of FGSCs. Blocking the Hh pathway by GANT61 depletes ovarian germ cells and FGSCs [68].

Furthermore, gene expression analysis inferred increased expression of proliferation-related genes c-Myc and Cyclin A in the OSE and cortical cells, while expression of the differentiation marker Zp3 was significantly decreased. Rapamycin inhibits the activation of primordial follicles, promotes FGSCs proliferation, and inhibits their differentiation, thus providing a new prospect for delaying ovarian senescence. Furthermore, the novel administration of Daidzein to FGSCs promoted the survival and proliferation of FGSCs by activating the Akt signaling pathway through Type C lectin domain family 11 member a, which functions as a growth factor [66].

8. Differentiation of FGSCs

Stem cell differentiation is critically dependent on the ability to differentiate stem cells into a specific cell type with a highly efficient and scalable system. Recent evidence has demonstrated that the differentiation signals are strongly modified by adhesive and mechanical factors. Furthermore, an environment that mimics the microenvironment in tissues is desired to stimulate stem cell potential and differentiation. Mimicking the cellular microenvironment *in vitro* is increasingly influential in guiding stem cell proliferation and differentiation [69]. Studies claim that cells from both menopausal and non-menopausal women may produce *in vitro* oocyte-like cells (OLCs-large spherical cells). This suggests that neo-oogenesis may take place during ovarian senescence. Tilly and colleagues discovered that OSCs developed *in vitro* to produce OLCs with gradual expansion up to 30–50 μm in diameter. These cells have expressed terminal markers such as zona pellucida (ZP) glycoproteins, GDF–9, NOBOX, YBX2, and SYCP3. In addition, they have examined that OLCs have the haploid karyotype [15]. OLCs generated by regulatory factors or spontaneous development lose expression of developmental pluripotency-associated genes, resulting in strong expression of oocyte makers and maintenance of germline markers. However, although many groups have been studied for the establishment of FGSCs, the differentiation potential of FGSCs in mammalian ovaries remains a controversial issue among germline biologists and stem cell researchers. To date, no mammals other than mice and rats have successfully produced offspring from FGSCs [14, 48]. Whether FGSCs can undergo growth, maturation, and fertilization to become functional oocytes is one of the crucial questions for us.

While most researchers have focused on the role of media (cytokines and growth factors) in regulating FGSCs differentiation into OLCs, the ideal culture system condition has yet to be established. We have investigated the effect of culture medium on FGSCs and further studied the effect of different culture systems (gelatin-coated dish, MEF feeder layer, co-culture with granulosa cells, and co-culture with MEF cells) on isolated FGSCs differentiation into OLCs. Co-culture of stem cells with a somatic cell population has been investigated as an alternative growth factor to induce stem cell differentiation [70]. This system provides growth factors and overcomes the requirement for exogenous growth factors to promote stem cell differentiation [71]. Somatic cells have been demonstrated to support oocyte development through cell-to-cell communication. This plays a vital role in oocyte growth and functioning via the transport of metabolites [72, 73]. Granulosa cells are one of the most important cell types which support oocyte development in the follicle. It was also proved that granulosa cells enhance oocyte development competence for *in vitro* culture [50, 74]. Another important cell type, MEF feeder cells, can produce and secrete growth factors and cytokines to provide an environment for stem cell migration, differentiation, and proliferation [75]. Hence, co-culture systems between FGSCs and somatic cells should be studied to improve the quality of OLCs.

2D culture system (monolayer) has been used as a conventional method for stem cell culture and differentiation. Recently, a new advanced 3D low attachment system has been developed. In this system, the primary cells exhibit a higher level of a specific function for a more extended period *in vitro* in 3D culture compared to monolayer culture. Furthermore, the 3D low attachment culture enhances the differentiation and stabilizes the functions of stem cells. [76]. During oocyte culture and development, it has been proved that improvement in ovarian culture systems increased the survival and growth of preantral follicles after the long-term culture period [77]. The majority of work on *in vitro* oocyte culture was undertaken using a conventional 2D culture system [72]. In recent years, further technical advancement has emerged form of 3D culture, which improved oocyte quality and development competence [78].

Our group also has successfully isolated FGSCs from adult pig ovaries and differentiated them into OLCs [50]. FGSCs were passed to the differentiation medium. Firstly, the old culture medium was gently removed from the tissue culture plate. Cells were then washed three times with PBS solution. After removing the PBS solution, 0.25% Trypsin-EDTA was added to the dish for 3–5 minutes. The dish was shaken slightly to separate FGSCs from the bottom of the surface. 10% FBS was added to stop trypsin action. The cell suspension was then divided into new tubes and centrifuged at 1000 rpm for 5 minutes twice using DMEM solution (Sigma). The supernatant was taken out, and

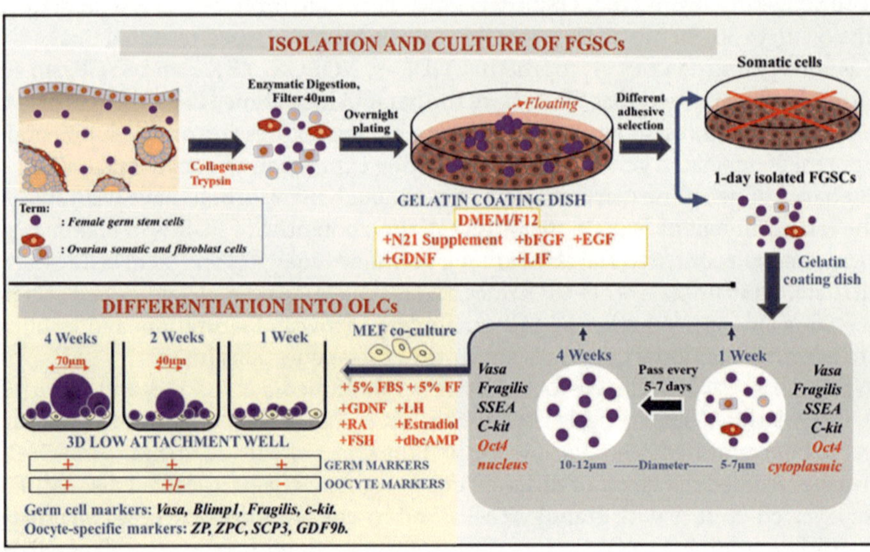

Figure 3.

The female germline stem cells (FGSCs) were isolated from porcine ovarian tissue and cultured in vitro, in DMEM/F-12, and N21 free-serum supplemented. These cells possessed spherical morphology and expressed specific germline characteristics (Vasa, Stella, Oct4, c-kit). For in vitro differentiation induction, using FGSCs 1-week after isolation, where ovarian somatic cells remained in the culture, or using 1-month cultured cells with less or no ovarian somatic cells. Co-culturing the isolated FGSCs with MEF cells under three-dimensional (3D) cell cultures supplemented with follicle fluid. After 1-month in differentiation culture, OLCs could reach about 70 μm in diameter, with a large number of surrounding somatic cells. OLCs expressed germ cell-specific markers (Vasa, Blimp1, Fragilis, C-kit) and oocyte-specific markers (ZP, ZPC, SCP3, GDF9b), contained large nuclei, about 25-30μm, with filamentous chromatin, similar to the oocyte.

the pellet was resuspended with a differentiation medium. This study evaluated four differentiation conditions to differentiate FGSCs into OLCs *in vitro* (**Figure 3**).

9. Application of FGSCs

Preservation of fertility is one of the most important qualities of life issues fertility preservation (FP) for young women with threatening premature ovarian insufficiency, especially for young cancer survivors who have not completed their family upon a cancer diagnosis. Among the currently available options for FP, especially in young patients, is cryopreservation of ovarian cortical tissue containing primordial follicles followed by autotransplantation [79]. Although ovarian tissue cryopreservation (OTC) currently represents an experimental approach, it offers not only FP but a restoration of endocrine function for women with cancer prior to undergoing gonadotoxic treatments. Re-implantation of OTC is currently the only option to use stored tissue, but in many cases, this procedure carries the potential risk of reintroducing the malignancy. However, in cancer patients, especially women with leukemia, re-transplantation is not an option due to the presence of malignant cells in the ovaries [80, 81].

FGSCs have the great clinical potential to be one of the options for the treatment in regenerative medicine for restoring declined female reproductive function caused by ovarian aging and perimenopausal-related diseases and preservation of fertility for patients with post-gonadotoxic therapy for ovarian cancer. In the former case, activation of FGSCs may restore ovarian function through their self-renewal and ability for committed differentiation into oocytes. The ovary is one of the most important female organs and reflects physiological signs of aging. Delay of ovarian age may avoid the negative consequences of menopause on one's health and its climacteric symptoms. It can also be applied to the fertility of women with POF to prolong their lives. Regarding the latter, FGSCs can be isolated and preserved for future use by cryopreservation after the biopsy of the ovarian cortex from the patient. Then, it might be able to use for *in vitro* fertilization by starting *in vitro* maturation to mature oocytes or injected back into the patient's ovaries to undergo neo-folliculogenesis. These will offer several advantages. First, the collection of ovarian cortex samples does not create the need to delay life-saving treatment, contrary to ovarian superovulation regimens. Additionally, more new follicles and oocytes may be obtained from FGSCs than from cryopreserved tissue or ovarian stimulation. However, the actual clinical application has not yet been achieved, due to technical and regulatory issues. Thus, research on FGSCs for therapeutic application has become an important topic.

Nowadays, OTCs and re-transplantation is a viable methods to preserve fertility in cancer patients. Therefore, most research focuses on technical aspects of OTCs, including follicle survival from freezing/thawing and fragment size, and duration of ovarian function after re-transplantation [82–84]. Study on *in vitro* culture for human OTCs is limited by the extreme difficulty of this technique and the unavailability of human tissue. Despite the undeniable advancements strengthening the protocols currently used for OTCs in domestic animals and endangered species has been achieved, this technique is still considered to be experimental for livestock such as swine [85], caprine [86], ovine [87], bovine [88], equine [89]. Using OTCs to establish FGSCs and differentiate into functional oocytes will answer the question about the possibility of creating gametes cells from adult mammalian ovaries. Gamete cells are ready for insemination with sperm to form embryos. This study will contribute significantly to the study of biological processes in human eggs in infertility treatment.

Moreover, FGSCs have great applicability in the basic science of stem cell models to study oocyte development and maturation, especially for treating human disease. Besides, FGSC is also very important in the production of transgenic animals. Transgenic animals are the animals with modified genome. A foreign gene is inserted into the genome of the animal to alter its DNA. This method is done to improve the genetic traits of the target animal. Until now, people have created many products of this type, and many products are suitable for use as food or medicine. Zhang and colleagues established FGSCs in mice and transplanted various genes [14]. FGSCs were transferred into the ovaries of infertile female mice. The result was to create transgenic mice after coordination with normal male mice. Scientists have suggested that the ability to produce transgenic animals in this method would be a great tool for biological reproduction in the future. Producing transgenic livestock can significantly improve human health, enhance nutrition, protect the environment, increase animal welfare, and decrease livestock disease. Especially, the creation of transgenic animals with biotechnology-based pharmaceuticals to produce precious protein for human or animal organ replacements, such as miniature pigs, due to the size of their organs being similar to humans.

In order to control these factors, a suitable culture system must be designed and optimized. From the basic research on the ability to generate eggs from FGSCs, applying this method to produce transgenic animals would be carried out efficiently. A study showed that FGSCs were established in mice and transplanted different genes into them. Then these FGSCs were transferred into the mouse ovaries of infertile females. The result was to create transgenic mice after coordination with normal male mice [14]. Recently, scientists have proven that FGSCs were a useful tool for the genetic manipulation of animals by creating transgenic rats [48]. Moreover, a study reported the success to restore ovarian function that suffers from cancer chemotherapy treatment and eventually produces offspring for the first time by transplanting the FGSCs [90]. Therefore, FGSCs played an essential role in the treatment of diseases caused by infertility females or in extending the period of menopause, as well as the application of stem cell therapy. In addition, FGSCs may play a significant role in treating diseases caused by infertile females or in extending the period of menopause, as well as the application of stem cell therapy.

10. Summary

FGSCs have a vital role in the treatment of diseases caused by infertility of females or in extending the period of menopause, as well as the application of stem cell therapy. Research on FGSCs opened up a new direction in reproductive biotechnology to treat infertility and produce biological drugs supported in pre-menopausal syndrome in women.

Our results support the theory that the ovary contains a small number of undifferentiated cells with stem cell characteristics. They may remain in the postnatal and adult ovary, but they are generally unable to proliferate due to inhibitory factors in the ovary. Under appropriate conditions, however, they can proliferate, differentiate into OLCs, and self-renewal of FGSCs. The presence of such FGSCs in mammalian ovaries and the depletion of ovarian reserve as the female reproductive system ages leads to the hypothesis that such "neo-oogenesis" was present in ancestors and is still present in insects, some fish, and mollusks. Nevertheless, it has been lost in terrestrial vertebrates during evolution. FGSCs are usually unable to proliferate in the ovary due to the presence of inhibitory factors unless placed under appropriate conditions. Although we have successfully established pFGSCs and differentiated them into

OLCs, it is still inconclusive whether FGSCs become functional oocytes through their growth, maturation, fertilization, and embryonic development in large animals.

In summary, FGSCs appear to exist in ovaries and have been independently isolated by different research groups and from various species (e.g., humans, pigs, mice, rats, etc.). Furthermore, these cells can be manipulated *in vitro* and transplanted to produce offspring. However, only mice and rats have successfully produced offspring from FGSCs. Although the biological significance of these cells remains controversial, their identification and isolation are expected to provide a valuable model for understanding germ cell development and represent a significant step forward in the future for reproductive biotechnology and infertility treatment. Thus, research on the isolation and culture of FGSCs from ovarian tissue before or after cryopreservation may be helpful in the treatment of fertility in women.

Acknowledgments: Our research is funded by Vietnam National University HoChiMinh City (VNU-HCM) under grant number B2022-28-01.

Author details

Hong-Thuy Bui[1,2*], Nhat-Thinh Nguyen[1,2,3], Truc Phuong Lam Do[1,2], Anh My Le Ba[1,2] and Nguyen Van Thuan[1,2]

1 Cellular Reprogramming Lab, School of Biotechnology, International University, Ho Chi Minh City, Vietnam

2 Vietnam National University, Ho Chi Minh City, Vietnam

3 School of Medicine-VNU, Ho Chi Minh City, Vietnam

*Address all correspondence to: bhthuy@hcmiu.edu.vn

References

[1] Brinster RL. Male germline stem cells: From mice to men. Science. 2007;**316**:404-405. DOI: 10.1126/science.1137741

[2] Zuckermann S. The numbers of oocytes in the mature ovary. Recent Progress in Hormone Research. 1951:**6**;63-108

[3] Johnson J, Canning J, Kaneko T, Pru JK, Tilly JL. Germline stem cells and follicular renewal in the postnatal mammalian ovary. Nature. 2004;**428**:145-150. DOI: 10.1038/nature02316

[4] Zou K, Yuan Z, Yang Z, Luo H, Sun K, Zhou L, et al. Production of offspring from a germline stem cell line derived from neonatal ovaries. Nature Cell Biology. 2009;**11**:631-636. DOI: 10.1038/ncb1869

[5] Wang X-F, Zhang L, Wu Q-H, Min J-X, Ma N, Luo L-C. Biological mechanisms of premature ovarian failure caused by psychological stress based on support vector regression. International Journal of Clinical and Experimental Medicine. 2015;**8**:21393-21399

[6] Hernández-Angeles C, Castelo-Branco C. Early menopause: A hazard to a woman's health. The Indian Journal of Medical Research. 2016;**143**:420. DOI: 10.4103/0971-5916.184283

[7] Dumesic DA, Oberfield SE, Stener-Victorin E, Marshall JC, Laven JS, Legro RS. Scientific statement on the diagnostic criteria, epidemiology, pathophysiology, and molecular genetics of polycystic ovary syndrome. Endocrine Reviews. 2015;**36**:487-525. DOI: 10.1210/er.2015-1018

[8] Oktay K, Buyuk E, Rodriguez-Wallberg KA, Sahin G. In vitro maturation improves oocyte or embryo cryopreservation outcome in breast cancer patients undergoing ovarian stimulation for fertility preservation. Reproductive Biomedicine Online. 2010;**20**:634-638. DOI: 10.1016/j.rbmo.2010.01.012

[9] Silber S. Ovarian tissue cryopreservation and transplantation: Scientific implications. Journal of Assisted Reproduction and Genetics. 2016;**33**:1595-1603. DOI: 10.1007/s10815-016-0814-1

[10] Johnson J, Bagley J, Skaznik-Wikiel M, Lee H-J, Adams GB, Niikura Y, et al. Oocyte generation in adult mammalian ovaries by putative germ cells in bone marrow and peripheral blood. Cell. 2005;**122**:303-315. DOI: 10.1016/j.cell.2005.06.031

[11] Kerr JB, Duckett R, Myers M, Britt KL, Mladenovska T, Findlay JK. Quantification of healthy follicles in the neonatal and adult mouse ovary: Evidence for maintenance of primordial follicle supply. Reproduction. 2006;**132**:95-109. DOI: 10.1530/rep.1.01128

[12] Bukovsky A, Svetlikova M, Caudle MR. Oogenesis in cultures derived from adult human ovaries. Reproductive Biology and Endocrinology. 2005;**3**:17. DOI: 10.1186/1477-7827-3-17

[13] Virant-Klun I, Zech N, Rožman P, Vogler A, Cvjetičanin B, Klemenc P, et al. Putative stem cells with an embryonic character isolated from the ovarian surface epithelium of women with no naturally present follicles and oocytes. Differentiation. 2008;**76**:843-856. DOI: 10.1111/j.1432-0436.2008.00268.x

[14] Zhang Y, Yang Z, Yang Y, Wang S, Shi L, Xie W, et al. Production

of transgenic mice by random recombination of targeted genes in female germline stem cells. Journal of Molecular Cell Biology. 2011;**3**:132-141. DOI: 10.1093/jmcb/mjq043

[15] White YAR, Woods DC, Takai Y, Ishihara O, Seki H, Tilly JL. Oocyte formation by mitotically active germ cells purified from ovaries of reproductive-age women. Nature Medicine. 2012;**18**:413-421. DOI: 10.1038/nm.2669

[16] Woods DC, Tilly JL. Isolation, characterization and propagation of mitotically active germ cells from adult mouse and human ovaries. Nature Protocols. 2013;**8**:966-988. DOI: 10.1038/nprot.2013.047

[17] Bui H-T, Van Thuan N, Kwon D-N, Choi Y-J, Kang M-H, Han J-W, et al. Identification and characterization of putative stem cells in the adult pig ovary. Development. 2014;**141**:2235-2244. DOI: 10.1242/dev.104554

[18] Parte S, Bhartiya D, Telang J, Daithankar V, Salvi V, Zaveri K, et al. Detection, characterization, and spontaneous differentiation In vitro of very small embryonic-like putative stem cells in adult mammalian ovary. Stem Cells and Development. 2011;**20**:1451-1464. DOI: 10.1089/scd.2010.0461

[19] Fereydouni B, Salinas-Riester G, Heistermann M, Dressel R, Lewerich L, Drummer C, et al. Long-term oocyte-like cell development in cultures derived from neonatal marmoset monkey ovary. Stem Cells International. 2016;**2016**:1-17. DOI: 10.1155/2016/2480298

[20] de Souza G, Costa J, da Cunha E, Passos J, Ribeiro R, Saraiva M, et al. Bovine ovarian stem cells differentiate into germ cells and oocyte-like structures after culture *in vitro*. Reproduction in Domestic Animals. 2017;**52**:243-250. DOI: 10.1111/rda.12886

[21] Clarkson YL, McLaughlin M, Waterfall M, Dunlop CE, Skehel PA, Anderson RA, et al. Initial characterisation of adult human ovarian cell populations isolated by DDX4 expression and aldehyde dehydrogenase activity. Scientific Reports. 2018;**8**:6953. DOI: 10.1038/s41598-018-25116-1

[22] Silvestris E, Cafforio P, D'Oronzo S, Felici C, Silvestris F, Loverro G. In vitro differentiation of human oocyte-like cells from oogonial stem cells: Single-cell isolation and molecular characterization. Human Reproduction. 2018;**33**:464-473. DOI: 10.1093/humrep/dex377

[23] Eggan K, Jurga S, Gosden R, Min IM, Wagers AJ. Ovulated oocytes in adult mice derive from non-circulating germ cells. Nature. 2006;**441**:1109-1114. DOI: 10.1038/nature04929

[24] Kujjo LL, Chang EA, Pereira RJG, Dhar S, Marrero-Rosado B, Sengupta S, et al. Chemotherapy-induced late transgenerational effects in mice. PLoS One. 2011;**6**:e17877. DOI: 10.1371/journal.pone.0017877

[25] Bristol-Gould SK, Kreeger PK, Selkirk CG, Kilen SM, Mayo KE, Shea LD, et al. Fate of the initial follicle pool: Empirical and mathematical evidence supporting its sufficiency for adult fertility. Developmental Biology. 2006;**298**:149-154. DOI: 10.1016/j.ydbio.2006.06.023

[26] Lei L, Spradling AC. Female mice lack adult germ-line stem cells but sustain oogenesis using stable primordial follicles. Proceedings of the National Academy of Sciences of the United States of America. 2013;**110**:8585-8590. DOI: 10.1073/pnas.1306189110

[27] Zhang H, Zheng W, Shen Y, Adhikari D, Ueno H, Liu K. Experimental evidence showing that no mitotically

active female germline progenitors exist in postnatal mouse ovaries. Proceedings of the National Academy of Sciences of the United States of America. 2012;**109**:12580-12585. DOI: 10.1073/pnas.1206600109

[28] Inoue N, Matsuda F, Goto Y, Manabe N. Role of cell-death ligand-receptor system of granulosa cells in selective follicular atresia in porcine ovary. The Journal of Reproduction and Development. 2011;**57**:169-175. DOI: 10.1262/jrd.10-198E

[29] Baker TG. Comparative aspects of the effects of radiation during oogenesis. Mutation Research/Fundamental and Molecular Mechanisms of Mutagenesis. 1971;**11**:9-22. DOI: 10.1016/0027-5107(71)90028-5

[30] Beaumont HM. Radiosensitivity of Oogonia and oocytes in the Foetal rat. International Journal of Radiation Biology and Related Studies in Physics, Chemistry and Medicine. 1961;**3**:59-72. DOI: 10.1080/09553006114550071

[31] Kanatsu-Shinohara M, Inoue K, Lee J, Yoshimoto M, Ogonuki N, Miki H, et al. Generation of pluripotent stem cells from neonatal mouse testis. Cell. 2004;**119**:1001-1012. DOI: 10.1016/j.cell.2004.11.011

[32] Izadyar F, Pau F, Marh J, Slepko N, Wang T, Gonzalez R, et al. Generation of multipotent cell lines from a distinct population of male germ line stem cells. Reproduction. 2008;**135**:771-784. DOI: 10.1530/REP-07-0479

[33] Schofield R. The relationship between the spleen colony-forming cell and the haemopoietic stem cell. Blood Cells. 1978;**4**:7-25

[34] Ermolaeva M, Neri F, Ori A, Rudolph KL. Cellular and epigenetic drivers of stem cell ageing. Nature Reviews. Molecular Cell Biology. 2018;**19**:594-610. DOI: 10.1038/s41580-018-0020-3

[35] Massasa E, Costa XS, Taylor HS. Failure of the stem cell niche rather than loss of oocyte stem cells in the aging ovary. Aging. 2010;**2**:1-2. DOI: 10.18632/aging.100119

[36] Niikura Y, Niikura T, Tilly JL. Aged mouse ovaries possess rare premeiotic germ cells that can generate oocytes following transplantation into a young host environment. Aging. 2009;**1**:971-978. DOI: 10.18632/aging.100105

[37] Satirapod C, Wang N, Mac Donald JA, Sun M, Woods DC, Tilly JL. Estrogen regulation of germline stem cell differentiation as a mechanism contributing to female reproductive aging. Aging. 2020;**12**:7313-7333. DOI: 10.18632/aging.103080

[38] Wu J, Tu D, Yuan L-Y, Yuan H, Wen L-X. T-2 toxin exposure induces apoptosis in rat ovarian granulosa cells through oxidative stress. Environmental Toxicology and Pharmacology. 2013;**36**:493-500. DOI: 10.1016/j.etap.2013.03.017

[39] Bhartiya D. Ovarian stem cells are always accompanied by very small embryonic-like stem cells in adult mammalian ovary. Journal of Ovarian Research. 2015;**8**:70. DOI: 10.1186/s13048-015-0200-0

[40] Patel H, Bhartiya D, Parte S. Further characterization of adult sheep ovarian stem cells and their involvement in neo-oogenesis and follicle assembly. Journal of Ovarian Research. 2018;**11**:3. DOI: 10.1186/s13048-017-0377-5

[41] Miyauchi H, Ohta H, Nagaoka S, Nakaki F, Sasaki K, Hayashi K, et al.

Bone morphogenetic protein and retinoic acid synergistically specify female germ-cell fate in mice. The EMBO Journal. 2017;**36**:3100-3119. DOI: 10.15252/embj.201796875

[42] Teletin M, Vernet N, Ghyselinck NB, Mark M. Roles of retinoic acid in germ cell differentiation. In: Current Topics in Developmental Biology. 2017;**125**:191-225. DOI: 10.1016/bs.ctdb.2016.11.013

[43] Yazdekhasti H, Hosseini MA, Rajabi Z, Parvari S, Salehnia M, Koruji M, et al. Improved isolation, proliferation, and differentiation capacity of mouse ovarian putative stem cells. Cellular Reprogramming. 2017;**19**:132-144. DOI: 10.1089/cell.2016.0054

[44] Hainaut M, Clarke HJ. Germ cells of the mammalian female: A limited or renewable resource? Biology of Reproduction. 2021;**105**:774-788. DOI: 10.1093/biolre/ioab115

[45] Paravati S, Rosani A, Warrington SJ. Physiology, Catecholamines. Treasure Island (FL): Stat Pearls Publishing; 2021

[46] Liu J, Shang D, Xiao Y, Zhong P, Cheng H, Zhou R. Isolation and characterization of string-forming female germline stem cells from ovaries of neonatal mice. Journal of Biological Chemistry. 2017;**292**:16003-16013. DOI: 10.1074/jbc.M117.799403

[47] Esmaeilian Y, Atalay A, Erdemli E. Putative germline and pluripotent stem cells in adult mouse ovary and their *in vitro* differentiation potential into oocyte-like and somatic cells. Zygote. 2017;**25**:358-375. DOI: 10.1017/S0967199417000235

[48] Zhou L, Wang L, Kang JX, Xie W, Li X, Wu C, et al. Production of fat-1 transgenic rats using a post-natal female germline stem cell line. MHR: Basic Science of Reproductive Medicine. 2014;**20**:271-281. DOI: 10.1093/molehr/gat081

[49] Bai Y, Yu M, Hu Y, Qiu P, Liu W, Zheng W, et al. Location and characterization of female germline stem cells (FGSCs) in juvenile porcine ovary. Cell Proliferation. 2013;**46**(5):516-528. DOI: 10.1111/cpr.12058

[50] Nguyen H-H, Nhu BLQ, Uyen NNP, Nguyen V-T, Bui H-T. Isolation of female germline stem cells from porcine ovarian tissue and differentiation into oocyte-like cells. The Journal of Reproduction and Development. 2019;**65**:423-432. DOI: 10.1262/jrd.2019-050

[51] Xie W, Wang H, Wu J. Similar morphological and molecular signatures shared by female and male germline stem cells. Scientific Reports. 2015;**4**:5580. DOI: 10.1038/srep05580

[52] Mikedis MM, Downs KM. Mouse primordial germ cells. International Review of Cell and Molecular Biology. 2014;**309**:1-57. DOI: 10.1016/B978-0-12-800255-1.00001-6

[53] Mikedis MM, Downs KM. STELLA-positive subregions of the primitive streak contribute to posterior tissues of the mouse gastrula. Developmental Biology. 2012;**363**:201-218. DOI: 10.1016/j.ydbio.2011.10.003

[54] Reizel Y, Itzkovitz S, Adar R, Elbaz J, Jinich A, Chapal-Ilani N, et al. Cell lineage analysis of the mammalian female germline. PLoS Genetics. 2012;**8**:e1002477. DOI: 10.1371/journal.pgen.1002477

[55] Abban G, Johnson J. Stem cell support of oogenesis in the human. Human Reproduction. 2009;**24**:2974-2978. DOI: 10.1093/humrep/dep281

[56] De Felici M. Germ stem cells in the mammalian adult ovary: Considerations

by a fan of the primordial germ cells. Molecular Human Reproduction. 2010;**16**:632-636. DOI: 10.1093/molehr/gaq006

[57] Shamblott MJ, Axelman J, Wang S, Bugg EM, Littlefield JW, Donovan PJ, et al. Derivation of pluripotent stem cells from cultured human primordial germ cells. Proceedings of the National Academy of Sciences USA. 1998;**95**:13726-13731. DOI: 10.1073/pnas.95.23.13726

[58] Turnpenny L, Brickwood S, Spalluto CM, Piper K, Cameron IT, Wilson DI, et al. Derivation of human embryonic germ cells: An alternative source of pluripotent stem cells. Stem Cells. 2003;**21**:598-609. DOI: 10.1634/stemcells.21-5-598

[59] Kerr CL, Hill CM, Blumenthal PD, Gearhart JD. Expression of pluripotent stem cell markers in the human fetal ovary. Human Reproduction. 2008;**23**:589-599. DOI: 10.1093/humrep/dem411

[60] Castrillon DH, Quade BJ, Wang TY, Quigley C, Crum CP. The human VASA gene is specifically expressed in the germ cell lineage. Proceedings of the National Academy of Sciences USA. 2000;**97**:9585-9590. DOI: 10.1073/pnas.160274797

[61] Ding X, Liu G, Xu B, Wu C, Hui N, Ni X, et al. Human GV oocytes generated by mitotically active germ cells obtained from follicular aspirates. Scientific Reports. 2016;**6**:28218. DOI: 10.1038/srep28218

[62] Parte S, Bhartiya D, Manjramkar DD, Chauhan A, Joshi A. Stimulation of ovarian stem cells by follicle stimulating hormone and basic fibroblast growth factor during cortical tissue culture. Journal of Ovarian Research. 2013;**6**:20. DOI: 10.1186/1757-2215-6-20

[63] Hu Y, Bai Y, Chu Z, Wang J, Wang L, Yu M, et al. GSK3 inhibitor-BIO regulates proliferation of female germline stem cells from the postnatal mouse ovary. Cell Proliferation. 2012;**45**:287-298. DOI: 10.1111/j.1365-2184.2012.00821.x

[64] Zhang X, Yang Y, Xia Q, Song H, Wei R, Wang J, et al. Cadherin 22 participates in the self-renewal of mouse female germ line stem cells via interaction with JAK2 and β-catenin. Cellular and Molecular Life Sciences. 2018;**75**:1241-1253. DOI: 10.1007/s00018-017-2689-4

[65] Zhang X, Wei R, Sun Y, Xia Q, Xie W, Song H, et al. AKT3 is a pivotal molecule of Cadherin-22 and GDNF family receptor-α1 signal pathways regulating self-renewal in female germline stem cells. Stem Cells. 2019;**37**:1095-1107. DOI: 10.1002/stem.3030

[66] Li X, Tian G, Wu J. Novel circ GFRα1 promotes self-renewal of female germline stem cells mediated by m6A writer METTL14. Frontiers in Cell and Development Biology. 2021;**9**:640402. DOI: 10.3389/fcell.2021.640402

[67] Ye H, Li X, Zheng T, Hu C, Pan Z, Huang J, et al. The hippo signaling pathway regulates ovarian function via the proliferation of ovarian germline stem cells. Cellular Physiology and Biochemistry. 2017;**41**:1051-1062. DOI: 10.1159/000464113

[68] Jiang Y, Zhu D, Liu W, Qin Q, Fang Z, Pan Z. Hedgehog pathway inhibition causes primary follicle atresia and decreases female germline stem cell proliferation capacity or stemness. Stem Cell Research & Therapy. 2019;**10**:198. DOI: 10.1186/s13287-019-1299-5

[69] Eyckmans J, Lin GL, Chen CS. Adhesive and mechanical regulation of

mesenchymal stem cell differentiation in human bone marrow and periosteum-derived progenitor cells. Biology Open. 2012;**1**:1058-1068. DOI: 10.1242/bio.20122162

[70] Hendriks J, Riesle J, van Blitterswijk CA. Co-culture in cartilage tissue engineering. Journal of Tissue Engineering and Regenerative Medicine. 2007;**1**:170-178. DOI: 10.1002/term.19

[71] Hwang NS, Varghese S, Elisseeff J. Controlled differentiation of stem cells. Advanced Drug Delivery Reviews. 2008;**60**:199-214. DOI: 10.1016/j.addr.2007.08.036

[72] Eppig JJ, Wigglesworth K, Pendola FL. The mammalian oocyte orchestrates the rate of ovarian follicular development. Proceedings of the National Academy of Sciences USA. 2002;**99**:2890-2894. DOI: 10.1073/pnas.052658699

[73] Kidder G, Mhawi A. Gap junctions and ovarian folliculogenesis. Reproduction. 2002;**123**:613-620. DOI: 10.1530/rep.0.1230613

[74] Zarate-Garcia L, Lane SIR, Merriman JA, Jones KT. FACS-sorted putative oogonial stem cells from the ovary are neither DDX4-positive nor germ cells. Scientific Reports. 2016;**6**:27991. DOI: 10.1038/srep27991

[75] Brizzi MF, Tarone G, Defilippi P. Extracellular matrix, integrins, and growth factors as tailors of the stem cell niche. Current Opinion in Cell Biology. 2012;**24**:645-651. DOI: 10.1016/j.ceb.2012.07.001

[76] Abu-Absi SF, Hansen LK, Hu W-S. Three-dimensional co-culture of hepatocytes and stellate cells. Cytotechnology. 2004;**45**:125-140. DOI: 10.1007/s10616-004-7996-6

[77] Gupta P, Nandi S. Isolation and culture of Preantral follicles for retrieving oocytes for the embryo production: Present status in domestic animals: Isolation and culture of Preantral follicles. Reproduction in Domestic Animals. 2012;**47**:513-519. DOI: 10.1111/j.1439-0531.2011.01904.x

[78] Belli M, Vigone G, Merico V, Redi CA, Zuccotti M, Garagna S. Towards a 3D culture of mouse ovarian follicles. The International Journal of Developmental Biology. 2012;**56**:931-937. DOI: 10.1387/ijdb.120175mz

[79] Kim S-Y, Kim SK, Lee JR, Woodruff TK. Toward precision medicine for preserving fertility in cancer patients: Existing and emerging fertility preservation options for women. Journal of Gynecologic Oncology. 2016;**27**:e22. DOI: 10.3802/jgo.2016.27.e22

[80] Rosendahl M, Greve T, Andersen CY. The safety of transplanting cryopreserved ovarian tissue in cancer patients: A review of the literature. Journal of Assisted Reproduction and Genetics. 2013;**30**:11-24. DOI: 10.1007/s10815-012-9912-x

[81] Medrano JV, Andrés MDM, García S, Herraiz S, Vilanova-Pérez T, Goossens E, et al. Basic and clinical approaches for fertility preservation and restoration in cancer patients. Trends in Biotechnology. 2018;**36**:199-215. DOI: 10.1016/j.tibtech.2017.10.010

[82] Donnez J, Martinez-Madrid B, Jadoul P, Van Langendonckt A, Demylle D, Dolmans M-M. Ovarian tissue cryopreservation and transplantation: A review. Human Reproduction Update. 2006;**12**:519-535. DOI: 10.1093/humupd/dml032

[83] Yding Andersen C, Mamsen LS, Kristensen SG. Fertility preservation:

Freezing of ovarian tissue and clinical opportunities. Reproduction. 2019;**158**:F27-F34. DOI: 10.1530/REP-18-0635

[84] Hoekman EJ, Louwe LA, Rooijers M, Westerlaken LAJ, Klijn NF, Pilgram GSK, et al. Ovarian tissue cryopreservation: Low usage rates and high live-birth rate after transplantation. Acta Obstetricia et Gynecologica Scandinavica. 2020;**99**:213-221. DOI: 10.1111/aogs.13735

[85] Borges EN, Silva RC, Futino DO, Rocha-Junior CMC, Amorim CA, Báo SN, et al. Cryopreservation of swine ovarian tissue: Effect of different cryoprotectants on the structural preservation of preantral follicle oocytes. Cryobiology. 2009;**59**:195-200. DOI: 10.1016/j.cryobiol.2009.07.003

[86] Luz VB, Santos RR, Pinto LC, Soares AAX, Celestino JJH, Mafezoli J, et al. Dimethyl sulfoxide perfusion in caprine ovarian tissue and its relationship with follicular viability after cryopreservation. Fertility and Sterility. 2009;**91**:1513-1515. DOI: 10.1016/j.fertnstert.2008.07.1778

[87] Oskam IC, Asadi BA, Santos RR. Histologic and ultrastructural features of cryopreserved ovine ovarian tissue: Deleterious effect of 1, 2-propanediol applying different thawing protocols. Fertility and Sterility. 2010;**93**:2764-2766. DOI: 10.1016/j.fertnstert.2010.02.003

[88] Kagawa N, Silber S, Kuwayama M. Successful vitrification of bovine and human ovarian tissue. Reproductive Biomedicine Online. 2009;**18**:568-577. DOI: 10.1016/S1472-6483(10)60136-8

[89] Gastal GDA, Aguiar FLN, Alves BG, Alves KA, de Tarso SGS, Ishak GM, et al. Equine ovarian tissue viability after cryopreservation and in vitro culture. Theriogenology. 2017;**97**:139-147. DOI: 10.1016/j.theriogenology.2017.04.029

[90] Xiong J, Lu Z, Wu M, Zhang J, Cheng J, Luo A, et al. Intraovarian transplantation of female germline stem cells rescue ovarian function in chemotherapy-injured ovaries. PLoS One. 2015;**10**:e0139824. DOI: 10.1371/journal.pone.0139824

Section 3

Clinical Application - Novel Indications and Present Ethical Concerns

Chapter 8

Ethics of International Stem Cell Treatments and the Risk-Benefit of Helping Patients

Neil H. Riordan, Luis Gerardo Jiménez Arias and Ramón Coronado

Abstract

Numerous and diverse participants are involved in the development of novel therapies: patients, physicians, scientists, sponsors, governing bodies, lawmakers, institutional review boards, and bioethics proponents. While the welfare of the patient must always and unquestionably be at the forefront of any intervention along with informed consent, their wishes, their requests, and their expectations should also be considered at every step. The availability of stem cell research in various countries with dissimilar regulatory agencies has opened the door for thought-provoking questions about their validity from an ethical, legal, and moral perspective, which will be addressed in this chapter, framed within the doctor-patient relationship.

Keywords: ethics, stem cell therapy, regenerative medicine, international bioethics, mesenchymal stem cells

1. Introduction

In 2010, the Ministry of Health of Costa Rica disallowed stem cell therapy, citing concerns about the experimental nature of these procedures. This abrupt decision left several patients unable to continue receiving treatment at the San José hospital – most notably, a young pilot suffering from spinal cord injury after a 2008 airplane accident that rendered him paraplegic with no perspective of ever regaining mobility, according to two independent physicians. Having experienced encouraging progress in muscle recovery and bowel and sexual function following stem cell therapy, [1] he promptly filed a legal remedy for protecting his constitutional rights to the Supreme Court of Costa Rica, along with other patients in a similar predicament. In their appeal, they argued that patients had *"a right to exhaust all technically feasible procedures to recover [their] health and quality of life"*, citing Article XI of the 1948 American Declaration of the Rights and Duties of Man (the right to the preservation of health and well-being) [2]. This right should not, they contended, be limited by a political authority. Additionally, they invoked the American Convention on Human Rights, namely Articles 5 (the right to physical integrity) and 2 (whereby States Parties undertake to adopt (...) legislative or other measures to grant the rights or freedoms enshrined in The Convention). Lastly, none of the plaintiffs declared having experienced any adverse effects from their therapy at the time of the appeal; they had, in fact, perceived improvements in their health condition.

The Supreme Court ruled that medical treatments had to be permitted by law before their implementation, as had been argued by the defendant (the Ministry of Health). While the judges acknowledged the principle of patient autonomy, the doctor/patient relationship, human dignity, and the right to health, selecting a treatment should rely on the law. However, the Court also recognized that no adverse events had been noted throughout these treatment cycles with stem cells and that they had no factual or legal arguments to halt the treatments that had already begun. The plaintiffs were therefore permitted to continue their therapy.

Six years later, presidential decree n°39,986-S authorized regenerative therapies with adult stem cells in Costa Rica, based on national scientific recommendations, under the human right to health access principles, and citing decades of past research on the safety of hematopoietic stem cells transplants [3]. Therein are outlined the requirements to be submitted by those seeking stem cell therapy administration: safety profile, scientific rationale, cellular characterization, administration, and requirements for the qualifications of health professionals and facilities. While the decree makes a distinction between minimally manipulated and more than minimally manipulated cells, both are permitted with different avenues for authorization.

This Costa Rican case is pivotal because the Supreme Court ruling marks the first legal precedent in a Latin American country that allows patients who are already being treated with stem cells to continue their treatment – a first step, perhaps, to implement international legislation on stem cells. Following the law to the letter in that instance proved to be rather impractical, falling almost into irrationality as it limited the right of access to health for patients who required treatment as a last resort, and going against international law, which stipulated that Costa Rica must have taken the necessary legislative, economic and political measures to facilitate access to the said right to health. The subsequent presidential decree is also a landmark case since it authorizes using a non-minimally manipulated therapeutical product currently not permitted in other jurisdictions, notably in the USA according to the Federal Drugs Administration (FDA) guidance. The compassionate use of or access to drugs and new therapies remains limited on the grounds of minimizing harm to terminally ill patients [4]. How much evidence is necessary to release a given drug or a given experimental therapy? Are Phase I studies sufficient? Can effectiveness be shown at this stage? Answers remain unclear [4].

This chapter will examine legal and moral issues arising from stem cell treatments, the patient/doctor relationship, and the right of access to health and patient welfare in the context of current international regulations and medical tourism – and how conflicting regulations between countries pose conflicting views on ethics regarding patient access to new therapeutics.

2. Stem cells: definitions, applications, and considerations

Human "stem cells" is a broad term that may refer to various types of cells differing in their origin, applications, and characteristics. Hematopoietic stem cells, multipotent, self-renewing, and capable of generating blood cells, were first described in the early 60s [5]. Soon after, another type of multipotent stem cells differing from those of hematopoietic lineages were observed in bone marrow [6] – these would later become known as mesenchymal stem cells [7] or mesenchymal stromal cells (MSCs), with a limited capacity to differentiate into specific types of adult cells. Pluripotent embryonic stem cells (ESCs), capable of giving rise to all cells in the body, were derived in the late 90s from blastocysts [8]. Finally, induced pluripotent stem cells (iPSCs) were introduced in 2006: somatic cells reprogrammed to have the embryonic capacity for differentiation [9]. While much of the research in the 20th century was focused on isolating and identifying techniques to culture stem cells, the early 21st century saw a growing number of case reports and clinical trials seeking to establish their therapeutic potential. This line of research was not without challenges or controversy (**Table 1**).

Most of the earlier bioethical controversy was focused, understandably, on the use of embryonic stem cells due to the loss of viable embryos in the process of isolation and derivation of the stem cells, which led to bans or limitations imposed on research in several nations. Mesenchymal stem cells, in contrast, share little of those ethical concerns related to their tissue origin since they can be derived from adult tissue, notably for autologous use, or from perinatal tissue, such as the umbilical cord that would routinely be discarded after a normal birth, for allogeneic use. Consequently, the ethical controversies surrounding the use of mesenchymal stem cells have shifted to considerations applicable to clinical trials and the development and commercialization of novel experimental therapies.

Recent clinical trials with mesenchymal stem cells cover a broad range of tissue sources and administration routes, complicating comparisons between published results. In brief, mesenchymal stem cells appear safe for administration at least in the short term, [21–23], and some long-term reports are also available [24, 25]. Adverse

Product	Biological Source	Biological concerns with administration or transplantation	Ethical concerns
Hematopoietic stem cells	Peripheral blood, bone marrow	Graft versus host disease, infections, neutropenia [10]	Need/availability ratio, donor / recipient ethical issues common to other organ and tissue donations [11]
Mesenchymal stem cells	Adult tissue (e.g., bone marrow, adipose tissue...), perinatal tissue (umbilical cord, amnion...)	Lack of standardization of source, isolation, preparation, expansion, doses, homing effectivity [12, 13], poor understanding of therapeutic mechanisms [14]	Donor/recipient ethical issues, issues common to clinical trial participation, medical misuse, commercialization [15]
Embryonic stem cells	Blastocysts	Tumorigenicity [16]	Right to life of the embryo, objectification of human life, commodification of oocytes [17, 18]
Induced pluripotent stem cells	Somatic cells	Genomic instability, tumorigenicity [16]	Personhood definitions, limited informed consent of donor for downstream research [19], risk-benefit assessment [20]

Table 1.
Biological and ethical concerns arising from the administration or transplantation of various stem cell types.

events are reportedly transient, usually fever or fatigue, and no tumorigenicity or malignancy has been observed thus far [23]. While the mechanisms of action of mesenchymal stem cells are not entirely understood, current research views stem cell secretions as key for their immuno-modulatory, anti-inflammatory, and therapeutic properties. The conditions treated in various clinical trials or case reports include cardiovascular, neurological, autoimmune, orthopedic, pulmonary, and graft-versus-host disease, [26] and more recently, COVID-19 [27]. At least ten products derived from mesenchymal stem cells have received regulatory approval in South Korea, Japan, India, Canada, Australia, and Europe [12]. The wide variety of conditions researched as well as the legitimacy of some products, may give the erroneous impression that mesenchymal stem cells are a "cure-all": more research is necessary to standardize isolation and culture techniques, as well as to establish risk-benefits and true efficacy for certain conditions to avoid these pitfalls. Most importantly, trials should be conducted ethically, ensuring that disappointing results are also published, that all adverse events are carefully documented and reported, and that patients duly consent with a proper understanding of the procedures and the potential benefits and limits.

Throughout this chapter, "stem cells" should be understood to mean mesenchymal stem cells or adult stem cells unless otherwise indicated.

3. International regulation and medical tourism

The use of stem cells in Costa Rica has been commercially permitted since 2016. The current law does not specify the effectiveness of a treatment or the amounts to be charged, if any. The broad wording was designed to position and cement the place of Costa Rica as a safe biotechnological or medical tourism hub, thereby disregarding ideological debates arising in other countries regarding the restrictions on the use of stem cells. In Turkey, the use of stem cells has been open and regulated as long as they are not embryonic cells, giving way to laws increasingly open to medical tourism [28]. South Korea is a pioneer in the production of stem cell-derived products, [29] notably a composite of allogeneic umbilical cord blood-derived mesenchymal stem cells and hyaluronate to treat cartilage defects in osteoarthritis [24, 30]. In 2014, Japan introduced a "fast-track" regulatory path for regenerative medical products in the Pharmaceutical and Medical Devices Act [31], whereby products can be commercially available following a short trial: patients recruited into that scheme are enrolled in a registry to be followed up for a period during which efficacy must be demonstrated. This law and the subsequent applications of this law were criticized in the international scientific community, [32, 33] using terms such as an *"obsession [with staying at the forefront of regenerative medicine]"* [34]. Nevertheless, Japan stood its ground [35] and continued research with induced pluripotent stem cells, announcing the start of prospective trials for spinal cord injury and heart conditions in Japanese university hospitals [36, 37], among others. Japanese scientists appear to have followed up on these developments and have published some relevant self-reflection [38–40].

Irrespective of stem cells, medical tourism is hardly a new phenomenon, and a wealth of literature has been written about ethical considerations and repercussions both in the country of origin and the destination country [41, 42]. Certain countries such as Thailand, Singapore, and India have developed this medical tourism industry to the extent that it represents a non-negligible percentage of their GDP. Contributors to the debate should examine their own biases when referring to foreign countries, avoiding terms such as 'third world' or assuming higher possibilities of substandard care, infections, or fraudulence as this may alienate so-called developing nations from much-needed transparency and open cooperation. The perspectives of physicians or researchers from "destination" nations must be included and perhaps should even be the primary source for scientific output, as they should be considered capable of producing critical literature from their own cultural and scientific perspectives. There is a marked inequality in availability and volume of research produced in more affluent countries compared to underprivileged ones, [43] though a process of "catching up" appears to be in place [44].

Nevertheless, medical tourism is a cause of concern, [45] particularly in the case of stem cell therapy, since prospective patients may not be sufficiently informed about the risk and benefits of the treatments they are receiving, or may be charged disproportionate amounts for dubious or unregulated procedures – a form of preying on their state of mind due to their illnesses. Many patients who travel abroad to be treated with stem cells do so as a last resort and in some cases desperately seek a cure or at least an improvement for their medical condition. However, medical malpractice or dishonesty can occur regardless of whether a country approves a given drug or procedure. While medical tourism may contribute or aggravate the proliferation of questionable practices, it can hardly be considered the only reason. Every country has an obligation to implement and uphold good medical practices promoting principles of medical ethics and the deontological standards of medicine. Moreover, the proliferation of fraudulent clinics in some destination clinics is likely hindering the development of legitimate stem cell research [45]. The case of Stamina in Italy is an example where internal regulations successfully identified a fraudulent clinical practice; this operation needed to be halted because of an individual holding no medical qualifications, poor quality standards for manufacturing and no sound scientific rationale, not because of an inherent evil or penchant for fraudulence in stem cell research.

Should medical tourism then be regulated? There is currently no consensus or homogeneity in criteria among countries on regulatory affairs of therapeutic stem cell treatment. The implantation of international regulations would not be a trivial exercise. The suggestion of a recognized agency such as the FDA or the European Medicines Agency (EMA) granting certificates of approval to international clinics, while desirable in theory, would not be possible in practice as it would interfere with the sovereignty of each territory to regulate and legislate their internal affairs. Similarly, the jurisdiction of national agencies is limited to their borders, and allocating a portion of their budget to oversee practices in other countries would not be economically feasible. Some international networking organizations such as the International Society for Cell and Gene Therapy (ISCT), have recently established working groups to tackle regulatory and ethical issues with committee members primarily based in North America, [46] but unlike charters of international intergovernmental organizations, binding for the undersigning nations, their advisories are ultimately non-binding. The Pan American Health Organization (PAHO) has issued recommendations for the use of "advanced therapy medicinal products", [47] acknowledging that regulatory bodies are still developing in many countries and stressing that ethical implications must be considered, that treatments must take place in authorized, specialized centers, and that a risk-based approach could be considered for establishing regulations. This document notably states that *"continual advances in scientific research generally keep it several steps ahead of regulatory mechanisms, a situation that occurs worldwide,"* and calls for open communication between scientists and regulatory bodies [47].

Another possibility to curb medical tourism would be to introduce restrictions into the national health network for those patients who have traveled and received treatment in clinics unrecognized by regulatory bodies of e.g., the USA – such punitive measures might, however, not be conducive to the best interests of the patients. Patients may feel they have to lie or be reluctant to request their physician's opinion, thereby losing an opportunity for education or more extensive research into the treatments they seek abroad. A more constructive solution would then be to actively educate prospective patients before their decision, ensuring that the cost-benefits of stem cell treatment are carefully considered, including the possibility that effectiveness may be small, limited, or non-existent for their particular pathology. The education of patients is hardly a novel idea [48]. The information should be presented in a clear, easily understandable manner rather than a dry scientific report, as this is likely not the format that prospective patients are used to consuming in an era where multimedia is pervasive. Neither should the tone of the information provided be perceived as scolding or belittling, as this may have the unintended effect of the valuable advice being dismissed altogether to the detriment of the patient; efforts should be focused on empowering the patients to make their own decisions (to "know best" or to "be in the know") – or any other desirable traits that a sociological study into the characteristics of this population may reveal, along with decision-aid studies. For example, they could be more receptive to someone they perceive as a peer than to a disengaged physician or a removed academic seemingly removed from the daily struggles of their conditions. The same narrative and visual techniques employed to attract patients can also be used to educate. Recent efforts, e.g., by the International Society for Stem Cell Research (ISSCR) [49], remain very text-heavy and potentially unengaging for the general public.

However, if after considering the risk/benefits, the patient still decides to seek treatment abroad, be it a clinical trial or a treatment approved by the destination country's regulatory bodies, at a cost or not, it then becomes the right of a patient to health and the right to access health, to improve their quality of life – this is, perhaps, far more complex to regulate or legislate in an international context. There is a delicate balance between governmental control (imposition or restriction) of therapeutics and the reach of these controls inflicting upon the freedom of the patients to choose what they (along with their medical doctors) want to pursue for their health, under the sole assumption that the patients are fully informed about the risk and possible benefits associated with the treatment.

4. Let doctors be doctors: the patient-doctor relationship

The 2018 Right to Try Act of the USA creates a legal framework for access to unapproved medical products by patients with life-threatening illnesses who have exhausted all their options and may not participate in a clinical trial, provided said product has completed a Phase I study [50, 51]. Opponents of this law have argued that it creates conditions for physicians to prey on desperate patients by creating false hope, that the burden of treatment costs is shifted to patients and manufacturers, and that existing health disparities may be exacerbated, ultimately leading to greater patient suffering [52, 53]. Additionally, valuable information or data collection about product development or adverse events may be lost due to the lack of FDA oversight [52]. The legislation does not compel a manufacturer to provide access to treatment; some manufacturers may outright refuse, [54] and early reports seem to indicate that drugs are still being requested in greater volume under Expanded Access rather than with the Right to Try provision [55, 56]. This legislation raised much controversy, often politically charged, sparking ethical debates about what it was trying to achieve, how much it would truly help, or how it would be implemented in practice. The ethical problems of stem cell therapy then seem no different from that of any other experimental therapy (such as those for cancer or rare diseases contemplated by this law) where the patient's autonomy, the cost/benefit of the treatment and any possible abuses or misuses must be weighed in. Advocates of the Right to Try Act emphasized the liberty of patients to choose a treatment and to eliminate bureaucracy; in a similar vein, Texas House Bill 810 (85R) authorized the *"provision of certain investigational stem cell treatments"* under investigation in clinical trials to patients *"with certain severe chronic diseases or terminal illnesses."* [57]. The patients were required to provide written informed consent, and the treatment was to be overseen by an Institutional Review Board (IRB), and administered by a certified physician in a hospital, surgical center, or medical school. The IRBs were to submit annual reports of treatments enacted under this law. *"Why can't someone that is of age have the ability to sign off, so to speak, with regard to a proper medical release on the ability to do something that can make such a dramatic difference in their life, and their lifespan, and their quality of life?"* argued the proponents [58]. This would also encourage *"medical innovation"* [58] in the state of Texas – not unlike the intent of countries who have passed similar laws allowing stem cell research or therapy. These two landmark legislations of the United States exemplify a movement to put the patient and their doctor's relationship at the forefront, which has not been without controversy [59, 60].

If a therapy exists, every effort should be undertaken to implement a way to access said therapy as a last resort for patients. The requirement of having successfully completed Phase I of scientific research may be a reasonable compromise, as long as the patients are sufficiently and objectively educated about the risks and cost/benefits before reaching a decision. And yet, as the example of Japan has shown, a national law, a patient registry, and clinical trials overseen by universities, have still not been considered enough by the academic community. When is it enough? What and whose criteria drive this quantification? Would the debate not be enriched from the participation of physicians and patients who are, so to speak, in the front lines of the battle? One may argue that a "desperate" patient cannot adequately provide informed consent due to their state of mind, but this is a particularly thorny argument that toes close to discrimination, paternalism, [61] or, to use a more modern term, ableism: is having an illness ever sufficient to render a person incapable of making decisions regarding their own welfare? Is a person's dignity and mental ability lessened or invalidated when faced with a significant loss of quality of life or an eventual end of life? This decision-making process perhaps belongs more in the sphere of a qualified psychiatrist or therapist for each particular case rather than a broad stroke ruling in an academic setting or legislated by a government body. Broadly qualifying a disadvantaged person (in this case, one with an illness) as intrinsically "vulnerable" or incapable of making informed decisions for themselves may be an insult that reinforces the injustices and stigma that they already face [62]. A decision to use or seek new treatments is not in itself irrational; if prospective patients would be considered competent enough to consent to enter a phase I/II trial in a research setting, why could they not consent in a non-research setting under a physician's supervision? [63]. And if patients have the right to refuse a treatment, surely from this right to refuse follows a reciprocal right to choose or access an intervention [59, 64, 65]. It is an infringement on the right of patient to procure treatment that they understand would be useful for their condition after being presented with clear and truthful information about said treatment. Instead of paternalistic protection, prospective patients need empowerment: a greater voice in setting research agendas and designing studies [62].

Perhaps the most common concern is the potential harm of stem cells themselves or the problems of their commercialization, as the non-maleficence principle must always be kept in mind. The relative safety of mesenchymal stem cells has been sufficiently covered, as reported by systematic reviews [21–23]. Still, more research is needed for standardization of dosage, culture methods, and source of the stem cells, as well as a need for quantifying effectiveness for clearly defined conditions and thorough documenting of adverse events. Regarding commercialization, as long as there is a demand, there will always be a market to fulfill that demand with various degrees of ethical and legal shades. Stem cell therapy is no stranger to such a conundrum in the face of market greed. A problem of commercialization may derive from health providers being unwilling to fully inform patients about the risk/benefits for fear of losing business; however, not all practices are the same, comparable to medical practices running legally in which their marketing strategies might present skewed information to capture more clients. Within legal boundaries, there could still be unethical behaviors and vice-versa. On the other hand, patients who have gone to great lengths to receive treatment may experience a placebo effect or convince themselves it was worthwhile. Yet incurring in costs to access a treatment is not inherently unethical or fraudulent; the FDA has published a guidance outlining requirements where this practice may be authorized, notably when the costs would be extraordinary to the sponsor because of "*manufacturing complexity, scarcity of a natural resource, the large quantity of the drug needed (e.g., based on the size or duration of the trial), or some combination of these or other extraordinary circumstances (e.g., resources available to a sponsor)*" [66].

The pressure to find new therapies for illnesses with limited, insufficient, or no current treatment options comes precisely from the physicians, the scientific and medical industry, and patients seeking relief for their conditions. Herein lies the more significant risk: a race for supply and greed of demand when faced with pain or eventual death. Suppose a particular country's laws or guidelines are restrictive enough to hinder the physician/patient relationship. In that case, doctors may find it impossible to consider alternative therapies, even under compassionate use, due to the lack of adequate protocols. At the same time, patients feel powerless in the face of government regulations. It is precisely at this point where, in desperation, abuses or misuses arise, not from the new therapies themselves, but from a lack of expectations or incomplete information about the possibility of a cure or relief. And one may ask: what's the rush? Why do some patients insist on seeking a treatment that is not readily available or approved instead of waiting for the due process of clinical research? "*The rush is the daily necessity to help sick people. (...) The 'rush' arises from our human compassion for our fellow man who needs immediate help,*" *as* "*their illnesses will not wait for a more convenient time*" [67].

If a legally qualified doctor in his professional authority, after having read results of recent advances in the field, well within the boundaries of the regulatory bodies of the country where they operate concludes that such treatment could help a particular patient, a patient who is willing and fully informed to the best of the current understanding – would they be morally justified in refusing said treatment? Can moral objections ever be a sufficient basis for denying the right to healthcare? Patients who seek stem cell treatments may have spiritual distress or therapeutic hope, [68] an aspect some medical doctors may not be equipped to manage. But what of compassion? This use of "compassionate" may be close to that employed by ecclesiastical authorities who do not oppose but promote treatments that offer at least a better quality of life: a physician's duty is not limited to knowledge and technical expertise, but also compassion [69]. The compassionate act and the treatment as compassion, coming from the good judgment of a doctor seeking the best for his patient, must therefore be left in that sphere of the doctor's relationship with his patient. Depersonalized rulings do not allow the physician to exercise his art, profession, and oath. The relationship between the physician, or a team of physicians, and their patients is thus an essential aspect of making an informed decision: one the one hand, the patient exercising his autonomy to make medical decisions, and on the other, the physician, upholding his medical oath and not creating false expectations by promising more than what is expected to be achieved with a given treatment or to create hopes beyond what can be offered. Thus, the final decision to access an intervention must lie in the hands of the patient and their physician, based on real world evidence for safety, and within a sound legal and ethical framework.

"*Ethics in both research and clinical settings is most effective when it is preventive*" [70]: indeed, bioethicists do not go ahead with scientific developments but discuss scientific issues that are already on the table. Conversely, neither should physicians regard ethical questions *as "removed from their daily work at bench or bedside,"* as the purpose of new treatments is a societal benefit [70]. *"Market will efficiently allocate the resources, but not always in an ethical manner"* concerning medical tourism; [41] ethical considerations should therefore be contemplated before the application of treatments by creating a legal framework that promotes scientific research and keeps the welfare of the patient at the forefront. The fear of possible misuse or medical misconduct should not deny the patients' right to health, particularly when their lives are at risk or when they are the most vulnerable to their condition. Accumulated clinical experience and evidence-based medicine about safety, dosage, and efficacy would be more appropriate when determining whether to offer or withhold access to treatment.

5. Legal and moral issues

The international legal system is derived from ethical principles, at the heart of which lies the dignity of the human person, and modern bio-law too draws from ethical and axiological foundations. Human dignity as a concept is fundamentally imprecise, as its definition necessitates defining first what is dignified or worthy, and what is unworthy – an anthropocentric, Judeo-Christian notion [71]. Subsequently, the western concept of human dignity has evolved and dissociated itself from any deity to accept the teleological interpretation of the end of man (Kant): each person is an end in itself and not a means to satisfy the end of another person. In any case, a more pertinent application would be establishing what makes an act respectful of the dignity of others. A philosophical question arises as to whether human dignity is opening up in the last century, whether it is facing threats as never before, or whether both are occurring simultaneously [72]. But if dignity is being threatened, it is first necessary to seek and recognize, that is to say, to pinpoint these threats. One of the possible threats to human dignity comes precisely from biotechnological development and the interference of politicized legal systems that could be restricting fundamental rights, such as the right to health and the right to life. International treaties regarding human dignity appear to have different intensities and interpretations when applied to concrete cases [73], in legal cases ranging from political disappearances to in-vitro fertilization [74]. Interestingly, in the latter, a court of law and not science has defined a biological fact. Are we then facing an ideological system of human rights? Do human rights serve to protect life and health? Are all lives of equal value or are some worth less than others? Are scientific truths at the service of the law or should the law adjust to the reality of biotechnological progress?

The Catholic Church has been a notable exponent of the ethics of stem cell research and therapy, with contributions from the various Congregations of the Holy See and various speeches and interviews of the Popes after the Second Vatican Council promoting scientific development hand in hand with ethics. *"Progress becomes true progress only if it serves the human person and if the human person grows: not only in terms of his or her technical power, but also in his or her moral awareness,"* Pope Benedict XVI declared in 2006. Science shows its usefulness most strongly and richly when its end is to alleviate human suffering through new findings and resources: the efforts of the researchers result in the improvement of the affected and the different conditions or diseases. Stem cell research was deemed deserving of encouragement when it combined scientific knowledge, the most advanced technology in the biological field, and the ethics advocating respect for the human being in all phases of his existence [75]. Man is the actor of scientific research, but he is often the object of this same research; he must consequently be the beneficiary of scientific research, but never a mere instrument. Man cannot be disposed of as an object of research or commercialization, especially when his state is more vulnerable, in accordance with all the principles of personalist ethics; the main interest is the well-being of all and, in particular, of each individual [76]. Research initiatives with adult stem cells were deemed to be free of ethical problems, and clinical use presented no moral objections as long as *"scientific rigor and prudence [reduced] any risks to the patient to the bare minimum and [facilitated] the interchange of information among clinicians and full disclosure to the public at large"* [77]. A repeated call is thus made for dialog between science and ethics, particularly when the fruits of research remain inaccessible to those who lack the means to access them. "*Advances in medical science,*" it was noted, *"go hand in hand with just and equitable provision of health-care services,"* [75] and one may add to this, with public health policies that are willing and open to enable access to those fruits of research. The path of advancement thus leads to the promotion structures and economic means conducive to scientific achievements.

The morality of medical tourism from the perspective of prospective patients has been examined before, considering whether it is moral to "jump the queue" in countries with socialized medicine to seek care in countries where ordinary citizens may not afford the facilities offered to medical tourists. While some patients understood the perspective of the greater good, they were more willing to solve their problems (pain) than to consider fairness or morality at the time of their decision [78]. At the heart of this conflict lie two divergent ethical frameworks: the rights-based and the communitarian frameworks. In rights-based approach, "the rights and dignity of the individual should never (or rarely) be sacrificed to the interests of the larger society", whereas the "common good" will be at the center of communitarian views, where policies will be shaped to promote these "shared values, ideals and goals of a community" [64]. Having a thorough regulatory process protects the needs of the community, or more precisely, of the future members of the community who will become sick – paradoxically to the detriment of those who are currently ill or suffering and consequently have good reasons to prefer a quicker process or an alternative approach [59]. To what extent can the goal of seeking relief be considered immoral? When a patient fully and objectively informed decides to try an experimental or unproven therapy in a country where it is regulated, within any means reasonably available to them, should their freedom to choose be curtailed? Is there a price to feeling well, to regain mobility? To life?

The case of Ashya King, [79, 80] a young patient from the UK suffering from a brain tumor in 2015, is a dramatic example of the difficult decisions patients and their families encounter when at odds with current regulations or laws. Unwilling to subject him to radiation and chemotherapy at such a young age, the parents expressed a desire to try proton beam therapy in Prague, which was at that time not approved in the UK. When they were denied this option, the parents signed the child out of the hospital and attempted to travel to Prague – but an arrest warrant was issued against them because of potential child endangerment. After a short judicial deliberation, they were able to reach Prague. The intervention was successful; as of 2018, the child was reportedly in good health with cancer in remission [81]. The UK approved proton beam therapy shortly after positive results from a clinical trial, as it was safe and had fewer side effects than chemotherapy [82]. Randomized trials comparing proton therapy (desired by the parents) and traditional radiation (proposed by the UK health system) were deemed to be "*unethical and not feasible*" [82] and this was "*likely to be the best evidence available*" [83]. Clinical trials are the gold standard of research, but of what good would that have been to the child in 2015? He did not have the time to wait for this approval as he needed treatment as a matter of life or death. Was it morally justified to withhold this treatment option from him? Was it fair to arrest parents for seeking treatment, unapproved in their country, that ultimately would save their child's life and preserve his quality of life? The intervention was successful, but even if it had not been, was it immoral for them to try? Those who are healthy and not living with a significant disability should, perhaps, strive to achieve a greater understanding of the mindset and the goals of someone who is in pain or significantly impaired before they cast moral judgments on their actions, or seek to limit their liberties as self-appointed moral arbitrators. In the late 80s, when genetic therapy was in its infancy, the parent of three sick children wrote in the context of sickle cell anemia research, "*I resent the fact that a few well-meaning individuals have presented arguments strong enough to curtail the scientific technology which promises to give some hope. Aren't they deciding what's best for me without any knowledge of my suffering?*" [84]. Prospective patients may feel similarly bewildered when being deprived of their autonomy and their capacity to make their own decisions, and so the detached intellectual debate should be balanced with compassion [67] and with respect for the self-determination of patients.

6. Future perspectives

The case of Japan merits more consideration rather than outright dismissal. Following a desire to regulate the proliferation of stem cell clinics with dubious practices, a law was passed after various committee reports, some including medical professionals, lawyers, and lay participants [38]. The development of this legislative action must be understood in the Japanese context, which recognizes the right to access to health care and protects the freedom of discretion and academic freedom (physician's discretion) [38]. Subsequently, pertinent arrests have been made in cases where cells were processed without proper regulatory authorization [38, 85]. The approval of one ophthalmological trial with induced pluripotent stem cells was subject to a committee review process, where risks to patients were examined, along with how information was to be presented to prospective patients (who additionally had significant visual impairments) for informed consent [40]. The process in the latter was not perfect, [40] and neither is the current law [38]. Still, as Japanese researchers note, there is potential for suggestions, amendments, and recommendations to strengthen the practice of regenerative medicine in Japan: *"If various case studies on the review processes of [stem cell trials] or other cutting-edge biotherapeutic trials from around the world were similarly discussed, such case studies would contribute to establishing or improving guidelines for review committees to further improve the quality of these discussions"* [40]. That is, perhaps, the most tragic aspect of the case of Japan: a country exercising its national sovereignty, enacted legislation with reasonable regulatory controls, but it was maligned by academics in other countries since its inception. Yet Japanese researchers have demonstrated they are capable of internal criticism and suggestions for improvement. Rather than jumping to conclusions of fraud, obsessions, or lack of oversight, countries should be encouraged and trusted to develop their regulatory laws with guidance from recognized international health bodies, and scientists of said countries can be held accountable for upholding ethical, medical, and scientific standards.

Some notable lessons may be learned from the history of excimer laser. More commonly known as "LASIK", it was developed in the 1970s and applied to ophthalmology in the late 80s. Compared to other countries, the FDA was notoriously slow to authorize this intervention, or more precisely, the manufacturing process for the laser technology. During the lengthy review process, interesting questions such as "*what complication rates are acceptable?*"; "*what length of follow-up is acceptable [for complications]?*"; "*what is an excellent study design for a clinical trial (...)?*"; "*how good must a surgical procedure (...) be before it will be made widely available for clinical use in this country?*" [86]. These questions have many similarities with the current debates regarding stem cell trials. Excimer laser investigators "*expressed frustration with the deliberate pace of the review process, while outside the United States large numbers of patients [were] having the surgery*" in Canada, Europe, Australia, and Asia [86]. Indeed, clinics from those countries established *"systems of referral in which Americans fly in for surgery and return home for post-operative care"* [86]. The FDA approved the procedure between 1995 and 1999 following the usual clinical trial phase format. Excimer laser surgeries were extremely popular in the 2000s in the USA, then declined notably, [87] with a resurgence during the pandemic presumably due to mask wearing. The reasons for this decline in popularity are not well understood, but some patients cite concerns about complications after reading about the experiences of other patients [88]. There are lessons to be learned here: patients WILL find a way within their means to pursue a novel treatment or procedure they believe will help them, particularly if regulatory bodies are slow on the uptake. The conclusion is not necessarily that regulatory bodies should be less careful or quicker, but given that patients will seek treatment where it is available, once again education and information are of utmost importance. Indeed, once more information about excimer laser became available, even with the low rate of complications, some patients decided on their own that the procedure was not worth it or that they were unwilling to risk it. Thus, patients too should be trusted in their capability to exercise their own educated decisions when it comes to their access to health provided that they are given sufficient guidance.

The development, introduction and "fine-tuning" of stem cell interventions are perhaps not so different from the history of other medical breakthroughs, but this debate has become peculiarly heated, sometimes emotional. All stakeholders, patients, patient advocates, scientists, regulators, and clinicians, even those currently offering unproven therapies (as not all are fraudulent or untrustworthy, and may desire fair regulation) must engage in a constructive dialog to define acceptable policies [89]. This dialog would ideally lead to humane, ethical, scientifically sound, and commercially viable regulations, as long as every party listens actively and does not react with frustration when presented with differing points of views. The Colombian Xaverian university recently collaborated with multiple countries to develop an ISO for "biobanking of human mesenchymal stromal cells derived from bone marrow," [90] with debates and meetings in Berlin, Tokyo and Toronto, [91] proving cross-border collaboration is possible to develop international standards for stem cells. We are past the time of pointing fingers or engaging in unproductive academic accusations: all stakeholders must arrive at a consensus with the ultimate goal of developing sound regulatory policies at the international and national level, from which the rest will derive: non-fraudulent, ethically-driven clinics, physicians exercising their medical criteria, and patients empowered to make their own informed decisions.

Acknowledgements

The authors would like to thank Ms. Dorita Avila for manuscript preparation for publication.

Conflict of interest

NHR is a shareholder of Medistem Panama. RC and LGJA declare no conflicts of interests.

Author details

Neil H. Riordan[1*], Luis Gerardo Jiménez Arias[2] and Ramón Coronado[3,4,5,6]

1 Medistem Panama, Panama, Panama

2 International Institute of Human Rights Americas Chapter, San José, Costa Rica

3 University of Texas Health Science Center (Transplant Department) at San Antonio, Texas, USA

4 Department of Obstetrics and Gynecology, Baylor College of Medicine, Houston, Texas, USA

5 Crown Scientific, San Antonio, Texas, USA

6 Signature Biologics, Irving, Texas, USA

*Address all correspondence to: neil@aidanresearch.com

References

[1] Ichim TE, Solano F, Lara F, Paris E, Ugalde F, Rodriguez JP, et al. Feasibility of combination allogeneic stem cell therapy for spinal cord injury: A case report. International Archives of Medicine. 2010;**3**:30

[2] Organization of American States, editor. American Declaration of the Rights and Duties of Man. Bogotá: Ninth International Conference of American States; 1948

[3] Executive Power of Costa Rica. Decreto Ejecutivo N° 39986-S: Autorización para las Terapias Regenerativas con Células Madre Adultas. San José, Costa Rica: La Gaceta Costa Rica; 2016

[4] Falit BP, Gross CP. Access to experimental drugs for terminally ill patients. Journal of the American Medical Association. 2008;**300**(23):2793-2795

[5] Till JE, McCulloch EA. A direct measurement of the radiation sensitivity of normal mouse bone marrow cells. Radiation Research. 1961;**14**(2):213-222

[6] Friedenstein AJ, Petrakova KV, Kurolesova AI, Frolova GP. Heterotopic transplants of bone marrow. Transplantation. 1968;**6**(2):230-247

[7] Caplan AI. Mesenchymal stem cells. Journal of Orthopaedic Research. 1991;**9**(5):641-650

[8] Thomson JA, Itskovitz-Eldor J, Shapiro SS, Waknitz MA, Swiergiel JJ, Marshall VS, et al. Embryonic stem cell lines derived from human blastocysts. Science. 1998;**282**(5391):1145-1147

[9] Takahashi K, Yamanaka S. Induction of pluripotent stem cells from mouse embryonic and adult fibroblast cultures by defined factors. Cell. 2006;**126**(4):663-676

[10] Smith LA, Wright-Kanuth MS. Complications and risks in hematopoietic stem cell transplant patients. Clinical Laboratory Science. 2001;**14**(2):118-124

[11] Liso A, Neri M, Maglietta F, La Russa R, Turillazzi E. Hematopoietic stem cell transplantation: A bioethical lens. Stem Cells International. 2017;**2017**:1286246

[12] Levy O, Kuai R, Siren EMJ, Bhere D, Milton Y, Nissar N, et al. Shattering barriers toward clinically meaningful MSC therapies. Science Advances. 2020;**6**(30):eaba6884

[13] Regmi S, Pathak S, Kim JO, Yong CS, Jeong JH. Mesenchymal stem cell therapy for the treatment of inflammatory diseases: Challenges, opportunities, and future perspectives. European Journal of Cell Biology. 2019;**98**(5-8):151041

[14] Gao F, Chiu SM, Motan DA, Zhang Z, Chen L, Ji HL, et al. Mesenchymal stem cells and immunomodulation: Current status and future prospects. Cell Death & Disease. 2016;**7**:e2062

[15] MacPherson A, Kimmelman J. Ethical development of stem-cell-based interventions. Nature Medicine. 2019;**25**(7):1037-1044

[16] Suman S, Domingues A, Ratajczak J, Ratajczak MZ. Potential clinical applications of stem cells in regenerative medicine. Advances in Experimental Medicine and Biology. 2019;**1201**:1-22

[17] de Miguel-Beriain I. The ethics of stem cells revisited. Advanced Drug Delivery Reviews. 2015;**82-83**:176-180

[18] Holland S, Lebacqz K, Zoloth L. The Human Embryonic Stem Cell Debate: Science, Ethics, And Public Policy. Cambridge, Massachusetts, USA: MIT Press; 2001. ISBN: 9780262082990

[19] Zheng YL. Some ethical concerns about human induced pluripotent stem cells. Science and Engineering Ethics. 2016;**22**(5):1277-1284

[20] Fung RK, Kerridge IH. Uncertain translation, uncertain benefit and uncertain risk: Ethical challenges facing first-in-human trials of induced pluripotent stem (ips) cells. Bioethics. 2013;**27**(2):89-96

[21] Lalu MM, McIntyre L, Pugliese C, Fergusson D, Winston BW, Marshall JC, et al. Safety of cell therapy with mesenchymal stromal cells (SafeCell): A systematic review and meta-analysis of clinical trials. PLoS One. 2012;**7**(10):e47559

[22] Thompson M, Mei SHJ, Wolfe D, Champagne J, Fergusson D, Stewart DJ, et al. Cell therapy with intravascular administration of mesenchymal stromal cells continues to appear safe: An updated systematic review and meta-analysis. EClinicalMedicine. 2020;**19**:100249

[23] Wang Y, Yi H, Song Y. The safety of MSC therapy over the past 15 years: A meta-analysis. Stem Cell Research & Therapy. 2021;**12**(1):545

[24] Park YB, Ha CW, Lee CH, Yoon YC, Park YG. Cartilage regeneration in osteoarthritic patients by a composite of allogeneic umbilical cord blood-derived mesenchymal stem cells and hyaluronate hydrogel: Results from a clinical trial for safety and proof-of-concept with 7 years of extended follow-up. Stem Cells Translational Medicine. 2017;**6**(2):613-621

[25] Petrou P, Kassis I, Ginzberg A, Halimi M, Yaghmour N, Abramsky O, et al. Long-term clinical and immunological effects of repeated mesenchymal stem cell injections in patients with progressive forms of multiple sclerosis. Frontiers in Neurology. 2021;**12**:639315

[26] Galderisi U, Peluso G, Di Bernardo G. Clinical trials based on mesenchymal stromal cells are exponentially increasing: Where are we in recent years? Stem Cell Reviews and Reports. 2022;**18**(1):23-36

[27] Lanzoni G, Linetsky E, Correa D, Messinger Cayetano S, Alvarez RA, Kouroupis D, et al. Umbilical cord mesenchymal stem cells for COVID-19 acute respiratory distress syndrome: A double-blind, phase 1/2a, randomized controlled trial. Stem Cells Translational Medicine. 2021;**10**(5):660-673

[28] Ozturk Turkmen H, Arda B. Ethical and legal aspects of stem cell practices in Turkey: Where are we? Journal of Medical Ethics. 2008;**34**(12):833-837

[29] Lee S, Kwon T, Chung EK, Lee JW. The market trend analysis and prospects of scaffolds for stem cells. Biomaterials Research. 2014;**18**:11

[30] Lim HC, Park YB, Ha CW, Cole BJ, Lee BK, Jeong HJ, et al. Allogeneic umbilical cord blood-derived mesenchymal stem cell implantation versus microfracture for large, full-thickness cartilage defects in older patients: A multicenter randomized clinical trial and extended 5-year clinical follow-up. Orthopaedic Journal of Sports Medicine. 2021;**9**(1):2325967120973052

[31] Japanese Law Translation Database System (Ministry of Justice of Japan). Act on Securing Quality, Efficacy and Safety of Products Including Pharmaceuticals

and Medical Devices. Available from: https://www.japaneselawtranslation.go.jp/en/laws/view/3213#je_ch6

[32] Cyranoski D. Japan's approval of stem-cell treatment for spinal-cord injury concerns scientists. Nature. 2019;**565**(7741):544-545

[33] Editorial. A stem-cell race that no one wins. Nature. 2019;**573**(7775):463

[34] Editorial. Stem the tide. Nature. 2015;**528**(7581):163-164

[35] Miyamoto S. Japan responds: Stem-cell therapy justified. Nature. 2019;**569**(7754):40

[36] Keio University School of Medicine. Clinical Research on First-In-Human Clinical Trial of Transplantation: Regenerative Medicine Using iPS Cell-Derived Neural Stem/Progenitor Cells to Treat Complete Subacute Spinal Cord Injury. Tokyo, Japan: Keio University Hospital Press Release; 2022. Available from: https://www.keio.ac.jp/en/press-releases/2022/Jan/14/49-92092/

[37] Graduate School of Medicine. Doctor-Initiated Clinical Trial Begun for Cell Spray Heart Failure Treatment Method. Suita, Japan: Osaka University Hospital Press Release; 2019. Available from: https://www.med.osaka-u.ac.jp/eng/archives/5618

[38] Takashima K, Morrison M, Minari J. Reflection on the enactment and impact of safety laws for regenerative medicine in Japan. Stem Cell Reports. 2021;**16**(6):1425-1434

[39] Matsushita S, Tachibana K, Kusakabe T, Hirayama R, Tsutsumi Y, Kondoh M. The roadmap to approval under Japan's two-track regulatory system: Comparing six regenerative medical products. Cell Stem Cell. 2020;**27**(4):515-518

[40] Takashima K, Inoue Y, Tashiro S, Muto K. Lessons for reviewing clinical trials using induced pluripotent stem cells: Examining the case of a first-in-human trial for age-related macular degeneration. Regenerative Medicine. 2018;**13**(2):123-128

[41] Badulescu D, Badulescu A. Medical tourism: Between entrepreneurship opportunities and bioethics boundaries: Narrative review article. Iranian Journal of Public Health. 2014;**43**(4):406-415

[42] NaRanong A, NaRanong V. The effects of medical tourism: Thailand's experience. Bulletin of the World Health Organization. 2011;**89**(5):336-344

[43] Harzing A-W, Giroud A. The competitive advantage of nations: An application to academia. Journal of Informetrics. 2014;**8**(1):29-42

[44] Radosevic S, Yoruk E. Are there global shifts in the world science base? Analysing the catching up and falling behind of world regions. Scientometrics. 2014;**101**(3):1897-1924

[45] Lyons S, Salgaonkar S, Flaherty GT. International stem cell tourism: A critical literature review and evidence-based recommendations. International Health. 2022;**14**(2):132-141

[46] ISCT Head Office. ISCT Establishes Working Group to Tackle Regulatory and Ethical Issues with Expanded Access. Vancouver, Canada: Telegraft Hub. ISCT News and Community; 2022. Available from: https://www.isctglobal.org/telegrafthub/blogs/isct-head-office1/2022/02/28/eaworkinggroup

[47] Pan American Health Organization. Regulation of advanced therapy medicinal products: Concept note and recommendations. In: Ninth Conference

of the Pan American Network for Drug Regulatory Harmonization (PANDRH). San Salvador, October 2018: PAHO/HSS/19-004; 2019

[48] Master Z, Robertson K, Frederick D, Rachul C, Caulfield T. Stem cell tourism and public education: The missing elements. Cell Stem Cell. 2014;**15**(3):267-270

[49] International Society for Stem Cell Research. A Closer Look at Stem Cells. Available from: https://www.closerlookatstemcells.org/

[50] 115th Congress (2017-2018). H.R.878 - 115th Congress (2017-2018): Right to Try Act of 2017. 2017. Available from: https://wwwcongressgov/bill/115th-congress/house-bill/878

[51] Agarwal R, Saltz LB. Understanding the right to try act. Clinical Cancer Research. 2020;**26**(2):340-343

[52] Coughlin C, King NMP, McKinney M. Regenerative medicine and the right to try. Wake Forest Journal of Bus & Intell Prop L. 2017;**18**:590

[53] Mahant V. "right-to-try" experimental drugs: An overview. Journal of Translational Medicine. 2020;**18**(1):253

[54] Cohen T. Exclusive: BrainStorm will not provide ALS therapy under U.S. right to try act. Reuters. Healthcare & Pharma. 2018. Available from: https://www.reuters.com/article/us-health-brainstorm-cell-als-exclusive/exclusive-brainstorm-will-not-provide-als-therapy-under-u-s-right-to-try-act-idUSKBN1JM1BE

[55] Zettler ME, Jeune-Smith Y, Feinberg BA, Phillips EG Jr, Gajra A. Expanded access and right to try requests: The community Oncologist's experience. JCO Oncology Practice. 2021;**17**(11):e1719-e1e27

[56] Mehzer M. GAO Reports on FDA, Drugmaker Efforts to Boost Access to Investigational Drugs. www.raps.org

[57] Texas Legislative Session 85(R). HB 810. Relating to the Provision of Certain Investigational Stem Cell Treatments to Patients with Certain Severe Chronic Diseases or Terminal Illnesses and Regulating the Possession, Use, and Transfer of Adult Stem Cells; Creating a Criminal Offense. 2017. Available from: https://capitoltexasgov/billlookup/Historyaspx?LegSess=85R&Bill=HB810

[58] Matthews K, Kunisetty B, Sprung K, Texas HB. 810: Increased access to stem cell interventions or an increase in unproven treatments? Stem Cells and Development. 2018;**27**(21):1463-1465

[59] Brodrick M. Free to choose: A moral defense of the right-to-try movement. The Journal of Medicine and Philosophy. 2020;**45**(1):61-85

[60] Servick K. Texas signals support for unproven stem cell therapies. Science. 2017;**356**(6344):1219

[61] Woods S, McCormack P. Disputing the ethics of research: The challenge from bioethics and patient activism to the interpretation of the declaration of Helsinki in clinical trials. Bioethics. 2013;**27**(5):243-250

[62] Debruin D. Reflections on "vulnerability". Bioethics Examiner. 2001;**5**(2):1-4

[63] Robertson JA. Controversial medical treatment and the right to health care. The Hastings Center Report. 2006;**36**(6):15-20

[64] Chahal M. Off-trial access to experimental cancer agents for the

terminally ill: Balancing the needs of individuals and society. Journal of Medical Ethics. 2010;**36**(6):367-370

[65] Brodrick M. Evolving Ethics of 'Right to Try' Unproven Drugs. Morrisville, North Carolina, USA. Medical Ethics Advisor: Relias Media; 2021. pp. 13-24

[66] United States Food and Drug Administration. Charging for Investigational Drugs Under an IND - Questions and Answers. 2016. Available from: https://www.fda.gov/regulatory-information/search-fda-guidance-documents/charging-investigational-drugs-under-ind-questions-and-answers

[67] Anderson WF. What's the rush? Human Gene Therapy. 1990;**1**(2):109-110

[68] Hyun I. Therapeutic hope, spiritual distress, and the problem of stem cell tourism. Cell Stem Cell. 2013;**12**(5):505-507

[69] Pope Francis. To Directors of the Orders of Physicians of Spain and Latin America. 2016. Available from: https://wwwvaticanva/content/francesco/en/speeches/2016/june/documents/papa-francesco_20160609_ordini-medici-spagna-america-latinahtml

[70] King NM, Perrin J. Ethical issues in stem cell research and therapy. Stem Cell Research & Therapy. 2014;**5**(4):85

[71] Pele A. Una aproximación al concepto de dignidad humana. Universitas: Revista de filosofía, derecho y política. 2004;**1**:9-13

[72] Spaemann R. Sobre el concepto de dignidad humana. Persona y Derecho. 1988;**19**:13-33

[73] Bohórquez Monsalve V, Aguirre RJ. As tensões da dignidade humana: conceituação e aplicação no direito internacional dos direitos humanos. Sur Revista Internacional de Direitos Humanos. 2009;**6**(11):40-63

[74] Corte Interamericana de Derechos Humanos. Ficha Técnica: Artavia Murillo y otros (Fertilización in Vitro) Vs. Costa Rica Sentencia de excepciones preliminares, fondo, reparaciones y costas N° 257. 2012. Available from: https://www.corteidh.or.cr/cf/Jurisprudencia2/ficha_tecnica.cfm?nId_Ficha=235

[75] Pope Benedict XVI. Address of His Holiness Benedict XVI to Participants in the International Conference Promoted by The Pontifical Council for Culture. 2011. Available from: https://wwwvaticanva/content/benedict-xvi/en/speeches/2011/november/documents/hf_ben-xvi_spe_20111112_stem-cellshtml

[76] Ratzinger J, Dio E, Mondo I. Essere Cristiani Nel Nuovo Millennio. Revista Agustiniana. 2002;**43**:43

[77] Pope Benedict XVI. Instruction Dignitas Personae on Certain Bioethical Questions. 2008. Available from: https://wwwvaticanva/roman_curia/congregations/cfaith/documents/rc_con_cfaith_doc_20081208_dignitas-personae_enhtml

[78] Snyder J, Crooks VA, Johnston R. Perceptions of the ethics of medical tourism: Comparing patient and academic perspectives. Public Health Ethics. 2012;**5**(1):38-46

[79] Keener AB. Arrests reveal debate about costs and benefits of proton therapy. Nature Medicine. 2014;**20**(10):1081

[80] O'Brien A, Sokol DK. Lessons from the Ashya King case. BMJ. 2014;**349**:g5563

[81] Adams J. Ashya King cleared of cancer three years after his parents abducted him from hospital for treatment abroad. The LA Times. The Telegraph. 2018. Available from: https://www.telegraph.co.uk/news/2018/03/03/ashya-king-cleared-cancer-three-years-parents-abducted-hospital/

[82] Yock TI, Yeap BY, Ebb DH, Weyman E, Eaton BR, Sherry NA, et al. Long-term toxic effects of proton radiotherapy for paediatric medulloblastoma: A phase 2 single-arm study. The Lancet Oncology. 2016;**17**(3):287-298

[83] NHS. Proton Beam Therapy 'effective' and 'Causes Fewer side Effects': Defunct Site. 2016. Available from: https://web.archive.org/web/20160201230211/http://www.nhs.uk/news/2016/02February/Pages/Proton-beam-therapy-effective-and-causes-fewer-side-effects.aspx

[84] Siegel B. More labs than patients : Desire to Be first colors gene studies. Los Angeles Times. 1987. Available from: https://www.latimes.com/archives/la-xpm-1987-12-14-mn-19227-story.html

[85] Sipp D, Okano H. Japan strengthens regenerative medicine oversight. Cell Stem Cell. 2018;**22**(2):153-156

[86] McDonnell PJ. Excimer laser photorefractive keratectomy. The Food and Drug Administration panel speaks. Archives of Ophthalmology. 1995;**113**(7):858-859

[87] Stuart A. A look at LASIK past, present and future. American Academy of Ophthalmology. San Francisco, California, USA: EyeNet Magazine. American Academy of Ophthalmology; June 2009. Available from: https://www.aao.org/eyenet/article/look-at-lasik-past-present-future?june-2009

[88] Schoenberg N. Lasik surgery falling out of favor with patients. chicagotribunecom. The Chicago Tribune. 2016. Available from: https://www.chicagotribune.com/lifestyles/health/sc-lasik-loses-luster-health-0525-20160526-story.html

[89] Matthews KR, Iltis AS. Unproven stem cell-based interventions and achieving a compromise policy among the multiple stakeholders. BMC Medical Ethics. 2015;**16**(1):75

[90] International Organization for Standardization. ISO 24651:2022. Biotechnology — Biobanking — Requirements for Human Mesenchymal Stromal Cells Derived from Bone Marrow. Geneva, Switzerland: ISO; 2022

[91] Martínez DM. Colombiana lidera reglamentación mundial para cierto tipo de células madre. Bogotá, Colombia: Pontificia Universidad Javeriana (a university in Colombia). 2022. Available from: https://www.javeriana.edu.co/pesquisa/celulas-madre-colombiana-reglamenta-investigacion/

Chapter 9

Therapeutic Features of Mesenchymal Stem Cells and Human Amniotic Epithelial Cells in Multiple Sclerosis

Reza ArefNezhad and Hossein Motedayyen

Abstract

Imbalance in immune responses plays an indispensable role in pathogenesis and development of multiple sclerosis (MS), as a neurodegenerative disorder. Current treatments are not always successful in preventing MS development and treating the disease. Stem cell-based cell therapy has provided a new window for treating neurodegenerative disorders. Stem cells can regulate the immune system and improve axonal remyelination. They can be isolated from different origins such as bone marrow, embryonic, neural, and adipose tissues. However, there is a challenge in choosing the best cell source for stem cell therapy. Mesenchymal stem cells (MSCs) derived from different origins have significant immunoregulatory impacts on different cells from the immune system. A growing body of evidence indicates that adipose tissue and umbilical cord can be a suitable source for obtaining MSCs. Moreover, human amniotic epithelial cell (hAEC), as a novel stem cell with immunoregulatory effects, regenerative effects, and low antigenicity, can be a candidate for MS treatment. This chapter discusses therapeutic impacts of MSCs and hAECs in MS disease.

Keywords: multiple sclerosis, mesenchymal stem cell, human amniotic epithelial cell, regenerative impacts, immunomodulatory effects

1. Introduction

Neurodegenerative disorders are considered as a chronic and progressive inflammatory condition resulting in the deposition of abnormal forms of specific proteins in the nervous system and destruction of neurons in motor, sensory, or cognitive systems [1, 2]. These disorders mainly involve women and are observed in subjects with age ranging from 20 to 30 years [3]. It is reported that more than 2.5 million individuals suffer from multiple sclerosis (MS) around the world who require effective therapeutic approaches to control disability and recover the central nervous system (CNS) functions [3]. The major problem of the management of the disease is the lack of successful regeneration of neurons [4].

Until now, several therapeutic approaches for MS have been suggested to control abnormal immune responses including natalizumab, interferon-β (IFN-β), glatiramer acetate, and fingolimod (FTY720). These treatments mainly exert their inhibitory effects on immune reactions and thereby reduce the number of relapses and modulate the progression of neurologic disability. However, they have not been consistently successful and are suitable in arresting the disease in approximately 30% of relapsing-remitting (RR) MS patients as the most common form of MS [5, 6]. It is reported that these treatments fail to control the degeneration of nerve tissue in an aggressive form of MS [7]. Among these approaches, stem cell-based therapies show a hopeful outlook for decreasing neural damages in the neurological diseases through regenerative roles for remyelination, the secretions of neurotrophic mediators with immunomodulatory impacts, and differentiation into astrocytes and oligodendrocytes effectively *in vivo* and *in vitro* [8]. Previous studies have shown some challenges for using stem cells as a curative treatment in clinical trials such as tumorigenicity and immunogenicity [9, 10]. However, extensive data of the literature have indicated that stem cell therapy exerts positive effects on animal models with neurological disorders [11, 12]. Clinical uses of adult stem cells, particularly mesenchymal stem cells (MSCs) and human amniotic epithelial cells (hAEC), have been recommended for the management of neurological diseases such as MS [13–15]. Several advantages have been reported for their therapeutic applications including the following: (1) their relative safety and low immunogenicity in comparison with other stem cell sources [16, 17]; (2) the ease of their accessibility, isolation, expansion, and manipulation *ex vivo* [18]; (3) their potency in differentiation into mesodermal lineages [16]; and (4) their capability to transport from the blood to damaged sites. Hereby, this chapter aimed to describe and discuss evidence regarding MSC- and hAEC-based therapies and their mechanisms for treating MS.

2. MS and its pathogenesis

Multiple sclerosis (MS) is the most common non-traumatic disabling disease, resulting in axonal loss and myelin disruption. The frequent features of MS are formations of lesions and sclerotic plaques in the central nervous system (CNS) and the cerebrospinal cord. The immune system plays a critical role in neural evolution through regulating oligodendrogenesis, neurogenesis, and synaptic organization. Therefore, immune cells can participate in the pathogenesis and development of MS [19, 20]. The pathogenesis of MS is largely related to hormone, environmental, and genetic factors. It is reported that alterations in expressions and functions of some immune agents such as major histocompatibility complex (MHC), immunoglobulin (Ig), T-cell receptor (TCR), and cytokines can contribute to the increased risk of MS [6, 21]. Today, studies on MS have indicated that autoreactive T-cell migration to the CNS occurs upon autoimmune cascade initiation and blood-brain barrier (BBB) disruption, which leads to destroy myelin sheath and creates sclerotic lesions and plaques [6, 22]. Destruction of the myelin sheath, which plays a significant role in survival and integration of axon, is a major reason for the development of MS [3]. T helper 1 (Th1) and T helper 17 (Th17) cells are the main effector cells that participate in the demyelination and destruction of the CNS [19, 20]. Th1 and Th17 produce some pro-inflammatory cytokines, including inerleukine-1 (IL-1), IL-17, interferon-gamma (IFN-γ), and tumor necrosis factor-alpha (TNF-α) [23]. Moreover, CD8+ T cells are found in MS lesions, especially around the blood vessels. Previous studies

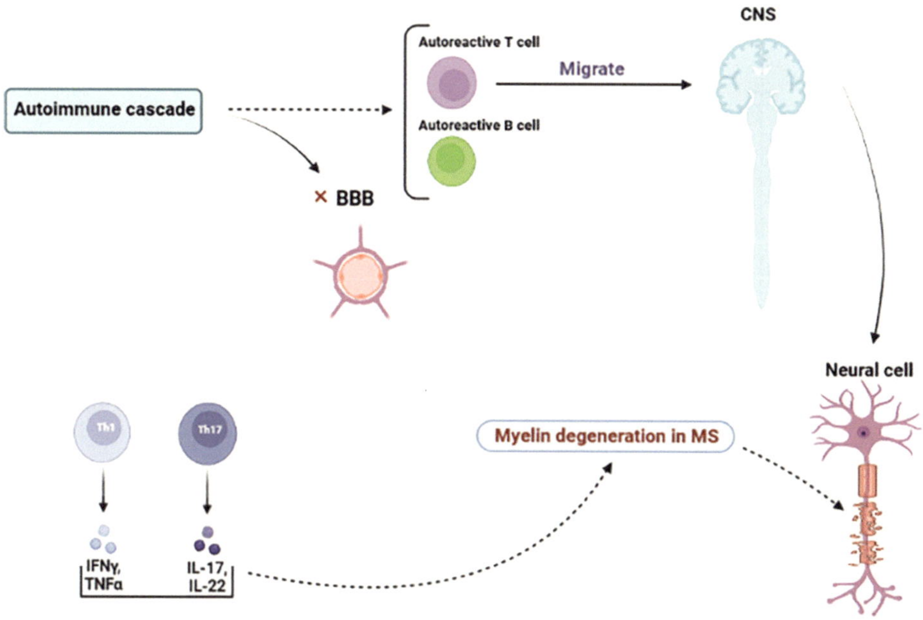

Figure 1.
The impacts of immune responses in MS pathogenesis.

have revealed that the proliferation of CD8+ T cell in patients with MS is more than CD4+ T cell, which is largely associated with axon injury [1]. Besides the roles of T cells in the pathogenesis of MS, other immune cells play important roles in the formations of lesions and plaques. The activation of macrophage by Th1 cytokines leads to the destruction of the myelin and thereby exposes more CNS antigens. Although it is demonstrated that autoreactive T cells are the major effector cells for the pathogenesis of MS, some reports have indicated that autoreactive B cells have critical roles in disappearing the myelin sheaths and axonal loss, through cytokine secretions, antigen presentations, and autoantibody productions [24]. Autoantibodies can be major immune mediators that can be found in MS plaques. There are some reports pointing toward the association of immunoglobulin G (IgG) with MS signs. Furthermore, it is shown that IgG, especially IgG against proteolipid proteins (PLP) and myelin basic proteins (MBP), can be considered as the features of the disease, although their roles in MS pathogenesis are not well identified yet (**Figure 1**) [25].

3. Mesenchymal stem cells

MSCs can be obtained from different tissues such as bone marrow, adipose tissue, umbilical cord, brain, dental tissue, and fetal lung [26–28]. MSCs can differentiate into monocytes and neurons *in vitro* and *in vivo* [29]. These cells can migrate to injured tissue *via* expressions of the receptors for chemokines such as CXCR4, CXCR5, CXCR6, CCR1, and some growth factors [7]. In line with potential therapeutic effects of MSCs, it is revealed that these cells possess anti-oxidant and anti-apoptotic impacts and are able to secrete trophic factors, which can contribute to support axon and increase neural stability [30]. They improve neural cell differentiation, promote angiogenesis, inhibit neuron apoptosis, and repair the CNS in MS patients [25].

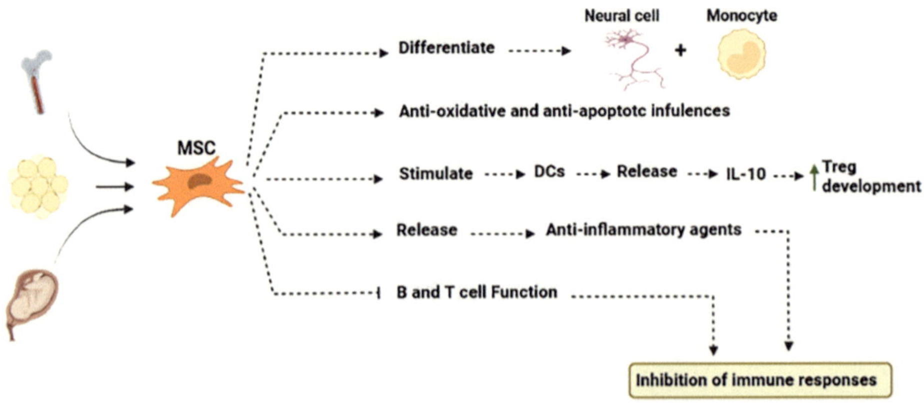

Figure 2.
Immunoregulatory and therapeutic effects of MSCs.

Furthermore, these cells can recruit oligodendrocyte precursors to the CNS and induce their differentiation into neuronal cells [31, 32]. MSCs exert immunomodulatory impacts through suppressing the activity of B, T, and other immune cells [33]. Intravascular MSC therapy improves CNS tissue repair through the induction of T-cell tolerance to myelin glycoproteins [34]. Studies on experimental autoimmune encephalomyelitis (EAE), an animal model of MS, indicated that intravenous injection of syngeneic MSCs induces tolerance in MOG-specific T cells and thereby reduces immune cell infiltrations to the CNS and increases the clinical course [35, 36]. Others have revealed that the immunoinhibitory effects of MSCs are mediated through the secretions of anti-inflammatory cytokines such as TGF-β, prostaglandinE-2 (PGE-2), and indoleamine-pyrrole 2, 3-dioxygenase (IDO) [37]. Previous studies have demonstrated that MSCs can impair B-cell proliferation and antibody production through inhibiting the activation and proliferation of Th1 cells [38]. These cells also improve the activation of suppressor of cytokine signaling 3 (SOCS3) and decrease the differentiation of Th17 via the IFN-γ pathway [39]. IDO, as a mechanism used by MSC for controlling immune responses, depletes tryptophan from the environment of lymphocytes, which plays a key role in lymphocyte activations [40]. MSC participates in the development of regulatory T cells (Treg) through inducing IL-10 secretion of peripheral dendritic cells (DCs) [40]. In line with the improvement of peripheral tolerance, it is reported that MSCs inhibit the differentiation and function of DCs, resulting in the inhibition of clonal expansion of autoreactive T cells via the reduction of antigen presentation [41]. Hepatocyte growth factor (HGF) produced by MSC increases tolerogenic DCs [42]. MSCs with HGF can reduce immune cell infiltrations and CNS inflammation in EAE mice [42]. Thus, HGF derived from MSC may be effective in MS treatment. Several studies on genetically modified MSCs have shown that over-expressed anti-inflammatory cytokines such as IL-10 and IL-4 can participate in suppressing immune responses, reducing BBB injury, and improving remyelination of neurons in EAE mice (**Figure 2**) [43].

4. Human bone marrow-derived MSCs (hBM-MSCs)

MSCs obtained from bone marrow have multiple properties, which make them an attractive cell source for therapeutic applications (**Table 1**). These cells are the most

Different sources of MSCs	Advantages	Disadvantages
hBM-MSCs	1. Colony formations that contribute to hemopoiesis [44]. 2. Differentiation into mesodermal lineage cells [44]. 3. Secretion of brain-derived neurotrophic factor (BDNF) and enhancement of oligodendrogenesis. 4. Immunomodulatory impacts in animal models of neurodegenerative disorders, especially in the early stages of MS [16].	1. Invasive and painful techniques with low efficiency for its isolation [45]. 2. Low efficiency in controlling disease progression in the stabilized stage of MS [16]. 3. The probability of malignant transformation and immune rejection after clinical applications [46, 47].
hUC-MSCs	1. Less-invasive techniques for its isolation [48]. 2. Less ethical issues for its isolation [48]. 3. High proliferation ability [48]. 4. Low immunogenicity potential [48]. 5. Differentiation capacity into various lineages [48] 6. Immunoregulatory impacts in animal models of autoimmune disorders [49]. 7. Production of nerve growth factors [50].	1. Tumorigenic potential and immune rejection after clinical use [51]. 2. Procoagulant properties that may contribute to pulmonary embolism [52]. 3. The risk of viral and prion transmission after administration [53, 54].
AD-MSCs	1. Easy isolation with high efficiency of adipose tissue [53]. 2. Adipogenic, cardiogenic, neurogenic, myogenic, chondrogenic, and osteogenic features [53]. 3. Migration to different organs through expression of α4 integrin [55]. 4. Production of different growth factors [56].	1. Tumorigenic potential and immune rejection after their clinical use [57]. 2. Nephrotoxicity potential [49, 57]. 3. Procoagulant properties [52].
hAECs	1. Easy isolation with high efficiency of the amniotic membrane [58]. 2. Pluripotency and self-renewal properties [18, 59]. 3. Differentiation into the cells originating from three germinal layers [18]. 4. Production of neurotrophic mediators [59, 60]. 5. Immunomodulatory impacts in animal models of inflammatory neurological disorders [61, 62]. 6. Low tumorigenic and immunogenicity potential [15, 18, 63].	1. The controversial formation of teratomas upon hAEC-derived neural cell engraftment [64]. 2. Are not described as stem cells because they do not show long-term self-renewal and fail to grow the cells from single-cell clones [17].

Table 1.
The pros and cons of MSC-based therapies.

frequent cell sources used in clinical settings [49]. Given therapeutic features of hBM-MSCs in neurological disorders, it is revealed that they can promote disease recovery in relapsing-remitting and chronic types of MS in EAE mice, due perhaps to reduce demyelination regions and inflammatory infiltrates, induce oligodendrogenesis, and enhance brain-derived neurotrophic factor (BDNF) production [65]. Several studies have reported that BM-MSCs have immunomodulatory effects in EAE through preventing the maturation of antigen-presenting cells (APCs) and proliferation of B and T cells [39]. These immunosuppressive impacts are mainly mediated by releasing various bioactive mediators [66]. Moreover, their neuroprotective effects can induce local progenitor cells and suppress scar creation, gliosis, and neuron apoptosis [67]. Besides having immunomodulatory impacts, they have the ability to differentiate into the neurons and improve the replacement of the cells [67]. Nonetheless, the isolation of BM-MSC is painful, invasive, and low efficiency [45], which may be considered a disadvantage in their clinical applications. In the first phase of clinical trial using autologous *ex vivo* expanded BM-MSCs on patients with advanced MS, it was reported that 30% of patients were unable to grow an acceptable number of these cells ($< 2 \times 10^6$) despite several bone marrow aspirations. This observation has reflected an inherent deficiency of MSCs in the bone marrow of participants [67]. Thus, MSCs derived from other tissues can be considered for MS treatment. In EAE, BM-MSCs are notable curative effects if they are used before disease initiation due to a significant suppression on effector T cells and the induction of peripheral tolerance. However, these cells fail to control disease development in the stabilized stage of MS [16].

5. Human umbilical cord (hUC)-MSCs

HUC-MSCs have significant characteristics, which distinguish them from other sources of MSCs (**Table 1**). Several lines of evidence suggest the administration of hUC-MSCs in autoimmune disorders such as encephalomyelitis, type 1 diabetes, and rheumatoid arthritis due to its immunoregulatory impacts [68–71]. Immunomodulatory effects of these cells have fundamental roles in tissue recovery [72]. They have the decreased expression of HLA-I, increased capacity of proliferation, and more rapid growth *in vitro*, compared with BM-MSCs [73]. *In vitro* and *in vivo* studies have indicated that hUC-MSCs have a positive effect on Treg proliferation [74]. hUC-MSCs can increase behavioral activities and reduce the histopathological impairments of EAE. Furthermore, they exert a positive effect on the productions of IL-4 and IL-10, unlike IL-1 and IL-6 [70]. *In vitro* studies have demonstrated that hUC-MSCs can enhance the frequency of Treg and secretion of anti-inflammatory

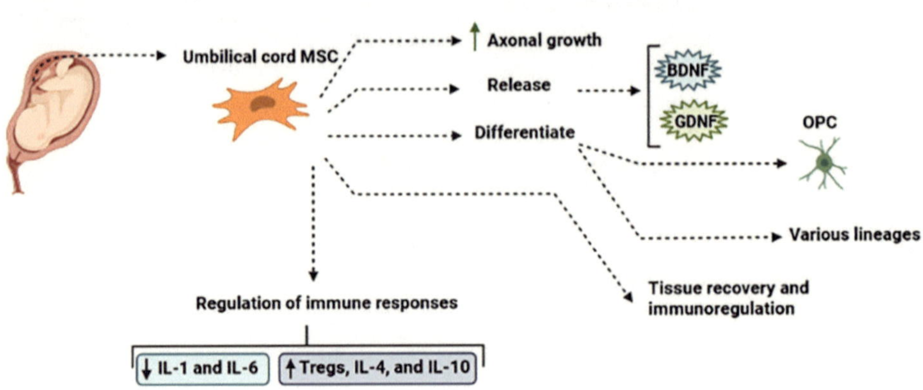

Figure 3.
Immunomodulatory and therapeutic impacts of hUC-MSCs.

cytokines of peripheral blood mononuclear cells (PBMCs) (**Figure 3**) [74, 75]. In addition to their immunoregulatory effects, this source of stem cells is able to release several nerve growth factors, for example, glial cell-derived neurotrophic factor (GDNF) and BDNF. Moreover, their capacity to differentiate into oligodendrocyte precursor cells can improve axonal growth [50].

6. Human adipose-derived MSCs (AD-MSCs)

AD-MSC can be obtained from adipose tissue by collagenase digests. Adipose tissue contains high levels of MSCs (approximately 100–1000 MSCs per gram of fat) and is easily accessible for use. Thus, this tissue is an important source of the cell for cellular therapy. AD-MSCs show the adipogenic, cardiogenic, neurogenic, myogenic, chondrogenic, and osteogenic features *in vitro*, which make them a fantastic cell source for stem cell therapy [8, 76]. Unlike BM-MSCs, these cells are able to migrate to different organs due to express α4 integrin, an adhesive molecule [55]. It is suggested that autologous and allogeneic AD-MSCs are effective in the treatment of the diseases with immunopathogenesis such as MS and autoimmune encephalomyelitis [77–80]. Study on EAE mice revealed that intravascular AD-MSC participates in the reduction of immune infiltration in the CNS and decreases demyelination and axonal loss [9]. Various growth factors released from AD-MSCs, such as anti-apoptotic, angiogenic, and neurotrophic mediators, play critical roles in cell differentiation, proliferation, and maturation [56]. It is thought that AD-MSCs have more capabilities for stem cell-based cell therapy, due perhaps to the expression of integrin α4β1; pass the BBB; and exert their anti-inflammatory, immunoregulatory, and neurodegenerative impacts (**Figure 4**) [7]. Until now, several studies have been performed to find a standard method for the treatment of MS by these cells [81, 82]. Nonetheless, there are some concerns regarding the clinical application of MSCs such as tumorigenesis and immune rejection after use that must be addressed in future studies.

7. Human amniotic epithelial cells (hAECs)

hAECs are easily isolated from the amniotic membrane, the inner layer of the fetal membranes, and possess some stem cell-like properties [83–86]. These cells express

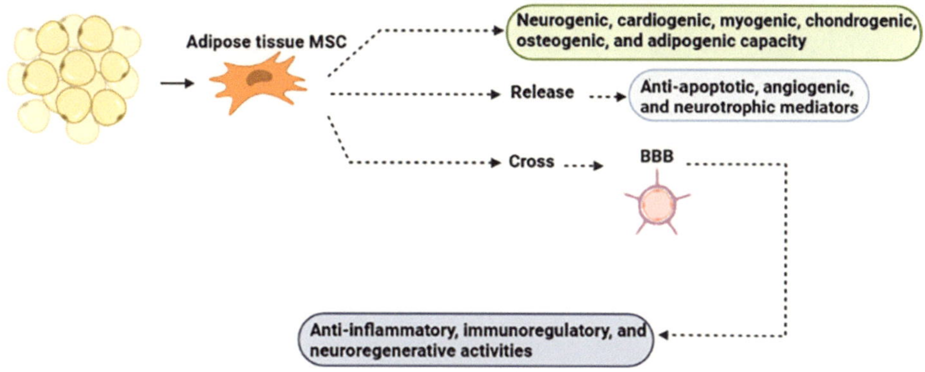

Figure 4.
Immunoregulatory and therapeutic impacts of AD-MSCs in degenerative disorders.

some markers of pluripotent stem cells, including FGF-T, Sox-2, Nanog, Rex-1, SSEA-4, and Oct4. Some of these markers have important roles in pluripotency and self-renewal properties in induced pluripotent stem (iPS) cells and embryonic stem cells (ESCs) [87]. hAEC can differentiate into different cells such as the pancreatic cells, neural cells, hepatocytes, cardiomyocytes, adipocytes, and myocytes, which originate from the endoderm, ectoderm, and mesoderm [18]. It is reported that hAECs have immunomodulatory impacts on adaptive and innate immune systems [17, 88, 89]. They exert suppressive effects on the activations of natural killer (NK) and CD4+ T cells, migrations of neutrophil and macrophage, secretions of pro-inflammatory cytokines of CD4+ T cells, and proliferation of B cells [23, 61, 90, 91]. These impacts are primarily mediated through the productions of immunoregulatory mediators, such as IL-4, PG-E2, and transforming growth factor-beta (TGF-β), which may participate in the increase of Tregs and Th2 cells, inhibition of pathogenic T-cell reactions, and protection of the peripheral naive CD4+ T-cell source [61, 63, 92–95]. These effects suggest that hAECs may be considered as an effective cell source for MS treatment [62, 95]. To support this notion, they can contribute to a shift from Th1-type responses to Th2-type responses [95]. EAE mice treated with hAECs experienced significant reductions in demyelination and immune infiltration into the CNS [95]. It is indicated that hAECs have a negative effect on Th17 differentiation through reducing the productions of TGF-β and IL-6, which play indispensable roles in the differentiation of these cells [88]. Studies on animal models of MS have revealed that alpha-fetoprotein (AFP) produced from hAECs participates in the reduction of lymphocyte function and neuroinflammation [96, 97]. Others have indicated that these cells can reduce gray and white matter damages through residing in inflammation locations such as the brain [98]. Furthermore, they can release neurotrophic agents such as neurotrophin-3 (NT-3), nerve growth factor (NGF), and brain-derived neurotrophic factor (BDNF) (**Figure 5**) [59, 60]. These features along

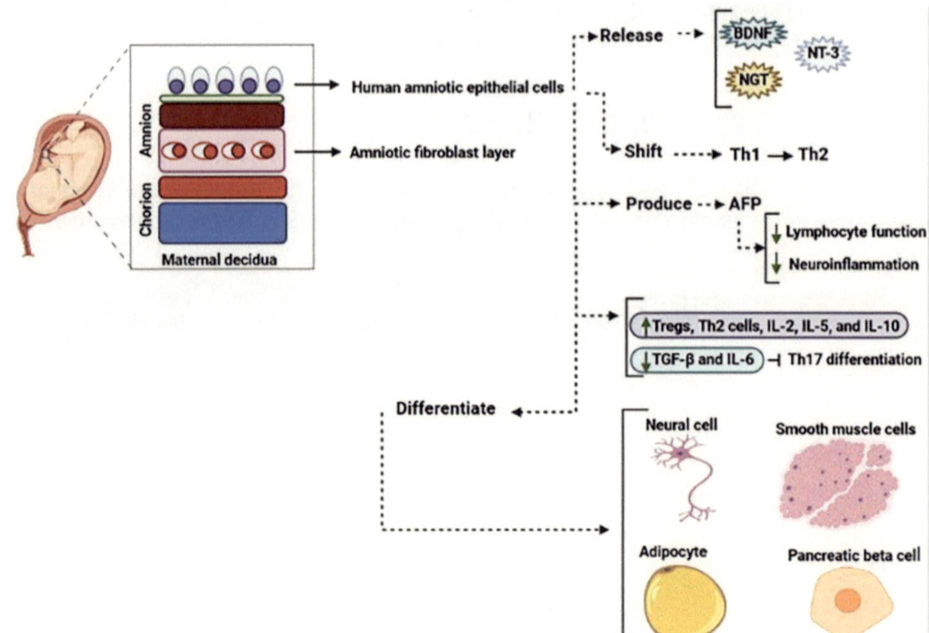

Figure 5.
Immunomodulatory and stem cell characteristics of hAECs.

with low antigenicity provide additional confirmations to clarify therapeutic properties of hAECs in the treatment and management of inflammatory neurological disorders such as MS [99]. hAECs possess a limited proliferative potential due perhaps to the lack of telomerase [87, 92, 100, 101], which helps to reduce potential tumorigenicity of stem cell-based therapies. Nevertheless, it should be noted that further works and more information are needed to illustrate the possible capability of these cells in treating diseases with immune pathophysiology.

8. Comparison of hAECs with MSCs derived from different sources

There are some differences and similarities between hAECs and MSCs derived from different sources, for example, morphologic and tumorigenic properties, immunoregulatory characteristics, angiogenesis capacities, and ethical issues associated with their isolations and applications [45, 102]. In line with morphology, hAECs show a cobblestone-like morphology, while the cultured hAMSCs have a spindle fibroblast-like morphology [103]. The morphologic feature of the cultured MSCs derived from BM can range from fibroblast-like spindle-shaped cells to large flat cells [104]. MSCs from other sources, such as AD-MSCs and hUC-MSCs, indicate spindle shapes in the culture [105, 106]. The amniotic membrane can be collected by standard isolation methods following cesarean section, which is not invasive and does not have unfavorable effects on human embryos and ethical issues [102]. The isolation of amniotic cells can simply be performed upon prenatal testing. However, there are some ethical problems in regard to clinical applications of MSCs and the isolation of some sources of MSCs [104, 107]. As mentioned above, the isolation of hBM-MSCs is done by invasive techniques with low efficiency, which is painful [45]. Today, there is no document pointing to the tumorigenicity of amnion membrane or membrane-originated cells after clinical applications [17]. However, some reports have shown that MSCs may raise tumor growth in some cancer mouse models [108]. Several lines of evidence propose that hAECs, BM-MSCs, and AD-MSCs participate in enhancing angiogenesis through the productions of some cytokines and angiogenic factors, such as VEGF, HGF, and EGF, and mechanisms associated with protease [103, 109]. According to evidence, hAECs possess better immunomodulatory impacts but lesser osteogenic effects than BM-MSCs and MSCs derived from the human amniotic fluid (hAF) [110]. hAECs express some MSC markers such as CD90, CD44, and CD105. However, the levels of SSEA4 and SSEA3 expressions are higher on hAECs than those on hBM-MSCs and hAFMSC, revealing more multipotent potential of these cells [17, 111]. Furthermore, hAECs and hAFMSC possess higher levels of PD-L1 and PD-L2 than hBM-MSCs, which may make them more successful in providing peripheral tolerance in immune cells [110, 112].

9. Mesenchymal stem cell-based cell therapy and clinical trials

Until now, several clinical trials were carried out using MSCs as a therapeutic approach for MS. In a phase II clinical trial, intravascular MSCs were employed in the treatment of nine relapsing-remitting multiple sclerosis (RRMS) patients. After 6 months, the results revealed a significant reduction in MS lesions in magnetic resonance imaging (MRI) [67]. In a phase IIa clinical trial, autologous BM-MSCs were injected to one RRMS and nine secondary progressive multiple sclerosis (SPMS) patients. After 3 months to 1 year, authors observed that BM-MSCs improved clinical features in the treated patients. In this clinical trial, 10 SPMS patients were treated with intravascular MSCs for 6 months, and the results revealed neuroprotection effect of MSCs and remyelination [67]. Furthermore, in a study conducted by Bonab et al., in 2007, therapeutic impacts of intrathecal injection of MSC were studied on 10 MS patients. This study indicated that the disease progression was gradually reduced in half of the participants [113]. Another study on 22 patients with primary progressive multiple sclerosis (PPMS) demonstrated that intravascular and intrathecal injections of BM-MSCs were effective in MS treatment [35]. In a triple-blind and placebo-controlled study on 30 patients with SPMS, the researchers indicated that AD-MSC injection is a possible and safe method in the treatment of SPMS patients [114]. Staff et al. reported the safety of intrathecal administration of AD-MSCs in amyotrophic lateral sclerosis (ALS) patients [115]. In a study conducted by Li et al., it was demonstrated that hUC-MSC transplantation is able to reduce MS symptoms and relapse occurrence in comparison with control individuals. In addition, the researchers observed that hUC-MSC administration results in a shift in Th1 responses toward Th2 immunity [116]. In line with the therapeutic impacts of MSCs, Riordan et al. indicated that hUC-MSC transplantation is safe and exerts suitable impacts on life quality and brain lesion in MS patients [117].

10. Conclusion

There are many documents pointing to stem cell-based cell therapy as a treatment for MS and other neurological disorders. However, an inconsistency in the results of these studies is observed. Among different types of stem cells, MSCs are more possible to consider as a therapeutic approach for MS treatment because they utilize different mechanisms involved in regulating immune responses and repairing CNS damages. Furthermore, MSCs have anti-oxidant and anti-apoptotic properties and trophic factor secretion, which exert positive effects on the axon and neural stability. Numerous studies have recommended that MSCs derived from umbilical cord and adipose tissue can be more effective for stem cell therapy. Moreover, hAECs are mentioned as a novel source of the cells, which have immunoregulatory effects and show a potential for differentiation into the cells originating from three germinal layers. Consequently, hAECs may be considered a therapeutic method to manage and control MS. However, more experimental studies should be done to illustrate their efficiency and mechanisms involved in the treatment of MS.

Finding

This study was not financially supported and was performed in personal capacity.

Conflicts of interest

The authors declare that there is no conflict of interest.

Author details

Reza ArefNezhad[1] and Hossein Motedayyen[2*]

1 Department of Anatomy, School of Medicine, Shiraz University of Medical Sciences, Shiraz, Iran

2 Autoimmune Diseases Research Center, Shahid Beheshti Hospital, Kashan University of Medical Sciences, Kashan, Iran

*Address all correspondence to: hmotedayyen@gmail.com

References

[1] Martin JB. Molecular basis of the neurodegenerative disorders. The New England Journal of Medicine. 1999;**340**(25):1970-1980

[2] ArefNezhad R, Motedayyen H, Mohammadi A. Therapeutic aspects of mesenchymal stem cell-based cell therapy with a focus on human amniotic epithelial cells in multiple sclerosis: A mechanistic review. International Journal of Stem Cells. 2021;**14**(3):241-251

[3] Bejargafshe MJ, Hedayati M, Zahabiasli S, Tahmasbpour E, Rahmanzadeh S, Nejad-Moghaddam A. Safety and efficacy of stem cell therapy for treatment of neural damage in patients with multiple sclerosis. Stem cell Investigation. 2019;**6**:44

[4] Minzenberg MJ. Dousing the flames to repair brains. Science Translational Medicine. 2010;**2**:27ec60-27ec60

[5] Yousefi F, Arab FL, Saeidi K, Amiri H, Mahmoudi MJ. Various Strategies to Improve Efficacy of Stem Cell Transplantation in Multiple Sclerosis: Focus on Mesenchymal Stem Cells and Neuroprotection. Journal of neuroimmunology. 2019;**328**:20-34

[6] Bejargafshe MJ, Hedayati M, Zahabiasli S, Tahmasbpour E, Rahmanzadeh S, Nejad-Moghaddam A. Safety and Efficacy of Stem Cell Therapy for Treatment of Neural Damage in Patients with Multiple Sclerosis. Stem Cell Investigation. 2019;**6**:44

[7] Ghasemi N. Therapeutic effects of adipose derived mesenchymal stem cells on remyelination process in inflammatory demyelinating diseases. Channels. 2015;**4**:6

[8] Abdallah AN, Shamaa AA, El-Tookhy OS. Evaluation of treatment of experimentally induced canine model of multiple sclerosis using laser activated non-expanded adipose derived stem cells. Research in Veterinary Science. 2019;**125**:71-81

[9] Nguyen H, Zarriello S, Coats A, Nelson C, Kingsbury C, Gorsky A, et al. Stem cell therapy for neurological disorders: A focus on aging. Neurobiology of Disease. 2019;**126**:85-104

[10] Pouryousefi-Koodehi T, Shayegan S, Hashemi S, Arefnezhad R, Roghani-Shahraki H, Motedayyen H, et al. Can mesenchymal stem cells derived from adipose tissue and their conditioned medium improve ovarian functions? A mini-review. Zygote. 2022;**2022**:1-4

[11] Nguyen H, Zarriello S, Coats A, Nelson C, Kingsbury C, Gorsky A, et al. Stem Cell Therapy for Neurological Disorders: A Focus on Aging. Neurobiology of disease. 2019;**126**:85-104

[12] Namchaiw P, Wen H, Mayrhofer F, Chechneva O, Biswas S, Deng W. Temporal and partial inhibition of GLI1 in neural stem cells (NSCs) results in the early maturation of NSC derived oligodendrocytes in vitro. Stem Cell Research & Therapy. 2019;**10**(1):272

[13] Mansoor SR, Zabihi E, Ghasemi-Kasman M. The potential use of mesenchymal stem cells for the treatment of multiple sclerosis. Life sciences. 2019;**235**:1-8

[14] Bitsch A, Bruhn H, Vougioukas V, Stringaris A, Lassmann H, Frahm J, et al. Inflammatory CNS demyelination: Histopathologic correlation with in vivo quantitative proton MR spectroscopy.

American Journal of Neuroradiology. 1999;**20**(9):1619-1627

[15] Akle C, Welsh K, Adinolfi M, Leibowitz S, McColl I. Immunogenicity of human amniotic epithelial cells after transplantation into volunteers. The Lancet. 1981;**318**(8254):1003-1005

[16] Darlington PJ, Boivin M-N, Bar-Or A. Harnessing the therapeutic potential of mesenchymal stem cells in multiple sclerosis. Expert Review of Neurotherapeutics. 2011;**11**(9):1295-1303

[17] Miki T, Lehmann T, Cai H, Stolz DB, Strom SC. Stem cell characteristics of amniotic epithelial cells. Stem Cells. 2005;**23**(10):1549-1559

[18] Motedayyen H, Zarnani A-H, Tajik N, Ghotloo S, Rezaei A. Immunomodulatory effects of human amniotic epithelial cells on naive CD4+ T cells from women with unexplained recurrent spontaneous abortion. Placenta. 2018;**71**:31-40

[19] Rahmati M, Ghannadian SM, Kasiri N, Ahmadi L, Motedayyen H, Shaygannejad V, et al. Modulation of Th17 proliferation and IL-17A gene expression by acetylated form of apigenin in patients with multiple sclerosis. Immunological Investigations. 2020;**2020**:1-14

[20] Sedaghat N, Motedayyen H, Alsahebfosoul F, Etemadifar M, Ostadi V, Kianpour F, et al. Increased expression of lymphocyte activation gene-3 by regulatory t cells in multiple sclerosis patients with fingolimod treatment multiple sikleroz hastalar ฤ ฑ nda fingolimod tedavisi sonucu d ร ผ zenleyici t h ร ผ crelerinde lenfosit aktivasyon gen-3 ekspresyonunda art ฤ ฑ ล. Turkish Journal of Immunology. 2019;7(1):31-39

[21] Imitola J. Regenerative neuroimmunology: The impact of immune and neural stem cell interactions for translation in neurodegeneration and repair. Journal of Neuroimmunology. 2019;**331**:1-3

[22] Yousefi F, Arab FL, Saeidi K, Amiri H, Mahmoudi M. Various strategies to improve efficacy of stem cell transplantation in multiple sclerosis: Focus on mesenchymal stem cells and neuroprotection. Journal of Neuroimmunology. 2019;**328**:20-34

[23] Motedayyen H, Rezaei A, Zarnani A-H, Tajik N. Human amniotic epithelial cells inhibit activation and pro-inflammatory cytokines production of naive CD4+ T cells from women with unexplained recurrent spontaneous abortion. Reproductive Biology. 2018;**18**(2):182-188

[24] Sedaghat NMH, Etemadifar M, Zarkesh H, Kianpour F, Vestri E, et al. Effect of Fingolimod on the frequency of regulatory T cells in patients with relapsing-remitting multiple sclerosis. Journal of Immune Research. 2018;**1**(5):1032

[25] Pirttilä T, Nurmikko T. CSF oligoclonal bands, MRI, and the diagnosis of multiple sclerosis. Acta Neurologica Scandinavica. 1995;**92**(6):468-471

[26] Markarian CF, Frey GZ, Silveira MD, Milani AR, Ely PB, Horn AP, et al. Isolation of adipose-derived stem cells: A comparison among different methods. Biotechnology Letters. 2014;**36**(4):693-702

[27] Fukuchi Y, Nakajima H, Sugiyama D, Hirose I, Kitamura T, Tsuji K. Human placenta-derived cells have mesenchymal stem/progenitor cell potential. Stem Cells. 2004;**22**(5):649-658

[28] Schwab KE, Hutchinson P, Gargett CE. Identification of surface markers for prospective isolation of human endometrial stromal

colony-forming cells. Human Reproduction. 2008;**23**(4):934-943

[29] Xu C, Diao Y-F, Wang J, Liang J, Xu H-H, Zhao M-L, et al. Intravenously infusing the secretome of adipose-derived mesenchymal stem cells ameliorates neuroinflammation and neurological functioning after traumatic brain injury. Stem Cells and Development. 2020;**29**(4):222-234

[30] Song C-G, Zhang Y-Z, Wu H-N, Cao X-L, Guo C-J, Li Y-Q, et al. Stem cells: A promising candidate to treat neurological disorders. Neural Regeneration Research. 2018;**13**(7):1294

[31] Samper Agrelo I, Schira-Heinen J, Beyer F, Groh J, Bütermann C, Estrada V, et al. Secretome analysis of mesenchymal stem cell factors fostering Oligodendroglial differentiation of neural stem cells In vivo. International Journal of Molecular Sciences. 2020;**21**(12):4350

[32] Jaramillo-Merchan J, Jones J, Ivorra J, Pastor D, Viso-León M, Armengol JA, et al. Mesenchymal stromal-cell transplants induce oligodendrocyte progenitor migration and remyelination in a chronic demyelination model. Cell Death & Disease. 2013;**4**(8):e779

[33] Forte A, Finicelli M, Mattia M, Berrino L, Rossi F, De Feo M, et al. Mesenchymal stem cells effectively reduce surgically induced stenosis in rat carotids. Journal of Cellular Physiology. 2008;**217**(3):789-799

[34] Peng Z, Gao W, Yue B, Jiang J, Gu Y, Dai J, et al. Promotion of neurological recovery in rat spinal cord injury by mesenchymal stem cells loaded on nerve-guided collagen scaffold through increasing alternatively activated macrophage polarization. Journal of Tissue Engineering and Regenerative Medicine. 2018;**12**(3):e1725-e1e36

[35] Brück W, Porada P, Poser S, Rieckmann P, Hanefeld F, Kretzschmarch HA, et al. Monocyte/macrophage differentiation in early multiple sclerosis lesions. Annals of Neurology: Official Journal of the American Neurological Association and the Child Neurology Society. 1995;**38**(5):788-796

[36] Freedman MS, Bar-Or A, Atkins HL, Karussis D, Frassoni F, Lazarus H, et al. The therapeutic potential of mesenchymal stem cell transplantation as a treatment for multiple sclerosis: Consensus report of the International MSCT Study Group. Multiple Sclerosis Journal. 2010;**16**(4):503-510

[37] Squillaro T, Peluso G, Galderisi U. Clinical trials with mesenchymal stem cells: An update. Cell Transplantation. 2016;**25**(5):829-848

[38] Najar M, Fayyad-Kazan M, Meuleman N, Bron D, Fayyad-Kazan H, Lagneaux L. Mesenchymal stromal cells of the bone marrow and natural killer cells: Cell interactions and cross modulation. Journal of Cell Communication and Signaling. 2018;**12**(4):673-688

[39] Liu X, Ren S, Qu X, Ge C, Cheng K, Zhao RCH. Mesenchymal stem cells inhibit Th17 cells differentiation via IFN-γ-mediated SOCS3 activation. Immunologic Research. 2015;**61**(3):219-229

[40] Wang D, Huang S, Yuan X, Liang J, Xu R, Yao G, et al. The regulation of the Treg/Th17 balance by mesenchymal stem cells in human systemic lupus erythematosus. Cellular & Molecular Immunology. 2017;**14**(5):423-431

[41] Selmani Z, Naji A, Gaiffe E, Obert L, Tiberghien P, Rouas-Freiss N, et al. HLA-G is a crucial immunosuppressive molecule secreted by adult

human mesenchymal stem cells. Transplantation. 2009;**87**(9S):S62-SS6

[42] Lu Z, Chang W, Meng S, Xu X, Xie J, Guo F, et al. Mesenchymal stem cells induce dendritic cell immune tolerance via paracrine hepatocyte growth factor to alleviate acute lung injury. Stem Cell Research & Therapy. 2019;**10**(1):1-16

[43] Duffy MM, Ritter T, Ceredig R, Griffin MD. Mesenchymal stem cell effects on T-cell effector pathways. Stem Cell Research & Therapy. 2011;**2**(4):34

[44] Uccelli A, Laroni A, Freedman MS. Mesenchymal stem cells for the treatment of multiple sclerosis and other neurological diseases. The Lancet Neurology. 2011;**10**(7):649-656

[45] Quirici N, Soligo D, Bossolasco P, Servida F, Lumini C, Deliliers GL. Isolation of bone marrow mesenchymal stem cells by anti-nerve growth factor receptor antibodies. Experimental Hematology. 2002;**30**(7):783-791

[46] Røsland GV, Svendsen A, Torsvik A, Sobala E, McCormack E, Immervoll H, et al. Long-term cultures of bone marrow–derived human mesenchymal stem cells frequently undergo spontaneous malignant transformation. Cancer Research. 2009;**69**(13):5331-5339

[47] Huang X-P, Sun Z, Miyagi Y, McDonald Kinkaid H, Zhang L, Weisel RD, et al. Differentiation of allogeneic mesenchymal stem cells induces immunogenicity and limits their long-term benefits for myocardial repair. Circulation. 2010;**122**(23):2419-2429

[48] Ding Y, Yang H, Feng JB, Qiu Y, Li DS, Zeng Y. Human umbilical cord-derived MSC culture: The replacement of animal sera with human cord blood plasma. In Vitro Cellular & Developmental Biology. Animal. 2013;**49**(10):771-777

[49] Musiał-Wysocka A, Kot M, Majka M. The pros and cons of mesenchymal stem cell-based therapies. Cell Transplantation. 2019;**28**(7):801-812

[50] Dong H, Li G, Shang C, Yin H, Luo Y, Meng H, et al. Umbilical cord mesenchymal stem cell (UC-MSC) transplantations for cerebral palsy. American Journal of Translational Research. 2018;**10**(3):901

[51] Herberts CA, Kwa MS, Hermsen HP. Risk factors in the development of stem cell therapy. Journal of Translational Medicine. 2011;**9**(1):1-14

[52] Tatsumi K, Ohashi K, Matsubara Y, Kohori A, Ohno T, Kakidachi H, et al. Tissue factor triggers procoagulation in transplanted mesenchymal stem cells leading to thromboembolism. Biochemical and Biophysical Research Communications. 2013;**431**(2):203-209

[53] Mazini L, Rochette L, Amine M, Malka G. Regenerative capacity of adipose derived stem cells (ADSCs), comparison with mesenchymal stem cells (MSCs). International Journal of Molecular Sciences. 2019;**20**(10):2523

[54] Chen G, Yue A, Ruan Z, Yin Y, Wang R, Ren Y, et al. Monitoring the biology stability of human umbilical cord-derived mesenchymal stem cells during long-term culture in serum-free medium. Cell and Tissue Banking. 2014;**15**(4):513-521

[55] Kashani IR, Hedayatpour A, Pasbakhsh P, Kafami L, Atlasi N, Mahabadi VP, et al. 17β-Estradiol enhances the efficacy of adipose-derived mesenchymal stem cells on remyelination in mouse model of multiple sclerosis. Acta Medica Iranica. 2012;**2012**:789-797

[56] Dai R, Wang Z, Samanipour R, Koo K-i, Kim K. Adipose-derived stem cells

for tissue engineering and regenerative medicine applications. Stem Cells International. 2016;**2016**:1-19

[57] Kim J-H, Jo CH, Kim H-R, Hwang Y-i. Comparison of immunological characteristics of mesenchymal stem cells from the periodontal ligament, umbilical cord, and adipose tissue. Stem Cells International. 2018;**2018**:1-13

[58] Motedayyen H, Esmaeil N, Tajik N, Khadem F, Ghotloo S, Khani B, et al. Method and key points for isolation of human amniotic epithelial cells with high yield, viability and purity. BMC Research Notes. 2017;**10**(1):1-8

[59] Uchida S, Suzuki Y, Araie M, Kashiwagi K, Otori Y, Sakuragawa N. Factors secreted by human amniotic epithelial cells promote the survival of rat retinal ganglion cells. Neuroscience Letters. 2003;**341**(1):1-4

[60] Xt M, Chen D, Dong, y Z, Liu J. Enhanced neural differentiation of neural stem cells and neurite growth by amniotic epithelial cell co-culture. Cell Biology International. 2007;**31**(7):691-698

[61] Taheri RA, Motedayyen H, Ghotloo S, Masjedi M, Mosaffa N, Mirshafiey A, et al. The effect of lipopolysaccharide on the expression level of immunomodulatory and immunostimulatory factors of human amniotic epithelial cells. BMC Research Notes. 2018;**11**(1):1-5

[62] McDonald CA, Payne NL, Sun G, Moussa L, Siatskas C, Lim R, et al. Immunosuppressive potential of human amnion epithelial cells in the treatment of experimental autoimmune encephalomyelitis. Journal of Neuroinflammation. 2015;**12**(1):1-14

[63] Xu H, Zhang J, Tsang KS, Yang H, Gao W-Q. Therapeutic potential of human amniotic epithelial cells on injuries and disorders in the central nervous system. Stem Cells International. 2019;**2019**:1-12

[64] Li J-Y, Christophersen NS, Hall V, Soulet D, Brundin P. Critical issues of clinical human embryonic stem cell therapy for brain repair. Trends in Neurosciences. 2008;**31**(3):146-153

[65] Ryu CH, Park KY, Hou Y, Jeong CH, Kim SM, Jeun S-S. Gene therapy of multiple sclerosis using interferon β-secreting human bone marrow mesenchymal stem cells. BioMed Research International. 2013;**2013**:1-10

[66] Patel SA, Sherman L, Munoz J, Rameshwar P. Immunological properties of mesenchymal stem cells and clinical implications. Archivum Immunologiae et Therapiae Experimentalis. 2008;**56**(1):1-8

[67] Yamout B, Hourani R, Salti H, Barada W, El-Hajj T, Al-Kutoubi A, et al. Bone marrow mesenchymal stem cell transplantation in patients with multiple sclerosis: A pilot study. Journal of Neuroimmunology. 2010;**227**(1-2):185-189

[68] Hu J, Yu X, Wang Z, Wang F, Wang L, Gao H, et al. Long term effects of the implantation of Wharton's jelly-derived mesenchymal stem cells from the umbilical cord for newly-onset type 1 diabetes mellitus. Endocrine Journal. 2012;**2021**:EJ12

[69] Li J-F, Yin H-L, Shuboy A, Duan H-F, Lou J-Y, Li J, et al. Differentiation of hUC-MSC into dopaminergic-like cells after transduction with hepatocyte growth factor. Molecular and Cellular Biochemistry. 2013;**381**(1):183-190

[70] Liu R, Zhang Z, Lu Z, Borlongan C, Pan J, Chen J, et al. Human umbilical cord stem cells ameliorate experimental autoimmune encephalomyelitis by

regulating immunoinflammation and remyelination. Stem Cells and Development. 2013;**22**(7):1053-1062

[71] Tavakoli A, Kahrizi MS, Safa K, ArefNezhad R, Rezaei-Tazangi F. Umbilical cord mesenchymal stem cells and breast cancer: A good therapeutic candidate or not? A minireview. Molecular Biology Reports. 2022;**2022**:1-6

[72] Rong L, Chi Y, Yang S, Chen D, Chen F, Xu S, et al. Effects of interferon-γ on biological characteristics and immunomodulatory property of human umbilical cord-derived mesenchymal stem cells. Zhongguo Shi Yan Xue Ye Xue Za Zhi. 2012;**20**(2):421-426

[73] Lu Z, Zhao H, Xu J, Zhang Z, Zhang X. Human umbilical cord mesenchymal stem cells in the treatment of secondary progressive multiple sclerosis. Journal of Stem Cell Research Theraphy. 2013;**6**:2

[74] Yang H, Yang H, Xie Z, Wei L, Bi J. Systemic transplantation of human umbilical cord derived mesenchymal stem cells-educated T regulatory cells improved the impaired cognition in AβPPswe/PS1dE9 transgenic mice. PLoS One. 2013;**8**(7):1-10

[75] Yang H, Sun J, Wang F, Li Y, Bi J, Qu T. Umbilical cord-derived mesenchymal stem cells reversed the suppressive deficiency of T regulatory cells from peripheral blood of patients with multiple sclerosis in a co-culture–a preliminary study. Oncotarget. 2016;**7**(45):72537

[76] Mazini L, Rochette L, Amine M, Malka GJIjoms. Regenerative capacity of adipose derived stem cells (ADSCs). Comparison with Mesenchymal Stem Cells (MSCs). 2019;**20**(10):2523

[77] Stepien A, Dabrowska NL, Maciagowska M, Macoch RP, Zolocinska A, Mazur S, et al. Clinical application of autologous adipose stem cells in patients with multiple sclerosis: Preliminary results. Mediators of Inflammation. 2016;**2016**:1-6

[78] Gadelkarim M, Abushouk AI, Ghanem E, Hamaad AM, Saad AM, Abdel-Daim MM. Adipose-derived stem cells: Effectiveness and advances in delivery in diabetic wound healing. Biomedicine & Pharmacotherapy. 2018;**107**:625-633

[79] Chi K, Fu R-H, Huang Y-C, Chen S-Y, Hsu C-J, Lin S-Z, et al. Adipose-derived stem cells stimulated with n-butylidenephthalide exhibit therapeutic effects in a mouse model of Parkinson's disease. Cell Transplantation. 2018;**27**(3):456-470

[80] Li J, Chen Y, Chen Z, Huang Y, Yang D, Su Z, et al. Therapeutic effects of human adipose tissue-derived stem cell (hADSC) transplantation on experimental autoimmune encephalomyelitis (EAE) mice. Scientific Reports. 2017;**7**(1):1-10

[81] Ghasemi N, Razavi S, Mardani M, Esfandiari E, Salehi H, Esfahani SHZ. Transplantation of human adipose-derived stem cells enhances remyelination in lysolecithin-induced focal demyelination of rat spinal cord. Molecular Biotechnology. 2014;**56**(5):470-478

[82] Tomita K, Madura T, Mantovani C, Terenghi G. Differentiated adipose-derived stem cells promote myelination and enhance functional recovery in a rat model of chronic denervation. Journal of Neuroscience Research. 2012;**90**(7):1392-1402

[83] Motedayyen H, Rezaei A, Zarnani A-H, Tajik NJRb. Human amniotic epithelial cells inhibit activation and pro-inflammatory cytokines production of naive CD4+ T cells from women with unexplained recurrent spontaneous abortion. 2018;**18**(2):182-188

[84] Li H, Niederkorn JY, Neelam S, Mayhew E, Word RA, McCulley JP, et al. Immunosuppressive factors secreted by human amniotic epithelial cells. 2005;**46**(3):900-907

[85] Insausti CL, Blanquer M, García-Hernández AM, Castellanos G, Moraleda JM, et al. Amniotic membrane-derived stem cells: immunomodulatory properties and potential clinical application. 2014;7:53

[86] Fathi F, Atapour A, Eskandari N, Keyhanmehr N, Hafezi H, Mohammadi S, et al. Regulatory T-cells and their impacts on cytokine profile of end-stage renal disease patients suffering from systemic lupus erythematosus. International Journal of Immunopathology and Pharmacology. 2019;**33**:2058

[87] Yang P-j, Yuan W-x, Liu J, Li J-y, Tan B, Qiu C, et al. Biological characterization of human amniotic epithelial cells in a serum-free system and their safety evaluation. Acta Pharmacologica Sinica. 2018;**39**(8):1305-1316

[88] McDonald CA, Short M, Jenkin G, Bernard CCA. The Potential of Human Amnion Epithelial Cells as an Immunomodulatory and Neuroregenerative Treatment for Multiple Sclerosis. Perinatal stem cells. Springer; 2014:231-242

[89] Li H, Niederkorn JY, Neelam S, Mayhew E, Word RA, McCulley JP, et al. Immunosuppressive factors secreted by human amniotic epithelial cells. Investigative Ophthalmology & Visual Science. 2005;**46**(3):900-907

[90] Alipour R, Motedayyen H, Sereshki N, Rafiee M, Alsahebfosul F, Pourazar A. Human amniotic epithelial cells affect the functions of neutrophils. International Journal of Stem Cells. 2020;**13**(2):212-220

[91] İnsan TGKOK, Öldürücü AEHD. Immunoregulatory effects of human amnion epithelial cells on natural killer and T cells in women with recurrent spontaneous abortion (RSA). Turkish Journal of Immunology. 2019;**7**(1):21-30

[92] McDonald CA, Payne NL, Sun G, Moussa L, Siatskas C, Lim R, et al. Immunosuppressive potential of human amnion epithelial cells in the treatment of experimental autoimmune encephalomyelitis. Journal of neuroinflammation. 2015;**12**(1):112

[93] Motedayyen H, Esmaeil N, Tajik N, Khadem F, Ghotloo S, Khani B, et al. Method and key points for isolation of human amniotic epithelial cells with high yield, viability and purity. BMC Research Notes. 2017;**10**(1):552

[94] Motedayyen H, Fathi F, Fasihi-Ramandi M, Taheri RA. The effect of lipopolysaccharide on anti-inflammatory and pro-inflammatory cytokines production of human amniotic epithelial cells. Reproductive Biology. 2018;**18**(4):404-409

[95] Liu YH, Vaghjiani V, Tee JY, To K, Cui P, Oh DY, et al. Amniotic epithelial cells from the human placenta potently suppress a mouse model of multiple sclerosis. PLoS One. 2012;**7**(4):e35758

[96] Han K, Lee J, Kwon S, Park S, Shim S, Kim H, et al. Human amnion-derived mesenchymal stem cells are a potential source for uterine stem cell therapy. Cell Proliferation. 2008;**41**(5):709-725

[97] Irony-Tur-Sinai M, Grigoriadis N, Tsiantoulas D, Touloumi O, Abramsky O, Brenner T. Immunomodulation of EAE by alpha-fetoprotein involves elevation of immune cell apoptosis markers and the transcription factor FoxP3. Journal of the Neurological Sciences. 2009;**279**(1-2):80-87

[98] Yawno T, Schuilwerve J, Moss TJ, Vosdoganes P, Westover AJ, Afandi E, et al. Human amnion epithelial cells reduce fetal brain injury in response to intrauterine inflammation. Developmental Neuroscience. 2013;**35**(2-3):272-282

[99] McDonald CA, Payne NL, Sun G, Moussa L, Siatskas C, Lim R, et al. Immunosuppressive potential of human amnion epithelial cells in the treatment of experimental autoimmune encephalomyelitis. Journal of Neuroinflammation. 2015;**12**(1):112

[100] Liu YH, Vaghjiani V, Tee JY, To K, Cui P, Oh DY, et al. Amniotic epithelial cells from the human placenta potently suppress a mouse model of multiple sclerosis. 2012;**7**(4):e35758

[101] Uchida S, Inanaga Y, Kobayashi M, Hurukawa S, Araie M, Sakuragawa NJ. Neurotrophic function of conditioned medium from human amniotic epithelial cells. Journal of neuroscience research. 2000;**62**(4):585-590

[102] Kim EY, Lee K-B, Kim MK. The potential of mesenchymal stem cells derived from amniotic membrane and amniotic fluid for neuronal regenerative therapy. BMB Reports. 2014;**47**(3):135

[103] Wu Q, Fang T, Lang H, Chen M, Shi P, Pang X, et al. Comparison of the proliferation, migration and angiogenic properties of human amniotic epithelial and mesenchymal stem cells and their effects on endothelial cells. International Journal of Molecular Medicine. 2017;**39**(4):918-926

[104] Motamedi B, Ibrahim T, Abdul A, Allaudin Z, Moshrefi M, Hajghani M, et al. Characteristics of human amniotic epithelial cells and bone marrow mesenchymal stem cells in a parallel study: Hanging drop colony formation and doubling time. Journal of Regenerated Medicine. 2015;**4**:2-10

[105] Wu K-C, Chang Y-H, Liu H-W, Ding D-C. Transplanting human umbilical cord mesenchymal stem cells and hyaluronate hydrogel repairs cartilage of osteoarthritis in the minipig model. Tzu-Chi Medical Journal. 2019;**31**(1):11

[106] Debnath T, Chelluri LK. Standardization and quality assessment for clinical grade mesenchymal stem cells from human adipose tissue. Hematology, Transfusion and Cell Therapy. 2019;**41**(1):7-16

[107] Serras ASM, Cipriano MZdRF, da Graça Silva PM, Miranda JPG. Challenges for Deriving Hepatocyte-like Cells from Umbilical Cord Mesenchymal Stem Cells for In Vitro Toxicology Applications. Novel Perspectives of Stem Cell Manufacturing and Therapies. 2020

[108] Xu Z, Wei W, Gagneur J, Perocchi F, Clauder-Münster S, Camblong J, et al. Bidirectional promoters generate pervasive transcription in yeast. Nature. 2009;**457**(7232):1033-1037

[109] Kachgal S, Putnam AJ. Mesenchymal stem cells from adipose and bone marrow promote angiogenesis via distinct cytokine and protease expression mechanisms. Angiogenesis. 2011;**14**(1):47-59

[110] Si J, Dai J, Zhang J, Liu S, Gu J, Shi J, et al. Comparative investigation of human amniotic epithelial cells and mesenchymal stem cells for application in bone tissue engineering. Stem Cells International. 2015;**2015**:1-15

[111] Parolini O, Alviano F, Bagnara GP, Bilic G, Bühring HJ, Evangelista M, et al. Concise review: Isolation and characterization of cells from human term placenta: Outcome of the first international workshop on placenta derived stem cells. Stem Cells. 2008;**26**(2):300-311

[112] Banas RA, Trumpower C, Bentlejewski C, Marshall V, Sing G, Zeevi A. Immunogenicity and immunomodulatory effects of amnion-derived multipotent progenitor cells. Human Immunology. 2008;**69**(6):321-328

[113] Mohyeddin Bonab M, Yazdanbakhsh S, Lotfi J, Alimoghaddom K, Talebian F, Hooshmand F, et al. Does mesenchymal stem cell therapy help multiple sclerosis patients? Report of a pilot study. Iranian Journal of Immunology. 2007;**4**(1):50-57

[114] Fernández O, Izquierdo G, Fernández V, Leyva L, Reyes V, Guerrero M, et al. Adipose-derived mesenchymal stem cells (AdMSC) for the treatment of secondary-progressive multiple sclerosis: A triple blinded, placebo controlled, randomized phase I/II safety and feasibility study. PLoS One. 2018;**13**(5):e0195891

[115] Staff NP, Madigan NN, Morris J, Jentoft M, Sorenson EJ, Butler G, et al. Safety of intrathecal autologous adipose-derived mesenchymal stromal cells in patients with ALS. Neurology. 2016;**87**(21):2230-2234

[116] Li J-F, Zhang D-J, Geng T, Chen L, Huang H, Yin H-L, et al. The potential of human umbilical cord-derived mesenchymal stem cells as a novel cellular therapy for multiple sclerosis. Cell Transplantation. 2014;**23**(1_suppl):113-122

[117] Riordan NH, Morales I, Fernández G, Allen N, Fearnot NE, Leckrone ME, et al. Clinical feasibility of umbilical cord tissue-derived mesenchymal stem cells in the treatment of multiple sclerosis. Journal of Translational Medicine. 2018;**16**(1):1-12

Chapter 10

From the Classification of Stem Cells to the Release of Potential in Cell Therapies: Limits, Considerations and Future Aspects in Regenerative Medicine

Arnaud Martino Capuzzo, Riccardo Ossanna, Lindsey Alejandra Quintero Sierra, Federica Virla, Alessandro Negri, Anita Conti, Andrea Sbarbati and Sheila Veronese

Abstract

Regenerative medicine aims to repair organs or tissues that have congenital abnormalities, or that have been damaged by disease, aging, or trauma, and to restore or at least improve their native function. One of the strategies used in regenerative medicine is stem cell therapy, due to the enormous regenerative potential of stem cells. A staminal cell line is a group of cells that can replicate for an extended period *in vitro*, that is outside the body. These cells are grown in incubators using a culture medium that should have a temperature and an oxygen/carbon dioxide composition that simulates the desired environment. This chapter describes the main characteristics of stem cells, the main fields of application, and outlines what could be the future developments of their use, also considering the ethical and technical problems that currently limit their use. There is still much to be done in the field of stem cell research, and researchers are working tirelessly to remain leaders and innovators in it. A struggle, step by step, will make it possible to have more information on current knowledge by expanding the scientific literature and push current limits ever further.

Keywords: stem cells, regenerative medicine, clinical studies, research strategies, therapeutic method

1. Introduction

In recent years, increasing attention has been paid to the study of various types of cells, with particular attention to their properties, to promote regenerative processes and/or to be used for the cellular treatment of many diseases [1]. Great interest in research has been aroused by stem cells; their ability to self-renew, and differentiate into mature adult cells has made them, since their discovery, extremely promising for the regeneration of human tissue [2]. It is no coincidence that the first results of their use have contributed to the real definition of "regenerative medicine" [3]. Today, stem cells and their differentiated derivates are increasingly used in an ever-widening field of cellular studies, often with the aim of treating the condition of cell loss related to various diseases [4].

Stem cell division can give birth to an asymmetrical offspring with an additional progenitor cell and a daughter stem cell. For this reason, they exhibit both self-renewal and regeneration capabilities. The differentiation capacity of stem cells depends on their specification potential.

Generally, the renewal of adult stem cells (ASCs) is limited because they can only differentiate into specific cells of a single tissue. Finding stem cells capable of differentiating into all tissue types is a challenge. In the event that all terminal cell populations can be reached the cells are said to have the property of totipotency, while the ability to pursue a more restricted pattern of phenotypes is the property of multipotency [5] (**Figure 1**).

Scientists have traditionally dealt with two types of animal and human stem cells: embryonic stem cells (ESCs) and non-embryonic "somatic" or "adult" stem cells. Almost 30 years ago, in 1981, researchers discovered how to obtain ESCs from early mouse embryos [7]. A method to extract stem cells from human embryos, and

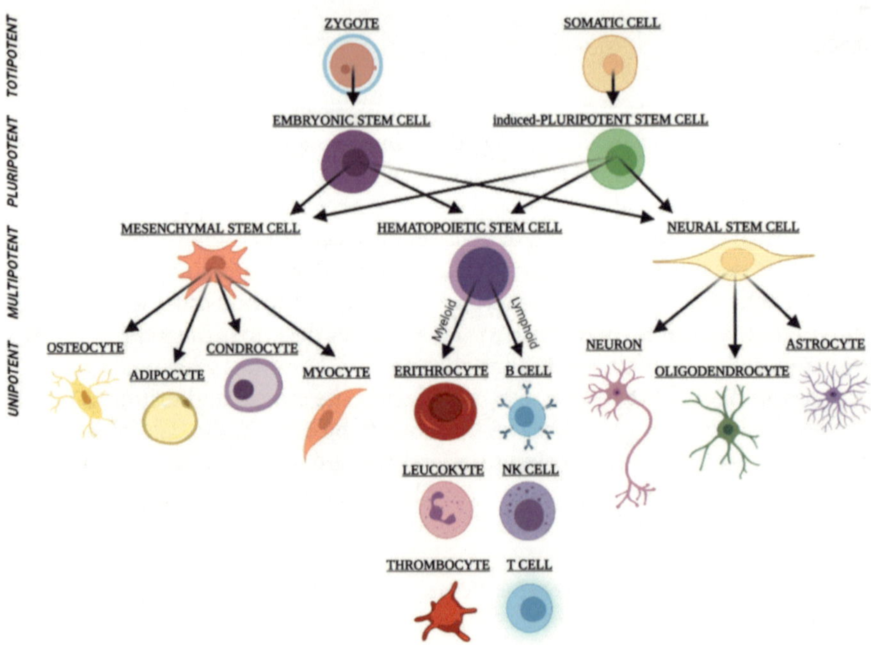

Figure 1.
Totipotent stem cells generate all the cell types of the organism (e.g., zygote or fertilized egg). Pluripotent stem cells produce all the embryonic germ layers (endoderm, ectoderm, and mesoderm). Multipotent stem cells generate a limited number of cell types based on their tissue of origin. Mesenchymal stem cells give rise to fat, bone, muscle, and cartilage. Hematopoietic stem cells give rise to different types of blood cells (for example platelets, and red and white cells). Neural stem cells give rise to neurons, oligodendrocytes, and astrocytes. Oligopotent stem cells generate some closely related cell types (for example myeloid stem cells). Unipotent stem cells generate a single cell type (e.g., epidermal stem cells or muscle stem cells) [6].

growing the cells in the laboratory was discovered in 1998, as a result of a deepen examination of mouse stem cell biology [8]. Human embryonic stem cells (hESCs) were the name given to these cells. The *in vitro* fertilization techniques utilized in these investigations were used to produce the embryos for use in reproduction. When they were no longer needed for that function, they were given to study with the informed consent of the donor. ESCs possess powerful properties, but their use goes against ethical principles [9], and for this reason it has been limited. Today there is already and will probably continue to be in the future, a debate on their use, on the possibility of keeping them in culture for more than 14 days, and on all the social, moral and ethical problems connected to them [10].

A possible way to overcome this obstacle was already found in 2006 by Takahashi and Yamanaka. Under unknown circumstances, some specialized adult cells were genetically "reprogrammed" to take on a stem cell-like form. The current name of this novel form of stem cell is induced pluripotent stem cell (iPSCs). iPSCs exhibit morphology, growth properties, and cell marker gene expression of ESCs, without ethical concerns. iPSCs can differentiate in various human tissues and exploit regenerate properties [11, 12], given their pluripotency capability. Unlike hESCs, iPSCs do not raise any ethical concerns regarding the timing of human personality initiation [13, 14].

However, both ESCs and IPSCs carry the risk of tumor formation, a risk related to both pluripotency and self-renewal. This is a critical factor of both cells types [15]. Furthermore, iPSCs still present some technical issues related to immune rejection after transplantation. This means that research in this field needs to be expanded because more studies are needed before using iPSCs as a viable tool for *in vivo* tissue regeneration [16, 17].

2. Origin of stem cells

Stem cells have the ability to self-renew, i.e. to create copies of themselves, and to differentiation into lineage populations, i.e. to develop into more specialized cells, which allow cell turnover in the respective tissues present in multicellular organisms [18]. The production of tissue-specific stem cells, which generally assist the cell renewal of all tissue types for the development of the organism, is a necessary step in the life cycle of a complex organism [19].

The core cells of the 3- to 5-day-old embryo, known as blastocysts, give rise to the complete body of an organism, including the numerous specialized cell types and organs such as the heart, lung, skin, sperm, eggs, and other tissues [20].

Distinct populations of ASCs provide replacements for cells lost due to natural wear and tear, injury, or disease in different adult tissues, such as bone marrow, muscle, and brain. Stem cells allow novel therapeutic possibilities for addressing these conditions thanks to their exceptional ability to regenerate [21].

Scientists can study the basic characteristics of stem cells and what distinguishes them from other specialized cell types through laboratory investigations. In laboratories, stem cells are already being used by researchers to test new drugs, create models/systems to study healthy growth, and pinpoint the origins of birth abnormalities. Understanding of how an organism grows from a single cell and how healthy cells replace damaged ones in adult creatures has been advanced through stem cell research [19, 22].

One of the most promising areas of modern biology is stem cell biological product research. It has recently been ascertained that miRNA-containing vesicles, such as exosomes, could induce a change in some pathologies [23–25] (**Figure 2**).

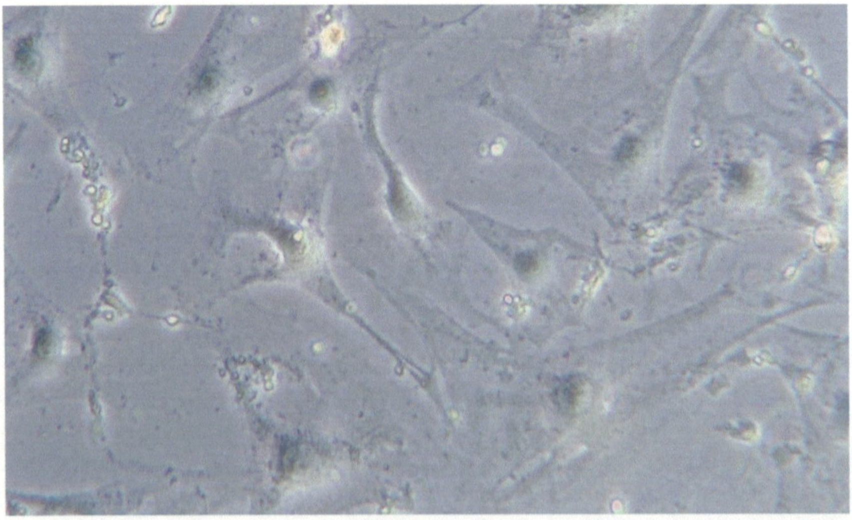

Figure 2.
Adult adipose tissue derived murine stem cells ADAS. Here shown, ADAS cultured with complete DMEM containing 10% FBS and 1% P/S mix 1:1, and incubated at 37°C in a 5% CO2 atmosphere. These cells have the characteristic of producing exosomes, which in recent studies have shown to have unique characteristics for some neuropathologies [23]. The image acquisition was done using a bright field optical microscope, Olympus BX-51 (Olympus, Tokyo, Japan) equipped with a digital camera (DKY-F58 CCD JVC, Yokohama, Japan) and connected with a PC endowed with image-pro plus 7.0 software.

There is still much work to be done in laboratories and clinics to improve the efficacy of using these cells and their byproducts in what are termed cellular treatments, or even regenerative or reparative medicine treatments, of a diverse pool of diseases [26].

3. Stem cells subpopulations

All stem cells can self-renew and develop, as described in the previous chapter, but they differ greatly in what they can and cannot become and in the conditions under which they can and cannot perform certain functions. This is one reason why scientists employ different kinds of stem cells in their study.

3.1 Adult stem cells

ASCs are undifferentiated cells found in some differentiated tissues of the body and have the possible property of self-renewing or producing new cells to replace damaged or dead tissue [27]. Alternatively, ASCs are sometimes referred to as "somatic stem cells", where the term "somatic" refers to the non-reproductive cells of the body (eggs or sperm). Some examples of ASCs are: Epithelial and Skin Stem Cells, Neural Stem Cells, Hematopoietic Stem Cells (Blood Stem Cells), Mesenchymal Stem Cells [28] (**Figure 3**).

ASCs are often insufficient in native tissues, making them difficult to study and harvest for research [29]. Distinct populations of ASCs, which are present in most tissues in the human body, produce new cells to replace those lost as a result of natural repair, disease, or damage.

All tissues in a person, including the umbilical cord, placenta, bone marrow, muscle, brain, adipose tissue and lipoaspirates, skin, stomach, etc. include ASCs.

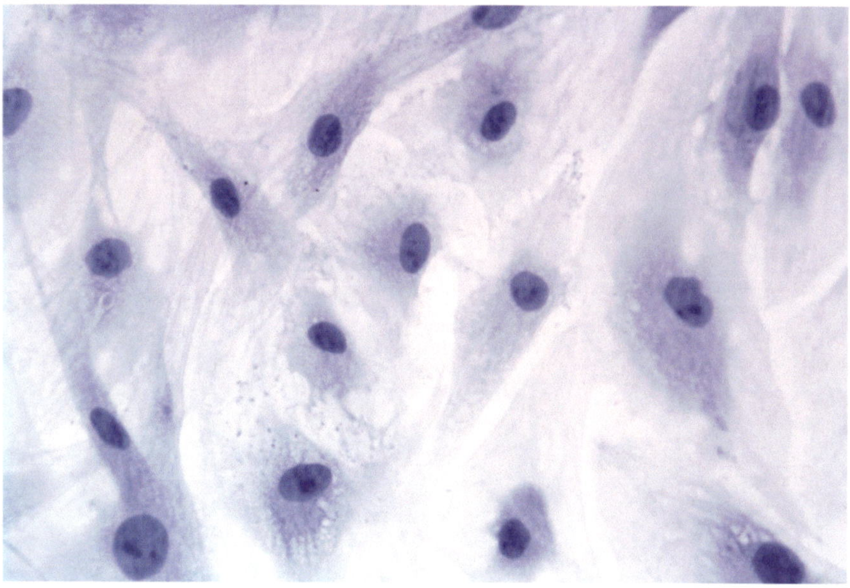

Figure 3.
Fat harvesting via liposuction. Here shown, human adipose derived mesenchymal stem cells cultured with DMEM complete medium for 2 weeks, fixed with PFA 3%, washed PBS 1x stained with hematoxylin, the protocol involved the seeding of ADSCs on a 12-wells plate with sterile slides on the bottom of each well. The cells were seeded and incubated with 1 ml of complete culture medium for 24 hours at 37° C and 5% CO2. At the end the cells were fixed with paraformaldehyde 4%, stained with Mayer's hematoxylin (bio-Optica, Milan, Italy) for 5 min. Finally, the cells were washed with tap water for 5 min and mounted with mount quick aqueous solution (bio-Optica, Milan, Italy). The image acquisition was done using a bright field optical microscope, Olympus BX-51 (Olympus, Tokyo, Japan) equipped with a digital camera (DKY-F58 CCD JVC, Yokohama, Japan) and connected with a PC endowed with image-pro plus 7.0 software. Slides were gently cleaned with ethanol, acquired using a 20X.

In 1948, the first ASCs were removed and utilized to create blood [30]. When the first adult bone marrow cells were employed in clinical therapy for blood disorders in 1968, this process was expanded [31].

For more than 40 years, treatments for blood disorders such as leukemia and lymphoma have included transplantation of peripheral blood stem cell and bone marrow [32].

There is an ongoing debate. According to some studies [33], ASCs can only produce the cell types of the tissue in which they reside. However, other studies suggest that ASCs may be able to produce cells of other tissue types [34]. More research is required to clarify this aspect.

Scientists have demonstrated that ASCs are present in most body tissues. Scientific research is looking for ways to locate, isolate, and multiply these cells for therapeutic use.

Most of the biological effects of ASC are probably mediated by extracellular vesicles, such as exosomes, which influence surrounding cells. The current development of exosome therapies requires efficient and non-invasive methods to localize, monitor, and trace exosomes [25] (**Figure 4**). The idea behind these therapies is that the exosomes and the chemicals released are the stem cells' way of manifesting their therapeutic function.

3.1.1 What are exosomes?

Exosomes are vesicles that include peptides, mRNA, and microRNAs [35, 36], range in size from 50 to 150 nm, and are essential for intercellular communication [37].

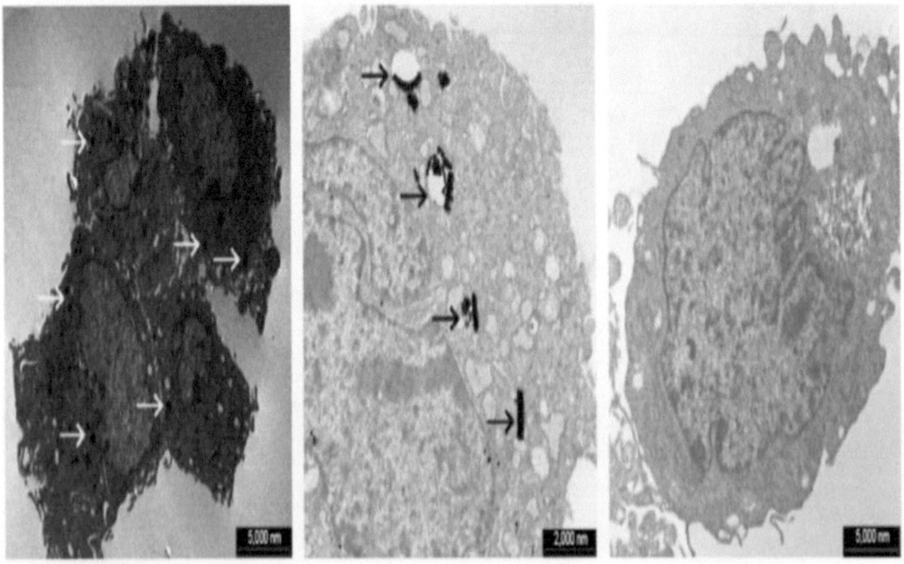

Figure 4.
TEM images of ADAS incubation with nanoparticles (NPs) were morphologically analyzed through a transmission electron microscope (TEM) in order to confirm the intracellular uptake of NPS and visualize their intracellular localization. The scale bar in the left and right pictures is 5000 nm, and the Centre picture is focused on the endocytic invagination containing nanoparticles and the internalized nanoparticles inside the endosome (scale bar 2000 nm). Cell pellets were fixed for 1 h in 2% glutaraldehyde in 0.1 M phosphate buffer (PB) and, after washed, postfixed for 1 h in 1% OsO4 diluted in 0.2 M K3Fe (CN)6. After rinsing in 0.1 M PB, the samples were dehydrated in graded concentrations of acetone and embedded in a mixture of Epon and araldite (electron microscopic sciences, Fort Washington, PA, USA). Ultrathin sections were cut at 70 nm thickness on a Ultracut E ultramicrotome (Reichert-Jung, Heidelberg, Germany), placed on Cu/Rh grids and contrasted with lead citrate. Samples were observed with Pa Philips Morgagni 268 D electron microscope (Fei company, Eindhoven, the Netherlands) equipped with a mega view II camera to acquire digital images [25].

They mimic the effects of stem cell transplantation by delivering physiologically active chemicals to recipient cells, which change their gene expression and behavior.

According to several studies, stem cell-derived exosomes may have a role in synaptic plasticity, nerve regeneration, neuronal protection, and neurological recovery [38, 39].

By using these vesicles as a treatment, rather than their generated parental cells, restrictions and dangers for cell transplantation are avoided.

3.2 Embryonic stem cells

The embryo, known as a blastocyst at this stage, contains an inner cell mass capable of growing all the specialized tissues that make up the human body, 3 to 5 days after fertilization and prior to implantation [40]. ESCs are produced from the inner cell mass of an *in vitro* fertilized embryo, donated for scientific research. ESCs are not made from eggs that have been fertilized inside a female's body [41].

Isolable only in the early stages of development, these pluripotent stem cells can develop into virtually any cellular form (**Figure 5**). One of the research objectives is to understand how these cells differentiate during development [42]. The progressive increase in knowledge about these stages of development, could allow researchers to use ESCs generated *in vitro* to rebuild different types of tissue, such as neurons, skin, gut, and liver for transplantation [43].

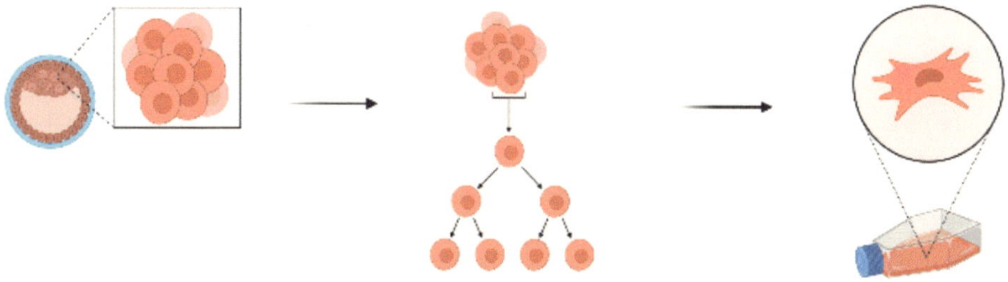

Figure 5.
Schematic representation of ESCs. ESCs are pluripotent, and are derived from the inner cell mass of a blastocyst. Human embryos reach the blastocyst stage 4–5 days after fertilization and at that time consist of approximately 50–150 cells [40].

In the future, ESCs could be used to treat a broader spectrum of disorders. It is hoped that once this technique is well understood, the information will be applied to vehicular ESCs, i.e., induce them to differentiate into the specific cell type required for patient therapy [44]. Currently, diseases treated with ESCs transplantation include diabetes, spinal cord injury, muscular dystrophy, heart disease, and vision/hearing loss [45].

3.3 Reprogrammed pluripotent stem cells

Halfway between ASCs and ESCs are iPSCs, which are stem cells produced in a laboratory, by introducing embryonic genes into a somatic cell, such as a skin cell, so that it returns to a "stem cell-like" state [46].

The production of these cells is an innovative technique of genetic reprogramming. First identified in 2006 [47], several years of study will be required before they can be used therapeutically.

The potential to alter recipient somatic cells into a "ESC-like" state undoubtedly makes therapies using iPSCs attractive [48]. The cells required for the therapies could be produced using appropriate differentiation processes of these iPSC cells. What makes this technique attractive is that it circumvents the need for lifelong histocompatibility immunosuppression, as is the case with transplanted cells of donor stem cells [49].

iPSC cells are regarded as pluripotent, making them similar to most ESCs. However, unlike for ESCs, manipulation of iPSCs has not been successful in growing the outer layer of an embryonic cell, which is required for the cell to develop into a full human individual [50]. But, iPSC research is rapidly moving towards translational and clinical applications [51].

4. Advantages of different stem cell lineages

It is possible to list some advantages and disadvantages associated with the three distinct stem cell types (ASCs, ESCs, and iPSC) previously described. If we start from the analysis of the disadvantages and ongoing debates on the use of different stem cells, we are immediately redirected towards their advantages.

ASCs – Among the positive characteristics of ASCs is their ability to transdifferentiate and reprogram themselves. Also they are less likely to be rejected when used in transplants [52]. Their efficiency in the therapeutic field has already been proven in numerous clinical applications [53].

ESCs – Among their advantageous properties, ESCs have the potential to be maintained in culture, and to grow even for more than a year. Numerous protocols have been established for their maintenance in culture, protocols that consider the ability of these cells to produce most cell types in the body [54–57]. There are numerous studies relating to these cells, which appear to be among the most investigated stem cells. A further increase in these studies may lead to more knowledge about how living things develop and thrive.

iPSCs – iPSCs are mostly derived from donor somatic cells. This means that they can be utilized in large quantities, avoiding histocompatibility issues in transplants. These cells have performed well in preclinical drug testing and research/development studies [58–60]. The definition of the new cellular "reprogramming" procedure, and the new knowledge deriving from it, could be applied to define *in vivo* therapies for the reprogramming of damaged or diseased cells and tissues.

The currently best-known cell therapy for the treatment of blood malignancies and other blood problems is bone marrow transplantation, which transplants blood stem cells [61].

Theoretically, Parkinson's disease, spinal cord damage, stroke, burns, heart disease, Type 1 diabetes, osteoarthritis, rheumatoid arthritis, muscular dystrophy, and liver disease are all possible candidates for stem cell therapy [62]. Additionally, regeneration of the retina using isolated ocular stem cells could one day contribute to the reversal of blindness, providing a potential treatment for distressed or injured eyes [63].

Cell therapy, which replaces unhealthy cells with healthy ones to treat disease, is one potential use of stem cells; it is comparable to organ transplantation, except that the cells rather than the organs are transplanted (**Figure 6**).

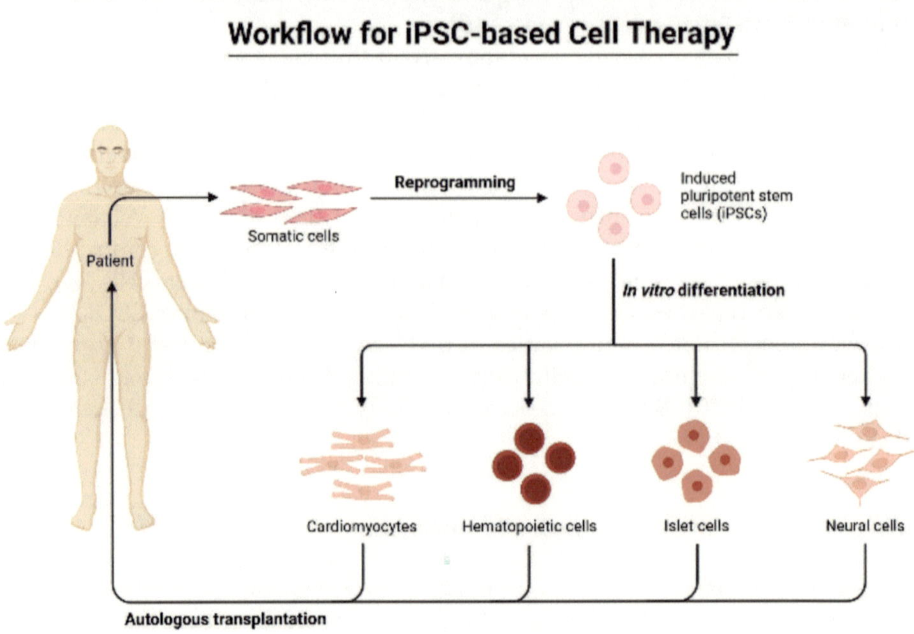

Figure 6.
Autologous human iPSCs can be derived from individual patients and differentiate into different cell types. To develop new therapies, isolated and cultured cells are used to observe specific disease phenotypes and identify possible new pathological mechanisms. Cell therapy includes therapies based on stem and non-stem cells, unicellular and multicellular, with different immunophenotypic profiles, isolation techniques, mechanisms of action and regulatory levels. The use of human iPSCs, autologous to the patient, offers an innovative approach for regenerative medicine ([35, 36]).

5. Limitations and obstacles to their use

Many factors still need to be considered and many studies are still needed to fully understand how to use stem cells.

ASCs – The differentiating capacity of ASCs are not yet fully elucidated [64]. This may mean that a different interpretation of their properties is needed, considering them multi- or unipotent.

Furthermore, ASCs cannot be grown for long periods of time in culture without an observable phenotypic change occurring and fail to prevent their immortalization.

Another limitation of the use of these cells is that they are present in small quantities in the tissues. This makes it difficult to identify the niche and the purification/isolation process is tedious. Currently few technologies are available to generate large quantities of cultured stem cells and keep them incubating. We have previously successfully tested the combination of CELLviewer [65] with Spin∞, a new bioreactor still in the prototype stage [66].

ESCs – The limitations and obstacles that can be found in the use of ESCs concern various aspects, among which the fundamental one is that the cell line generation process is inefficient. Moreover, their use is strictly regulated [67].

It is uncertain whether they would be rejected if used in transplants. Therapies that use ESC pathways are largely new, and much more research and testing is needed to ascertain their effectiveness as alternative pathways to conventional clinical treatments [68].

Finally, when used directly from the undifferentiated culture, ESCs for tissue transplants can generate and cause tumors (teratomas) or the development of cancer.

iPSCs – The reasoning on iPSCS turns out to be more complex. Since the transcription and reprogramming factors of stem cells were identified in 2006 [47], several methods to obtain iPSCs have been certified. These methods have been duly included in protocols that ensure cell reproducibility and maintenance [49]. New discoveries are not uncommon, given that the tissues into which they differentiated are not known *a priori* [69].

An important limit to the use of these cells are the viruses that are currently used to introduce embryonic genes into somatic cells. Studies in mouse models have shown that these viruses can cause tumors.

The use of iPSCs has given rise to several controversies because, however, there could be numerous application possibilities in the research field and in the market.

As far as ethical aspects are concerned, for the use of ASCs has not raised any significant issue. For the use of ESC, to acquire the inner cell mass, the embryo is destroyed. Therefore, donor consent is required. Relevant problem may concern iPSC cells because, when exposed to the right conditions, they have the potential to become embryos.

It should be noted that there can be many possible hitches related to the cellular therapy. For example, in the presence of insufficient synthesis of stem cells in the bone marrow, a poor transfusion is obtained, i.e. a transfusion with few cells. This is typical of elderly individuals who, compared to younger individuals, have a lower capacity of the bone marrow to produce stem cells. Regardless of the number of cells, the quality of these cells is also crucial.

Finally, not all existing health problem can be solved with stem cell treatment. Each person responds differently; the fundamental goal of this method is to provide the body with the means and the optimal environment to repair damaged tissue on its own. The knowledge of the type of patients to be treated in the different modalities to optimize therapeutic outcomes has significantly improved. For some pathologies/

patients, on the other hand, there has been no progress and it is not yet known what the most suitable cell therapy might be. Additionally, the outcome of any procedure relies on the participant's individual body resilience. Each body can respond differently. Therefore, it is currently not possible to generalize the results of a treatment or to determine it as the ideal treatment for a given pathological condition. There is no universal treatment plan.

6. Future perspectives of culturing stem cells

Faced with the possibility of using different stem cells, it should be remembered that the cells can be used in abundant quantities of donor somatic cells, thus making autologous treatments possible. In other words, the histocompatibility problems typical of transplants from donors other than the recipient, which are normally the main reasons for rejection, could be avoided.

Stem cells can be useful for several areas of interest such as drug development or development studies.

Finding a suitable stem cell source is the first step. Finding, isolating, and cultivating the proper type of stem cell, such as a rare cell in adult tissue, requires painstaking work.

Tissue-specific stem cells are thought to be less adaptable than embryonic and foetal stem cells, in general. The correct environment must be created after the stem cells have been identified and separated, an environment where the cells can differentiate into the specialized cells needed for a specific therapy.

A transport/migration system for cells to the area of the body where they are to act must be created. It is only in this area that the cells must perform their action, integrating with the body's native cells. However, although cells can chemically recognize injured tissue, there may be a physical barrier, such as a blocked artery, preventing these cells from "traveling" to damaged regions. The rate of tissue regeneration will be poor, slow, or non-existent if the root cause of the blockage is not removed.

Like organ transplants, it may be necessary to suppress the immune system of the body to lessen the immune response triggered by the donated cells. It should be considered that the body may react negatively to the addition of progenitor cells causing the formation of tumors or accelerating their growth rate. The oncological field is, without a doubt, a field in which cell therapy research must be strengthened, given its potential, but with the appropriate measures.

Since 2001, hundreds of stem cell lines have been developed; these lines are much more adaptable and easier to deal with than lines established nearly a decade ago [70]. They also have the advantage of not being "contaminated" by being produced from other cells. Therefore, the National Institutes of Health (NIH) or other competitive funding organizations are unlikely to support experiments limited to older cell lines, given the advantages of newer ones [71]. The study of stem cells and their potential uses for the treatment of various human diseases is still in its infancy, despite encouraging results from animal models. To guarantee long-term efficacy and safety, a thorough research process should be followed.

7. Conclusion and final considerations

Fundamental knowledge of how organisms grow and develop, as well as how tissues are maintained throughout adult life, is aided by stem cell research.

Understanding what goes wrong during disease and damage and, ultimately, how to treat these diseases, is necessary. In the future, researchers will have the means to simulate diseases, test medications, and create increasingly effective treatments with the help of developing human tissue and stem cell lines, and related biological products.

In conclusion, there are two factors that make the creation of disease- or patient-specific pluripotent stem cells extremely therapeutically promising. First, these cells may offer a powerful new tool for researching the causes of human diseases and for developing new drugs. Secondly, the generated ESCs could be transformed into a specific cell type and, if transplanted into the original donor, would be recognized as "autologous," eliminating the issues with immunosuppression and rejection that arise with transplants from unrelated donors. The research must go on and unlock the potential of stem cells to further advance regenerative medicine.

Acknowledgements

We gratefully acknowledge the Department of Pathology and Diagnostics at the University of Verona, which allow to perform research.

Conflict of interest

The author declares no conflict of interest.

Thanks

To Natasha, who in her modest comprehension of things knows how to pique my curiosity about the areas of my life that have not yet been thoroughly explored.

Thanks to all my dearest supporters who in the dark moment know how to encourage and spur me to work more and more. I would also like to thank all dear colleagues in the Department of Diagnostics and Public Health for providing knowledge and insights into some of the considerations in the chapter.

A special thanks to Sheila who took care to correct and revise the chapter several times, leaving no mistakes.

Special thanks to the IntechOpen Team for giving me the opportunity to write this chapter.

Author details

Arnaud Martino Capuzzo[1*], Riccardo Ossanna[1], Lindsey Alejandra Quintero Sierra[2], Federica Virla[2], Alessandro Negri[1], Anita Conti[1], Andrea Sbarbati[2] and Sheila Veronese[2]

1 Department of Pathology and Diagnostics, Nanoscience and Advanced Technologies, University of Verona, Verona, Italy

2 Department of Neurosciences, Biomedicine and Movement Sciences, University of Verona, Verona, Italy

*Address all correspondence to: arnaudmartino.capuzzo@univr.it

References

[1] Mahla RS. Stem cells applications in regenerative medicine and disease therapeutics. International Journal of Cell Biology. 2016;**2016**:e6940283. DOI: 10.1155/2016/6940283

[2] Stoltz J-F, de Isla N, Li YP, Bensoussan D, Zhang L, Huselstein C, et al. Stem cells and regenerative medicine: Myth or reality of the 21th century. Stem Cells International. 2015;**2015**:e734731. DOI: 10.1155/2015/734731

[3] Haseltine WA. The emergence of regenerative medicine: A new field and a new society. e-biomed: The Journal of Regenerative Medicine. 2001:17-23. DOI: 10.1089/152489001753309652

[4] Hyun I. The bioethics of stem cell research and therapy. The Journal of Clinical Investigation. 2010;**120**(1):71-75. DOI: 10.1172/JCI40435

[5] Cofano F, Boido M, Monticelli M, Zenga F, Ducati A, Vercelli A, et al. Mesenchymal stem cells for spinal cord injury: Current options, limitations, and future of cell therapy. International Journal of Molecular Sciences. 2019;**20**(11):2698. DOI: 10.3390/ijms20112698

[6] Totipotent Stem Cell-An overview Science Direct Topics. (s.d.). Recuperato 21 gennaio 2023, da. Available from: https://www.sciencedirect.com/topics/medicine-and-dentistry/totipotent-stem-cell

[7] Paige CJ, Kincade PW, Shinefeld LA, Sato VL. Precursors of murine B lymphocytes. Physical and functional characterization, and distinctions from myeloid stem cells. Journal of Experimental Medicine. 1981;**153**(1):154-165. DOI: 10.1084/jem.153.1.154

[8] Nichols J, Zevnik B, Anastassiadis K, Niwa H, Klewe-Nebenius D, Chambers I, et al. Formation of pluripotent stem cells in the mammalian embryo depends on the POU transcription factor Oct4. Cell. 1998;**95**(3):379-391. DOI: 10.1016/S0092-8674(00)81769-9

[9] de Wert G, Mummery C. Human embryonic stem cells: Research, ethics and policy. Human Reproduction. 2003;**18**(4):672-682. DOI: 10.1093/humrep/deg143

[10] Adashi EY, Cohen IG. Who will oversee the ethical limits of human embryo research? Nature Biotechnology. 2022;**40**(4):463-464. DOI: 10.1038/s41587-022-01274-6. PMID: 35361998

[11] Kim C. Disease modeling and cell based therapy with iPSC: Future therapeutic option with fast and safe application. Blood Research. 2014;**49**(1):7. DOI: 10.5045/br.2014.49.1.7

[12] Steeg R, Mueller SC, Mah N, Holst B, Cabrera-Socorro A, Stacey GN, et al. EBiSC best practice: How to ensure optimal generation, qualification, and distribution of iPSC lines. Stem Cell Reports. 2021;**16**(8):1853-1867. DOI: 10.1016/j.stemcr.2021.07.009

[13] Scesa G, Adami R, Bottai D. iPSC preparation and epigenetic memory: Does the tissue origin matter? Cells. 2021;**10**(6):Art 6. DOI: 10.3390/cells10061470

[14] Tukker AM, de Groot MWGDM, Wijnolts FMJ, Kasteel EEJ, Hondebrink L, Westerink R. Is the time right for in vitro neurotoxicity testing using human iPSC-derived neurons? ALTEX-Alternatives to Animal Experimentation. 2016;**33**(3):Art. 3. DOI: 10.14573/altex.1510091

[15] Knoepfler PS. Deconstructing stem cell Tumorigenicity: A roadmap to safe regenerative medicine. Stem Cells. 2009;**27**(5):1050-1056. DOI: 10.1002/stem.37

[16] Hong SG, Winkler T, Wu C, Guo V, Pittaluga S, Nicolae A, et al. Path to the clinic: Assessment of iPSC-based cell therapies In vivo in a nonhuman primate model. Cell Reports. 2014;7(4):1298-1309. DOI: 10.1016/j.celrep.2014.04.019

[17] Taguchi J, Yamada Y. In vivo reprogramming for tissue regeneration and organismal rejuvenation. Current Opinion in Genetics & Development. 2017;**46**:132-140. DOI: 10.1016/j.gde.2017.07.008

[18] Bongso A, Lee EH. Stem Cells: From Bench to Bedside. World Scientific; 2005

[19] Ramalho-Santos M, Willenbring H. On the origin of the term "stem cell". Cell Stem Cell. 2007;**1**(1):35-38. DOI: 10.1016/j.stem.2007.05.013

[20] Rossant J. Stem cells from the mammalian blastocyst. Stem Cells. 2001;**19**(6):477-482. DOI: 10.1634/stemcells.19-6-477

[21] Bourne DA, Thomas RD, Bliley J, Haas G, Wyse A, Donnenberg A, et al. Amputation-site soft-tissue restoration using adipose stem cell therapy. Plastic and Reconstructive Surgery. 2018;**142**(5):1349. DOI: 10.1097/PRS.0000000000004889

[22] Lajtha LG. Stem cell concepts. Differentiation. 1979;**14**(1):23-33. DOI: 10.1111/j.1432-0436.1979.tb01007.x

[23] Bonafede R, Turano E, Scambi I, Busato A, Bontempi P, Virla F, et al. ASC-exosomes ameliorate the disease progression in SOD1 (G93A) murine model underlining their potential therapeutic use in human ALS. International Journal of Molecular Sciences. 2020;**21**(10):art. 10. DOI: 10.3390/ijms21103651

[24] Busato A, Bonafede R, Bontempi P, Scambi I, Schiaffino L, Benati D, et al. Magnetic resonance imaging of ultrasmall superparamagnetic iron oxide-labeled exosomes from stem cells: A new method to obtain labeled exosomes. International Journal of Nanomedicine. 2016;**11**:2481-2490. DOI: 10.2147/IJN.S104152

[25] Busato A, Bonafede R, Bontempi P, Scambi I, Schiaffino L, Benati D, et al. Labeling and magnetic resonance imaging of exosomes isolated from adipose stem cells. Current Protocols in Cell Biology. 2017;**75**(1):3.44.1-3. 44.15. DOI: 10.1002/cpcb.23

[26] Kolios G, Moodley Y. Introduction to stem cells and regenerative medicine. Respiration. 2013;**85**(1):3-10. DOI: 10.1159/000345615

[27] Clevers H. What is an adult stem cell? Science. 2015;**350**(6266):1319-1320. DOI: 10.1126/science.aad7016

[28] Young HE, Black AC Jr. Adult stem cells. The Anatomical Record Part A: Discoveries in Molecular, Cellular, and Evolutionary Biology. 2004;**276A**(1):75-102. DOI: 10.1002/ar.a.10134

[29] Barker N, Bartfeld S, Clevers H. Tissue-resident adult stem cell populations of rapidly self-renewing organs. Cell Stem Cell. 2010;7(6):656-670. DOI: 10.1016/j.stem.2010.11.016

[30] Schepers A, Clevers H. Wnt signaling, stem cells, and cancer of the gastrointestinal tract. Cold Spring Harbor Perspectives in Biology. 2012;**4**(4):a007989. DOI: 10.1101/cshperspect.a007989

[31] Ikehara S. Grand challenges in stem cell treatments. Frontiers in Cell and Developmental Biology. 2013;**1**:2

[32] Serakinci N, Keith WN. Therapeutic potential of adult stem cells. European Journal of Cancer. 2006;**42**(9):1243-1246. DOI: 10.1016/j.ejca.2006.01.036

[33] What is an adult stem cell? Science. (s.d.). Recuperato 20 gennaio 2023, da. Available from: https://www.science.org/doi/full/10.1126/science.aad7016

[34] Alison M, Islam S. Attributes of adult stem cells. The Journal of Pathology. 2009;**217**(2):144-160. DOI: 10.1002/path.2498

[35] De Francesco F, Gravina P, Busato A, Farinelli L, Soranzo C, Vidal L, et al. Stem cells in autologous microfragmented adipose tissue: Current perspectives in osteoarthritis disease. International Journal of Molecular Sciences. 2021a;**22**(19):art. 19. DOI: 10.3390/ijms221910197

[36] De Francesco F, Matta C, Riccio M, Sbarbati A, Mobasheri A. Reevolution of tissue regeneration: From recent advances in adipose stem cells to novel therapeutic approaches. Stem Cells International. 2021b;**2021**:e2179429. DOI: 10.1155/2021/2179429

[37] Bonafede R, Brandi J, Manfredi M, Scambi I, Schiaffino L, Merigo F, et al. The anti-apoptotic effect of ASC-exosomes in an In vitro ALS model and their proteomic analysis. Cells. 2019;**8**(9):art. 9. DOI: 10.3390/cells8091087

[38] Exosomes: Immune properties and potential clinical implementations SpringerLink. (s.d.). Recuperato 7 gennaio 2023, da. Available from: https://link.springer.com/article/10.1007/s00281-010-0233-9

[39] Xiong M, Zhang Q, Hu W, Zhao C, Lv W, Yi Y, et al. Exosomes from adipose-derived stem cells: The emerging roles and applications in tissue regeneration of plastic and cosmetic surgery. Frontiers in Cell and Developmental Biology. 2020;**8**:574223

[40] Rippon HJ, Bishop AE. Embryonic stem cells. Cell Proliferation. 2004;**37**(1):23-34. DOI: 10.1111/j.1365-2184.2004.00298.x

[41] Waldby C. Oocyte markets: Women's reproductive work in embryonic stem cell research. New Genetics and Society. 2008;**27**(1):19-31. DOI: 10.1080/14636770701843576

[42] Odorico JS, Kaufman DS, Thomson JA. Multilineage differentiation from human embryonic stem cell lines. Stem Cells. 2001;**19**(3):193-204. DOI: 10.1634/stemcells.19-3-193

[43] Keller G. Embryonic stem cell differentiation: Emergence of a new era in biology and medicine. Genes & Development. 2005;**19**(10):1129-1155. DOI: 10.1101/gad.1303605

[44] Elisseeff JH. Embryonic stem cells: Potential for more impact. Trends in Biotechnology. 2004;**22**(4):155-156. DOI: 10.1016/j.tibtech.2004.02.006

[45] Menasché P. Human embryonic stem cells still have a place in the cell therapy landscape. Cardiovascular Research. 2022;**118**(13):e96-e97. DOI: 10.1093/cvr/cvac117

[46] Cherry ABC, Daley GQ. Reprogrammed cells for disease modeling and regenerative medicine. Annual Review of Medicine. 2013;**64**:277-290. DOI: 10.1146/annurev-med-050311-163324

[47] Takahashi K, Yamanaka S. Induction of pluripotent stem cells from mouse embryonic and adult fibroblast

cultures by defined factors. Cell. 2006;**126**(4):663-676. DOI: 10.1016/j.cell.2006.07.024

[48] Malik N, Rao MS. A review of the methods for human iPSC derivation. In: Lakshmipathy U, Vemuri MC (A c. Di). Pluripotent Stem Cells: Methods and Protocols. Humana Press; 2013. pp. 23-33. DOI:10.1007/978-1-62703-348-0_3

[49] Wiegand C, Banerjee I. Recent advances in the applications of iPSC technology. Current Opinion in Biotechnology. 2019;**60**:250-258. DOI: 10.1016/j.copbio.2019.05.011

[50] Nicholson MW, Ting C-Y, Chan DZH, Cheng Y-C, Lee Y-C, Hsu C-C, et al. Utility of iPSC-derived cells for disease modeling, drug development, and cell therapy. Cells. 2022;**11**(11):Art. 11. DOI: 10.3390/cells11111853

[51] MacArthur CC, Pradhan S, Switalski S, Landon M, Lakshmipathy U. Establishing iPSC cell banks derived using reagents and workflows optimized for cell therapy manufacturing. Cytotherapy. 2019;**21**(5 Supplement):S13. DOI: 10.1016/j.jcyt.2019.03.574

[52] Körbling M, Estrov Z. Adult stem cells for tissue repair—A new therapeutic concept? New England Journal of Medicine. 2003;**349**(6):570-582. DOI: 10.1056/NEJMra022361

[53] Squillaro T, Peluso G, Galderisi U. Clinical trials with mesenchymal stem cells: An update. Cell Transplantation. 2016;**25**(5):829-848. DOI: 10.3727/096368915X689622

[54] Meisner LF, Johnson JA. Protocols for cytogenetic studies of human embryonic stem cells. Methods. 2008;**45**(2):133-141. DOI: 10.1016/j.ymeth.2008.03.005

[55] Pouton CW, Haynes JM. Embryonic stem cells as a source of models for drug discovery. Nature Reviews Drug Discovery. 2007;**6**(8):Art. 8. DOI: 10.1038/nrd2194

[56] Tamm C, Galitó SP, Annerén C. A comparative study of protocols for mouse embryonic stem cell culturing. PLoS One. 2013;**8**(12):e81156. DOI: 10.1371/journal.pone.0081156

[57] Turksen K. Embryonic Stem Cells: Methods and Protocols. Springer Science & Business Media; 2008

[58] Appelt-Menzel A, Oerter S, Mathew S, Haferkamp U, Hartmann C, Jung M, et al. Human iPSC-derived blood-brain barrier models: Valuable tools for preclinical drug discovery and development? Current Protocols in Stem Cell Biology. 2020;**55**(1):e122. DOI: 10.1002/cpsc.122

[59] Harding J, Mirochnitchenko O. Preclinical studies for induced pluripotent stem cell-based therapeutics *. Journal of Biological Chemistry. 2014;**289**(8):4585-4593. DOI: 10.1074/jbc.R113.463737

[60] van Amerongen RA, Morton LT, Chaudhari UG, Remst DFG, Hagedoorn RS, van den Berg CW, et al. Human iPSC-derived preclinical models to identify toxicity of tumor-specific T cells with clinical potential. Molecular Therapy-Methods & Clinical Development. 2023;**28**:249-261. DOI: 10.1016/j.omtm.2023.01.005

[61] Chiu RC-J. Bone-marrow stem cells as a source for cell therapy. Heart Failure Reviews. 2003;**8**(3):247-251. DOI: 10.1023/A:1024769617018

[62] Kim JY, Nam Y, Rim YA, Ju JH. Review of the current trends in clinical trials involving induced pluripotent stem cells. Stem Cell Reviews and Reports. 2022;**18**(1):142-154. DOI: 10.1007/s12015-021-10262-3

[63] Menon A, Vijayavenkataraman S. Novel vision restoration techniques: 3D bioprinting, gene and stem cell therapy, optogenetics, and the bionic eye. Artificial Organs. 2022;**46**(8):1463-1474. DOI: 10.1111/aor.14241

[64] Stocum DL, Zupanc GKH. Stretching the limits: Stem cells in regeneration science. Developmental Dynamics. 2008;**237**(12):3648-3671. DOI: 10.1002/dvdy.21774

[65] Capuzzo AM, Vigo D. Microfluidic live-imaging technology to perform research activities in 3D models. 2021. Available from: https://www.researchgate.net/publication/349923987_Microfluidic_Live-Imaging_Technology_to_Perform_Research_Activities_in_3D_Models

[66] Capuzzo AM, Vigo D. Updated modifications about spin∞ bioreactor. Journal of Engineering Technology and Applied Sciences. 2020;**5**:53-56

[67] Findikli N, Candan N, Kahraman S. Human embryonic stem cell culture: Current limitations and novel strategies. Reproductive Biomedicine Online. 2006;**13**(4):581-590. DOI: 10.1016/S1472-6483(10)60648-7

[68] Murry CE, Keller G. Differentiation of embryonic stem cells to clinically relevant populations: Lessons from embryonic development. Cell. 2008;**132**(4):661-680. DOI: 10.1016/j.cell.2008.02.008

[69] All Roads Lead to Induced Pluripotent Stem Cells: The Technologies of iPSC Generation Stem Cells and Development. (s.d.). Recuperato 7 gennaio 2023, da. Available from: https://www.liebertpub.com/doi/full/10.1089/scd.2013.0620

[70] Bongso A, Richards M. History and perspective of stem cell research. Best Practice & Research Clinical Obstetrics & Gynaecology. 2004;**18**(6):827-842. DOI: 10.1016/j.bpobgyn.2004.09.002

[71] Tagle DA. The NIH microphysiological systems program: Developing in vitro tools for safety and efficacy in drug development. Current Opinion in Pharmacology. 2019;**48**:146-154. DOI: 10.1016/j.coph.2019.09.007

Chapter 11

A New Cell Stem Concept for Pelvic Floor Disorders Prevention and Treatment – Endometrial Mesenchymal Stem Cells

Manuela Cristina Russu

Abstract

High rate complications and recurrences in reconstructive surgery using *in situ* synthetic/polypropylene (PP) meshes have driven to a new concept based on mesenchymal stem cells (MSCs) for homeostasis repair in pelvic floor disorders (PFD). Prevention and therapy with MSCs are up to date analyzed on small and large animal models, less in women trials. Cell based-vaginal/intraurethral, or systemically introduced, tissue engineering (TE) with new generation meshes/scaffolds MSCs seeded-bone marrow, adipose tissue and recently proposed the endometrial/menstrual MSCs (eMSCs/MenSCs) for PFDs, management. Easy collected, isolated with specific markers, cultured for number harvesting, without ethic and immune compatibility issues, with unique biologic properties eMSCs/MenSCs differentiate in many cellular types—smooth muscle, and fibroblast-like cells, preserving cell shape, and phenotype, without oncogenic risks, and collagen, elastin fibers; eMSCs/MenSCsare appropriate for PFDs management, respecting good protocols for human safety. The quick appeared regenerative effect-mediated by angiogenesis, apoptosis inhibition, cell proliferation, no chronic inflammation and low/no foreign body reactions, less thick collagen fibers, and fibrosis improve connective/neuromuscular tissues; less pelvic structures stiffness with more elasticity are advantages for new meshes/scaffolds generation in TE. Human eSMCs/MenSCs deliver bioactive factors by their exosomes/microvesicles/secretome for paracrine effects to injury site, facilitating in vivo tissue repair.

Keywords: pelvic floor disorders, endometrial/menstrual mesenchymal stem cells, cell based therapy, tissue engineering, new generation meshes/scaffolds

1. Introduction

Pelvic Floor Disorders (PFDs) are a continuous gynecologist's challenge, reason for this paper to review their correction attempts, by better knowledge of pathophysiology, pelvic connective tissue structure and biology with the new concepts of regenerative medicine by mesenchymal stem cells (MSCs) and/or their bioactive mollecules, sometimes crucial. All these objectives are after the era of reconstructive surgery with native tissue or *in situ* synthetic/polypropylen (PP), non-absorbable meshes for fascia repair. The surgical procedures have immediate, and subsequent (within 5 years) complications (10–30% rate) [1], disorder's recurrence or a new disorder, and host reactions to mesh—as a foreign body reaction (FBR), with fibrosis, mainly to transvaginal type, explaining abnormal functions after interventions. FDA (USA) had two public health notifications, warning since 2010 against vaginal mesh use in POP, because complications, recurrences, and litigations [2]; PP mesh use is prohibited for transvaginal POP surgery in Australia, New Zealand, USA, and UK [3]. Stem cells, as endometrial/menstrual mesenchymal stem cells (eMSCs/MenSCs) with their remarkable unique biological properties are considered a new potential tool for PFDs prevention and therapy, in a properly response to pathophysiology. MSCs are key regulatory components in the regenerating stem cell local microenvironment termed "stem cell niche" or for MSCs culture conditions, *per se* or by their secretome, facts that positively influence altered pelvic floor connective tissue, and contribute to tissue homeostasis restoration [4]. Allogeneic and autologus MSCS/MenSCs are proposed for these aims, being easy collected, isolated, purified with known perivascular surface markers for pericytes [5], and with the novel single perivascular marker Sushi Domain Containing 2 (SUSD2) for purifying eMSCs [6], and maintaining cells clonogenicity, reduced by flow cytometry used for their sorting [5, 6]. Different to other MSCs culture expandation imposing presence in culture medium of some constituents of their secretome [7, 8], for eMSCs and ERCs one adds transforming growth factor-β receptor inhibitor (TGF-β R inhibitor), that limits cells spontaneous differentiation to fibroblasts, and maintains the undifferentiated cells status in days following administration, to ensure efficacy [9]. eMSCs and ERCs have no ethic issue, or incompatibility risk, no need of immunosuppressive adjuvant drug, they intervene in repair and regeneration by new blood vessels formation, modulating host immune system, reducing chronic inflammation, FBRs, and fibrosis, with no scar as endometrium regenerates after each menstruation from the menarcha to menopause, under a normal estrogen/progesterone balance, and in postmenopause when on menopausal hormone therapy (MHT). One proposes cell based therapy-eMSCs/MenSCs direct placement in vaginal walls or intraurethral, or systemic administration [10, 11] to prevent postpartum POP or SUI, and tissue engineering (TE)/tissue grafts with eMSCc/MenSCs seeded in a new meshes/scaffold generation preferable biodegradable, obtained by a variety of technologies, as knitting (like old ones) from alternative synthetic and natural polymers, electrospinning, and three-dimensional printing [12], and a composite mesh of polyamide plus a gelatin layer [13], instead of PP meshes, with no ectopic tissue formation, or malignant tumor [14]. There are many small and large animal models, and few human clinical studies on MSCs for tissue restoration and repair when PFDs.

2. Contemporary burden of women's pelvic floor disorders

PFDs named also Pelvic Floor Dysfunctions represent a women old pathology, with high incidence and prevalence in the last 70 years, associated to worldwide lifespan increase, independently to their different definitions and diagnosis criteria, last updated in 2010 by the International Urogynecological Association and International Continence Society [15], with over 250 definitions for clinical categories and subclassifications, and alphanumeric code for each definition. PFDs is an umbrella term for a group of clinical conditions caused by pelvic floor supportive tissue weakening, sometimes degenerative, occurring independently or simultaneously, connected to genetics, childbirth and aging, and it includes pelvic organs prolapse (POP) into vagina, as a hernia in the endopelvic fascia [16], and alterations in sensory or emptying abnormalities of the lower urinary and gastrointestinal tracts: urinary incontinence (UI) with clinical manifestations of stress urinary incontinence (SUI), overactive bladder syndrome (ORB), detrusor instability, urinary retention, and fecal incontinence, and sexual dysfunctions. PFDs are diagnosed in middle-aged women as stress urinary incontinence (SUI)—incidence peak around 46 years, and in elderly women when a second SUI peak between 70–71 years, being described in young women, also, around 18 years, with annual risk of 3.8 and 3.9 per 1000 women at that bimodal peaks; the risk for pelvic organ prolapse (POP) is increasing progressively from 50 to 60 years in Swedish population [17], to ages of 71 and 73 years, when annual risk is 4.3 per 1000 North American women evaluated for lifetime risks for SUI, or POP, or both surgery between 2007 and 2011 [18], and higher in institutionalized, and when dementia [19]. The recent Swedish national register-based cohort study [20] estimated a 12–19% lifetime risk for surgery for POP, similar to the earliear 19%-for Australian women [21]. PFDs have serious negative influences on women's quality of life at any age [22]. SUI and POP are highly related to childbirth-associated pelvic floor injury [23], less influenced by route of delivery (vaginal, cesarean section), sometimes inherited, as a familial disorder, associated to other hereditary conditions—joint hypermobility and even Ehlers–Danlos syndrome, a mutation in the gene for collagen III [24, 25]. One recommends the extension of familial conditions lower than first degree generation relationship [26] for familial risk reduction and prolapse prevention, many cases being more frequently in postmenopause, without any pregnancy/delivery, when the association of ovarian aging hypoestrogenism, plus vasomotor syndrome, depressive mood, chronic constipation and cough, obesity, characteristics for this life period, are increasing PDFs medical and financial burden [27].

2.1 Synthetic mesh unfavorable results in surgery for pelvic floor disorders

In front of PFDs burden, one tried to correct the supportive tissue of cardinal, utero-sacral ligaments, levatorani muscle, and urethral sphincter damaged by gestation, parturition, and aging initially by non-surgical interventions: physical exercises (after delivery), pessaries—first choice for POP symptomatic women [28], or as a tool for decision of mid-urethral slim mesh, when one decides POP surgery, to avoid over and under-treatment if vaginal mesh is used for an nonexistent SUI or a urinary leakage when surgery for POP [29], pharmacological therapies, laser, nonablative monopolar radiofrequency [30], and surgery, when all these failure; either vaginal or abdominal route, or both with native tissue repair, through open surgery or laparoscopy, or reconstructive surgery with synthetic, monofilament PP mesh respecting the principles for different PFDs manifestations, classified in anterior and posterior compartment, and vaginal vault disorders. It was evident that the addition of mesh as reinforcement to vaginal walls provides better prolapse correction, compared with colporrhaphy using native tissue alone, by both objective and subjective criteria [31], as it was demonstrated for abdominal wall defects/hernia repair [32]. Unfortunately medical staff and patients wishes for pelvic floor normal functions restoration with PP meshes were not accomplished, or the results were not better than after native tissue repair, being reported high risks—immediate, subsequent (within first 5 years) readmissions for later postoperative complications, such as intractable pain, or mesh erosion or extrusion into the bladder, bowel or vagina [33], requiring surgical excision in ≥10% [28], or deterioration of vaginal biomechanical properties by high stiffness mesh implanted for prolapse [34], and further recurrences of previous disorder, or a *de novo* one, as incontinence surgery-incidence of 9.9% after surgery for POP without occult SUI (apical and non-apical) [35] or further prolapse surgery, sacro-spinous fixation with synthetic mesh being a risk factor for POP recurrence [36], andtransvaginal PP mesh having a higher risk than vaginal vault sacro–colpopexy [37, 38]. The population-based cohort study from Scotland [39] compared the primary outcome—immediate postoperative complications and subsequent (within 5 years) readmissions for later postoperative complications, further incontinence surgery, or further prolapse surgery in women older than 20 years, after first, single PP mesh (retropubic mesh, or single prolapse mesh procedure) to non-mesh procedures during 20 years. Immediate complications were lower after mesh procedures [adjusted relative risk (aRR) 0.44 (95%CI 0.36–055) and subsequent prolapse surgery [adjusted incidence rate ratio (air) 0.30 (0.24–0.39)], and a similar risk for further incontinence surgery [0.90 (0.73–1.11)], and later complications [1.12 (0.98–1.27)]. Anterior compartment prolapse PP mesh repair was associated to a similar risk of immediate complications as non-mesh surgery [aRR 0.93 (95%CI 0.49–1.79)], but with an increased risk of further incontinence [air 3.20 (2.06–4.96)], and prolapse surgery [1.69 (1.29–2.20)], and a substantially higher risk of later complications [3.15 (2.46–4.04)]. Posterior compartment mesh repair was associated to a similarly increased risk of repeat prolapse surgery and later complications as non-mesh surgery. Vaginal vault prolapse had similar outcome when vaginal and, separately, abdominal mesh repair were compared with vaginal non-mesh repair. Both vaginal and abdominal mesh procedures for vaginal vault prolapse repair are associated with similar effectiveness and complication rates to non-mesh repair.

The synthetic meshes used for pelvic floor reinforcement or reconstruction may provide the necessary mechanical support for damaged tissue, but implants' biological actions interfere with host biology, inducing the growth of a fibrous tissue layer, as an additional physical support, but the scar may contract the mesh, and surrounding tissue up to 60% [40], Mesh parameters influence host tissues: microstructure-porosity (permeability for host's cells, mainly immune cells, fibroblasts, macrophages, metabolites, oxygen at repair place, and also bacteria effects, by chronic inflammation), fiber filament type (monofilament: polyamide and polyetheretherketone monofilaments or multifilaments; synthetic/natural/composite), and diameter, mesh weight (light/ultralight/heavy), mechanical properties as stiffness, and elasticity, chemical properties, materials biodegradability, and integration in host organ by new blood vessels formation, facts that contributed to the new concept of stem cells in cells based therapy, and tissue engineering (TE) for PFDs prevention or repair [13, 41].

3. A new concept of stem cell use in prevention and therapy of women's pelvic floor disorders

Tissue engineering is a new option in the field of pelvic floor repair when soft tissue reinforcement or reconstruction, or normal function are necessary. MSCs and/or their secretome are proposed after many studies regarding pelvic tissue biology, materials properties associated to stem cells designed to restore the anatomical functions, to provide a real pelvic floor mechanical support after damage, usually to be like a hammock, and to offer both the lost stiffness when under tension and the flexibility under bending [42].

3.1 Tissue homeostasis/remodeling behind repair in pelvic floor disorders

Pelvic floor—a complex resistance piece, keeps pelvic organs within the body, still allowing passage through urethra, vagina and anal canal, around which are designated striated muscles—levatorani, with its three fascicles, and superficialperinealmuscles or urogenital diaphragm [43], forming a functional neuromuscular unit, and fibrous connective tissue, generating endopelvic fascia, cardinal and utero-sacral suspensory ligaments, and vaginal dense fibromuscular-connective tissue. Pelvic connective tissue maintains the position of organs adjacent to vagina, and the close anatomical relationship among vagina, bladder, and rectum may contribute to the emergence of anatomical-functional failure of adjacent organs/systems, in PFDs [44], according to their normal different stiffness/elasticity. Animal models indicate that molecular changes in tissue composition, mainly protein content, coincide to altered biomechanical properties, in PFDs, mainly in POP, as one cites in Australian [45], European [46], North American analyses [47, 48]. Human and mouse pelvic floor provided similarities of pathological changes, centered on deep and superficial muscles, ligaments, and connective tissue, mainly on vaginal walls. Pelvic floor connective tissue contains stromal cells and a very complex extracellular matrix (ECM). The balance between ECM synthesis and degradation during women/females life is a key in pelvic floor properties, and the vaginal structures are strongly influenced during women, small and large animals life, ovine models demonstrating architectural and functionaldifferences according the reproductive status. The Australian ovine model [45] revealed the lowest total collagen content in virgin vaginal tissue, in contrast to parous tissue with highest total collagen and lowest elastin content with concomitant high maximum stress, in contrast to pregnant sheep with lowest collagen and highest elastin contents, and thickest smooth muscle layer and low maximum stress, and poor dimensional recovery following repetitive gestational loading. The vaginal tissue is anisotropic with some biomechanical properties—loading pressure, deformation rates, resistance to rupture, which were tested in ewes [49], and in vaginal specimens collected during surgery for POP [50] compared to specimens from cadavers without noticed PFDs (non-pelvic organ prolapse)—the first experimental study providing vaginal tissue mechanical behavior. The results highlight the non-linear relationship between stress (force per unit of surface) and strain, the vagina being hyperelastic and supporting very large deformation before rupture appearance, as in labor, and fetal expulsion. The vaginal walls tissue is stiffer in patients with POP than non-POP [51]. Comparison of biomechanical properties of the crucial organs of pelvic support [52], showed significant differences at large strain levels: vagina is more rigid, and less extendible than rectum, which, is more rigid than the bladder. The anterior and posterior vaginal walls have different stiffness, and the bladder tissue was anisotropic at large strain levels, facts very important for tissue repair: region with dysfunction/disorder, or procedure type.

ECM molecules are arranged in a matrix/scaffold, surrounding stromal cells (fibroblasts, myofibroblasts, smooth muscle) that synthesize collagen, as tropocollagen and elastin, as tropoelastin to form the fibers complex network, plus proteoglycans, and matricellular proteins, enzymes, according to their genes. All forming the complex network of pelvic floor support, recently updated [53]. EMC contains:

- tropocollagen, self-assembled into fibrils, aggregating to form a collagen type I (for tensile strength) and type III collagen fiber (for elastic properties, and increased collagen III reduces mechanical strength [54]; collagen form a cross-linked network intertwined with elastin—the elastic fibers core component, secreted by elastogenic cells as the monomer tropoelastin, and undergoes self aggregation, cross-linking and deposition on to microfibrils assemble into insoluble elastin polymers. A microfibril scaffold-primarily formed by the protein fibrilin-1, is required for elastic fibers formation [53]. Collagen and elastin fibers are surrounded by a viscous substance of proteoglycans-consisting of a core protein to which one or more glycosaminoglycan (GAG) chains—ashyaluronan or hyaluronic acid, heparan, dermatan sulfate and the small leucine-rich repeat proteoglycans (SLRPs)-decorin, lumican and fibromodulin [17] are covalently attached; SLRPs cover the surface of collagen fibers, contributing to fiber optimal formation [55]. Proteoglycans have key roles in controlling gradients and availability of potent growth factors, chemokines, cytokines, and morphogens, very important in tissue's homeostasis, mechanical strength, development, and repair. One or more proteoglycans are cell surface or transmembrane receptors for adhesion molecules in all mammalian extracellular matrices [56], contributing to progenitor stem cells microenvironment/niche [57, 58].

- matricellular proteins or elastic fibers associated [59]: fibrilin-1, fibrilin-2, fibulin-3, -4, -5, are involved in elastic fibers synthesis and assembly. Fibulin-5 is a pivotal molecule with dual functions involving MMP-9 enzyme regulation, and in tethering polymerized monomeric form of elastin to surrounding cells in vaginal wall, and positively regulating coacervation, but negatively regulating maturation of coacervated elastin in vitro [60]. When pre pregnancy POP fibulin-5 knockout mice increases in severity after vaginal delivery [61–63].

- adhesion molecules: fibronectin; integrins—α, β or α-β integrins interaction with fibulin-5 is essential in vascular development, but dispensable for fibronectin fibrils assembly.

- enzymes:

  - matrix proteases as matrix metalloproteinase (MMP)-2, MMP-9 involved indisruption of collagen and elastin fibers, and particularly increased in POP, and in postmenopause comparative to premenopauseal asymptomatic cases; estrogens withdrawal or antiestrogenic therapy upregulates MMP-9; and TIMP-1, TIMP-2

  - lysyl oxidase-like–1 (LOXl-1)—predominantly catalyzes elastin cross-linking, its inhibition associated to increased MMP-9 led to subclinical POP [27, 64], because tropoelastin accumulation, according to the theory of antielastase-elastase inbalance in mice lacking LOXl-1, and the lack of deposit with normal elastic fibers in the uterine tract, and an abnormal postpartum heal of elastin, In LOXL1 knockout mice, smooth muscle cells stiffness and cells adhesion are altered, being proved the interplay between smooth muscle mechanics and ECM remodeling, mainly in postpartum [65, 66].

- water, very important for all body composition.

The content, aspect and cross-link of collagen and elastic fibers, matricellular proteins, proteoglycans, and enzymes specially MMPs, LOXl-1 are negatively changed when POP; elastic and mechanical strength are decreased during gestation, and with age, beingconceivable that a loss of elastic fiber–associated proteins in pelvic floor connective tissues with aging, may disrupt the optimal balance between synthesis and degradation of vaginal elastic fibers, and lead to POP, fact associated to the critical negative proteases role in POP progression [48]. The fibers amount is less interested initially after vaginal delivery; their histomorphology is first changed, regarding length (shorter), and cross-linking in net-work [61], fibers density decreases later, by aging [67]. The quantification of collagen and elastic fibers shows a more important decrease of elastic fibers in superficial epithelial layers near vaginal epithelium, and less in the deepness of pelvic cavity, around muscles, and thin, irregular and disrupted collagen bundles, higher levels of collagen type III in the vaginal wall, and fragmentation of collagen fibers [68], being appreciated that epithelial-stromal interactions, and fibulin-5-integrin interactions—that suppress ROS generation, are critical in regulation of MMP-9 in mice vaginal wall [48], with an increased level of MMP-2, -9 in advanced prolapse [69, 70]. It is sure that such molecular changes are not corrected by surgical techniques with native tissue or with PP mesh, and the procedures of tissue engineering by different MSCs types, and/or their secretome may change pelvic floor future histomorphology and functionality with normal/near normal connective tissue appearance, that will be discussed in Section 2.5.

3.2 Genetics, gestational and postmenopausal influences on pelvic floor connective tissue disorders

Genes, sexual steroid hormones with their receptors, ligands and co-activators modulate pelvic floor structures, and volumes entire women's life. One analyses Homeobox genes (*HOXA-11*involved in utero sacral ligaments fibroblasts proliferation and p53 regulation [71]), gene encoding *LOXl1*-generating a primarily failure of elastin postpartum healing in knockout mice, the decrease gene signal for production of three SLRPs-decorin, lumican and fibromodulin, which are collagen fiber assembly regulators with affected collagen fibrilogenesis and collagen fibrils shape, and impairment in elastic fiber assembly by down regulation of fibulin-5 in POP [17]. Genes encoding*fibulin-5*, *fibulin-3*, *Upii-sv40t*—involved in elastin fiber structure, are analyzed in PFDs associated to knockout mice aging [27].

Pregnancy induces adaptations in pelvic floor structures for vaginal delivery to withstand deformations with minimum damages, but vaginal delivery leads to floor disorders, damaging nerves, connective tissue, pelvic smooth and striated muscles. Pelvic connective tissue reduced stiffness and elasticity is essential, being demonstrated that the load carrying response (other than the functional response to the pelvic organs) of each fascia component, pelvic organ, smooth muscle, and ligament are assumed to be isotropic, hyperelastic, and incompressible [72]. There are parallel gestational changes in levatorani muscle: sarcomerogenesis, fiber elongation, and an increased ECM collagen content, with muscular stiffness [73], to avoid sarcomere hyperelongation resulted from mechanical strains imposed by vaginal delivery. Sometimes delivery related strains lead to acute sarcomere hyperelongation, and pregnancy pelvic floor muscles (PFMs) adaptation is exceeded [74]; with pelvic floor muscles (PFMs) avulsions discovered postpartum [73]. Human parturition needs PFMs elongation of 300% in resting muscle length to achieve fetal vaginal delivery, as computational models revealed [75]. An instrumental vaginal delivery with forceps may induce important damages of levator ani muscle visible at 3D/4D postpartum ultrasound [76], and one considers that the majority of vaginal deliveries are followed by subclinical damages. The postpartum pelvic floor repair is different from *cervix uteri* repair, with loss of pregestational EMC composition restoration. Each vaginal delivery, in special genes and familial heredity conditions, and also without these risk factors, may contribute to EMC damages, with changes of pelvic floor shape and in biochemical structure in pregnancy, and post delivery versus nullipara [77], gestation has the greatest impact on vaginal tissue composition and biomechanical properties proved in animal models (mice, sheep), and women. Important differences of pelvic floor EMC structures are described [45] between virgin, pregnant and parous females regarding total collagen, collagen III/I + III ratios, GAGs, and elastin, and in passive biomechanical properties-compliance and elasticity, and maximum stress and strain, with permanent strain following cyclic loading after each gestation. Vaginal tissue of virgin sheep had the lowest total collagen content and permanent strain, and parous tissue had the highest total collagen, and lowest elastin content with concomitant high maximum stress in contrast to pregnant sheep, that had the highest elastin and lowest collagen contents, and thickest smooth muscle layer, situation associated with low maximum stress, and poor dimensional recovery following repetitive loading. Vaginal biomechanical properties do not recover after pregnancy to those of virgins [44], and tensile strength appears to be linked to vaginal content: total collagen, elastin, and smooth muscles show a direct influence on tissue compliance—reduced after ovine consecutive pregnancies, different to rectum and bladder compliance which are stiffer than vaginal walls after many deliveries [78]. Vaginal distensibility pregnancy-induced and along vaginal delivery by tissue vaginal pressure is not recovered in late postpartum rats [79]. It was demonstrated in mice vaginal culture, the POP appearance and progression after each pregnancy and vaginal delivery, caused by a combination of inhibited elastin linking with tropoelastin accumulation [47], because inability to initiate damaged tissue necessary clear and replacement, with poor elastin properly self repair after each delivery, through abnormal enzymatic actions of MMP-2 (decrease)—TMP-4 (rise) [48], after an initial total elastin amount preservation.

Ageing associated hypoestrogenism is worsening pelvic floor condition. Sexual steroid hormones have receptors in all pelvic organs, not only in genitalia (**Table 1**) [80]. Uterine prolapse, but not SUI is diagnosed in nuliparous postmenopausal women.

	ERs	P4Rs	ARs
Vagina	+	+	+
Urethra	+		
Urethral sphincter	+		
Periurethal venous system	+		
Bladder trigon	+	+	+
Pelvic floor	+	+	
Pubo cervical muscle	?(+/−)	+	
Levator Ani muscle	+	−	
Cardinal ligaments	+ (α, β)	−	
Utero-sacral ligaments	+ (α, β)		
Periurethral Fascia	+		
Perivaginal tissue			

Adapted from Rechberger and Skorupski [80] (creative common: License CC BY4.0 for adapt). ERs: estrogen receptors; P4Rs: progesterone receptors; ARs: androgens receptors.

Table 1.
Sexual steroid hormones receptors distribution in pelvic floor structures.

Aging is associated to intrinsic aging stem cells-meaning self renewal reducing, through their genes aging [81] in the general dysfunctional frailty syndrome, where pro-inflammatory cytokines-TNF-α, IL-6 and C-reactive, are increased [82], and one may delay aging effects by menopausal hormone therapy (MHT), started from perimenopause for urogenital aging, with local estriol to reduce vaginal atrophy, and some symptoms of bladder aging [83, 84], and it is adjuvant in pre and long term postoperative care, associated to systemic MHT, around age of 50′ as new protocols advice, and different MSCs, types treatments, for frailty delay to safe health and function of organs/tissues, and one speaks about safe proper frailty treatment with MSCs, to increase health and function of organs/tissues [85, 86].

3.3 Mesenchymal stem/stromal cells for pelvic tissue repair and regeneration

High number surgical gynecological procedures proposed along 100 years showed limitations, low adequacy to PFDs pathophysiology, no restoring organs' normal positions and functions, and better understanding by three dimensional digital models combining DeLancey JO's theory [16] to Petros P integral theory of continence [interdependence between pelvic organ support systems, linking ligament fascia lesions, and clinical expression, with less critics after TVT (tension free vaginal tape) technique] [87], and tissue structure continuing to deteriorate by aging after correction, being far to be restored by surgery either by native tissue or by PP meshes. MSCs, and their secretome, are discussed since more than 10 years, specially after techniques for their potency enhancement by specific culture systems, including three-dimensional culture conditions, or their priming preconditioning with some molecules of their secretome [7, 65]. MSCs represent a pathophysiologic correction, and limitation of PDFs progression [88]. Tissue engineering is a new option to restore and maintain micturition normality via direct effects on damaged or dysfunctional tissues, or pelvic floor repair when soft tissue reinforcement is necessary [89, 90], to improve outcomes in POP management [91]. MSCs also referred as mesenchymal stromal cells belong to the pool of progenitor and adult stem cells (ASCs) family, from all postnatal organs and tissues, with specific properties for each one, ensuring the capacity of renewal after damage, and in aging [7]. Collected and isolated from various anatomical sites, more or less easy accessible, few ethics-related issues, MSCs are actually easily separated/purified [92], with specific markers, and cultured. Some consider multipotent MSCs to have a limited self-renewal capacity [93], in a specific microenvironment termed "stem cells niche", first described by Schofield R [94] as "adult stem cell niche hypothesis", which is reconsidered to be more dynamically than originally appreciated, with a bilateral influence from healthy or damaged tissue to MSCs, mainly by immunological and inflammatory signals in conjunction to MSCs' paracrine effects. Others [95] consider that MSCs unique properties-high proliferative ability, self-renewal, differentiation to mesodermal lineages, appropriate to their location, are supporting tissue regeneration in physiologic and pathologic conditions, and are contributing to tissue homeostasis. MSCs are key regulatory components in the regenerating stem cell niche, by the increase of their own compound, or increasing physiologic cells turn-over [96] to support tissue regeneration after injury, or are activated in injured tissue, where they are inactive [97], or are attracting supporting cells to niche [4], or are activating tissue's own cells to facilitate repair [98], capabilities that are different according to tissue type. These biological properties determined the change of "stem" cell nomination to "stromal" for a more appropriate connotation [7], and earlier Caplan AI [99] proposed the name of "Medicinal Signaling Cells" for a more accurate presentation:

when systemically administered MSCs home in on sites of injury/disease, and exhibit a paracrine action, by secreting bioactive molecules as regulatory and growth factors, chemokines, cytokines, nucleic acids, packaged into extracellular vesicles or MSCs-derived exosomes, with trophic and immunomodulatory actions, reflecting that MSCs make therapeutic drugs *in situ* that are medicinal, important for tissue repair. MSCs fate in a tissue is influenced by local microenvironment or niche fixed compartment, where ASCs are in a dormant state (G_0) through signaling pathways inhibitory for growth and differentiation, often involving transforming growth factor-ß (TGF-ß), and bone morphogenetic protein (BMP) family members [57], being anchored to niche cells by adhesion molecules—cadherins, integrins [100] during stem cells' periods of inactivity, and niche cells differentiation signals to resident stem cells [58]. The bioactive molecules produced through MSCs homing when systemic administrated, or when are added in cultures to potentiate MSCs action, as MSCs primers [65], are considered more important than cell engraftment and replacement. MSCs and their bioactive molecules have proangiogenic [101], antifibrotic, anti-inflammatory, and pro-inflammatory actions, which sustain proliferation [102], and stimulate effect of resident progenitor cells, in relationship to disease/organ/tissue type [103].

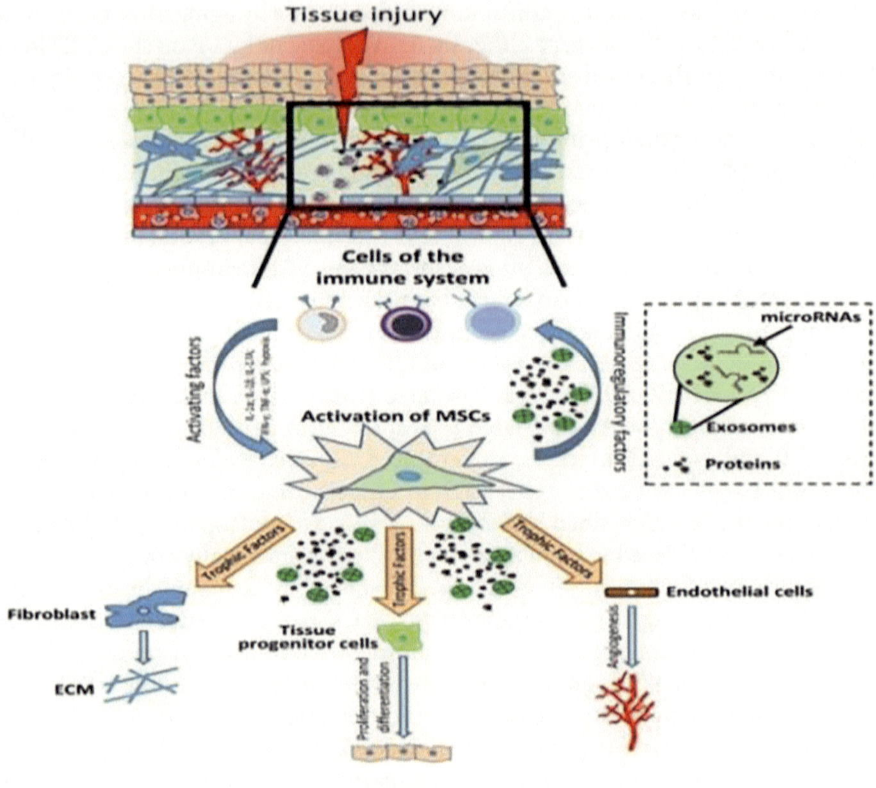

Figure 1.
MSCs role in damaged connective tissue repair. MSCs activation after tissue injury. The damaged tissue activates MSCs after injury through different inflammatory signals (hypoxia, cytokines asIL-1β, IFN-γ, TNF-α, LPS). MSCs activation leads microenvironment to coordinate the production of immunomodulatory factors, that sustain inflammation progression, and of growth factors to stimulate endothelial cells, fibroblasts, and tissue progenitor cells to differentiate, and all contribute to tissue repair, in an orderly action by angiogenesis, EMC remodeling, and functional tissue restoration. Adapted from Miceli et al. [7]. It is an open access article, distributed under the terms and conditions of the creative commons attribution (CCBY), licensee MDPI, Basel, Switzerland.

The damaged, ischemic tissue activates MSCs after injury, through different inflammatory signals (hypoxia, proinflammatory cytokines as IL-1β, IFN-γ, TNF-α, lipopolysaccharide) from host innate immune system and leads microenvironment to coordinate the production of immunomodulatory factors to sustain inflammation progression and rapid remove of allogenic MSCs, and production of growth factors to stimulate endothelial cells, fibroblasts, and tissue progenitor cells' differentiation from MSCs niche, all contributing to tissue repair, in an orderly action by angiogenesis, EMC remodeling, and functional tissue restoration [7], MSCs have the ability to home to injured tissues to exert their paracrine actions when systemically administered, a very attractive feature for this chapter discussions (**Figure 1**) [104].

MSCs do not impose immunosuppresion, being immune-privileged due to their low expression of *CD40*, *CD80*, *CD86*, and major histocompatibility complex I (*MHC I*), and the lack of *MHC II* [92], or because they are immune evasive [105].

4. Endometrial and menstrual mesenchymal stem cells for prevention and therapy of pelvic floor disorders

MSCs from different tissues exhibit many common characteristics, their biological activity and some markers are different and depend on tissue origin: bone marrow (obtained by aspirate), adipose tissue (obtained by liposuction), placenta (for maternal MSCs), and umbilical cord, amniotic membranes and liquid (for fetal MSCs) collected at birth. All these MSCs have some limitations, as in vitro expansion [106] for their rarity in original tissue, invasive methods to harvest bone marrow aspirates, donor aging affects MSCs proliferative capacity [86], and the necessity to add their secretome's active molecules in culture medium [7, 8, 107].

4.1 Unique characteristics of endometrial mesenchymal stromal/stem cells and menstrual mesenchymal stromal cells

In the last 15–20 years, a MSCs subpopulation of stem/progenitor/regenerative cells has been identified and characterized in human endometrium (eMSCs) and in menstrual blood (MenMSCs or ERCs), comparable to bone marrow and adipose tissue MSCs [95], but with unique biologic characteristics [1, 108], and knowledge plus new technique capabilities made them a very promising MSCs category in autologous and allogeneic cellular therapy, and TE. Uterine fragments of shedding endometrial tissue with their remarkable cells turn-over—like hematopoietic bone marrow, intestinal epithelium, epidermis, contribute to endometrial repair and renewal without scar, and gene profiling has demonstrated that the lysed stroma is enriched in genes involved in EMC dynamics, biosynthesis and degradation [109, 110], very promising in pelvic floor connective tissue repair/restoration, and other endometrial fragments from menstrual blood contain MenSCs. The concept of endometrial renewing after each menstruation (~400 cycles in woman's life) by endometrial stem/progenitor cells located perivascular [6], in the basalis of endometrial glands near myometrium was first hypothetized by Prianishnikov VA, 1978 [111], reloaded by Padykula HA, 1989 [112], and it is continued in Australia at the Department of Obstetrics—Gynecology, Monash University and Centre for Women's Health Research (Melbourrne, Victoria) by a team led by Gargett CE [113], who presented the first direct evidence that human endometrium contains rare populations of epithelial (0.22%), mesenchymal/stromal-eMSCs (1.25%), and endothelial progenitor cells, which exhibit the adult stem cells behavior in vitro, as clonogenicity, and later their differentiation potential (reviewed in 2016) (**Figure 2**) [115].

One may add a small side population (SP) cells which enriches endometrial stem cells fractions, according to their identity and differentiation potential [116]. eMSCs are responsible for cyclic regeneration of human, mice, and ovine endometrium [117–120]. Other research group [121] demonstrated that the low number of human endometrial stromal/stem cells seems to belong to the family of MSCs, by possessing the minimal criteria of MSCs assessment [122] clonogenity, self-renewal, plastic adherence in culture, high proliferative potential and capacity and ability to differentiate into at least one type of mature functional progeny, but eMSCs have multilineage differentiation capacity [123]. eMSCs have proliferative capability to undergo 30 populations doubling before reaching senescence, generating 6.5×10^{11} cells (**Table 2**) [124].

Taylor HS [125] at Yale University (USA) presented bone marrow (BM) as an exogenous source of eMSCs, which appear histologically as epithelial and stromal endometrial cells, expressing appropriate markers of endometrial cell differentiation, and cyclic mobilization of BM-derived stem cells is considered a normal physiologic process [126]. Menstrual blood, an usual waste tissue, but a "bio-waste" as recently reconsidered [127] (endometrial functionalis layer shed during menstruation) is an easy obtained source of MSCs, with no ethic issue, and isolated, cultured similarly to bone marrow aspirated.

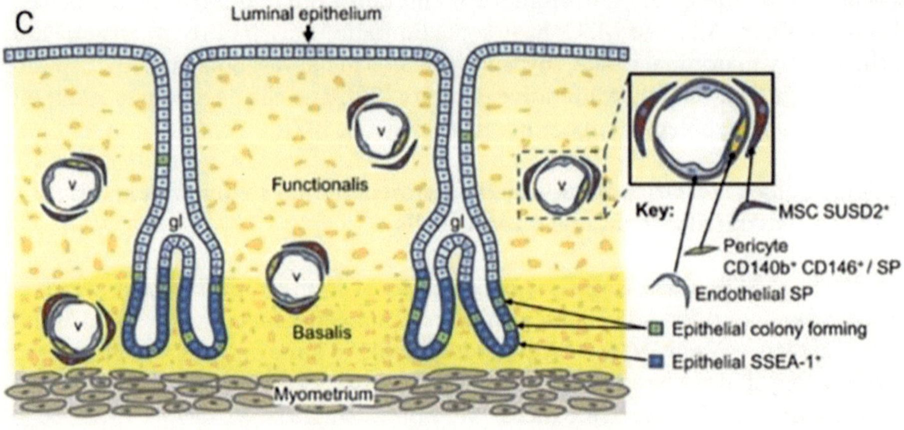

Figure 2.
Schematic perivascular localisation of human eMSCs. Co-expressing CD146 and PDGFRβ/CD140b and SUSD2⁺ eMSC in the endometrial basalis and functionalis layer, indicating eMSC will be shed into menstrual blood. Reprinted from Gargett and Masuda [114] with license permission: 5351991932 77 for Copyright Oxford University Press, 2022, July.

- MSCs must be plastic-adherent when maintained in standard culture conditions.
- MSCs must express CD105, CD73 & CD90, and lack expression of CD45, CD34, CD14 or CD11b, CD79α or CD19 & HLA-DR surface molecules.
- MSCs must differentiate to osteoblasts, adipocytes and chondroblasts in vitro

International Society for Cellular Therapy position statement [122].

Table 2.
Minimal criteria for defining multipotent MSCs.

Menstrual stem cells (MenSCs), named also endometrial regenerative cells (ERCs) by the team from Bio-Communications Research Institute (Wichita, USA), who first isolated and cultured them [128], have the classic properties, and pattern of MSCs surface markers, are multipotent [129], retain a stable karyotype in culture [130], proved at more than 68 doublings without any karyotype or functional abnormalities [128, 131]. ECRs have similar capabilities for tissue repair and restoration as eMSCs, according to their secretome that ensure cells paracrine actions on endometrium (after menstruation, and Asherman syndrome) [132], ovaries [133], and on different organ [134, 135]. There are many controversies on eMSCs and ERCs makers [108], being demonstrated that endometrial/menstrual MSCs clones express MCSs markers [ITGB1 (CD29), CD44, NT5E (CD73), THY1 (CD90), ENG (CD105), PDGFRB (CD140B), MCAM (CD146)], but not endothelial or hemopoietic markers PECAM1 (CD31), CD34, PTPRC (CD45), and the pan-leukocyte marker CD45 [128]. The Australian team used these markers for eMSCs isolation and culture [124], plus two perivascular cell surface markers—CD146, and platelet-derived growth factor-receptor β (PDGF-Rβ) or CD140B, and have determined eMSCs location near blood vessels in human endometrium [114, 136, 137], blood vessel wall is considered the eMSCs niche, as it is for all MSCs [138, 139]. These markers were used in association to Sushi Domain Containing 2 (SUSD2) or W5C5, a special marker for eMSCs and MenSCs isolation, proposed as a novel single marker for purifying eMSCs, and to reconstitute endometrial stromal tissue in vivo from endometrial biopsies [6]. MenSCs need a selective marker enrichment to be consistent and efficacious as eMSCs obtained by endometrial biopsy. It is known that perivascular MSCs or pericytes are rare cells, difficult to harvest from adult tissues, and necessitate substantial ex-vivo culture expansion to achieve a sufficient number of potent cells, and prolonged culture of MSCs determine a spontaneous differentiation to fibroblasts, which limits culture expansion, and these significant limits challenged the special add in culture medium of a novel small molecule-transforming growth factor-β receptor inhibitor, namely A83-01, that limits the inconvenient and maintains eMSCs and other MSCs undifferentiated in the days following administration and ensure the therapeutic efficacy of a small proportion (2%) of cells which are estimated to remain in vivo in the days following administration [9, 91, 114, 140]. MenSCs secretome needs a special attention for its exceptional therapeutic effects, due to extracellular vesicles (EV) [135], including microvesicles, exosomes and apoptotic bodies transporting bioactive molecules. A total of 895 molecules are identified in exosomes [141], as micro RNA, lipids, growth factors (vascular endothelial growth factor, insulin-like growth factor-1, hepatocyte growth factor), chemokines, cytokines as regulators of immune response in different tissues [142], which can be isolated from menstrual blood as are MenSCs isolated [143]. The human MenSCs transcriptome and methylome profiles showed their most distinctive expression and epigenetic signature compared to human bone marrow and adipose MSCs [144]. MenSCs trandifferentiation capacity is extensively discussed, associated to their possibility to differentiate into mesodermal lineage (including chondrogenic, osteogenic, adipogenic, and cardiomyogenic fate), endodermal lineage (hepatocyte), and ectodermal lineage (neural and glial) [145], processes that varies considerably between each type, and it is different when one compares them to bone marrow and adipocyte MSCs [128]. Recently there were proved the beneficial effects of MenSCs and their secretome in pulmonary healing in severe acute adults lung cells injury (ARDS) from COVID, by increasing number of CD4 lymphocytes, reducing expression of inflammatory markers (C Reactive Proteine, ferritine, LDH), absorption of bilateral pleural exudates, better than other MSCs types (BM, adipose tissue) when systemic transplanted [146], having also the advantage of easy collection in emergency.

4.2 Potential application of endometrial and menstrual mesenchymal stem cells in pelvic floor disorders prevention and therapy

eMSCs with autologous and allogeneic origins are easily procured from endometrial biopsies—during reproductive years, under contraceptives, in postmenopause [115], without anesthesia or from menstrual blood (source that can be repeatedly used at every menstrual cycle, much easier obtained vs. other sources of ASC). with remarkable differentiation capabilities, plus paracrine actions from their secretome, proved by preclinical and some phase III trials—much discussed and criticized in USA [147], because industry commercial enthusiasm [148]. FDA (USA) approved in 2011 clinical trials with ERCs, which can be used for PFDs prevention and therapy. Actually eMSCs are not accessible for human trials all over the world, or there are no legally approved banks for eMSCs/MenMSCs, these cells being used in some countries only for academic/scientific health centers or not-for profit public institutions. European Medicine Agency has a "hospital exemption" clause, with existence of many unknowns/controversies on eMSCs/MenMSCs-based therapies in human (allogeneic or autologous), as they are used in animal models. Actually one must consider that amalgamation of highly specialized disciplines such as tissue engineering, stem cell therapy and personalized medicine provide important approaches and tools to respond to these challenges in PFDs prevention and therapy, as there were found regulatory approval and deployment for disorders with unmet medical needs [149].

4.2.1 Endometrial/menstrual mesenchymal stem cells based for pelvic floor disorders prevention and therapy

Cell therapy is an emerging field in clinical practice, bone marrow, adipose tissue being subjects for trials in chronic and degenerative disease. Basic evidences provide the preventive role of MSCs autologous cell based therapy in rats SUI by local-urethral administration [150], which is a minimally invasive procedure, and intravenous (i.v.) route [10]. MSCs are homing in damaged pelvic organs, where they are attracted by cytokines [151] and chemokines [152]. After i.v. administration, post vaginal distension (VD)—a model for childbirth injury, GFP-labeled cells were depicted at 4 to 10 days in urethra, vagina, rectum, and levator ani muscle of sacrificed animals, with significantly more MSCs homing at 4 days *versus* after sham VD, and reduction of GFP intensity at 10 days after VD [153]. It is discussed PFDs prevention with MSCs in high risk women by genetic predisposition, with postpartum SUI and/or POP, and the possibility to induce homing a long time after injury or to increase homing after an acute injury *via* stem cells genetic modification to express a greater number of homing ligands [154] or an electrical stimulation to the paravaginal region, which induces neural stem cells migration [155], or by MSCs paracrine effects in damaged tissues [growth factors as Granulocyte-Macrophage Colony-Stimulating Factor (GM-CSF), angiopoietin-2, and Platelet-Derived Growth Factor (PDGF-BB)], which were proved to be in a rate of about 10–100,000 times more for MenMSCs than the control mesenchymal cell lines derived from umbilical cord blood, and by stimulating metalloproteinases involved in elastin postpartum [128]. Post infusion febrile reaction is the sole adverse event associated to bone marrow MSCs i.v. [156]. One discusses the i.v. dose differentiation between small (rodents), large (ovine) animals, and human: 50 million MSCs/kg for rodents, and 1–2 million cells/kg and never more than 12 million cells/kg for human. MSCs pulmonary entrapment, with their quick phagocytisation, by lung tissue macrophages [157], and the genetic instability and tumorigenicity [156], not valuable for eMSCs/MenMSCs [14]. ERCs presenting more than 68 doublings without any karyotype or functional abnormalities [128, 131].

Australia has programs based on techniques to purify eMSCs by magnetic beads and special markers, for production of large number of cells under Good Manufacturing Practice (GMP) conditions to offer women's own eMSCs when they need for PFDs, for TE. Actually there are only animal studies, no human trial on eMSCs, /MenSCs for PFDs, being discussed other MSCs types. The Cochrane Database Syst Rev. (2017) [158] found in Cochrane Incontinence Group Specialized Trials Register only one small RCT for SUI, on injection of autologous MSCs with fat origin *vs placebo*, terminated early because of safety concerns, and afterwards there are mentioned some clinical trials for SUI treated with autologous muscle derived MSCs compared to *placebo* [159], or transurethral and periurethral intrasphincteric injections of cellular suspension for SUI, with limited accuracy of results [160], or a preliminary randomized study of stem cells from adipose tissue implanted for fecal incontinence [161].

4.3 Tissue engineering with endometrial and menstrual mesenchymal stromal cells for pelvic floor disorders

Tissue engineering (TE) for pelvic floor disorders treatment combines principles of eMSCs/MenSCs biology with materials science (scaffolds, meshes for pelvic implantation), and biomedical engineering [162]. The most discussed PFDs beneficiary of eMSCs therapy are POP and SUI—*de novo*, persistent or recurrent after failure of surgery with native tissue or reconstruction. There are necessary multidisciplinary trained teams. and special protocols, as are presenting Ichim T (2008) CEO at Medistem Laboratory (San Diego, USA), and Gargett CE (2013)—head of the Australian Stem Cell Centre.

The key to safe and efficacious TE in PFDs was the generation of some tissue substitute by materials with nano-architecture/nanofiber technology [163] or 3D printing, mimicking EMC pelvic floor topography, mainly vaginal EMC, or to induce favorable tissue mechanical responses, or to add some EMC constituents (as tropoelastin-the elastin core in EMC network, besides collagen, lost in POP) [164], for production of mesh/scaffold new generation, which allows entrapment, and persistence of seeded eMSCs/MenSCs up to 14 days after implant [41]. eMSCs are in vitro optimized in serum free conditions—fibronectine is the optimal substrate for human eMSCs attachment [91], and eMSCs transcriptome reveals improved potential for cell based therapy after adding TGF-β receptor inhibitor in culture medium (to prolong their undifferentiated status after implantation, by eMSCs apoptosis, and senescence prevention, and maintaining the percentage of $SUSD2^+$ cells to more than 90% for all samples) [165]. Another proposal for eMSCs tissue repair efficacy augmentation is the add of a protective delivery system, as a compatible bio-hydrogel carrier that encapsulates eMSCs in mesh/scaffold and improves cells retention at site from host immune system actions to rapid their remove, due to loss of vascular niches [107], and by their encapsulation in hydrogel MSCs can promote endogenous cellular repair [140]. There are many composite meshes produced from different materials: nondegradable polyamide meshes, as those of polyamide/gelatin seeded with 100,000 human or ovine MenSCs/cm^2, which stimulate angiogenesis, host synthesis type 1 and type III collagen, lower leukocytes infiltrate at 90 days postimplantation (when tested on rats) [166], andthe new biomimetic tissue generation of degradable nano/microstructured meshes [167], meshes obtained through new technologies of electrospinning and 3 D bioprinted endometrial stem cells on an aloe vera–alginate (AV-ALG) injectable hydrogel or on melt electrospun poly epsilon-caprolactone mesh, with the largest open pore diameter and the lowest thickness that promotes eMSCs encapsulated in the hydrogel attachment, which reduces FBRs associated to eMSCs same action [168, 169]. The meshes designed with nanoscale fibers using electrospinning techniques promote cell–cell and cell-biomaterial interactions, being appreciated in Australia [91, 167, 170], and Nederland studies [171] that biometric properties of this nanostructured mesh can improve the integration, overcome erosion, and offer good outcomes in POP reconstructive surgery. The mesh type added to eMSCs have different persistence time after implantation, the natural ones have the shortest "life" duration after implant; the non-degradable polyamide/gelatin mesh plus autologuse MSC persisted 3 days in immunocompetent mice, 1–2 weeks in immunocompromised rodent model when xenogenic human eMSCs, and 90 days in ovine; the longest duration is at 3 D printed mesh [168]. The Chinese study [12] shows that a composite mesh based on synthetic and natural polymers seems to provide the best combination for an ideal pelvic floor mesh material, because natural polymers can provide ligands for cell adhesion and growth factors that promote tissue remodeling, while synthetic polymers provide mechanical strength. New scaffolds proposed for POP provide a three-dimensional environment, and are mimicking EMC network, specially by their micro/nanoscale architecture [172], offering a larger area for EMC constituent proteins, and more binding sites for cell membrane receptors, and adhesion molecules, growth factors, genes, immunomodulatory agents, and external stimuli (electrical, or magnetic pulses), which are delivered simultaneously to target sites after scaffold implantation for promotion of new healthy tissue [173]. No ideal mesh/scaffold for PFDs exists to day, and one may choose the mesh/scaffold according to patient history, and implants' properties, such as material type: natural (as purified human/animal collagen, chitosan, gelatin, elastin, fibrin, silk, and fibronectin for human eSMCs [174, 175], or synthetic, or by the old criteria used for PP meshes, as pore sizes, mesh's weight, and the host tissue response to mesh/scaffold material or to eMSCS/ERCs, understanding that reaction to implant materials are crucial for balance between their own elasticity and stiffness, vaginal tissue capacity of strain, which are absent when PP meshes are used. Novel blends of electrospun synthetic and natural polymers combined with eMSC in new generation of implants show that this approach promotes host cell infiltration and slows biomaterial degradation that has potential to strengthen the vaginal wall during healing [164, 165], actioning like intrinsic ECM, but with some limitation regarding small pore size of electrospun nanofiber meshes and toxic organic solvents used for their production [12]. /Human eMSCs/MenSCs modulate host tissue response to implanted materials, by stimulating tissue proper stem cells proliferation, their own high proliferation rate [128], and scaffolds' eMSCs infiltration, and constituents of their secretome as growth factors, enzymes as MMPs—important for elastin postpartum recovery [128], influence mesh mechanical behavior after implantation [166], by fibrosis and FBRs reducing, when nondegradable polyamide mesh implant, through influencing macrophage polarization switching from an M1 to M2 phenotype, as in rodent and ovine models [166, 176, 177]. eMSCs seeded in degradable nano/microstructured meshes improve mesh tissue integration, eMSCs are entraped over 6 weeks in vivo, by cells with immunomodulatory effects, and by increasing local angiogenesis reduce FBRs to mesh implanted in mice with POP [166, 167, 169], and induce an up-regulation of M2 markers—as CD206 and *Arg1*, *Mrc1*, and *Il10* genes in host tissue macrophages, parallel to reduction of cellular infiltration and secretion of inflammatory cytokines Il-1β and Tnf-α [175].

The Australian researchers [140] appreciate that tissue engineered mesh inserted transvaginally in large animal models will aid the validation of these constructs prior to clinical translation by assessing their integration with host tissue, and FBRs through histological analysis, immunoassays and gene profiling. Research has commenced with the completion of multiple xenogenic small and large animals studies assessing eMSC/PAG constructs, as mentioned above [166]. These animal models will be crucial in further assessing the efficacy of locally delivered eMSC and further determination of their action mechanisms. Based on these findings, multiple heterologous small and large animal studies are underway to assess the efficacy of other biomimetic degradable materials such as PLCL and 3D-PCL meshes seeded with eMSC/MenSCs.

5. Future proposals for stem cell prevention and therapy in pelvic floor disorders

The Australian research teams recognize some unknown data about eMSCs/MenSCs, mediation on cellular migration and recruitment by their paracrine effects, how eMSCs mediate M2 immunomodulatory responses during the FBR after implantation of bioengineered constructs. One discusses eMSCs secretome constituents as future associated to new generation implants for tissue engineering in PFDs. The Canadian and North American and Chinese researchers recognize the high financial burden for the studies and introduction in human clinical practice according to legislation issues which are incomplete resolved in Europe and North America, in comparison to Australia and Japan.

Conflict of interest

None.

Author details

Manuela Cristina Russu
"Dr I. Cantacuzino" Discipline of Obstetrics and Gynecology, "Carol Davila" University of Medicine and Pharmacy, Bucharest, Romania

*Address all correspondence to: manuela_russu@yahoo.com

References

[1] Emersson JS, Gargett EC. Endometrial mesenchymal stem cells as a cell based therapy for pelvic organ prolapse. World Journal of Stem Cells. 2016;**8**(5):202-215. DOI: 10.4252/wjsc.v8.i5.202

[2] Macowski SJ, Jurnalov C, Phelps JY. Use of vaginal mesh in the face of recent FDA warning and litigation. American Journal of Obstetrics & Gynecology. 2010;**203**(2):103.e1-103.e4

[3] Holt E. US FDA rules manufacturers to stop selling mesh devices. Lancet. 2019;**393**:1686. DOI: 10.1016/S0140-6736(19)30938-9

[4] Wosczyna MN, Konishi CT, Perez Carbajal EE, Wang TT, Rando TA, et al. Mesenchymal stromal cells are required for regeneration and homeostatic maintenance of skeletal muscle. Cell Reports. 2019;**27**:2029-2035.e5. DOI: 10.1016/j.celrep.2019.04.074

[5] Schwab KE, Gargett CE. Co-expression of two perivascular cell markers isolates mesenchymal stem-like cells from human endometrium. Human Reproduction. 2007;**22**(11):2903-2911

[6] Masuda H. A novel marker of human endometrial mesenchymal stem like cells. Cell Transplantation. 2012;**21**(10):2201-2214. DOI: 10.3727/096368911X37362

[7] Miceli V, Bulati M, Iannolo G, Zito G, Gallo A, Conaldi PG. Therapeutic properties of mesenchymal stromal/stem cells: the need of cell priming for cell-free therapies in regenerative medicine. International Journal of Molecular Sciences. 2021;**22**(2):763. DOI: 10.3390/ijms22020763

[8] Alvites R, Branquinho M, Sousa CA, Lopes B, Sousa P, Mauricio AC. Mesenchymal stem/stromal cells and their paracrine activity-immunomodulation mechanisms and how to influence the therapeutic potential. Pharmaceutics. 2022;**14**(2):381. DOI: 10.3390/pharmaceutics14020381

[9] Gurung S, Werkmeister JA, Gargett CE. Inhibition of transforming growth factor-beta receptor signaling promotes culture expansion of undifferentiated human endometrial mesenchymal stem/stromal cells. Scientific Reports. 2015;**5**:15042. DOI: 10.1038/srep15042

[10] Cruz M, Dissaranan C, Cotleur A, Kiedrowski M, Damaser M, et al. Pelvic organ distribution of mesenchymal stem cells injected intravenously after simulated childbirth injury in female rats. Obstetrics and Gynecology International. 2012;**2012**:612946. DOI: 10.1155/2012/612946

[11] Ben Menachem-Zidon O, Gropp M, Ben Shushan E, Reubinoff B, Shveiky D. Systemically transplanted mesenchymal stem cells induce vascular-like structure formation in a rat model of vaginal injury. PLoS One. 2019;**14**(6):e0218081

[12] Yuan M, Hu M, Dai F, Fan Y, Cheng Y, et al. Application of synthetic and natural polymers in surgical mesh for pelvic floor reconstruction. Materials & Design. 2021;**209**:109984

[13] Edwards SL, Werkmeister JA, Rosamilia A, Ramshaw JAM, White JF, et al. Characterisation of clinical and newly fabricated meshes for pelvic organ prolapse repair. Journal of the Mechanical Behavior of Biomedical Materials. 2013;**23**:53-61

[14] von Bahr L, Batsis I, Moll G, Hagg M, Le Blanc K, et al. Analysis of tissue

following mesenchymal stromal cell therapy in humans indicates limited long term engraftment and no ectopic tissue formation. Stem Cells. 2012;**30**(7):1575-1578. DOI: 10.1002/stem.1118

[15] Haylen BT, de Ridder D, Freeman RM, Swift SE, Berghmans B, et al. An international Urogynecological association (IUGA)/international continence society (ICS) joint report on the terminology for female pelvic floor dysfunction. Neurourology and Urodynamics. 2010;**29**:4-20

[16] DeLancey JO. Anatomy and biomechanics of genital prolapse. Clinical Obstetrics and Gynecology. 1993;**36**:897-909

[17] Westergren Söderberg M, Byström B, Kalamajski S, Malmstörm A, Ordeberg G-E. Gene expression of small leucine-rich repeat proteoglycans and fibulin are decreased in pelvic organ prolapse. Molecular Human Reproduction. 2009;**15**(4):251-257

[18] Wu MJ, Matthews AC, Conover MM, Pate V, Jonsson Funk M. Lifetime risk of stress urinary incontinence or pelvic organ prolapse surgery. Obstetrics and Gynecology. 2014;**123**(6):1201-1206. DOI: 10.1097/AOG.0000000000000286

[19] National Institute of health consensus: Consensus development conference. Urinary incontinence in adults. Journal of the American Geriatrics Society. 1990;**38**:265-276

[20] Nussler E, Granäsen G, Bixo M, Löfgren M. Long term outcome after routine surgery for pelvic organ prolapse. A national register-based cohort study. International Urogynecology Journal. 2022. DOI: 10.1007/s00192-022-05156-y

[21] Smith FJ, Holman CD, Moorin RE, Tsokos N. Lifetime risk of undergoing surgery for pelvic organ prolapsed. Obstetrics and Gynecology. 2010;**116**(5):1096-1100. DOI: 10.1097?AOG.0b01363181f3729

[22] Dieter AA, Wilkins MF, Wu JM. Epidemiological trends and future care needs for pelvic floor disorders. Current Opinion in Obstetrics and Gynecology. 2015;**27**:380-384

[23] Dooley Y et al. Urinary incontinence prevalence: Results from the National Health and nutrition examination survey. The Journal of Urology. 2008;**179**(2):656-671

[24] McIntosh LJ, Mallett VT, Frahm JD, Richardson DA, Evans MI. Gynecologic disorders in women with ehlers-danlos syndrome. Journal of the Society for Gynecologic Investigation. 1995;**2**:559-564

[25] Norton PA, Baker JE, Sharp HC, Warenski JC. Genitourinary prolapse and joint hypermobility in women. Obstetrics and Gynecology. 1995;**85**:225-228

[26] Allen-Brady K, Bortolini MAT, Damaser MS. Mouse knockout models for pelvic organ prolapse, a systematic review. International Urogynecology Journal. 2022. DOI: 10.1007/s00192-021-05066-5

[27] Subak LL, Waetjen LE, van den Eeden S, Thom DH, Brown JS, et al. Cost of pelvic organ prolapse surgery in the United States. Obstetrics and Gynecology 2001;**98**(4):646-651. DOI: 10.1016/S0029-7844(01)01472-7

[28] Van Geelen JM, Dwyer PL. Where to for pelvic organ prolapsed treatment after FDA pronouncements? A systematic review of the recent literature. International Urogynecology Journal. 2013;**5**:707-718. DOI: 10.1007/s00192-012-2025-3

[29] Goessens E, Deriemaeker H, Cammu H. The use of vaginal pessary to decide whether a mid urethral sling should be added to prolapsed surgery. The Journal of Urology. 2020;**203**(3):593-603. DOI: 10.1097/JU.0000000000000575

[30] Dobrokhotova EY, Nagieva T, Kareva EN. Focused radiofrequency treatment of postpartum vaginal relaxation syndrome. Obstetrics Gynecology and Reproduction. 2020;**14**(4):437-446

[31] Altman D, Vayrynen T, Engl ME, Axelsen S, Falconer C. For Nordic transvaginal mesh group: Anterior colporrhaphy versus transvaginal mesh for pelvic-organ prolapse. New England Journal of Medicine. 2011;**364**(19):1826-1836. DOI: 10.1056/NEJMoa1009521

[32] Chapple CR, MacNeil S. The use of implanted materials for treating women with pelvic organ prolapse and stress urinary incontinence. Current Opinion in Urology. 2019;**29**(4):431-436. DOI: 10.1097/MOU.0000000000000619

[33] Abhari ER, Izett-Kay LM, Morris LH, Cartwright R, Snelling JBS. Host-biomaterial interactions in mesh complications after pelvic floor reconstructive surgery. Nature Reviews Urology. 2021;**18**(12):725-738. DOI: 10.1038/s41585-021-00511-y

[34] Feola A, Abramowitch S, Jallah Z, Stein S, Moalli P, et al. Deterioration in biomechanical properties of the vagina following implantation of a high-stiffness prolapse mesh. BJOG. 2013;**120**:224-232

[35] Alas AN, Chinthakanan O, Espaillat L, Davila GW, Aguilar VC, et al. De novo stress urinary incontinence after pelvic organ prolapse surgery in women without occult incontinence. International Urogynecology Journal. 2017;**28**(4):583-590. DOI: 10.1007/s00192-016-3149-7

[36] Weemhoff M, Vergeldt TF, Notten K, Serroyen J, Roumen FJ, et al. Avulsion of puborectalis muscle and other risk factors for cystocele recurrence: A 2-year follow-up study. International Urogynecology Journal. 2012;**23**(1):65-71. DOI: 10.1007/s00192-011-1524-y

[37] Quiroz LH, Gutman RE, Shippey S, Cundiff GW, Handa VL, et al. Abdominal sacrocolpopexy: Anatomic outcomes and complications with Pelvicol, autologous and synthetic graft materials. American Journal of Obstetrics and Gynecology. 2008;**198**(5):557. e1-557. e5

[38] Maher C, Feiner B, Baessler K, Adams EJ, Hagen S, Glazener CM. Surgical management of pelvic organ prolapse in women. Cochrane Database of Systematic Reviews. 2013;**4**:CD004014. DOI: 10.1002/14651858.CD004014.pub5

[39] Morling JR, Mc Allister AD, Agur W, Fischbacher MC, Wood R, et al. Adverse events after first, single, mesh and non-mesh surgical procedures for stress urinary incontinence and pelvic organ prolapse in Scotland, 1997-2016: A population-based cohort study. Lancet. 2017;**389**(10069):629-640

[40] Pascal G, Hernandez Gascon B, Rodriguez M, Bellon JM, et al. The long-term behavior of lightweight and heavyweight meshes used to repair abdominal wall defects is determined by the host tissue repair process provoked by the mesh. Surgery. 2012;**152**:886-895

[41] Ulrich D, Edwards SL, Su K, Tan KS, White JF, et al. Human endometrial mesenchymal stem cells modulate the tissue response and mechanical behaviour of polyamide mesh implants

for pelvic organ prolapse repair. Tissue Engineering. 2013;**20**:785-798

[42] Vashaghian M, Zaat SJ, Smit TH, Roovers JP. Biomimetic implants for pelvic floor repair. Neurourology and Urodynamics. 2018;**37**:566-580

[43] Thompson P. The myology of the pelvic floor, Newton: McCorquodale, 1899, cited by Raizada V, Mittal KM. Pelvic floor anatomy and applied physiology. Gastroenterology Clinics of North America. 2008;**37**(3):493-vii. DOI: 101016jgec.2008.06.003

[44] Rynkevic R, Martins P, Hympanova L, Deprest J, et al. Biomechanical and morphological properties of the multiparous ovine vagina and effect of subsequent pregnancy. Journal of Biomechanics. 2017;**57**:94-102. DOI: 10.1016/j.biomech.2017.03.023

[45] Ulrich D, Edwards SL, Su K, White JE, Gargett CE, et al. Influence of reproductive status on tissue composition and biomechanical properties of ovine vagina. PLoS One. 2014;**9**(4):e93172. DOI: 10.1371/journal;pone.0093172

[46] de Landsheere L, Brieu M, Blacher S, Munaut C, Cosson M, et al. Elastin density. Link between histological and biomechanical properties of vaginal tissue in women with pelvic organ prolapse? International Urogynecology Journal. Apr 2016;**27**(4):629-635. DOI: 10.1007/s00192

[47] Jameson S, Swaminathan G, Dahal S, Court B, Damaser MS, et al. Elastin homeostasis is altered with pelvic organ prolapsed in cultures of vaginal cells from lysiloxydase like-1 knockout mice model physiological reports. 2020;**8**(11):e14436

[48] Dahal S, Kuang M, Rietsch A, Damaser MS, et al. Quantitative morphometry of elastic fibers in pelvic organ prolapse. Annals of Biomedical Engineering. 2021;**49**(8):1909-1922. DOI: 10.1007/s10439-021-02760-9

[49] Rubod C, Boukerrou M, Brieu M, Dubois P, Cosson M. Biomechanical properties of vaginal tissue. Part 1: New experimental protocol. The Journal of Urology. 2007;**178**:320-325

[50] Rubod C, Boukerrou M, Brieu M, Jean-Charles C, Dubois P, et al. Biomechanical properties of vaginal tissue: Preliminary results. International Urogynecology Journal. 2008;**19**:811-816

[51] Jean-Charles C, Rubod C, Brieu M, Boukerrou M, Fasel J, et al. Biomechanical properties of prolapsed or non-prolapsed vaginal tissue: Impact on genital prolapse surgery. International Urogynecology Journal. 2010;**21**:1535-1538

[52] Rubod C, Brieu M, Cosson M, Rivaux G, Clay J-C, et al. Biomechanical properties of human pelvic organs. The Journal of Urology. 2012;**79**(968):e917-e922

[53] Shin SJ, Yanagisawa H. Recent updates on the molecular network of elastic fiber formation. Essays in Biochemistry. 2019;**63**(3):395-376. DOI: 10.1042/EBC20180052

[54] Hulmes DJS. Building collagen molecules, fibrils, and suprafibrillar structures. Journal of Structural Biology. 2002;**137**:2-10. DOI: 10.1006/jsbi.2002.4450

[55] Svensson L, Narlid I, Oldberg A. Fibromodulin and lumican bind to the same region on collagen type I fibrils. FEBS Letters. 2000;**470**:178-182

[56] Couchman JR, Pataki CA. An introduction to proteoglycans and their

localization. Journal of Histochemistry & Cytochemistry. 2012;**60**:885-897

[57] Li L, Xie T. Stem cell niche: Structure and function. Annual Review of Cell and Developmental Biology. 2005;**21**:605-631

[58] Moore KA, Lemischka IR. Stem cells and their niches. Science. 2006;**311**:1880-1885

[59] de Vega S, Iwamoto T, Yamada Y. Fibulins: Multiple roles in matrix structures and tissue functions. Cellular and Molecular Life Sciences. 2009;**66**(11-12):1890-1902. DOI: 10.1007/s00018-009-8632-6

[60] Yanagisawa H, Davis EC. Unraveling the mechanism of elastic fiber assembly: The roles of short fibulins. The International Journal of Biochemistry & Cell Biology. 2010;**42**(7):1084-1093. DOI: 10.1016/j.biocel.2010.03.009

[61] Drewes PG, Yanasgisawa H, Starcher B, Hornstra I, Word RA, et al. Pelvic organ prolapse in fibulin-5 knockout mice: Pregnancy-induced changes in elastic fiber homeostasis in mouse vagina. The American Journal of Pathology. 2007;**170**(2):578-589. DOI: 10.2353/ajpath.2007.060662

[62] Rahn DD, Ruff MD, Brown SA, Tibbals HF, Word RA. Biomechanical properties of the vaginal wall: Effect of pregnancy, elastic fiber deficiency, and pelvic organ prolapse. American Journal of Obstetrics and Gynecology. 2008;**198**(590):e591-e596

[63] Rahn DD, Acevedo FJ, Roshanravan S, Keller PW, Word RA, et al. Failure of pelvic organ support in mice deficient in fibulin-3. The American Journal of Pathology. 2009;**174**(1):206-215. DOI: 10.2353/ajpath.2009.080212

[64] Liu X, Zhao Y, Gao J, Pawlyk B, Li T, et al. Elastic fiber homeostasis requires lysyl oxidase-like 1 protein. Nature Genetics. 2004;**36**(2):178-182. DOI: 10.1038/ng1297

[65] Ferreira JR, Teixeira GQ, Santos SG, Barbosa MA, Goncalves RM, et al. Mesenchymal stromal cell secretome: Influencing therapeutic potential by cellular pre-conditioning. Frontiers in Immunology. 2018;**9**:2837. DOI: 10.3389/fimmu.2018.02837

[66] Svensson K, Bortolini MAT, Damaser MS. Mouse knockout models for pelvic organ prolapse, a systematic review. International Urogynecology Journal. 2022. DOI: 10.1007/s00192-021-05066-5

[67] Bruel A, Ortoft G, Oxlund H. Inhibition of cross-links in collagen is associated with reduced stiffness of the aorta in young rats. Atherosclerosis. 1998;**140**(1):135-145. DOI: 10.1016/S0021-9150(98)00130-

[68] Edwall L, Carlström K, Fianu Jonasson A. Markers of collagen synthesis and degradation in urogenital tissue and serum from women with and without uterovaginal prolapse. Molecular Human Reproduction. 2008;**14**(3):193-197

[69] Moalli PA, Shand SH, Zyczynski HM, Gordy SC, Meyn LA. Remodeling of vaginal connective tissue in patients with prolapse. Obstetrics and Gynecology. 2005;**106**(5 pt 1):953-963

[70] Phillips CH, Anthony F, Benyon C, Monga AK. Collagen metabolism in the uterosacral ligaments and vaginal skin of women with uterine prolapse. BJOG : An International Journal of Obstetrics and Gynaecology. 2006;**113**(1):39-46

[71] Connell KA, Guess MK, Chen HW, Lynch T, Taylor HS, et al. HOXA11

promotes fibroblast proliferation and regulates p53 in uterosacral ligaments. Reproductive Sciences. 2009;**16**(7):694-700

[72] Bhattarai A, Staat M. Modeling of soft connective tissue to investigate female pelvic floor dysfunction. Computational and Mathematical Methods in Medicine. 2018:9518076

[73] Catanzarite T, Bremner S, Barlow C, Bou-Malham L, Alperin M, et al. Pelvic muscles' mechanical response to strains in the absence and presence of pregnancy-induced adaptations in a rat model. American Journal of Obstetrics & Gynecology. 2018;**210**(5):512.e1-512.e9. DOI: 10.1016/j.ajog.2018.02.001

[74] Bartolini MA, Drutz HP, Lovatsis D, Alarab M. Vaginal delivery and pelvic floor dysfunction, current evidence and implications to future research. International Urogynecology Journal. 2010;**21**(8):1025-1030

[75] Lien KC, Mooney B, DeLancey JO, Aston-Miller JA. Levatorani muscle stretch induced by simulated vaginal birth. Obstetrics and Gynecology. 2004;**103**(1):31-40

[76] Kroffta L, Otcenasek M, Kasikova E, Feyereisl J. Pubococcgygeus-puborectalis trauma after forceps delivery-evaluation of levatorani muscle with 3D/4D ultrasound. International Urogynecology Journal and Pelvic Floor Dysfunction. 2009;**20**:1175-1181

[77] Rautzang RM, Rostaminia G, Moalli AP, Abramovitch S. Pelvic floor shape variations during pregnancy and after vaginal delivery. Computer Methods and Programs in Biomedicine. 2020;**194**:105516. DOI: 10.1016/J.cmpb.2020.105516

[78] Rynkevic R. The effect of consecutive pregnancies on the ovine pelvic soft tissue. Link between biomechanical and histological components. Annals of Anatomy. 2019;**222**:166-172. DOI: 10.1016/j.aanat.2018.12.002

[79] Alperin M, Feola A, Duerr R, Moalli P, Abramavowitch S. Pregancy and delivery-induced biomechanical changes in rat vagina persist postpartum. International Urogynecology Journal. 2010;**21**:1169-1174

[80] Rechberger T, Skorupski P. The controversies regarding the role of estrogens in urogynecology. Folia Histochemica et Cytobiologica. 2007;**45**(Supp. 1):S:17-S:21. DOI: 10.2478/4476

[81] Yu KR, Kang KS. Aging-related genes in mesenchymal stem cells: A mini review. Gerontology. 2013;**59**:557-563. DOI: 10.1159/000353857

[82] Schulman IH, Balkan W, Hare JM. Mesenchymal stem cell therapy for aging frailty. Frontiers in Nutrition. 2018;**5**:108. DOI: 10.3389/fnut.2018.00108

[83] Hillard T. The postmenopausal bladder. Menopause International. 2010;**16**(2):74-80. DOI: doi; 10.1258/mi.2010.010020

[84] Robinson D, Toozs Hobson P, Cardozo L. The effect of hormones on the lower urinary tract. Menopause International. 2013;**19**(1):155-162. DOI: doi; 10.1177/1754045511398

[85] Sousa-Victor P, Gutarra S, Garcia-Prat L, Rodriguez-Ubreva J, Serrano AL, et al. Geriatric muscle stem cells switch reversible quiescence into senescence. Nature. 2014;**506**:316-321. DOI: 10.1038/nature13013

[86] Beane OS, Fonseca VC, Cooper LL, et al. Impact of aging on the regenerative properties of bone marrow, muscle-, and

adipose-derived mesenchymal stem/stromal cells. PLoS One. 2014;**9**:e115963

[87] Petros P, Woodman JP. The integral theory of continence. International Urogynecology Journal. 2008;**19**(1):35-40. DOI: 10.1007/s00192-007-0475-9

[88] Tran C, Damaser SM. The potential role of stem cells in the treatment of urinary incontinence. Therapeutic Advances in Urology. 2015;7(1):22-40

[89] Furuta A, Jankowski R, Honda M, Pruchnic R, Chancellor M, et al. State of the art of where we are at using stem cells for stress urinary incontinence. Neurourology and Urodynamics. 2007;**26**:966-971

[90] Wang H, Chuang Y, Chancellor M. Development of cellular therapy for the treatment of stress urinary incontinence. International Urogynecology Journal. 2011;**22**:1075-1083

[91] Gargett CE, Gurung S, Darzi S, Werkmeister JA, Mukherjee S. Tissue engineering approaches for treating pelvic organ prolapse using a novel source of stem/stromal cells and new materials. Current Opinion in Urology. 2019;**29**:450-457. DOI: doi; 10.1097/mou.0000000000000634

[92] Le Blanc K, Tammik C, Rosendahl K, Zetterberg E, Ringden O. HLA expression and immunologic properties of differentiated and undifferentiated mesenchymal stem cells. Experimental Hematology. 2003;**31**:890-896. DOI: 10.1016/S0301-472X(03)00110-3

[93] Zhao Q, Gregory CA, Lee RH, Reger RL, McNeill E, et al. MSCs derived from iPSCs with a modified protocol are tumor-tropic but have much less potential to promote tumors than bone marrow MSCs. Proceedings of the National Academy of Sciences of the United States of America. 2015;**112**:530-535. DOI: 10.1073/pnas.1423008112

[94] Schofield R. The relationship between the spleen colony-forming cell and the haemopoietic stem cell. Blood Cells. 1978;**4**:7-25

[95] Darzi S, Werkmeister JA, Deane JA, Gargett CE. Identification and characterization of human endometrial mesenchymal stem/stromal cells and their potential for cellular therapy. Stem Cells Translational Medicine. 2016;**5**:1127-1132. DOI: 10.5966/sctm.2015-0190

[96] Barker N. Adult intestinal stem cells: Critical drivers of epithelial homeostasis and regeneration. Nature Reviews. Molecular Cell Biology. 2014;**15**:19-33. DOI: 10.1038/nrm3721

[97] Brack AS, Rando TA. Tissue-specific stem cells: Lessons from the skeletal muscle satellite cell. Cell Stem Cell. 2012;**10**:504-514. DOI: 10.1016/j.stem.2012.04.001

[98] Joe AW, Yi L, Natarajan A, Le Grand F, Rossi FM, et al. Muscle injury activates resident fibro/adipogenic progenitors that facilitate myogenesis. Nature Cell Biology. 2010;**12**:153-163. DOI: 10.1038/ncb2015

[99] Caplan AI. Mesenchymal stem cells: Time to change the name! Stem Cells Translational Medicine. 2017;**6**(6):1445-1451. DOI: 10.1002/sctm.17-0051

[100] Wilson A, Trumpp A. Bone-marrow haematopoietic-stem-cell niches. Nature Reviews. Immunology. 2006;**6**:93-106

[101] Maacha S, Sidahmed H, Jacob S, Gentilcore G, Cugno C, et al. Paracrine mechanisms of mesenchymal stromal cells in angiogenesis. Stem

Cells International. 2020:4356359. DOI: 10.1155/2020/4356359

[102] Fu X, Liu G, Halim A, Ju Y, Song AG, et al. Mesenchymal stem cell migration and tissue repair. Cell. 2019;**8**:784. DOI: 10.3390/cells8080784

[103] Xi J, Yan X, Zhou J, Yue W, Pei X. Mesenchymal stem cells intissue repairing and regeneration: Progress and future. Burns & Trauma. 2013;**1**:13-20

[104] Karp JM, LengTeo GS. Mesenchymal stem cell homing: The devil is in the details. Cell Stem Cell. 2009;**4**:206-216

[105] Ankrum JA, Ong JF, Karp JM. Mesenchymal stem cells: Immune evasive, not immune privileged. Nature Biotechnology. 2014;**32**(3):252-260

[106] Rastegar F, Shenaq D, Huang J, et al. Mesenchymal stem cells: Molecular characteristics and clinical applications. World Journal of Stem Cells. 2010;**2**:67-80

[107] Caplan AI. MSCs: The sentinel and safe-guards of injury. Journal of Cellular Physiology. 2016;**231**:1413-1416. DOI: 10.1002/jcp.25255

[108] Sanchez-Mata A, Gonzales-Muňoz E. Understanding menstrual blood-derived stromal/stem cells: Definition and properties. Are we rushing into their therapeutic applications? Science. 2021;**24**(12):103501

[109] Evans J, Kaitu'u-Lino T, Salamonsen LA. Extracellular matrix dynamics in scar-free endometrial repair: Perspectives from mouse in vivo and human in vitro studies. Biology of Reproduction. 2011;**85**:511-523

[110] Henriet P, Gaide Chevronnay HP, Marbaix E. The endocrine and paracrine control of menstruation. Molecular and Cellular Endocrinology. 2012;**358**:197-207

[111] Prianishnikov VA. On the concept of stem cell and a model of functional-morphological structure of the endometrium. Contraception. 1978;**18**(3):213-223

[112] Padykula HA, Coles LG, Kaiserman-Abramof IR, et al. The basalis of the primate endometrium: A bifunctional germinal compartment. Biology of Reproduction. 1989;**40**:681-690

[113] Chan RW, Schwab KE, Gargett CE. Clonogenicity of human endometrial epithelial and stromal cells. Biology of Reproduction. 2004;**70**:1738-1750

[114] Gargett CE, Masuda H. Adult stem cells in the endometrium. Molecular Human Reproduction. 2010;**16**:818-834. DOI: 10.1093/molehr/gaq061

[115] Gargett CE, Schwab KE, Deane JA. Endometrial stem/progenitor cells: The first 10 years. Human Reproduction Update. 2016;**22**:137-163

[116] Kato K, Yoshimoto M, Kato K, Adachi S, Wake N, et al. Characterization of side-population cells in human normal endometrium. Human Reproduction. 2007;**22**:1214-1223

[117] Schwab KE, Chan RW, Gargett CE. Putative stem cell activity of human endometrial epithelial and stromal cells during the menstrual cycle. Fertility and Sterility. 2005;**84**(Suppl. 2):1124-1130

[118] Chan RW, Gargett CE. Identification of label-retaining cells in mouse endometrium. Stem Cells. 2006;**24**:1529-1538

[119] Cervello I, Martinez-Conejero JA, Horcajadas JA, Pellicer A, Simón C.

Identification, characterization and co-localization of label-retaining cell population in mouse endometrium with typical undifferentiated markers. Human Reproduction. 2007;**22**(1):45-51

[120] Letouzey V, Tan KS, Deane JA, Ulrich D, Gargett CE, et al. Isolation and characterisation of mesenchymal stem/stromal cells in the ovine endometrium. PLoS One. 2015;**10**:e0127531

[121] Dimitrov R, Timeva T, Kyurkchiev D, Stamenova M, Kyurkchiev S, et al. Characterization of clonogenic stromal cells isolated from human endometrium. Reproduction. 2008;**135**:551-558

[122] Dominici M, Le Blanc K, Mueller I, Slaper-Cortenbach I, Horwitz E, et al. Minimal criteria for defining multipotent mesenchymal stromal cells. The International Society for Cellular Therapy position statement. Cytotherapy. 2006;**8**(4):315-317

[123] Wolff FE, Wolff BA, Du H, Taylor SH. Demonstration of multipotent stem cells in the adult human endometrium by in vitro chondrogenesis. Reproductive Sciences. 2007;**14**(6):524-533

[124] Gargett CE, Schwab KE, Zillwood RM, Nguyen HP, Wu D. Isolation and culture of epithelial progenitors and mesenchymal stem cells from human endometrium. Biology of Reproduction. 2009;**80**:1136-1145

[125] Taylor HS. Endometrial cells derived from donor stem cells in bone marrow transplant recipients. JAMA. 2004;**292**:81-85

[126] Du H, Taylor SH. Stem cells and female reproduction. Reproductive Sciences. 2009;**16**(2):126-139. DOI: 10.1177/1933719108329956

[127] Najafi-Ghalehlou N, Feizkhah A, MobayenM,Pourmohammadi-BejarpasiZ, Roudkenar MH, et al. Plumping up cushion of human biowaste in regenerative medicine, novel insights into a state-of the art reserve arsenal. Stem Cell Reviews and Reports. 2022

[128] Meng X, Ichim ET, Zhong J, Rogers A, Riordan HN, et al. Endometrial regenerative cells: A novel stem cell population. Journal of Translational Medicine. 2007;**5**:57. DOI: 10.1186/1479-5876-5- 57

[129] Ulrich D, Muralitharan R, Gargett CE. Toward the use of endometrial and menstrual blood mesenchymal stem cells for cell-based therapies. Expert Opinion on Biological Therapy. 2013;**13**(10):1387-1400. DOI: 10.157/147125982013.826187

[130] DarziS,ZarnaniAH,Jeddi-TehraniM, Entezami K, Kazemnejad S, et al. Osteogenic differentiation of stem cells derived from menstrual blood versus bone marrow in the presence of human platelet releasate. Tissue Engineering. 2012;**18**:1720-1728

[131] Patel A, Silva F. Menstrual blood stromal cells: The potential for regenerative medicine. Regenerative Medicine. 2008;**3**(4):43-44

[132] Zhang S, Li P, Yuan Z, Tan J. Platelet-rich plasma improves therapeutic effects of menstrual blood-derived stromal cells in rat model of intrauterine adhesion. Stem Cell Research & Therapy. 2019;**10**:61

[133] Feng P, Li P, Tan J. Human menstrual blood-derived stromal cells promote recovery of premature ovarian insufficiency via regulating the ECM-dependent FAK/AKT signaling. Stem Cell Reviews and Reports. 2019;**15**:241-255

[134] Chen L, Qu J, Cheng T, Chen X, Xiang C. Menstrual blood-derived stem cells: Toward therapeutic mechanisms, novel strategies, and future perspectives in the treatment of diseases. Stem Cell Research & Therapy. 2019;**10**:406

[135] Dalirfardouei R, Jamialahmadi K, Jafarian AH, Mahdipour E. Promising effects of exosomes isolated from menstrual blood-derived mesenchymal stem cell on wound-healing process in diabetic mouse model. Journal of Tissue Engineering and Regenerative Medicine. 2019;**13**:555-568

[136] Schwab KE, Hutchinson P, Gargett CE. Identification of surface markers for prospective isolation of human endometrial stromal colony-forming cells. Human Reproduction. 2008;**23**:934-943

[137] Crişan M, Corselli M, Chen CWW, Péault B. Perivascular cells for regenerative medicine. Journal of Cellular and Molecular Medicine. 2012;**16**(12):2851-2860. DOI: 10.1111/j.1582-4934.2012.01617.x

[138] Zhu X, Yu F, Yan G, Hu Y, Sun H, Ding L. Human endometrial perivascular stem cells exhibit a limited potential to regernerate endometrium after xenotransplantation. Human Reproduction. 2021;**36**(1):145-159. DOI: 10.1093/humrep/deaa261

[139] Gurung S, Ulrich D, Sturm M, Rosamilia A, Werkmeister JA, Gargett CE. Comparing the effect of TGF-β receptor inhibition on human perivascular mesenchymal stromal cells derived from endometrium, bone marrow and adipose tissues. Journal of Personalized Medicine. 2020;**10**:261. DOI: 10.3390/jpm10040261

[140] Hennes DMZB, Rosamilia A, Werkmeister JA, Gargett CE, Mukherjee S. Endometrial SUSD2$^+$ mesenchymal stem/stromal cells in tissue engineering: Advances in novel cellular constructs for pelvic organ prolapse. Journal of Personalized Medicine. 2021;**11**(9):840. DOI: 10.3390/jpm11090840

[141] Marinaro F, Gomez-Serrano M, Jorge I, Silla-Castro JC, Casado JG, et al. Unraveling the molecular signature of extracellular vesicles from endometrial-derived mesenchymal stem cells: Potential modulatory effects and therapeutic applications. Frontiers in Bioengineering and Biotechnology. 2019;**7**:431

[142] Marinaro F, Macias-Garcia B, Sanchez-Margallo FM, Blazquez R, Vazquez J, et al. Extracellular vesicles derived from endometrial human mesenchymal stem cells enhance embryo yield and quality in an aged murine modeldagger. Biology of Reproduction. 2019;**100**:1180-1192

[143] Dalirfardouei R, Jamialahmadi K, Mahdipour E. A feasible method for the isolation of mesenchymal stem cells from menstrual blood and their exosomes. Tissue & Cell. 2018;**55**:53-62

[144] Lopez-Caraballo L, Martorell-Marugan J, Carmona-Saez P, Gonzalez-Munoz E. Analysis of menstrual blood stromal cells reveals SOX15 triggers oocyte-based human cell reprogramming. iScience. 2020;**23**:101376

[145] Bozorgmehr M, Gurung S, Darzi S, Nikoo S, Kazemnejad S, Zarnani AH, et al. Endometrial and menstrual blood mesenchymal stem/stromal cells: Biological properties and clinical application. Frontiers in Cell and Development Biology. 2020;**8**:497

[146] Tang L, Jiang Y, Zhu M, Chen L, Zhon X, Zhou C, et al. Clinical study using mesenchymal stem cells for

the treatment of patients with severe COVID-19. Frontiers in Medicine. 2020;**14**(5):664-673

[147] Bianco P, Cao X, Frenette PS, Mao JJ, Robey PG, Simmons PJ, et al. The meaning, the sense and the significance: Translating the science of mesenchymal stem cells into medicine. Nature Medicine. 2013;**19**:35-42

[148] Turner L, Knoepfler P. Selling stem cells in the USA: Assessing the direct-to-consumer industry. Cell Stem Cell. 2016;**19**:154-157

[149] Galipeau J, Sensebe L. Mesenchymal stromal cells: Clinical challenges and therapeutic opportunities. Cell Stem Cell. 2018;**22**(6):824-833

[150] Herschorn S, Carr L, Birch C, et al. Autologous muscle derived cells as therapy fa or stress urinary incontinence: A randomized trial. Neurology and Urodynamics. 2010;**12**(2):307-312

[151] Woo L, Hijaz A, Kuang M, Penn M, Damaser M, Rackley R. Over expression of stem cell homing cytokines in urogenital organs following vaginal distention. The Journal of Urology. 2007;**177**:1568-1572

[152] Wood H, Kuang M, Woo L, Hijaz A, Butler R, Penn M, et al. Cytokine expression after vaginal distention of different durations in virgin Sprague-Dawley rats. The Journal of Urology. 2008;**180**:753-759

[153] Sarkar D, Vemula PK, Zhao W, Gupta A, Karp M, et al. Engineered mesenchymal stem cells with self assembled vesicles for systemic cell targeting. Biomaterials. 2010;**31**(19):5266-5274

[154] Li L, El-Hayek YH, Lin B, et al. Direct current electrical field guide neuronal stem/progenitor cell migration. Stem Cells. 2008;**26**(8):2193-2200

[155] Lalu MM, Mcintyre L, Pugliese C, Fergusson D, Stewart DJ, et al. Safety of cell therapy with mesenchymal stromal cells (SafeCell): A systematic review and meta-analysis of clinical trials. PLoS One. 2012;**2012**(7):e47559

[156] Nemeth K, Leelahavanichkul A, Yuen PS, Mayer B, Mezey E, et al. Bone marrow stromal cells attenuate sepsis via prostaglandin E(2)-dependent reprogramming of host macrophages to increase their interleukin-10 production. Nature Medicine. 2009;**15**:42-49

[157] Sensebe L, Fleury-Cappellesso S. Biodistribution of mesenchymal stem/stromal cells in a preclinical setting. Stem Cells International. 2013:678063. DOI: 10.1155/2013/678063

[158] Kirchin V, Page T, Keegan EP, Om Atiemo K, Aluko P, et al. Urethral injection therapy for urinary incontinence in women. Cochrane Database of Systematic Reviews. 2017;7(7):CD003881. DOI: 10.1002/14651858. CD003881.pub4

[159] Jankowski RJ, Tu LM, Carlson C, Robert M, Carr L, et al. A double-blind, randomized, placebo-controlled clinical trial evaluating the safety and efficacy of autologous muscle derived cells in female subjects with stress urinary incontinence. International Urology and Nephrology. 2018;**50**(12):2153-2165

[160] Burdzinska A, Dybowski B, Zarvchta-Wiśniewska W, Paczek L, et al. Limited accuracy of transurethral and periurethralintrasphincteric injections of cellular suspension. Neurourology and Urodynamics. 2018;**37**(5):1612-1622

[161] SarveazadA,NewsteadLG,MirzaerR, Taghi-Joghataer M, Mahjoubi BL, et al. A new method for treating fecal incontinence by implanting stem cells derived from human adipose tissue: Preliminary findings of a randomized

double-blind clinical trial. Stem Cell Research & Therapy. 2017;**8**(1):40

[162] Atala A. Tissue engineering of human bladder. British Medical Bulletin. 2011;**97**:81-104. DOI: 10.1093/bmb/ldr003

[163] Mukherjee S, Venugopal JR, Ravichandran R, Rmalingam M, Ramakrishna S, et al. Nanofiber technology for controlling stem cell functions and tissue engineering. Micro and Nanotechnologies in Engineering Stem Cells and Tissues. 2013;**1**:27-51. DOI: 10.1002/9781118574775.ch2

[164] Aghaei-Ghareh-Bolagh B, Mukherjee S, Lockley KM, Gargett CE, Weiss AS, et al. A novel tropoelastin-based resorbable surgical mesh for pelvic organ prolapse repair. Materials Today Bio. 2020;**8**:100081. DOI: doi.org/10.1016/j.mtbio.2020.100081

[165] Gurung S, Williams S, Deane JA, Westmeister JA, Gargett CE. The transcriptome of human endometrial mesenchymal stem cells under TGF betaR inhibition reveals improved potential for cell-based therapies. Frontiers in Cell and Development Biology. 2018;**6**:164. DOI: 10.3389/fcell.2018.00164

[166] Ulrich D, Edwards SL, Su K, Tan KS, Gargett CE, et al. Human endometrial mesenchymal stem cells modulate the tissue response and mechanical behavior of polyamide mesh implants for pelvic organ prolapse repair. Tissue Engineering. Part A. 2014;**20**:785-798. DOI: 10.1089/ten.TEA.2013.0170

[167] Mukherjee S, Darzi S, Rosamilia A, Kadam V, Werkmeister JA, et al. Blended nanostructured degradable mesh with endometrial mesenchymal stem cells promotes tissue integration and anti-inflammatory response in vivo for pelvic floor application. Biomacromolecules. 2019;**20**:454-468. DOI: 10.1021/acs.biomac.8b01661

[168] Paul K, Darzi S, Mcphee G, Del Borgo MP, Gargett CE, et al. 3D bioprinted endometrial stem cells on melt electrospun poly epsilon-caprolactone mesh for pelvic floor application promote anti-inflammatory responses in mice. Acta Biomaterialia. 2019;**97**:162-176. DOI: 10.1016/j.actbio.2019.08

[169] Paul K, Darzi S, Werkmeister JA, Gargett CE, Mukherjee S. Emerging nano/micro-structured degradable polymeric meshes for pelvic floor reconstruction. Nanomaterials (Basel). 2020;**10**(6):1120. DOI: 10.3390/nano10061120

[170] Mukherjee S, Darzi S, Paul K, Cousins FL, Werkmeister JA, Gargett CE. Electrospun nanofiber meshes with endometrial MSCs modulate foreign body response by increased angiogenesis, matrix synthesis, and anti-inflammatory gene expression in mice: Implication in pelvic floor. Frontiers in Pharmacology. 2020;**11**:353. DOI: 10.3389/fphar.2020.00353

[171] Vashaghian M, Ruiz-Zapata AM, Kerhof MH, Smit TH, et al. Toward a new generation of pelvic floor implants with electrospunnanofibrous matrices. A feasibility study. Neurourology and Urodynamics. 2017;**36**(3):565-575. DOI: 10.1002/nau.22969

[172] Sridharan I, Ma Y, Kim T, Kobak W, Wang R, et al. Structural and mechanical profiles of native collagen fibers in vaginal wall connective tissue. Biomaterials. 2012;**33**:1520-1527. DOI: 10.1016/j.biomaterials.2011.11.005

[173] Muzzio N, Moya S, Romero G. Multifunctional scaffolds and synergistic strategies in tissue engineering and

regenerative medicine. Pharmaceutics. 2021;**13**(6):792. DOI: 10.3390/pharmaceutics13060792

[174] Rajaraman G, White J, Tan KS, Ulrich D, Gargett CE, et al. Optimization and scale-up culture of human endometrial multipotent mesenchymal stromal cells: Potential for clinical application. Tissue Engineering. Part C, Methods. 2013;**19**:80-92

[175] Su K, Edwards SL, Tan KS, White JF, Gargett CE, Werkmeister JA, et al. Induction of endometrial mesenchymal stem cells into tissue-forming cells suitable for fascial repair. Acta Biomaterialia. 2014;**10**:5012-5020

[176] Darzi S, Deane JA, Nold CA, Edwards SE, Mukherjee S, et al. Endometrial mesenchymal stem/stromal cells modulate the macrophage response to implanted polyamide/gelatin composite mesh in immunocompromised and immunocompetent mice. Scientific Reports. 2018;**8**:6554. DOI: 10.1038/s41598-018-24919-6

[177] Emmerson SJ, Mukherjee S, Melendez-Munoz J, Cousins F, Edwards SL, Karjalainen P, et al. Composite mesh design for delivery of autologous mesenchymal stem cells influences mesh integration, exposure and biocompatibility in an ovine model of pelvic organ prolapse. Biomaterials. 2019;**225**:119495. DOI: 10.1016/j.biomaterials.2019.119495

Chapter 12

Therapeutic Approaches Targeting Cancer Stem Cells

Shin Mukai

Abstract

Cancer stem cells (CSCs) have been identified in many types of cancer since their discovery in leukemia in the 1990s. CSCs have self-renewal and differentiation capacity, and are thought to be a key driver for the establishment and growth of tumours. Several intracellular signalling pathways are reported to play a significant role in the regulation of the biological activities of CSCs. Thus, many researchers have considered CSCs to be a compelling therapeutic target for cancer, and blockade of CSC-related signalling pathways can be efficacious for the treatment of multiple cancer types. This chapter succinctly summarises the recent progress in the development of treatments targeting signalling pathways related to the functions of CSCs.

Keywords: therapeutic modalities, drug development, signalling pathways, self-renewal, cancer stem cells

1. Introduction

Cancer is a life-threatening disease in which abnormal cells grow and divide, resulting in the destruction of normal body tissues. The last several decades have seen advancements in cancer treatments [1]. However, the conventional therapeutic methods often fail due to cancer recurrence and metastasis [2]. This can be explained by the existence of cancer stem cells (CSCs), which are a minor population in tumours and can survive most traditional cancer therapies killing cancer cells with proliferative properties [3]. Cancer relapse, metastasis, multidrug resistance, and radiation resistance can be induced by the transient arrest of CSCs at the G0 phase, leading to the production of new malignant tumours [4]. The ability of CSCs to self-renew and differentiate into multiple cellular subtypes allows them to generate tumours [4]. Thus, researchers have regarded CSCs as a promising target for the treatment of cancer since they were discovered in leukemia in the 1990s [5, 6]. CSCs have also been found to be a subpopulation of many types of tumours, and tissue-specific expression of CSC markers has been reported [7]. It is known that the biological functions of CSCs are controlled by several signalling pathways [8]. This chapter focuses on the research status in cancer therapies targeting the signalling pathways that are believed to control the properties of CSCs.

2. General biology of CSCs

Since CSCs were found in leukemia in the 1990s, they have been studied intensively [5, 6]. However, the origin of CSCs remains to be elucidated [9]. It has been reported that (a) CSCs possess self-renewal capacity and high proliferation rate, (b) CSCs are able to generate and maintain tumours, and (c) cancer recurrence may be induced by the unlimited self-renewal capacity of CSCs [10]. CSCs are shown to be a distinct subpopulation in haematologic malignancies and solid tumours, and cell surface markers of CSCs in various types of cancer have been reported and can be used for the identification and isolation of CSCs (**Figure 1**) [11–15]. Signalling pathways such as Wnt/β-catenin, Notch and Hedgehog signalling pathways are thought to regulate the properties of CSCs [8]. Emerging evidence supports the clinical relevance of CSCs [16]. In particular, CSCs are shown to be resistant to conventional chemotherapy and radiation therapy, and they are believed to be a key player for cancer recurrence and metastasis [16]. Thus, understanding of signalling pathways that control the functions of CSCs can be useful for the creation of novel therapeutic interventions for cancer.

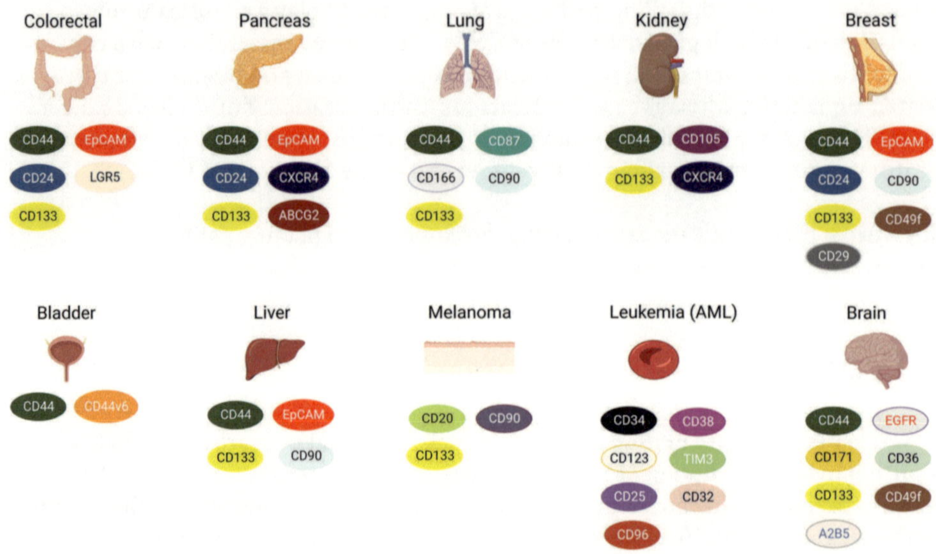

Figure 1.
Cell surface markers for CSCs in various organs. This figure was created with BioRender.

3. Development of therapeutic modalities targeting CSCs

3.1 Notch signalling

The Notch signalling pathway is a highly conserved pathway in mammals and controls a variety of cellular functions [17]. In mammals, the Notch pathway comprises four Notch receptors (Notch 1–4) and five Notch ligands (Jagged 1, Jagged 2, Delta-like ligand (DLL)-1, DLL-3, and DLL-4), and its primary effector is the transcription factor CSL (CBF1/RBP, Su(H)/Lag-1), which is crucial for the activation of the genes downstream of the Notch signalling pathway [18]. The Notch signalling pathway is divided into canonical and noncanonical pathways: the CSL-dependent pathway (canonical Notch signalling pathway) and non-CSL-dependent pathway (non-canonical Notch signalling pathway) [19]. Evidence suggests that these pathways play an instrumental role in preserving the existence of stem cells and initiating embryonic or foetal cell differentiation [20].

In the canonical Notch signalling pathway (**Figure 2**), the binding of receptor and ligand is induced through metalloproteinase- and γ-secretase-mediated proteolytic activation of the Notch receptor, leading to the release of the intracellular NOTCH domain (NICD) [21]. Subsequently, NICD migrates to the nucleus and forms a complex with CSL. As a result, the transcription and expression of the downstream target genes are triggered, leading to self-renewal, differentiation and proliferation [21].

Recent studies suggest that there is non-canonical Notch signalling, which can be activated either with or without ligand interaction [22]. In addition, the activation of non-canonical Notch signalling can occur in a γ-secretase-dependent or -independent manner [22]. In the case where non-canonical Notch signalling occurs in a γ-secretase-independent way, Notch remains bound to membrane [22]. Non-canonical Notch signalling does not require CSL [23, 24]. Instead, NICD or membrane bound Notch interacts with (a) Wnt, PI3K, mTORC2 and/or AKT pathways in the cytoplasm, and (b) NF-κB, YY1 and/or HIF-1α pathways in the nucleus [23, 24]. It is suggested that non-canonical Notch signalling plays a role in cell survival, metabolism and differentiation [23, 24]. Compared with the canonical Notch signalling pathway, there is less information on the non-canonical Notch signalling pathway [25]. Thus, more work will be needed for the identification of potential therapeutic targets in the non-canonical Notch signalling pathway [25].

Notch inhibition is believed to be a promising therapeutic approach to target CSCs, which are resistant to conventional methods such as chemotherapy and radiation [26]. As γ-secretase is a key player in Notch signalling, a great deal of effort has been invested in the development of γ-secretase inhibitors (GSIs) (**Figure 3**) [27]. It should be noted that GSIs show anti-CSC effects and that they were the first Notch inhibitors to reach clinical development [27]. However, one of the major concerns is the toxicity of GSIs [28]. In particular, serious toxicity in the gastrointestinal tract

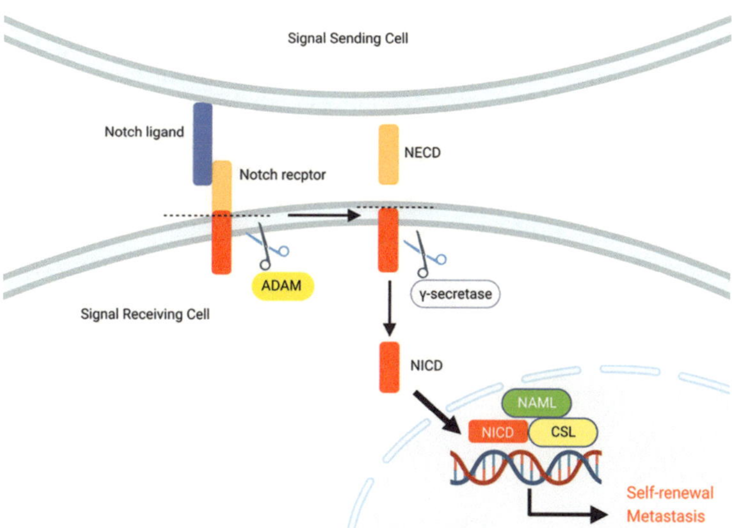

Figure 2.
Brief diagram of the canonical Notch signalling pathway. This figure was created with BioRender.

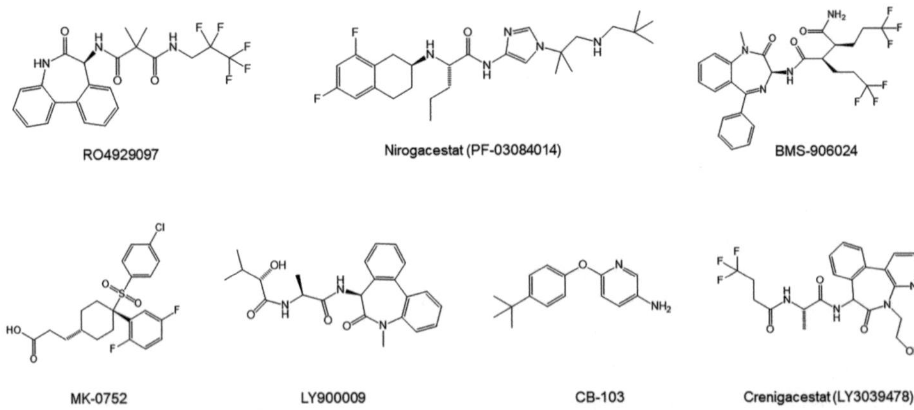

Figure 3.
Structures of Notch signalling inhibitors in clinical development.

can be caused by GSIs [28]. Several GSIs have entered clinical trials thus far. Data from a Phase II clinical trial suggest that RO4929097 did not show sufficient efficacy for the treatment of metastatic melanoma and platinum-resistant ovarian cancer as monotherapy [29, 30]. Nirogacestat (PF-3084014) is another GSI undergoing clinical trials for desmoid tumours [31]. The result of Phase II clinical trials indicates that (a) treatment of desmoid tumour fibromatosis patients with Nirogacestat could be a promising approach, (b) the objective response rate was 71.4% and (c) relatively low doses and high tolerability were achieved, resulting in prolonged disease control [31, 32]. Another study shows the antitumour and antimetastatic effects of Nirogacestat in hepatocellular carcinoma [33]. The potent pan-Notch Inhibitor BMS-906024 has advanced into clinical trials, and the data suggest that it could be effective for the treatment of leukemia and solid tumours [34]. The clinical development status of other small molecule inhibitors is as follows: (a) The GSI MK-0752 is in a Phase I clinical trial for the treatment of pancreatic ductal adenocarcinoma [27], (b) the GSI LY900009 is in a Phase I clinical trial for ovarian cancer [35], (c) CB-103, which inhibits the interaction between NICD and CSL, is in a Phase I clinical trial for adenoid cystic carcinoma (ACC), colorectal cancer, breast cancer and prostate cancer [36], and (d) Crenigacestat (LY3039478), which is an oral Notch and GSI inhibitor, is in a Phase I clinical trial for solid tumours [37]. Recently, cryoelectron microscopy (cryo-EM) structures of γ-secretase in complex with each of the two GSI clinical candidates Semagacestat and Avagacestat have been reported, and these pieces of information might be useful for the design of novel GSIs [38].

A study suggests that the expression of the Notch ligand DLL4 is increased in gastric cancer, enhancing self-renewal ability of CSCs [39]. Therefore, inhibition of DLL4 can be an effective approach to treating cancer [39]. Currently, Enoticumab (REGN421), a fully human IgG1 monoclonal antibody against DLL4, is in a phase I clinical trail for the treatment of solid tumours [40].

3.2 Wnt signalling

The Wnt signalling pathway is known to play a pivotal role in embryogenesis and tissue repair by controlling proliferation, differentiation, apoptosis and cell-to-cell interactions [41, 42]. The Wnt signalling pathway is classified into the canonical

Wnt pathway (β-catenin dependent) and the noncanonical Wnt pathway (β-catenin independent) [43]. Wnt ligands are required for the activation of Wnt signalling, and the acyltransferase Porcupine is known to be essential for the production of Wnt ligands [44]. In canonical Wnt signalling (**Figure 4**), the absence of Wnt ligands leads to the degradation of β-catenin due to phosphorylation by glycogen synthase kinase 3β (GSK3β), and thereby translocation of β-catenin from the cytoplasm to the nucleus does not occur [45]. In the presence of Wnt ligands, their ligation to Frizzled proteins and LRP5/6 receptors induces the activation of the cytoplasmic protein DVL and the subsequent suppression of GSK3β [46]. It enables β-catenin to migrate to the nucleus and trigger target gene transcription by binding to TCF/LEF transcription factors [46]. Noncanonical Wnt signalling does not require the cytoplasmic stabilisation of β-catenin or its translocation into the nucleus [47]. The noncanonical Wnt signalling is subdivided into the following two well-characterised pathways: the planar cell polarity (Wnt/PCP) pathway and the Wnt-Calcium (Wnt/Ca^{2+}) pathway [48]. In the Wnt/PCP pathway, Wnt ligands bind to the Frizzled receptor, leading to the activation of Dishevelled (DVL) protein [49]. Activated DVL forms the DVL-Rac complex and DVL-Rho complex [49]. The former stimulates the Rho kinase (ROCK), and the latter stimulates the c-Jun N-terminal Kinase (JNK) [50]. JNK is known to translocate into the nucleus and trigger the transcription of target genes [51]. The Wnt/Ca^{2+} pathway involves the activation of PLC and PKC, and an increase in intracellular Ca^{2+} [52]. The phosphatase calcineurin is stimulated by Ca^{2+} and dephosphorylates the transcriptions factor NF-AT, resulting in the migration of NF-AT to the nucleus [52].

Dysregulation of Wnt signalling is observed in many types of cancer [53]. The Wnt signalling cascade is reported to play an important role in controlling the properties of CSCs [53]. Thus, many scholars have been striving to create therapeutic modalities to target Wnt signalling for safe and effective elimination of CSCs, and several small molecule inhibitors and monoclonal antibodies have entered clinical trials (**Figure 5**) [53]. Evidence suggests that pharmacological inhibition of Porcupine can selectively block Wnt signalling and suppress tumour growth [54]. Thus, Porcupine is considered to be a promising therapeutic target for cancer [54]. LGK-974, a small molecule disrupting the enzymatic activity of Porcupine, is currently in Phase I clinical trials for pancreatic cancer, melanoma and triple-negative breast cancer [55]. Cryo-electron microscopy (cryo-EM) structures of Porcupine in complex with LGK-974 are available in the PDB database (PDB ID: 7URD) [55], and this information could be useful for creation of novel Porcupine inhibitors. Another Porcupine inhibitor ETC-159 is

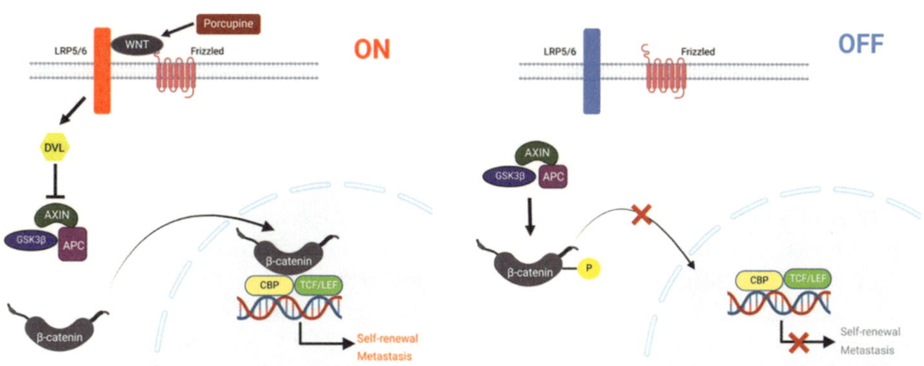

Figure 4.
Brief diagram of the canonical Wnt signalling pathway. This figure was created with BioRender.

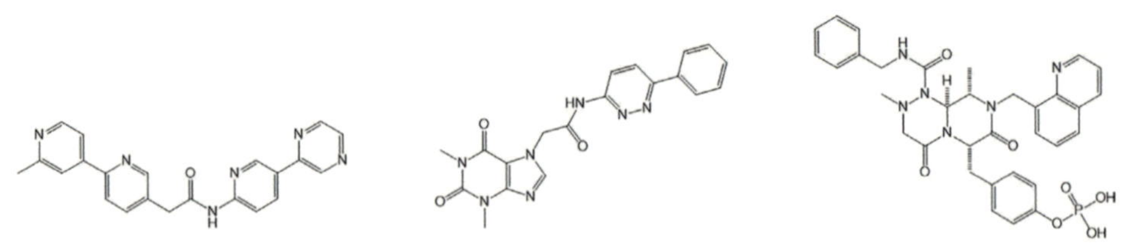

Figure 5.
Structures of Wnt signalling inhibitors in clinical development.

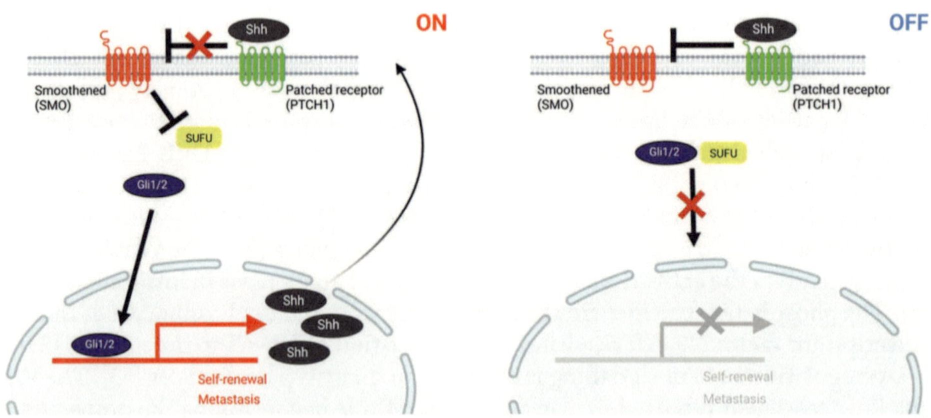

Figure 6.
Brief diagram of the canonical Hedgehog signalling pathway. This figure was created with BioRender.

in a Phase I clinical trial for the treatment of solid tumours [56]. The small molecule inhibitor PRI-724 is reported to block canonical Wnt signalling by preventing the interaction between β-catenin and its coactivator CREB binding protein (CBP) [57]. PRI-724 is in a Phase II clinical trial for advanced myeloid malignancies [58]. The therapeutic monoclonal antibody OMP-18R5 (vantictumab), which can target the Frizzled receptors, is now in a Phase I clinical trial for the treatment of non-small-cell lung cancer (NSCLC), pancreatic cancer and breast cancer [59–62]. The recombinant fusion protein ipafricept (OMP-54F28) can inhibit Wnt signalling by binding to Wnt ligands, and its safety and effectiveness are being evaluated in clinical trials [63–65].

3.3 Hedgehog signalling

The Hedgehog (Hh) signalling pathway contributes to the control of cell proliferation, cell survival, cell differentiation, and stem cell maintenance and development [66]. In Hh signalling (**Figure 6**), the autoproteolytic cleavage of Hh ligand precursor proteins leads to the production of an N-terminal protein, followed by dual lipid modification [67]. Subsequently, active Hh ligands are released through mediation of Dispatched and Scube2 [67]. In the absence of Hh ligand, Patched (PTCH) prevents the activation and ciliary localisation of Smoothened (SMO) [68, 69]. As a result, the Glioma-Associated Oncogene Homolog (GLI) forms a complex with Suppressor of Fused (SUFU), which precludes GLI from translocating into the nucleus [68, 69]. In the presence of Hh ligand, the binding of Hh to PTCH allows SMO to interact with β-arrestin (Arrb2) and kinesin family member 3A (KIF3A), leading to the ciliary localisation of SMO [68, 69]. As a result, GLI is released from SUFU and subsequently migrates to the nucleus, triggering the transcription of Hh target genes [68, 69].

Evidence suggests that the dysregulation of Hh signalling is associated with detrimental events such as the self-renewal and metastasis of cancer stem cells [67]. Thus, therapeutic targeting of Hh signalling has accorded a great deal of attention from many researchers, and SMO has been regarded as the most promising pharmacological target [68]. Indeed, several SMO inhibitors have been approved by the Food and Drug Administration (FDA) or are undergoing clinical trials (**Figure 7**) [70]. Vismodegib (GDC-0449) was the first SMO inhibitor that was granted FDA approval to treat basal cell carcinoma in 2011 [71]. Mounting evidence suggests the inhibitory activity of Vismodegib against self-renewal and mammosphere formation of breast CSCs [72]. Data from Phase II clinical trials demonstrate that (a) Vismodegib could be used as a neoadjuvant chemotherapy agent for patients with triple-negative breast cancer (NCT02694224) [73], (b) Vismodegib could be efficacious for the treatment of pancreatic cancer by suppressing self-renewal, proliferation and survival of pancreatic CSCs [74] and (c) Vismodegib could be effective for untreated metastatic colorectal cancer by reducing the stem cell markers of colon CSCs [75, 76]. These results support the notion that Vismodegib can inhibit the activities of CSCs by blocking the Hh signalling pathway. The SMO inhibitor Sonidegib (LDE225) was approved by FDA for the treatment of advanced basal cell carcinoma in 2015 [77]. A recent study indicates that Sonidegib can make triple-negative breast cancer more sensitive to Paclitaxel and improve clinical outcomes by reducing the expression of CSC markers [78]. Glasdegib (PF-04449913) was an FDA-approved SMO antagonist for the treatment of acute myeloid leukemia and launched in the USA in 2018 [79]. Glasdegib is also undergoing a Phase II clinical trial for the treatment of myelodysplastic syndrome and chronic myelomonocytic leukemia [80]. Although the development of SMO inhibitors is beneficial for cancer patients, monotherapy with each of the FDA-approved antagonists can cause SMO mutations in tumour tissues, leading to drug resistance [80]. Hence, novel therapeutic methods inhibiting SMO will need

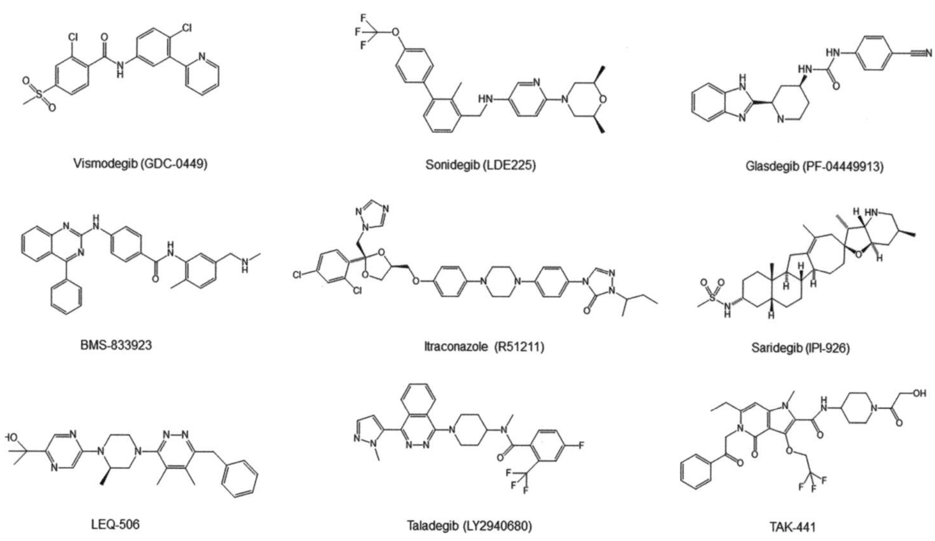

Figure 7.
Structures of Hedgehog signalling inhibitors in clinical development.

to be created in order to overcome this setback. The clinical development status of other SMO inhibitors is as follows [81]: (a) BMS-833923 is in a Phase I clinical trial for extensive-stage small cell lung cancer [82], (b) Itraconazole is in a Phase II clinical trial for prostate cancer [83], (c) Saridegib (IPI-926) is in a Phase I clinical trial for advanced and/or metastatic solid tumours [84], (d) LEQ-506 is in a Phase I clinical trial for advanced solid tumours [85], (e) Taladegib (LY2940680) is in a Phase I clinical trial for advanced solid tumours [86] and (f) TAK-441 is in a Phase I clinical trial for advanced nonhematologic malignancies [87]. In addition to the SMO inhibitors, arsenic trioxide is reported to inhibit the Hh signalling pathway and tumour growth by binding to GLI [81]. Arsenic trioxide is in a Phase II clinical trial for the treatment of advanced neuroblastoma or other childhood solid tumours [88]. Although there are no other GLI inhibitors undergoing clinical trials, *in vitro* and *in vivo* preclinical investigation into the GLI inhibitor GANT61 suggests its inhibitory activity against pancreatic cancer stem cells [89].

3.4 NF-κB signalling

The transcription factor nuclear factor kappa B (NF-κB) is a rapidly inducible transcription factor and a family of heterodimers or homodimers [90]. The heterodimers or homodimers are produced from different combinations of the five related proteins: p65/RelA, RelB, c-Rel, p105/p50 (NF-κB1) and p100/p52 (NF-κB2) [90]. The p50/p65 complex is thought to the most abundant form of NF-κB and serve main physiological functions [91]. The activation of NF-κB signalling induces the translocation of the transcription factor complexes from the cytoplasm to the nucleus [92]. The NF-κB signalling pathway diversifies into the canonical NF-κB signalling pathway and the noncanonical NF-κB signalling pathway [92]. The canonical NF-κB pathway is activated by the binding of ligands to their receptors such as the binding of (a) bacterial cell components to toll-like receptors (TLRs), (b) TNF-α to the TNF receptor (TNFR), (c) lipopolysaccharides to their respective receptors such as TLRs and (d) IL-1β to the IL-1 receptor (IL-1R) [93]. In response to the stimulation of these receptors, the kinase TGFβ-activated kinase 1 (TAK1) is activated, leading to the subsequent phosphorylation and activation of the IκB kinase (IKK) proteins [94]. The activated IKK proteins then phosphorylate IκB proteins. It induces the degradation of IκB, leading to the translocation of the activated p50/p60 complex into the nucleus [95]. The noncanonical NF-κB pathway is initiated by stimulation of receptors such as CD40, receptor activator for NF-κB (RANK), B cell activation factor (BAFF), TNFR2 and lymphotoxin β-receptor (LTBR) [94]. Subsequently, the kinase NF-κB-inducing kinase (NIK) is activated, resulting in the phosphorylation and activation of IKKα [94], which induces the phosphorylation of carboxy-terminal serine residues of p100 [94]. As a result, the C-terminal IκB-like structure of p100 is selectively degraded, which generates p52 and causes the p52-RelB complex to migrate to the nucleus [94].

A previous study shows that the production of cytokines, growth and angiogenic factors and proteases is promoted in the tumour development and progression, activating NF-κB signalling [96]. The NF-κB pathway is reported to contribute to self-renewal, maintenance and metastasis of CSCs [11]. Ovarian CSCs can display the enhanced capability of self-renewal, metastasis and maintenance due to the increased expression of RelA, RelB and IKKα [97]. In breast cancer, NIK expression is augmented, and the noncanonical NF-κB pathway is activated, leading to the self-renewal and metastasis of breast CSCs [98]. Regarding the development of therapeutic modalities to inhibit self-renewal, proliferation and metastasis of CSCs by targeting the NF-κB pathway (**Figure 8A**), Disulfiram is known to inhibit the activity of NF-κB in breast CSCs and is in a Phase II clinical trial for the treatment of metastatic breast cancer [99, 100]. Sulforaphane is suggested to prevent the translocation of p50/p65 and reduce the expression and transcriptional activity of p52/RelB, resulting in the inhibition of self-renewal of triple-negative breast CSCs [101]. Sulforaphane is in a Phase II clinical trial for the treatment of breast cancer [102]. The NF-κB pathway inhibitor curcumin is reported to impede self-renewal and metastasis of CSCs and is undergoing clinical trials for the treatment of breast cancer [103–105].

3.5 mTOR signalling

The mammalian target of rapamycin (mTOR) signalling pathway plays a pivotal role in cellular growth and metabolism in mammalian cells within various environments [106]. The activation of mTOR signalling can promote cell survival by (a) increasing cellular metabolism and the synthesis of proteins and lipids and (b) blocking apoptotic pathways [106]. The mTOR pathway is initiated by the binding of growth factors to tyrosine kinase receptors in the cell membrane, leading to the phosphorylation and activation of phosphatidylinositol 3-kinase (PI3K) [107]. Subsequently, protein kinase B (AKT) is phosphorylated and activated, leading to the activation of mTOR [107]. It activates a variety of transcription factors and enables them to migrate to the nucleus in order to promote the transcription of the target genes [108]. mTOR is divided into two protein complexes. mTORC1 is a complex of RAPTOR, mLST8, PRAS40 and DEPTOR, and mTORC2 is a complex of RICTOR, mLST8, DEPTOR, mSin1 and PROCTOR [109, 110]. mTORC1 is reported to (a) activate the ribosomal protein S6K, which promotes protein synthesis, (b) facilitate lipid synthesis and mitochondrial biogenesis and (c) reduce autophagy [109, 110].

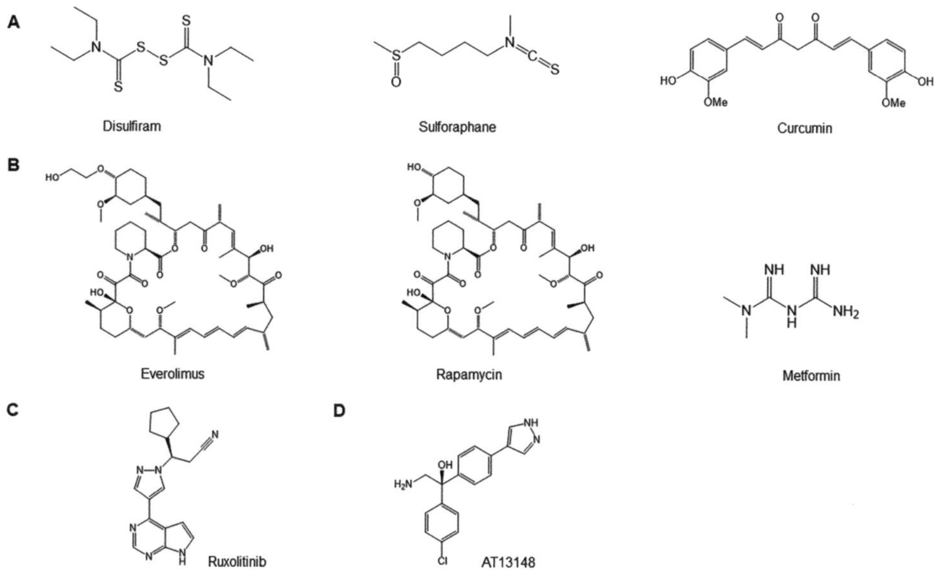

Figure 8.
Structures of CSC-related signalling inhibitors in clinical development. (A) NF-κB, (B) mTOR, (C) JAK/STAT, (D) ROCK.

mTORC2 is known to promote actin cytoskeleton and cell migration by phosphorylating various proteins [109, 110].

Evidence suggests that the mTOR signalling pathway is correlated with the functions of CSCs. According to previous studies, activation of mTOR signalling contributes to (a) prostate cancer radioresistance due to the enhancement of CSC properties and (b) the tumourigenicity of breast CSCs [111, 112]. Dysregulation of the PI3K/Akt/mTOR signalling pathway is reported to enhance the expression of chemokine (C-X-C motif) receptor 4 (CXCR4), and CXCR4-mediated STAT3 signalling is then activated, promoting the self-renewal of CSCs in non-small-cell lung cancer [113]. Furthermore, the activity of mTOR through the PI3K feedback loop is associated with the survival of prostate CSCs [114].

The mTOR signalling pathway has been regarded as a promising candidate for therapeutic modalities targeting CSCs, and several mTOR inhibitors have been approved for the treatment of cancer or undergoing clinical development (**Figure 8B**) [115]. As demonstrated by a series of *in vitro* and *in vivo* experiments, the mTOR inhibitor Everolimus can suppress the expression and phosphorylation of AKT-1, and thereby inhibit the activity of HER2-overexpressing breast CSCs [116]. Everolimus has been approved by FDA for the treatment of advanced breast cancer [117]. The mTOR inhibitor Rapamycin is shown to suppress the properties of colon CSCs [118] and inhibit the stemness of haemangioma stem cells [119]. Rapamycin is undergoing clinical trials for advanced or metastatic colorectal cancer and infantile hepatic haemangioendothelioma [120, 121]. Previous reports show the inhibitory effects of the mTOR antagonist Metformin on breast and pancreatic CSCs [122, 123]. Clinical trials of Metformin are in progress or completed for the treatment of breast and pancreatic cancer [124, 125]. Considering previous studies show that mTOR signalling plays an important role in controlling the functions of CSCs, further investigation into a link between mTOR signalling and CSCs could lead to the development of better therapeutic strategies for cancer.

3.6 JAK/STAT signalling

The Janus kinase/signal transducer and activator of transcription (JAK/STAT) signalling pathway contributes to a variety of biological processes such as embryonic development, stem cell maintenance, haematopoiesis and inflammatory response [126]. It is known that the JAK family in mammals comprises four members (JAK1, JAK2, JAK3 and tyrosine kinase 2 (TYK2)) and that the STAT family in mammals consists of seven members (STAT1, STAT2, STAT3, STAT4, STAT5a, STAT5b and STAT6) [127]. The JAK/STAT signalling pathway is initiated by the binding of cytokines to their corresponding receptors [128]. The receptors then undergo dimerisation, and the JAKs bound to the receptors come close to each other, leading to the activation of the JAKs through the interaction of tyrosine phosphorylation [128]. The activated JAKs induce the phosphorylation of the tyrosine residues of the catalytic receptor, resulting in the formation of a docking site with the SH2 domain of the STAT protein [128]. Subsequently, the STAT protein bound to the receptor is phosphorylated and dimerised, culminating in the translocation of the activated STAT to the nucleus and the transcription of the target genes [128].

Literature precedent indicates that JAK/STAT signalling plays a role in controlling the properties of CSCs [129]. It has been suggested that (a) the self-renewal and survival of breast CSCs are facilitated by the persistent activation of JAK/STAT signalling [112], (b) IL-10-mediated JAK1/STAT1 signalling promotes the self-renewal and migration of non-small-cell lung cancer and colorectal CSCs [130, 131], (c) the OCT4-activated JAK1/STAT6 pathway is associated with the functions of ovarian CSCs [132], (d) OCT4, which is a gene downstream of IL-6-mediated JAK1/STAT6 signalling, is involved in the transformation of bulk cancer cells to CSCs in breast cancer [133] and (e) JAK2/STAT3 signalling plays a role in the regulation of the properties of breast and colorectal CSCs [134–136].

With respect to the development of therapeutic strategies to target JAK/STAT signalling for the treatment of cancer, the JAK1/2 inhibitor Ruxolitinib is shown to inhibit the functions of CSCs, leading to a decrease in the number of CSCs (**Figure 8C**) [137]. Ruxolitinib is undergoing clinical development for the treatment of solid tumours [138]. Many studies have shown that the JAK/STAT signalling pathway is correlated with the survival, self-renewal and metastasis of CSCs. Thus, more endeavours will be needed to develop novel therapeutic modalities targeting CSCs through the inhibition of JAK/STAT signalling.

3.7 ROCK signalling

The Rho-associated coiled-coil-containing protein kinase (ROCK) signalling pathway plays a pivotal role in various cellular activities such as cell survival and apoptosis [139]. It is known that there are two types of ROCK in mammals: ROCK1 and ROCK2 [139]. Diverse extracellular stimuli activate guanine nucleotide exchange factors (GEFs), leading to the conversion of Rho-GDP to Rho-GTP [140]. Rho-GTO subsequently activates ROCK1 and ROCK2, resulting in the phosphorylation of their substrates and the induction of a range of cellular responses [140].

Dysregulation of ROCK signalling is reported to be involved in the pathogenesis of a variety of diseases such as cancer [139]. According to an *in vitro* study, the properties of CSCs can be reduced by pharmacological inhibition of ROCK with the ROCK inhibitors ML7 or Y-27632, supporting the involvement of ROCK signalling in controlling the functions of CSCs [141]. With regard to the clinical development of ROCK inhibitors, the dual ROCK1/2 antagonist AT13148 is undergoing clinical trials for patients with advanced cancer [142, 143]. There is still sparce information on the roles of the ROCK signalling pathway in the regulation of the functions of CSCs. Thus, further investigation into a correlation between ROCK signalling and CSCs could be beneficial for the development of novel cancer treatments.

3.8 TGF-β signalling

The transforming growth factor β (TGF-β) signalling pathway contributes to diverse biological processes including cell proliferation and differentiation [11]. When dimeric TGF-β binds to the TGF-β receptor type-2 (TβRII), TβRII is dimerised with the TGF-β receptor type-1 (TβRI), leading to the phosphorylation and activation of the receptor-regulated SMADs (R-SMADs) SMAD2 and SMAD3 [11]. Subsequently, SMAD2 and SMAD3 undergo trimerisation with the common-partner SMAD, SMAD4 [11]. The trimer migrates to the nucleus and promotes the transcription of target genes [11].

It has been suggested that TGF-β may have contradictory functions in the properties of CSCs [144]. A study using a breast cancer xenograft model indicates that the activation of TGF-β signalling can reduce the number and the self-renewal potential of breast CSCs [145]. In addition, another *in vivo* study suggests that the number of CSCs in diffuse-type gastric carcinoma can be reduced by the activation of TGF-β signalling, leading to the suppression of tumour formation [146]. In contrast, it is reported that the activation of TGF-β signalling leads to an increase in CSC counts and the enhancement of CSC properties in various types of cancer such as breast cancer liver cancer, gastric cancer, skin cancer, glioblastoma and leukaemia [147–156]. Considering these pieces of evidence, the therapeutic targeting of the TGF-β signalling pathway might be a promising strategy to eliminate CSCs. However, more work will be required to gain a better understanding of a link between TGF-β signalling and CSCs for the development of novel therapeutic modalities.

4. Conclusion

CSCs are a minor population in tumours, and their self-renewal capacity and differentiation potential contribute to tumour relapse, metastasis, chemoresistance and radioresistance. This chapter has provided a succinct summary of (a) major signalling pathways that are reported to be associated with the functions of CSCs and (b) clinical development of inhibitors targeting CSC-related signalling pathways for the purpose of encouraging research scientists (medicinal chemists, biologists, immunologists and others) to create new treatments (**Table 1**). Therapeutic targeting of CSCs *via* these signalling pathways has been considered to be a compelling strategy, and several small molecule inhibitors such as Vismodegib (GDC-0449), Sonidegib, Glasdegib (PF-04449913) and Everolimus have been approved by FDA for the treatment of cancer in the clinic. However, the development of such therapeutic interventions is challenging, and there is still vast scope for improvement. It is in part because signalling pathways interact with each other and because the CSC properties are thought to be controlled by the signalling network. AL/ML has been applied to the drug discovery, and it is reported that AL/ML is highly beneficial for target discovery, drug design and so on. These pieces of evidence can provide a scientific rational for applying AL/ML to the development of new therapeutic interventions targeting CSCs. The signal regulatory mechanisms of CSCs remain to be elucidated, and continuing studies of CSC-related pathways will lead to the creation of novel therapeutic modalities for various types of cancer.

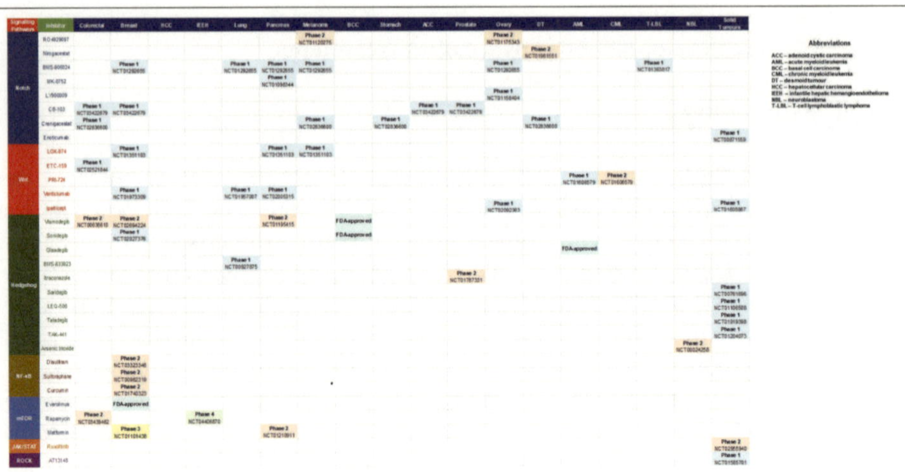

Signaling Pathways	Inhibitor	Colorectal	Breast	HCC	IEEH	Lung	Pancreas	Melanoma	BCC	Stomach	ACC	Prostate	Ovary	DT	AML	CML	T-LBL	NBL	Solid Tumours
Notch	RO4929097							Phase 2 NCT01120275					Phase 2 NCT01175343						
	Nirogacestat													Phase 2 NCT01981551					
	BMS-906024		Phase 1 NCT01292655			Phase 1 NCT01292655	Phase 1 NCT01292655	Phase 1 NCT01292655					Phase 1 NCT01292655				Phase 1 NCT01363817		
	MK-0752						Phase 1 NCT01098344												
	LY900009												Phase 1 NCT01158404						
	CB-103	Phase 1 NCT03422679	Phase 1 NCT03422679								Phase 1 NCT03422679	Phase 1 NCT03422679							
	Crenigacestat	Phase 1 NCT02836600						Phase 1 NCT02836600		Phase 1 NCT02836600				Phase 1 NCT02836600					
	Enoticumab																		Phase 1 NCT00871559
Wnt	LGK-974		Phase 1 NCT01351103				Phase 1 NCT01351103	Phase 1 NCT01351103											
	ETC-159	Phase 1 NCT02521844																	
	PRI-724														Phase 1 NCT01606579	Phase 2 NCT01606579			
	Vantictumab		Phase 1 NCT01973309			Phase 1 NCT01957007	Phase 1 NCT02005315												
	Ipafricept												Phase 1 NCT02092363						Phase 1 NCT01608867
Hedgehog	Vismodegib	Phase 2 NCT00636610	Phase 2 NCT02694224				Phase 2 NCT01195415		FDA approved										
	Sonidegib		Phase 1 NCT02027376						FDA approved										
	Glasdegib														FDA approved				
	BMS-833923					Phase 1 NCT00927875													
	Itraconazole											Phase 2 NCT01787331							
	Saridegib																		Phase 1 NCT00761696
	LEQ-506																		Phase 1 NCT01106508
	Taladegib																		Phase 1 NCT01919398
	TAK-441																		Phase 1 NCT01204073
	Arsenic trioxide																	Phase 2 NCT00024258	
NF-κB	Disulfiram		Phase 2 NCT03323346																
	Sulforaphane		Phase 2 NCT00982319																
	Curcumin		Phase 2 NCT01740323																
mTOR	Everolimus		FDA approved																
	Rapamycin	Phase 2 NCT03439462			Phase 4 NCT04406870														
	Metformin		Phase 3 NCT01101438				Phase 2 NCT01210911												
JAK/STAT	Ruxolitinib																		Phase 2 NCT02955940
ROCK	AT13148																		Phase 1 NCT01585701

Abbreviations

ACC – adenoid cystic carcinoma
AML – acute myeloid leukemia
BCC – basal cell carcinoma
CML – chronic myeloid leukemia
DT – desmoid tumour
HCC – hepatocellular carcinoma
IEEH – infantile hepatic hemangioendothelioma
NBL – neuroblastoma
T-LBL – T-cell lymphoblastic lymphoma

Table 1.
Landscape of clinical development of inhibitors targeting CSC-related pathways.

DOI: http://dx.doi.org/10.5772/intechopen.108963

Author details

Shin Mukai
New Wind Therapeutics L3C, Boston, United States of America

*Address all correspondence to: shin.mukai@new-wind-ther.com

References

[1] Advancing cancer therapy. Nature Cancer. 2021;**2**:245-246. DOI: 10.1038/s43018-021-00192-x

[2] Batlle E, Clevers H. Cancer stem cells revisited. Nature Medicine. 2017;**23**:1124-1134. DOI: 10.1038/nm.4409

[3] Reya T, Morrison SJ, Clarke MF, Weissman IL. Stem cells, cancer, and cancer stem cells. Nature. 2001;**414**:105-111. DOI: 10.1038/35102167

[4] Chen W, Dong J, Haiech J, Kilhoffer M-C, Zeniou M. Cancer stem cell quiescence and plasticity as major challenges in cancer therapy. Stem Cells International. 2016;**2016**:1740936. DOI: 10.1155/2016/1740936

[5] Lapidot T, Sirard C, Vormoor J, Murdoch B, Hoang T, Caceres-Cortes J, et al. A cell initiating human acute myeloid leukaemia after transplantation into SCID mice. Nature. 1994;**367**:645-648. DOI: 10.1038/367645a0

[6] Bonnet D, Dick JE. Human acute myeloid leukemia is organized as a hierarchy that originates from a primitive hematopoietic cell. Nature Medicine. 1997;**3**:730-737. DOI: 10.1038/nm0797-730

[7] Zhao W, Li Y, Zhang X. Stemness-related markers in cancer. Cancer Translational Medicine. 2017;**3**:87-95. DOI: 10.4103/ctm.ctm_69_16

[8] Matsui WH. Cancer stem cell signaling pathways. Medicine. 2016;**95**:S8-S19. DOI: 10.1097/MD.0000000000004765

[9] Bjerkvig R, Tysnes BB, Aboody KS, Najbauer J, Terzis AJA. The origin of the cancer stem cell: Current controversies and new insights. Nature Reviews Cancer. 2005;**5**:899-904. DOI: 10.1038/nrc1740

[10] Yu Z, Pestell TG, Lisanti MP, Pestellb RG. Cancer stem cells. The International Journal of Biochemistry & Cell Biology. 2012;**44**:2144-2151. DOI: 10.1016/j.biocel.2012.08.022

[11] Yang L, Shi P, Zhao G, Xu J, Peng W, Zhang J, et al. Targeting cancer stem cell pathways for cancer therapy. Signal Transduction and Targeted Therapy. 2020;**5**:8. DOI: 10.1038/s41392-020-0110-5

[12] Sun J-H, Luo Q, Liu L-L, Song G-B. Liver cancer stem cell markers: Progression and therapeutic implications. World Journal of Gastroenterology. 2016;**22**:3547-3557. DOI: 10.3748/wjg.v22.i13.3547

[13] Zhou Y, Xia L, Wang H, Oyang L, Su M, Liu Q, et al. Cancer stem cells in progression of colorectal cancer. Oncotarget. 2018;**9**:33403-33415. DOI: 10.18632/oncotarget.23607

[14] Ding Y, Gao H, Zhang Q. The biomarkers of leukemia stem cells in acute myeloid leukemia. Stem Cell Investigation. 2017;**4**:19. DOI: 10.21037/sci.2017.02.10

[15] Abou-Antoun TJ, Dombrowski SM, Hale JS, Lathia JD. Brain cancer stem cells in adults and children: Cell biology and therapeutic implications. Neurotherapeutics. 2017;**14**:372-384. DOI: 10.1007/s13311-017-0524-0

[16] Lathia J, Liu H, Matei D. The clinical impact of cancer stem cells. The Oncologist. 2020;**25**:123-131. DOI: 10.1634/theoncologist.2019-0517

[17] Zhou B, Lin W, Long Y, Yang Y, Zhang H, Wu K, et al. Notch signaling pathway: Architecture, disease, and therapeutics. Signal Transduction and Targeted Therapy. 2022;**24**:95. DOI: 10.1038/s41392-022-00934-y

[18] Hori K, Sen A, Artavanis-Tsakonas S. Notch signaling at a glance. Journal of Cell Science. 2013;**126**:2135-2140

[19] Fiúza U-M, Arias AM. Cell and molecular biology of Notch. Journal of Endocrinology. 2007;**194**:459-474. DOI: 10.1677/JOE-07-0242

[20] Nakata T, Shimizu H, Nagata S, Ito G, Fujii S, Suzuki K, et al. Indispensable role of Notch ligand-dependent signaling in the proliferation and stem cell niche maintenance of APC-deficient intestinal tumors. Biochemical and Biophysical Research Communications. 2017;**482**:1296-1303. DOI: 10.1016/j.bbrc.2016.12.031

[21] Wang Z, Li Y, Banerjee S, Sarkar FH. Emerging role of Notch in stem cells and cancer. Cancer Letters. 2009;**279**:8-12. DOI: 10.1016/j.canlet.2008.09.030

[22] Andersen P, Uosaki H, Shenje LT, Kwon C. Non-canonical Notch signaling: Emerging role and mechanism. Trends in Cell Biology. 2012;**22**:257-265. DOI: 10.1016/j.tcb.2012.02.003

[23] Jeffries S, Capobianco AJ. Neoplastic transformation by Notch requires nuclear localization. Molecular and Cellular Biology. 2000;**20**:3928-3941. DOI: 10.1128/mcb.20.11.3928-3941.2000

[24] Dumont E, Fuchs KP, Bommer G, Christoph B, Kremmer E, Kempkes B. Neoplastic transformation by Notch is independent of transcriptional activation by RBP-J signalling. Oncogene. 2000;**19**:556-561. DOI: 10.1038/sj.onc.1203352

[25] Ayaz F, Osborne BA. Non-canonical notch signaling in cancer and immunity. Frontiers in Oncology. 2014;**4**:345. DOI: 10.3389/fonc.2014.00345

[26] Purow B. Notch inhibition as a promising new approach to cancer therapy. Advances in Experimental Medicine and Biology. 2012;**727**:305-319. DOI: 10.1007/978-1-4614-0899-4_23

[27] Cook N, Basu B, Smith D-M, Gopinathan A, Evans J, Steward WP, et al. A phase I trial of the γ-secretase inhibitor MK-0752 in combination with gemcitabine in patients with pancreatic ductal adenocarcinoma. British Journal of Cancer. 2018;**118**:793-801. DOI: 10.1038/bjc.2017.495

[28] Takebe N, Nguyen D, Yang SX. Targeting notch signaling pathway in cancer: Clinical development advances and challenges. Pharmacology & Therapeutics. 2014;**141**:140-149. DOI: 10.1016/j.pharmthera.2013.09.005

[29] Diaz-Padilla I, Wilson MK, Clarke BA, Hirte HW, Welch SA, Mackay HJ, et al. A phase II study of single-agent RO4929097, a gamma-secretase inhibitor of Notch signaling, in patients with recurrent platinum-resistant epithelial ovarian cancer: A study of the Princess Margaret, Chicago and California phase II consortia. Gynecologic Oncology. 2015;**137**:216-222. DOI: 10.1016/j.ygyno.2015.03.005

[30] Lee SM, Moon J, Redman BG, Chidiac T, Flaherty LE, Zha Y, et al. Phase 2 study of RO4929097, a gamma-secretase inhibitor, in metastatic melanoma: SWOG 0933. Cancer. 2015;**121**:432-440. DOI: 10.1002/cncr.29055

[31] Villalobos VM, Hall F, Jimeno A, Gore L, Kern K, Cesari R, et al. Long-term follow-up of desmoid fibromatosis

treated with PF-03084014, an oral gamma secretase inhibitor. Annals of Surgical Oncology. 2018;**25**:768-775. DOI: 10.1245/s10434-017-6082-1

[32] Kummar S, Coyne GOS, Do KT, Turkbey B, Meltzer PS, Polley E, et al. Clinical activity of the γ-secretase inhibitor PF-03084014 in adults with desmoid tumors (aggressive fibromatosis). Journal of Clinical Oncology. 2017;**35**:1561-1569. DOI: 10.1200/JCO.2016.71.199

[33] Wu CX, Xu A, Zhang CC, Olson P, Chen L, Lee TK, et al. Notch inhibitor PF-03084014 inhibits hepatocellular carcinoma growth and metastasis via suppression of cancer stemness due to reduced activation of Notch1-Stat3. Molecular Cancer Therapeutics. 2017;**6**:1531-1543. DOI: 10.1158/1535-7163.MCT-17-0001

[34] Morgan KM, Fischer BS, Lee FY, Shah JJ, Bertino JR, Rosenfeld J, et al. Gamma secretase inhibition by BMS-906024 enhances efficacy of paclitaxel in lung adenocarcinoma. Molecular Cancer Therapeutics. 2017;**16**:2759-2769. DOI: 10.1158/1535-7163.MCT-17-0439

[35] Pant S, Jones SF, Kurkjian CD, Infante JR, Moore KN, Burris HA, et al. A first-in-human phase I study of the oral Notch inhibitor, LY900009, in patients with advanced cancer. European Journal of Cancer. 2016;**56**:1-9. DOI: 10.1016/j.ejca.2015.11.021

[36] Miranda EL, Stathis A, Hess D, Racca F, Quon D, Rodon J, et al. Phase 1 study of CB-103, a novel first-in-class inhibitor of the CSL-NICD gene transcription factor complex in human cancers. Journal of Clinical Oncology. 2021;**39**:3020. DOI: 10.1200/JCO.2021.39.15_suppl.3020

[37] Doi T, Tajimi M, Mori J, Asou H, Inoue K, Benhadji KA, et al. A phase 1 study of crenigacestat (LY3039478), the Notch inhibitor, in Japanese patients with advanced solid tumors. Investigational New Drugs. 2021;**39**:469-476. DOI: 10.1007/s10637-020-01001-5

[38] Yang G, Zhou R, Guo X, Yan C, Lei J, Shi Y. Structural basis of gamma-secretase inhibition and modulation by small molecule drugs. Cell. 2021;**184**:521-533. DOI: 10.1016/j.cell.2020.11.049

[39] Miao ZF, Xu H, Xu HM, Wang ZN, Zhao TT, Song YX, et al. DLL4 overexpression increases gastric cancer stem/progenitor cell self-renewal ability and correlates with poor clinical outcome via Notch-1 signaling pathway activation. Cancer Medicine. 2017;**6**:245-257. DOI: 10.1002/cam4.962

[40] Chiorean EG, LoRusso P, Strother RM, Diamond JR, Younger A, Messersmith WA, et al. A phase I first-in-human study of Enoticumab (REGN421), a fully human delta-like ligand 4 (Dll4) monoclonal antibody in patients with advanced solid tumors. Clinical Cancer Research. 2015;**21**:2695-2703. DOI: 10.1158/1078-0432.CCR-14-2797

[41] Murdoch B, Chadwick K, Martin M, Shojaei F, Shah KV, Gallacher L, et al. Wnt-5A augments repopulating capacity and primitive hematopoietic development of human blood stem cells in vivo. Proceedings of the National Academy of Sciences of the United States of America. 2003;**100**:3422-3427. DOI: 10.1073/pnas.0130233100

[42] Ferrer-Mayorga G, Niell N, Cantero R, González-Sancho JM, Peso LD, Muñoz A, et al. Vitamin D and Wnt3A have additive and partially overlapping modulatory effects on gene expression and phenotype in human colon fibroblasts. Scientific Reports. 2019;**9**:8085. DOI: 10.1038/s41598-019-44574-9

[43] Katoh M. Canonical and non-canonical WNT signaling in cancer stem cells and their niches: Cellular heterogeneity, omics reprogramming, targeted therapy and tumor plasticity (Review). International Journal of Oncology. 2017;**51**:1357-1369. DOI: 10.3892/ijo.2017.4129

[44] Torres VI, Godoy JA, Inestrosa NC. Modulating Wnt signaling at the root: Porcupine and Wnt acylation. Pharmacology & Therapeutics. 2019;**198**:34-45. DOI: 10.1016/j.pharmthera.2019.02.009

[45] Latres E, Chiaur DS, Pagano M. The human F box protein beta-Trcp associates with the Cul1/Skp1 complex and regulates the stability of beta-catenin. Oncogene. 1999;**18**:849-854. DOI: 10.1038/sj.onc.1202653

[46] Pan T, Xu J, Zhu Y. Self-renewal molecular mechanisms of colorectal cancer stem cells. International Journal of Molecular Medicine. 2017;**39**:9-20. DOI: 10.3892/ijmm.2016.2815

[47] Arnsdorf EJ, Tummala P, Jacobs CR. Non-canonical Wnt signaling and N-cadherin related bCatenin signaling play a role in mechanically induced osteogenic cell fate. PLoS One. 2009;**4**:e5388. DOI: 10.1371/journal.pone.0005388

[48] MacDonald BT, Tamai K, He X. Wnt/β-catenin signaling: Components, mechanisms, and diseases. Developmental Cell. 2009;**17**:9-26. DOI: 10.1016/j.devcel.2009.06.016

[49] Tree DRP, Shulman JM, Rousset R, Scott MP, Gubb D, Axelrod JD. Prickle mediates feedback amplification to generate asymmetric planar cell polarity signaling. Cell. 2002;**109**:371-381. DOI: 10.1016/s0092-8674(02)00715-8

[50] Topol L, Jiang X, Choi H, Garrett-Beal L, Carolan PJ, Yang Y. Wnt-5a inhibits the canonical Wnt pathway by promoting GSK-3-independent beta-catenin degradation. Journal of Cell Biology. 2003;**162**:899-908. DOI: 10.1083/jcb.200303158

[51] Kikuchi A, Yamamoto H, Sato A, Matsumoto S. New insights into the mechanism of Wnt signaling pathway activation. International Review of Cell and Molecular Biology. 2011;**291**:21-71. DOI: 10.1016/B978-0-12-386035-4.00002-1

[52] Kim J, Kim DW, Chang W, Choe J, Kim J, Park C-S, et al. Wnt5a is secreted by follicular dendritic cells to protect germinal center B cells via Wnt/Ca2+/NFAT/NF-κB–B cell lymphoma 6 signaling. The Journal of Immunology. 2012;**188**:182-189. DOI: 10.4049/jimmunol.1102297

[53] e Melo F d S, Vermeulen L. Wnt signaling in cancer stem cell biology. Cancers. 2016;**8**:60. DOI: 10.3390/cancers8070060

[54] Liu Y, Qi X, Donnelly L, Elghobashi-Meinhardt N, Long T, Zhou RW, et al. Mechanisms and inhibition of Porcupine-mediated Wnt acylation. Nature. 2022;**607**:816-822. DOI: 10.1038/s41586-022-04952-2

[55] Rodon J, Argilés G, Connolly RM, Vaishampayan U, Jonge M d, Garralda E, et al. Phase 1 study of single-agent WNT974, a first-in-class Porcupine inhibitor, in patients with advanced solid tumours. British Journal of Cancer. 2021;**125**:28-37. DOI: 10.1038/s41416-021-01389-8

[56] Madan B, Ke Z, Harmston N, Ho SY, Frois AO, Alam J, et al. Wnt addiction of genetically defined cancers reversed by PORCN inhibition. Oncogene. 2016;**35**:2197-2207. DOI: 10.1038/onc.2015.280

[57] Sasaki T, Kahn M. Inhibition of β-catenin/p300 interaction proximalizes mouse embryonic lung epithelium. Translational Respiratory Medicine. 2014;**2**:8. DOI: 10.1186/s40247-014-0008-1

[58] Safety and Efficacy Study of PRI-724 in Subjects With Advanced Myeloid Malignancies. Available from: https://clinicaltrials.gov/ct2/show/NCT01606579

[59] Pavlovic Z, Adams JJ, Blazer LL, Gakhal AK, Jarvik N, Steinhart Z, et al. A synthetic anti-frizzled antibody engineered for broadened specificity exhibits enhanced anti-tumor properties. MAbs. 2018;**10**:1157-1167. DOI: 10.1080/19420862.2018.1515565

[60] A Study of Vantictumab (OMP-18R5) in Combination With Docetaxel in Patients With Previously Treated NSCLC. Available from: https://clinicaltrials.gov/ct2/show/NCT01957007

[61] A Study of Vantictumab (OMP-18R5) in Combination With Nab-Paclitaxel and Gemcitabine in Previously Untreated Stage IV Pancreatic Cancer. Available from: https://clinicaltrials.gov/ct2/show/NCT02005315

[62] A Study of Vantictumab (OMP-18R5) in Combination With Paclitaxel in Locally Recurrent or Metastatic Breast Cancer. Available from: https://clinicaltrials.gov/ct2/show/NCT01973309

[63] Le PN, McDermott JD, Jimeno A. Targeting the Wnt pathway in human cancers: Therapeutic targeting with a focus on OMP-54F28. Pharmacology & Therapeutics. 2015;**146**:1-11. DOI: 10.1016/j.pharmthera.2014.08.005

[64] A Dose Escalation Study of OMP-54F28 in Subjects With Solid Tumors. Available from: https://clinicaltrials.gov/ct2/show/NCT01608867

[65] Dose Escalation Study of OMP-54F28 in Combination With Paclitaxel and Carboplatin in Patients With Recurrent Platinum-Sensitive Ovarian Cancer. Available from: https://clinicaltrials.gov/ct2/show/NCT02092363

[66] Babu D, Fanelli A, Mellone S, Muniswamy R, Wasniewska M, Prodam F, et al. Novel GLI2 mutations identified in patients with Combined Pituitary Hormone Deficiency (CPHD) Evidence for a pathogenic effect by functional characterization. Clinical Endocrinology. 2019;**90**:449-456. DOI: 10.1111/cen.13914

[67] Cochrane CR, Szczepny A, Watkins DN, Cain JE. Hedgehog signaling in the maintenance of cancer stem cells. Cancers. 2015;**7**:1554-1585. DOI: 10.3390/cancers7030851

[68] Lee D-H, Lee S-Y, Oh SC. Hedgehog signaling pathway as a potential target in the treatment of advanced gastric cancer. Tumor Biology. 2017;**39**:1-10. DOI: 10.1177/1010428317692266

[69] Han Y, Wang B, Cho YS, Zhu J, Wu J, Chen Y, et al. Phosphorylation of Ci/Gli by fused family kinases promotes Hedgehog signaling. Developmental Cell. 2019;**50**:610-626. DOI: 10.1016/j.devcel.2019.06.008

[70] Yang Y, Li X, Wang T, Guo Q, Xi T, Zheng L. Emerging agents that target signaling pathways in cancer stem cells. Journal of Hematology and Oncology. 2020;**13**:60. DOI: 10.1186/s13045-020-00901-6

[71] Ingram I: Vismodegib Granted FDA Approval for Treatment of Basal Cell Carcinoma. 2012. Available from: https://www.cancernetwork.com/view/vismodegib-granted-fda-approval-treatment-basal-cell-carcinoma

[72] Li W, Yang H, Li X, Han L, Xu N, Shi A. Signaling pathway inhibitors target breast cancer stem cells in triple-negative breast cancer. Oncology Reports. 2019;**41**:437-446. DOI: 10.3892/or.2018.6805

[73] Addition of Vismodegib to Neoadjuvant Chemotherapy in Triple Negative Breast Cancer Patients (SHH-CM). Available from: https://clinicaltrials.gov/ct2/show/NCT02694224

[74] Singh BN, Fu J, Srivastava RK, Shankar S. Hedgehog signaling antagonist GDC-0449 (Vismodegib) inhibits pancreatic cancer stem cell characteristics: Molecular mechanisms. PLoS One. 2011;**6**:e27306. DOI: 10.1371/journal.pone.0027306

[75] Wu C, Hu S, Cheng J, Wang G, Tao K. Smoothened antagonist GDC-0449 (Vismodegib) inhibits proliferation and triggers apoptosis in colon cancer cell lines. Experimental and Therapeutic Medicine. 2017;**13**:2529-2536. DOI: 10.3892/etm.2017.4282

[76] Berlin J, Bendell JC, Hart LL, Firdaus I, Gore I, Hermann RC, et al. A randomized phase II trial of vismodegib versus placebo with FOLFOX or FOLFIRI and bevacizumab in patients with previously untreated metastatic colorectal cancer. Clinical Cancer Research. 2013;**19**:258-267. DOI: 10.1158/1078-0432.CCR-12-1800

[77] Casey D, Demko S, Shord S, Zhao H, Chen H, He K, et al. FDA approval summary: sonidegib for locally advanced basal cell carcinoma. Clinical Cancer Research. 2017;**23**:2377-2381. DOI: 10.1158/1078-0432.CCR-16-2051

[78] Cazet AS, Hui MN, Elsworth BL, Wu SZ, Roden D, Chan C-L, et al. Targeting stromal remodeling and cancer stem cell plasticity overcomes chemoresistance in triple negative breast cancer. Nature Communications. 2018;**9**:2897. DOI: 10.1038/s41467-018-05220-6

[79] Norsworthy KJ, By K, Subramaniam S, Zhuang L, Valle PLD, Przepiorka D, et al. FDA approval summary: glasdegib for newly diagnosed acute myeloid leukemia. Clinical Cancer Research. 2019;**25**:6021-6025. DOI: 10.1158/1078-0432.CCR-19-0365

[80] Du F-Y, Zhou Q-F, Sun W-J, Chen G-L. Targeting cancer stem cells in drug discovery: Current state and future perspectives. World Journal of Stem Cells. 2019;**11**:398-420. DOI: 10.4252/wjsc.v11.i7.398

[81] Xie H, Paradise BD, Ma WW, Fernandez-Zapico ME. Recent advances in the clinical targeting of Hedgehog/GLI signaling in cancer. Cell. 2019;**8**:394. DOI: 10.3390/cells8050394

[82] A Study of BMS-833923 With Carboplatin and Etoposide Followed by BMS-833923 Alone in Subjects With Extensive-Stage Small Cell Lung Cancer. Available from: https://clinicaltrials.gov/ct2/show/NCT00927875

[83] Itraconazole in Treating Patients With Biochemically Relapsed Prostate Cancer. Available from: https://clinicaltrials.gov/ct2/show/NCT01787331

[84] A Phase 1 Study of IPI-926 in Patients With Advanced and/or Metastatic Solid Tumor Malignancies. Available from: https://clinicaltrials.gov/ct2/show/NCT00761696

[85] A Dose Finding and Safety Study of Oral LEQ506 in Patients With Advanced Solid Tumors. Available from: https://clinicaltrials.gov/ct2/show/NCT01106508

[86] A Study of LY2940680 in Japanese Participants With Advanced Cancers. Available from: https://clinicaltrials.gov/ct2/show/NCT01919398

[87] A Study of TAK-441 in Adult Patients With Advanced Nonhematologic Malignancies. Available from: https://clinicaltrials.gov/ct2/show/NCT01204073

[88] Arsenic Trioxide in Treating Patients With Advanced Neuroblastoma or Other Childhood Solid Tumors. Available from: https://clinicaltrials.gov/ct2/show/NCT00024258

[89] Fu J, Rodova M, Roy SK, Sharma J, Singh KP, Srivastava RK, et al. GANT-61 inhibits pancreatic cancer stem cell growth in vitro and in NOD/SCID/IL2R gamma null mice xenograft. Cancer Letters. 2013;**330**:22-32. DOI: 10.1016/j.canlet.2012.11.018

[90] Oeckinghaus A, Ghosh S. The NF-κB family of transcription factors and its regulation. Cold Spring Harvor Perspectives in Biology. 2009;**1**:a000034. DOI: 10.1101/cshperspect.a00003410.1101/cshperspect.a000034

[91] Siebenlist U, Franzoso G, Brown K. Structure, regulation and function of NF-kappa B. Annual Review of Cell Biology. 1994;**10**:405-455. DOI: 10.1146/annurev.cb.10.110194.002201

[92] Novack DV. Role of NF-κB in the skeleton. Cell Research. 2011;**21**:169-182. DOI: 10.1038/cr.2010.159

[93] Perkins ND, Gilmore TD. Good cop, bad cop: The different faces of NFkappaB. Cell Death & Differentiation. 2006;**13**:759-772. DOI: 10.1038/sj.cdd.4401838

[94] Sun S-C. The non-canonical NF-κB pathway in immunity and inflammation. Nature Reviews Immunology. 2017;**17**:545-558. DOI: 10.1038/nri.2017.52

[95] Hayden MS, Ghosh S. Shared principles in NF-kappaB signaling. Cell. 2008;**132**:344-362. DOI: 10.1016/j.cell.2008.01.020

[96] Greten FR, Eckmann L, Greten TF, Park JM, Li Z-W, Egan LJ, et al. IKKbeta links inflammation and tumorigenesis in a mouse model of colitis-associated cancer. Cell. 2004;**118**:285-296. DOI: 10.1016/j.cell.2004.07.013

[97] Gonzalez-Torres C, Gaytan-Cervantes J, Vazquez-Santillan K, Mandujano-TinocoEA,Ceballos-CancinoG, Garcia-Venzor A, et al. NF-kappaB participates in the stem cell phenotype of ovarian cancer cells. Archives of Medical Research. 2017;**48**:343-351. DOI: 10.1016/j.arcmed.2017.08.001

[98] Vazquez-Santillan K, Melendez-Zajgla J, Jimenez-Hernandez LE, Gaytan-Cervantes J, Muñoz-Galindo L, Piña-Sanchez P, et al. NF-kappaBeta-inducing kinase regulates stem cell phenotype in breast cancer. Scientific Reports. 2016;**6**:37340. DOI: 10.1038/srep37340

[99] Han D, Wu G, Chang C, Zhu F, Xiao Y, Li Q, et al. Disulfiram inhibits TGF-β-induced epithelial–mesenchymal transition and stem-like features in breast cancer via ERK/NF-κB/Snail pathway. Oncotarget. 2015;**6**:40907-40919. DOI: 10.18632/oncotarget.5723

[100] Phase II Trial of Disulfiram With Copper in Metastatic Breast Cancer (DISC). Available from: https://clinicaltrials.gov/ct2/show/NCT03323346

[101] Burnett JP, Lim G, Li Y, Shah RB, Lim R, Paholak HJ, et al. Sulforaphane enhances the anticancer activity of taxanes against triple negative breast cancer by killing cancer stem cells. Cancer Letters. 2017;**394**:52-64. DOI: 10.1016/j.canlet.2017.02.023

[102] Study to Evaluate the Effect of Sulforaphane in Broccoli Sprout Extract on Breast Tissue. Available from: https://clinicaltrials.gov/ct2/show/NCT00982319

[103] Marquardt JU, Gomez-Quiroz L, Camacho LOA, Pinna F, Lee Y-H, Kitade M, et al. Curcumin effectively inhibits oncogenic NF-κB signaling and restrains stemness features in liver cancer. Journal of Hepatology. 2015;**63**:661-669. DOI: 10.1016/j.jhep.2015.04.018

[104] Phase II Study of Curcumin vs Placebo for Chemotherapy-Treated Breast Cancer Patients Undergoing Radiotherapy. Available from: https://clinicaltrials.gov/ct2/show/NCT01740323

[105] A "Window Trial" on Curcumin for Invasive Breast Cancer Primary Tumors. Available from: https://clinicaltrials.gov/ct2/show/NCT03980509

[106] Laplante M, Sabatini DM. mTOR signaling at a glance. Journal of Cell Science. 2009;**122**:3589-3594. DOI: 10.1101/cshperspect.a011593

[107] Li J, Kim SG, Blenis J. Rapamycin: One drug, many effects. Cell Metabolism. 2014;**19**:373-379. DOI: 10.1016/j.cmet.2014.01.001

[108] Inoki K, Corradetti MN, Guan K-L. Dysregulation of the TSC-mTOR pathway in human disease. Nature Genetics. 2005;**37**:19-24. DOI: 10.1038/ng1494

[109] Karagianni F, Pavlidis A, Malakou LS, Piperi C, Papadavid E. Predominant role of mTOR signaling in skin diseases with therapeutic potential. International Journal of Molecular Sciences. 2022;**23**:1693. DOI: 10.3390/ijms23031693

[110] Papadopoli D, Boulay K, Kazak L, Pollak M, Mallette F, Topisirovic I, Hulea L: mTOR as a central regulator of lifespan and aging. F1000Research: 2019: 8: F1000 Faculty Rev-1998. doi: 10.12688/f1000research.17196.1

[111] Chang L, Graham PH, Hao J, Ni J, Bucci J, Cozzi PJ, et al. Acquisition of epithelial-mesenchymal transition and cancer stem cell phenotypes is associated with activation of the PI3K/Akt/mTOR pathway in prostate cancer radioresistance. Cell Death & Disease. 2013;**4**:e875. DOI: 10.1038/cddis.2013.407

[112] Zhou J, Wulfkuhle J, Zhang H, Gu P, Yang Y, Deng J, et al. Activation of the PTEN/mTOR/STAT3 pathway in breast cancer stem-like cells is required for viability and maintenance. Proceedings of the National Academy of Sciences of the United States of America. 2007;**104**:16158-16163. DOI: 10.1073/pnas.0702596104

[113] Jung M-J, Rho J-K, Kim Y-M, Jung JE, Jin YB, Ko Y-G, et al. Upregulation of CXCR4 is functionally crucial for maintenance of stemness in drug-resistant non-small cell lung cancer cells. Oncogene. 2013;**32**:209-221. DOI: 10.1038/onc.2012.37

[114] Marhold M, Tomasich E, El-Gazzar A, Heller G, Spittler A, Horvat R, et al. HIF-1alpha regulates mTOR signaling and viability of prostate cancer stem cells. Molecualr Cancer Research. 2015;**13**:556-564. DOI: 10.1158/1541-7786.MCR-14-0153-T

[115] Zhu Y, Zhang X, Liu Y, Zhang S, Liu J, Ma Y, et al. Antitumor effect of the mTOR inhibitor everolimus in combination with trastuzumab on human breast cancer stem cells in vitro and in vivo. Tumor Biology. 2012;**33**:1349-1362. DOI: 10.1007/s13277-012-0383-6

[116] Kim W-T, Ryu CJ. Cancer stem cell surface markers on normal stem cells.

BMB Reports. 2017;**50**:285-298. DOI: 10.5483/bmbrep.2017.50.6.039

[117] Leach B. FDA Approves Everolimus for Advanced Breast Cancer. 2012. Available from: https://www.onclive.com/view/fda-approves-everolimus-for-advanced-breast-cancer

[118] Wang Y, Liu Y, Lu J, Zhang P, Wang Y, Xu Y, et al. Rapamycin inhibits FBXW7 loss-induced epithelial-mesenchymal transition and cancer stem cell-like characteristics in colorectal cancer cells. Biochemical and Biophysical Research Communications. 2013;**434**:352-356. DOI: 10.1016/j.bbrc.2013.03.077

[119] Greenberger S, Yuan S, Walsh LA, Boscolo E, Kang K-T, Matthews B, et al. Rapamycin suppresses self-renewal and vasculogenic potential of stem cells isolated from infantile hemangioma. Journal of Investigative Dermatology. 2011;**131**:2467-2476. DOI: 10.1038/jid.2011.300

[120] ABI-009 (Nab-rapamycin) in Combination With FOLFOX and Bevacizumab as First-line Therapy in Patients With Advanced or Metastatic Colorectal Cancer. Available from: https://clinicaltrials.gov/ct2/show/NCT03439462

[121] Sirolimus in the Treatment for Infantile Hepatic Hemangioendothelioma(IEEH) (PRL-SRL-IHHE). Available from: https://clinicaltrials.gov/ct2/show/NCT04406870

[122] Iliopoulos D, Hirsch HA, Struhl K. Metformin decreases the dose of chemotherapy for prolonging tumor remission in mouse xenografts involving multiple cancer cell types. Cancer Research. 2011;**71**:3196-3201. DOI: 10.1158/0008-5472.CAN-10-3471

[123] Mohammed A, Janakiram NB, Brewer M, Ritchie RL, Marya A, Lightfoot S, et al. Antidiabetic drug metformin prevents progression of pancreatic cancer by targeting in part cancer stem cells and mTOR signaling. Translational Oncology. 2013;**6**:649-659. DOI: 10.1593/tlo.13556

[124] A Phase III Randomized Trial of Metformin vs Placebo in Early Stage Breast Cancer. Available from: https://clinicaltrials.gov/ct2/show/NCT01101438

[125] Metformin Combined With Chemotherapy for Pancreatic Cancer (GEM). Available from: https://clinicaltrials.gov/ct2/show/NCT01210911

[126] Villarino AV, Kanno Y, O'Shea JJ. Mechanisms and consequences of Jak-STAT signaling in the immune system. Nature Immunology. 2017;**18**:374-384. DOI: 10.1038/ni.3691

[127] Boehi F, Manetsch P, Hottiger MO. Interplay between ADP-ribosyltransferases and essential cell signaling pathways controls cellular responses. Cell Discovery. 2021;**7**:104. DOI: 10.1038/s41421-021-00323-9

[128] Morris R, Kershaw NJ, Babon JJ. The molecular details of cytokine signaling via the JAK/STAT pathway. Protein Science. 2018;**27**:1984-2009. DOI: 10.1002/pro.3519

[129] Wan S, Zhao E, Kryczek I, Vatan L, Sadovskaya A, Ludema G, et al. Tumor-associated macrophages produce interleukin 6 and signal via STAT3 to promote expansion of human hepatocellular carcinoma stem cells. Gastroenterology. 2014;**147**:1393-1404. DOI: 10.1053/j.gastro.2014.08.039

[130] Yang L, Dong Y, Li Y, Wang D, Liu S, Wang D, et al. IL-10 derived from M2 macrophage promotes cancer stemness via JAK1/STAT1/NF-kappaB/Notch1 pathway

in non-small cell lung cancer. International Journal of Cancer. 2019;**145**:1099-1110. DOI: 10.1002/ijc.32151

[131] Jia H, Song L, Cong Q, Wang J, Xu H, Chu Y, et al. The LIM protein AJUBA promotes colorectal cancer cell survival through suppression of JAK1/STAT1/IFIT2 network. Oncogene. 2017;**36**:2655-2666. DOI: 10.1038/onc.2016.418

[132] Ruan Z, Yang X, Cheng W. OCT4 accelerates tumorigenesis through activating JAK/STAT signaling in ovarian cancer side population cells. Cancer Management and Research. 2018;**11**:389-399. DOI: 10.2147/CMAR.S180418

[133] Kim S-Y, Kang JW, Song X, Kim BK, Yoo YD, Kwon YT, et al. Role of the IL-6-JAK1-STAT3-Oct-4 pathway in the conversion of non-stem cancer cells into cancer stem-like cells. Cellular Signalling. 2013;**25**:961-969. DOI: 10.1016/j.cellsig.2013.01.007

[134] Marotta LLC, Almendro V, Marusyk A, Shipitsin M, Schemme J, Walker SR, et al. The JAK2/STAT3 signaling pathway is required for growth of CD44(+)CD24(−) stem cell-like breast cancer cells in human tumors. Journal of Clinical Investigation. 2011;**121**:2723-2735. DOI: 10.1172/JCI44745

[135] Zhang X, Hu F, Li G, Li G, Yang X, Liu L, et al. Human colorectal cancer-derived mesenchymal stem cells promote colorectal cancer progression through IL-6/JAK2/STAT3 signaling. Cell Death & Disease. 2018;**9**:25. DOI: 10.1038/s41419-017-0176-3

[136] Zhou B, Damrauer JS, Bailey ST, Hadzic T, Jeong Y, Clark K, et al. Erythropoietin promotes breast tumorigenesis through tumor-initiating cell self-renewal. Journal of Clinical Investigation. 2014;**124**:553-563. DOI: 10.1172/JCI69804

[137] Dolatabadi S, Jonasson E, Lindén M, Fereydouni B, Bäcksten K, Nilsson M, et al. JAK-STAT signalling controls cancer stem cell properties including chemotherapy resistance in myxoid liposarcoma. International Journal of Cancer. 2019;**145**:435-499. DOI: 10.1002/ijc.32123

[138] An Open-Label Study to Enable Continued Treatment Access for Subjects Previously Enrolled in Studies of Ruxolitinib. Available from: https://clinicaltrials.gov/ct2/show/NCT02955940

[139] Julian L, Olson MF. Rho-associated coiled-coil containing kinases (ROCK). Small GTPases. 2014;**5**:e29846. DOI: 10.4161/sgtp.29846

[140] Mulherkar S, Tolias KF. RhoA-ROCK signaling as a therapeutic target in traumatic brain injury. Cell. 2020;**9**:245. DOI: 10.3390/cells9010245

[141] Srinivasan S, Ashok V, Mohanty S, Das A, Das S, Kumar S, et al. Purwar R: Blockade of Rho-associated protein kinase (ROCK) inhibits the contractility and invasion potential of cancer stem like cells. Oncotarget. 2017;**8**:21418-21428. DOI: 10.18632/oncotarget.15248

[142] Papadatos-Pastos D, Kumar R, Yap TA, Ruddle R, Decordova S, Halbert PJ, et al. A first-in-human study of the dual ROCK I/II inhibitor, AT13148, in patients with advanced cancers. Journal of Clinical Oncology. 2015;**33**:2566-2566. DOI: 10.1200/jco.2015.33.15_suppl.2566

[143] Phase I Study of AT13148, a Novel AGC Kinase Inhibitor. Available from: https://clinicaltrials.gov/ct2/show/NCT01585701

[144] Bellomo C, Caja L, Moustakas A. Transforming growth factor β as regulator of cancer stemness and metastasis. British Journal of Cancer.

2016;**115**:761-769. DOI: 10.1038/bjc.2016.255

[145] Tang B, Yoo N, Vu M, Mamura M, Nam J-S, Ooshima A, et al. Transforming growth factor-beta can suppress tumorigenesis through effects on the putative cancer stem or early progenitor cell and committed progeny in a breast cancer xenograft model. Cancer Research. 2007;**67**:8643-8652. DOI: 10.1158/0008-5472.CAN-07-0982

[146] Ehata S, Johansson E, Katayama R, Koike S, Watanabe A, Hoshino Y, et al. Transforming growth factor-β decreases the cancer-initiating cell population within diffuse-type gastric carcinoma cells. Oncogene. 2011;**30**:1693-1705. DOI: 10.1038/onc.2010.546

[147] Bruna A, Greenwood W, Quesne JL, Teschendorff A, Miranda-Saavedra D, Rueda OM, et al. TGFβ induces the formation of tumour-initiating cells in claudinlow breast cancer. Nature Communications. 2012;**3**:1055. DOI: 10.1038/ncomms2039

[148] Lo P-K, Kanojia D, Liu X, Singh UP, Berger FG, Wang Q, et al. CD49f and CD61 identify Her2/neu-induced mammary tumor-initiating cells that are potentially derived from luminal progenitors and maintained by the integrin-TGFβ signaling. Oncogene. 2012;**31**:2614-2626. DOI: 10.1038/onc.2011.439

[149] Bhola NE, Balko JM, Dugger TC, Kuba MG, Sánchez V, Sanders M, et al. TGF-β inhibition enhances chemotherapy action against triple-negative breast cancer. Journal of Clinical Investigation. 2013;**123**:1348-1358. DOI: 10.1172/JCI65416

[150] You H, Ding W, Rountree CB. Epigenetic regulation of cancer stem cell marker CD133 by transforming growth factor-beta. Hepatology. 2010;**51**:1635-1644. DOI: 10.1002/hep.23544

[151] Mima K, Okabe H, Ishimoto T, Hayashi H, Nakagawa S, Kuroki H, et al. CD44s regulates the TGF-β-mediated mesenchymal phenotype and is associated with poor prognosis in patients with hepatocellular carcinoma. Cancer Research. 2012;**72**:3414-3423. DOI: 10.1158/0008-5472.CAN-12-0299

[152] Hasegawa T, Yashiro M, Nishii T, Matsuoka J, Fuyuhiro Y, Morisaki T, et al. Cancer-associated fibroblasts might sustain the stemness of scirrhous gastric cancer cells via transforming growth factor-β signaling. International Journal of Cancer. 2014;**134**:1785-1795. DOI: 10.1002/ijc.28520

[153] Oshimori N, Oristian D, Fuchs E. TGF-β promotes heterogeneity and drug resistance in squamous cell carcinoma. Cell. 2015;**160**:963-976. DOI: 10.1016/j.cell.2015.01.043

[154] Ikushima H, Todo T, Ino Y, Takahashi M, Miyazawa K, Miyazono K. Autocrine TGF-beta signaling maintains tumorigenicity of glioma-initiating cells through Sry-related HMG-box factors. Cell Stem Cell. 2009;**5**:504-514. DOI: 10.1016/j.stem.2009.08.018

[155] Peñuelas S, Anido J, Prieto-Sánchez RM, Folch G, Barba I, Cuartas I, et al. TGF-beta increases glioma-initiating cell self-renewal through the induction of LIF in human glioblastoma. Cancer Cell. 2009;**15**:315-327. DOI: 10.1016/j.ccr.2009.02.011

[156] Naka K, Hoshii T, Muraguchi T, Tadokoro Y, Ooshio T, Kondo Y, et al. TGF-beta-FOXO signalling maintains leukaemia-initiating cells in chronic myeloid leukaemia. Nature. 2010;**463**:676-680. DOI: 10.1038/nature08734

Chapter 13

Induced Pluripotent Stem Cells: Advances and Applications in Regenerative Medicine

Igor Kizub, Andrii Rozhok and Ganna Bilousova

Abstract

Reprogramming adult somatic cells into induced pluripotent stem cells (iPSCs) through the ectopic expression of reprogramming factors offers truly personalized cell-based therapy options for numerous human diseases. The iPSC technology also provides a platform for disease modeling and new drug discoveries. Similar to embryonic stem cells, iPSCs can give rise to any cell type in the body and are amenable to genetic correction. These properties of iPSCs allow for the development of permanent corrective therapies for many currently incurable disorders. In this chapter, we summarize recent progress in the iPSC field with a focus on potential clinical applications of these cells.

Keywords: cell differentiation, cell-based therapy, genetic correction, induced pluripotent stem cells, iPSCs, regenerative medicine, stem cell reprogramming

1. Introduction

Direct reprogramming of somatic cells into induced pluripotent stem cells (iPSCs) through the ectopic expression of reprogramming factors [1–5] has had a dramatic impact on the field of regenerative medicine and has opened a new era in research and therapy. Human iPSCs represent an unprecedented source of patient-specific pluripotent stem cells suitable for disease modeling and tissue replacement therapy.

Stem cells (SCs) have the ability to self-renew through cell division and can differentiate into various cell types. Based on their origin, SCs are divided into embryonic SCs (ESCs), induced pluripotent SCs (iPSCs), and adult SCs. ESCs are pluripotent cells derived from the inner cell mass of the blastocyst. They can give rise to tissues of the three germ layers and are regarded as a renewable potent cell source for the regeneration of all bodily tissues [6]. Adult SCs are multipotent cells of adult tissues that are also essential for regenerative medicine [7]. iPSCs share many similarities with ECSs, including pluripotency, differentiation potential, and the capability to form teratomas and viable chimeras [8].

Human ESCs are isolated by the use of surplus *in vitro* fertilization embryos [9]. Therefore, unlike the iPSC technology, ESC-based techniques do not allow for the generation of genetically diverse patient-specific cells. Additionally, the use of ESCs is obstructed by the need to destroy human embryos in the process of cell isolation, which raises ethical considerations. Primary human ESCs, therefore, are a suboptimal

 IntechOpen

SC source for therapy and tissue engineering. ESC-based cell therapies may also result in immune rejection, which theoretically can be avoided if autologous iPSC-derived cells are used instead.

Similar to ESCs, iPSCs can proliferate indefinitely and differentiate into all three germ layers. Thus, the iPSC technology solves many problems associated with the use of ESCs and provides an unlimited source of autologous pluripotent SCs, which can be genetically corrected, differentiated into adult lineages, and returned to the same patient as an autograft [10]. Significant advances have been achieved in recent years in improving the safety of the iPSC technology; thus, expanding the opportunities for its clinical application. However, despite the tremendous potential of iPSCs, extensive analyses of their safety and reliability are still required. This chapter discusses the current progress and prospects of using the iPSC technology in tissue replacement therapy and as a tool for studying human pathologies.

2. Reprograming of somatic cells into iPSC

The first attempts to derive pluripotent SCs from adult cells were stimulated by early experiments that demonstrated the feasibility of reprogramming adult frog somatic cell nuclei by the cytoplasm of an enucleated unfertilized frog oocyte [11]. Later findings showed that reprogramming of somatic cells back to the pluripotent state is possible by transferring somatic cell nuclei into oocytes or by fusing somatic cells with pluripotent SCs [12–14]. Finally, the successful cloning of Dolly the sheep showed the feasibility of complete reprogramming of a mammalian somatic nucleus back to a pluripotent state from which it can develop a new animal [15].

2.1 Main reprograming factors

The success of animal cloning has demonstrated that unfertilized eggs and ESCs contain a set of factors that can confer pluripotency to somatic cells. Oct4, Sox2, c-Myc, and Klf4 were later identified by Takahashi and Yamanaka as sufficient to induce pluripotency in mouse somatic cells, resulting in iPSCs that were functionally equivalent to mouse ESCs [1, 2].

OCT4 is a key transcriptional factor, which maintains pluripotency in both early embryos and ESCs [16]. The level of OCT4 expression is vital for regulating pluripotency and it can both activate or repress the promoter of the *REX1* gene, also a critical regulator of pluripotency [17]. The transcription factor SOX2 (sex determining region Y (SRY)-box2) is also essential for maintaining cell pluripotency. It comprises a regulatory complex with OCT4 and REX1 that cooperatively binds to DNA to activate transcription of other pluripotency factors [18]. The proto-oncogene c-MYC has multiple downstream targets that enhance cell proliferation, resulting in SC self-renewal [19]. The c-MYC protein can also induce global histone acetylation allowing OCT4 and SOX2 to bind to otherwise inaccessible sites [20]. Krüppel-like factor 4 (KLF4) is an oncogene that contributes to the long-term maintenance of the ESC phenotype. The role of KLF4 in cell reprogramming is probably to downregulate the expression of the tumor suppressor protein p53 [21]. KLF4 can also activate transcription by interacting with histone acetyltransferases, suppressing cell proliferation, and reciprocally acting with c-MYC [22].

The combination of OCT4, SOX2, KLF4, and c-MYC (abbreviated as OSKM), is widely used to reprogram various types of somatic cells into a pluripotent state. Under specific conditions, reprogramming can also be achieved without c-MYC or with only one or two factors from the OSKM set [23–25]. An alternative combination of OCT4, SOX2, NANOG, and LIN28 has also been shown to be sufficient to reprogram human cells into iPSCs [5]. Additional factors can be used in combination with OSKM to enhance reprogramming efficiencies, such as LIN28, human telomerase reverse transcriptase (hTERT), and SV40 large T antigen [26–28]. Regardless of their combination, reprogramming factors remodel the epigenetic configuration of somatic cells in a way that allows for the conversion of these mature somatic cells into immature pluripotent SCs. Mechanisms of reprogramming into iPSCs are reviewed in detail by Meir and Li [29].

2.2 Reprograming approaches

Early approaches to obtain iPSCs relied on the use of integrating retro- and lentiviral vectors to deliver reprogramming factors into somatic cells [1]. However, the expression of these exogenous factors is only essential in the initial step of reprogramming, and their silencing must occur to ensure the stability of iPSCs [2]. The use of retroviral vectors can not only result in the reactivation of exogenous reprogramming factors and in turn destabilization of iPSCs [2] but also increase the risk of insertional mutagenesis and cancer transformation in iPSC-derived tissues. The development of integration-free reprogramming approaches made the production of iPSCs safer for potential clinical applications. Somatic cells have been successfully reprogrammed into iPSCs using episomal plasmids encoding the reprogramming factors and adenoviral vectors [30, 31]. A number of other non-integrating vectors of viral origins have been utilized for reprogramming, such as those based on Sendai virus [32], Epstein–Barr virus [26], and various circular plasmid constructs [33, 34]. While these methods produce integration-free iPSCs, many of these approaches suffer from extremely low efficiency, which hampers their potential clinical application [35].

The generation of iPSCs has been recently accomplished with defined chemicals that can functionally replace reprogramming factors [36]. iPSCs can also be generated by fusing reprogramming factors with cell-penetrating proteins that allow for the efficient transport of reprogramming proteins through cell membrane [37]. Despite their safety, these DNA-free approaches also suffer from low reprogramming efficiency.

A more promising approach for the transgene-free generation of iPSCs can be the use of synthetic modified mRNAs encoding the reprogramming factors. This approach has been shown to reprogram a variety of cell lines with an efficiency superior to that of other integration-free approaches [38]. The disadvantage of this method is that RNAs have to be delivered into the cells daily during the reprogramming process. MicroRNA (miRNAs), such as miR-200c, miR-291-3p, miR-294, miR-295, and miR-302a-d, have also been shown to significantly enhance the efficiency of pluripotency induction [39]. When combined with modified mRNAs encoding reprogramming factors, miR-367 and miR-302a-d have been shown to increase the reprogramming efficiency of human primary fibroblasts to an unprecedented level, and reprogramming up to 90.7% of individually plated cells [40].

2.3 Reprogramming process

Different cell types have been used for the generation of iPSCs, albeit with different reprogramming efficiencies. Fibroblasts are the most commonly used cells due to their availability and easy culture conditions. Keratinocytes, melanocytes, blood cells, hepatocytes, and gastric epithelial cells are also suitable for reprogramming [41].

The reprogramming process has been extensively studied in fibroblasts and has been shown to follow an organized sequence of events, which begins with the downregulation of somatic gene expression [42]. The first step requires a phenotype transition initiated by the activation of the early pluripotency stage-specific embryonic antigen (SSEA1) and alkaline phosphatase, and the inactivation of the differentiation-related antigen Thy-1 (CD90), followed by the activation of *NANOG* and *OCT4* [31, 42]. OCT4, SOX2, and NANOG further induce the expression of stemness genes, such as *STAT3* and *ZIC3*, and repress differentiation-associated genes [17, 43]. The expression level and balance of reprogramming factors are also important for iPSC generation. For example, the increased relative expression of OCT4 enhances reprogramming efficiency [44].

Reprogramming somatic cells often results in the generation of heterogeneous iPSCs with different molecular phenotypes and differentiation potentials [45]. The duration of the reprogramming process also affects the characteristics of the resulting iPSC. Prolonged cultivation of iPSCs yields phenotypes closer to those of ESCs as compared to cells in the early phase of reprogramming. This suggests that reprogramming continues even after iPSCs have been established [46]. Alterations in epigenetic modifications, such as DNA methylation, are also important for iPSC induction [47]. Interestingly, epigenetic profiling of iPSCs has revealed that reprogrammed cells retain epigenetic marks of the cell type of origin [48] although these marks disappear upon continued passaging [42].

3. Clinical applications of iPSCs

The rapid progress of the iPSC technology increases efforts to translate autologous iPSC-based therapies into the clinic. While only a few clinical trials have directly tested the delivery of iPSC-derived cells into patients, the technology continues to develop and researchers from virtually every field develop iPSC-based cell therapies for relevant diseases. These therapies are still at different stages of development and rely on the feasibility to derive functional somatic cell types from iPSCs. Examples of the efforts toward the development of iPSC-based therapies for a variety of tissues and organs are summarized for selected fields below.

3.1 Dermatology

The skin may represent an ideal tissue for testing novel iPSC-based therapies. It is readily accessible, highly proliferative, and can be easily monitored. Multiple skin cell lineages have been generated from iPSCs, such as keratinocytes [49], melanocytes [50], fibroblasts [51], and ectodermal precursor cells [52]. Mouse iPSC-derived keratinocytes display characteristics similar to those of primary keratinocytes and can regenerate differentiated epidermis and skin appendages when grafted together with mouse fibroblasts into athymic mice [49]. Human iPSC-derived keratinocytes have also been shown to establish functional organotypic skin in culture and 3D skin models [53]. These and other studies have demonstrated the potential of iPSCs to generate autologous donor cells for cell-based therapies for skin diseases.

As an essential step toward future clinical use, *in vitro* 3D skin equivalents have been generated using iPSC-derived components. These equivalents exhibit normal skin morphology, stratification, and terminal differentiation [54], and can potentially be used for drug screening. Skin organoids with stratified epidermal and dermal skin layers that are able to spontaneously produce *de novo* hair and sebaceous glands have also been generated from murine [55] and human [56] iPSCs.

Many of the most devastating forms of inherited skin diseases that are caused by known mutations, such as Epidermolysis Bullosa (EB), can be treated with genetically corrected iPSC-derived cells [54]. EB is a group of inherited skin blistering diseases that results in severe blistering and scarring [57]. Precise gene editing techniques, such as those based on CRISPR (clustered regularly interspaced short palindromic repeats)/CRISPR-associate (Cas) systems, can be used for the generation of these patient-specific genetically corrected iPSCs. These corrected iPSCs can then be differentiated into skin cells and transplanted back to the same patient in need of treatment. Similar strategies can be implemented for a variety of other diseases affecting many other organs (**Figure 1**). In fact, the generation of iPSCs, coupled with gene targeting, can solve many obstacles that are associated with gene correction in somatic cells. Unlike somatic cells, iPSCs can be expanded indefinitely, allowing for easier selection and expansion of corrected clones. In addition, iPSCs derived from very old patients can differentiate into "rejuvenated" cells [58].

While currently there are no approved clinical trials to test iPSC-based therapies for the treatment of skin diseases, EB is likely to be the first skin disease to benefit from the iPSC-based therapy due to the severity of this disorder. For

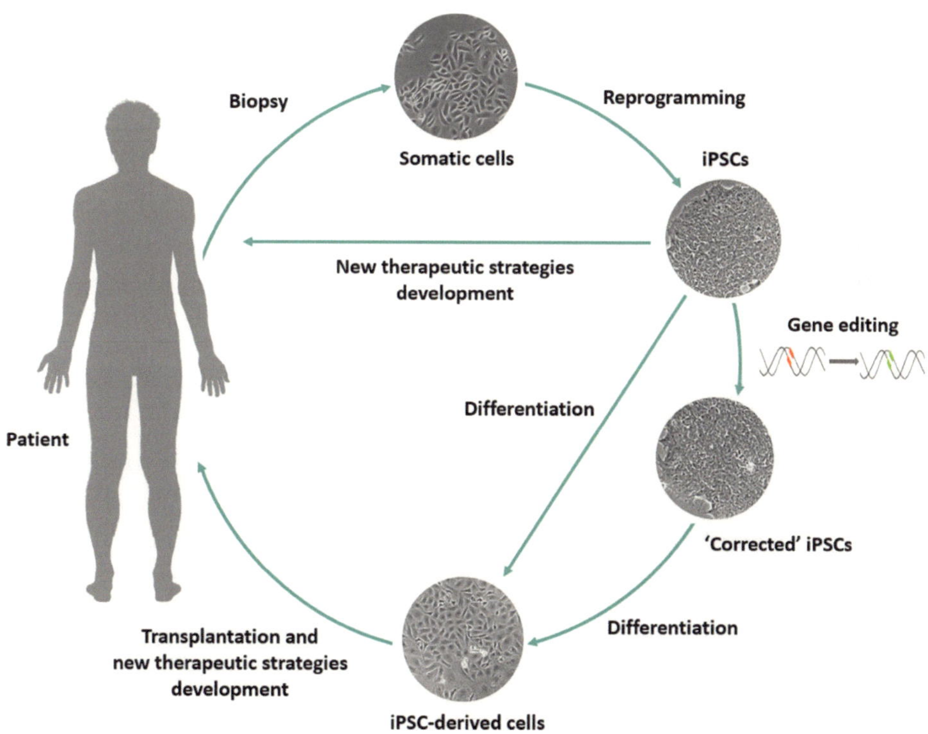

Figure 1.
Cell therapy strategies using iPSC-derived cells. Patient-specific primary cells of different origin can be isolated, cultured in vitro, reprogrammed to iPSCs, and if needed, genetically corrected. Genetically corrected or unmodified iPSC clones can be differentiated into desired cell lineages. These iPSC-derived cells can then be used for either autologous transplantation or the development of new therapeutic strategies, for example, by testing new drug modalities.

example, Anthony Oro's team at Stanford University has received an award from the California Institute for Regenerative Medicine (CIRM) to translate an iPSC-based gene correction therapy for the severe recessive dystrophic form of EB into the clinic by producing transplantable epidermal sheets from genetically corrected EB iPSCs [59]. Dr. Oro's team is currently generating data for an investigational new drug application (IND) with the Food and Drug Administration (FDA) to initiate a clinical trial.

3.2 Vascular therapy

Endothelial cells, pericytes, and vascular smooth muscle cells have been derived from animal and human iPSCs [60, 61]. iPSCs can also be directed into cord-blood endothelial colony-forming cells that can be used to derive highly proliferative blood vessel-forming cells applicable for the restoration of endothelial function in patients with vascular diseases [62].

Coronary artery disease continues to be the leading cause of death and morbidity around the world, with the existing therapy being not always efficient. Studies have shown that transplanted iPSCs can promote angiogenesis and effective tissue revascularization [63].

In diabetes, prolonged hyperglycemia causes aberrant angiogenesis in both micro- and macro-vessels, resulting in deficient functionality of endothelial progenitor cells and leading to decreased neovascularization. iPSC-derived endothelial cells have been widely explored as a model to study the mechanisms and novel treatments for endothelial dysfunction in type 1 and 2 diabetes and maturity-onset diabetes of the young (MODY) [64, 65].

iPSC-derived endothelial cells have also been used to study the mechanisms of macular degeneration and ischemic retinopathies [66, 67]. iPSC-derived spinal motor neurons and cerebral microvascular endothelial cells seeded into the spinal cord lead to vascular-neural interaction with specific maturation effects of endothelial cells on the neural tissue [68].

Vascular grafts have been successfully developed from iPSC-derived cells, recapitulating the cellular composition and orientation, as well as the anti-inflammatory properties, of functional blood vessels [69, 70].

Endothelial derived from iPSCs are yet to be evaluated in clinical trials. However, these cells are now successfully used for drug testing [71].

3.3 Cardiology

The recent advances in iPSC reprogramming into cardiomyocytes and other types of cardiac cells have provided potential avenues for cardiac repair, and functional cardiomyocytes have been successfully generated from iPSCs [72]. iPSC-derived cardiomyocyte-like cells demonstrate spontaneous contractility and exhibit molecular and structural similarities to cardiomyocytes.

Many studies have focused on testing iPSC-derived cells for post-myocardial infarction repair [73], which is one of the leading causes of morbidity and death throughout the world. In patients with extensive myocardial infarction, more than a billion cardiomyocytes can be lost, overwhelming the heart's repair capacity. Such massive cell death in the myocardium initiates the replacement of cardiomyocytes with fibrous tissue, resulting in heart failure. Beating iPSC-derived cardiomyocytes have been generated from patients with hypertrophic cardiomyopathy associated with diastolic dysfunction to study the cellular mechanisms and potential therapeutic targets of diastolic dysfunction [74].

Patient-specific iPSC-derived cardiomyocytes have also been generated from patients with different types of diabetes and used for modeling and studying the molecular mechanisms underlying diabetic cardiomyopathies [75]. Human iPSC-derived cardiomyocytes, endothelial cells, and cardiac fibroblasts have been generated and integrated in beating 3D cardiac microtissues as a platform for cardiovascular disease modeling [76].

Human iPSC lines have also been generated from patient-derived cells to study ventricular and atrial arrhythmias, which often lead to sudden cardiac death [77].

A few clinical trials to test the efficacy of iPSC-derived cardiomyocytes have been initiated. For example, an Osaka University spin-off company, Cuorips, Inc. has recently initiated a clinical trial to determine the efficacy and safety of a human allogeneic iPSC-derived cardiomyocyte sheet for ischemic cardiomyopathy patients (NCT04696328). Heartseed, Inc. is currently testing iPSC-derived cardiomyocyte spheroids in patients with severe heart failure in Phase I/II study (NCT04945018). Since both studies are still ongoing, no results have been reported.

3.4 Skeletal muscle regeneration

Converting iPSCs into skeletal muscle cells can offer a tool for *in vitro* modeling of muscular diseases and potential hope for patients afflicted with skeletal muscle dystrophic diseases. Myogenic progenitor cells have been derived from iPSCs in many studies [78, 79].

Skeletal muscle satellite cells, which drive skeletal muscle regeneration, have been shown to play an important role in the early regeneration of damaged skeletal muscles in muscular dystrophies and have been generated from human iPSCs for the identification of new therapeutic targets for the treatment of these disorders [80]. 3D functional skeletal muscle tissues have also been successfully generated from human iPSC-derived skeletal myotubes with sarcomeric structures as an *in vitro* model of contractile myofibrils for disease modeling and drug screening to study muscular and neuromuscular diseases [78].

Developing treatments for muscular dystrophy is a priority topic for researchers. Vita Therapeutics, Inc. has recently received an orphan drug designation from the FDA to initiate a clinical study that will test the efficacy of iPSC-derived myogenic stem cells to treat Duchenne muscular dystrophy [59]. More iPSC-based therapies for muscular dystrophies are currently in development [81].

3.5 Neurology

Several types of well-differentiated and functional populations of neural cells and neuronal multipotent progenitors have been generated from human and murine iPSCs. These progenitors have also been tested in cell replacement studies in rodent models with promising results [82]. Spinal neural progenitors have been differentiated from human iPSCs and together with human iPSC-derived brain microvascular endothelial cells were included in the dual-channel spinal cord chip system as an *in vitro* model of human vascularized motor neuron tissue [68]. Similarly, the blood–brain barrier chip system has been created for modeling neurological disorders and drug screening [83]. Cells in such systems can be generated from the same iPSC donor source, producing an isogenic *in vitro* model [84].

iPSCs are becoming an important source for the development of personalized therapeutic and preclinical strategies for research focusing on neurodegeneration. Parkinson's disease is one of the most common neurodegenerative disorders, resulting from the loss of dopamine neurons in substantia nigra. Cell replacement therapy, such as the transplantation of iPSC-derived neural progenitors, provides an alternative treatment strategy for Parkinson's disease (PD) [85]. An ongoing clinical trial in Japan is testing iPSC-derived dopaminergic neurons for PD [86]. In 2018, midbrain dopaminergic progenitor cells derived from autologous iPSCs were successfully transplanted into the brain of a patient with PD, and clinical symptoms improved in this patient at 18 to 24 months after implantation [85]. Aspen Neuroscience is currently developing two iPSC-based therapies for PD: an autologous iPSC-derived dopaminergic neuron therapy for idiopathic PD and an autologous gene-corrected iPSC-derived dopaminergic neuron therapy for genetic PD [87]. These and other therapies currently in development provide hope in the treatment of many neurological conditions.

3.6 Hearing loss

Hearing loss is a common impairment in humans that mainly results from the irreversible loss of sensory hair cells and auditory neurons. Patient-specific iPSCs are a promising tool for the regeneration of sensory hair cells and spiral ganglion neurons of the affected cochlea. iPSCs can be successfully reprogrammed into otic epithelial progenitors and otic neuroprogenitors that can subsequently be differentiated into inner ear hair cells [88, 89]. Functional cochlear supporting cells that can be important therapeutic targets for the treatment of hereditary deafness have also been successfully generated from mouse iPSCs [90]. Human iPSC-derived neural progenitors have been shown to innervate early postnatal cochlear hair cells *in vitro*, forming functional synapses [91].

Sensorineural hearing loss is a prevalent form of deafness, commonly arising from damage to the cochlear sensory hair cells and degeneration of the spiral ganglion neurons. iPSCs can serve as an autologous source of replacement neurons in an injured cochlea for the treatment of sensorineural hearing loss and as a model system to develop therapies to treat hereditary hearing loss [89]. While many iPSC-based treatment options are being developed in research settings, there are no approved clinical trials for hearing loss using iPSCs.

3.7 Ophthalmology

Transplantation of ocular cells derived from both autologous and allogeneic iPSCs in animal models and clinical trials showed great promise for cell-based therapies and disease modeling in ophthalmology. Using patient's own cells and the ability to correct disease-related gene mutations in patient-derived iPSCs provided a powerful approach for the treatment of ophthalmologic diseases. Various ocular cells have been generated from iPSCs, including corneal epithelial progenitor cells capable of terminal differentiation toward mature corneal epithelial-like cells [92], conjunctival epithelial cells, and conjunctival goblet-like cells [93], retinal pigment epithelial cells [92], photoreceptors, and retinal ganglion cells among others [94]. Human iPSCs can, in an autonomous manner, recapitulate the main steps of retinal development and form 3D retinal cups containing all major retinal cell types arranged in layers via retinal progenitors [95].

iPSCs hold a promise for the treatment of various degenerative eye disorders by filling clinical gaps in the use of adult limbal SCs or ESCs [96]. iPSC-derived photoreceptor cells and retinal pigment epithelium cells provide a cell replacement therapy for visual impairment associated with inherited retinal degeneration and age-related degeneration of photoreceptors [97].

Similar to the skin, the eye may represent an ideal tissue for testing iPSC therapies: it is relatively easy to monitor and access. Unsurprisingly, retinal pigment epithelium (RPE) cells derived from iPSCs were the first autologous iPSC-derived cell type to be transplanted into a human patient [85]. These RPE cells were used to treat age-related macular degeneration (AMD), the leading cause of vision loss in the elderly. A 4-year follow-up demonstrated that the iPSC-derived RPE sheets transplanted into the right eye of a 77-year-old patient had survived post-engraftment. While no improvement in vision was noted, the patient's vision remained stable, emphasizing the safety of the iPSC-based therapy to treat eye diseases [98]. Additional iPSC-based therapies are being assessed for the treatment of AMD in clinical trials in Japan [99] and in the United States by the team at the National Institute of Health (NCT04339764). The use of iPSC-derived photoreceptors to cure blindness is also being tested in preclinical research [59].

3.8 Bone and cartilage regeneration

While autologous bone grafting remains the main approach to reconstruct bone defects, the risk of bone resorption, infection, and insufficient amount of tissue available for transplantation is high. Therefore, iPSC technologies may provide a suitable alternative to grafting autologous bone/cartilage-forming cells. Functional osteoblasts, osteocytes, and chondrocytes have been generated from iPSCs [100, 101].

Human and animal iPSCs, as well as iPSC-derived mesenchymal stem/stromal cells (MSCs), have been differentiated into osteoblast- and osteocyte-like cells, which could be transplanted to achieve bone formation or regeneration of calvarial bone defects in *in vivo* animal models [102]. For bone tissue regeneration, engineered bioactive scaffolds provide mechanical support and components that mimic the extracellular matrix for iPSC-derived osteogenic cell grafting, increasing their adhesion, growth, and survival [103].

iPSCs also represent a potential source of viable chondroprogenitors for articular cartilage repair and engineering [104]. The regenerative potential of iPSCs is attractive for the therapy of intervertebral disc degeneration—a common cause of musculoskeletal diseases, such as low back pain, which is often attributed to a reduced number of nucleus pulposus cells that form the intervertebral disc. Current treatment strategies fail to replenish nucleus pulposus cells. The latter, however, have been successfully differentiated from iPSCs [105].

The development of iPSC-based therapies for bone and cartilage diseases is still ongoing. However, Australian Cynata Therapeutics has already tested allogeneic MSCs derived from iPSCs in a clinical trial for graft vs. host disease (GvHD) (NCT02923375). The infusion of these iPSC-derived MSCs was well tolerated by patients and promoted encouraging improvement in symptoms of GvHD [106]. Cynata therapeutics is also initiating a Phase III trial that will use iPSC-derived MSCs for the treatment of osteoarthritis [107].

3.9 Dentistry

Human iPSCs can potentially be a source to derive human odontoblasts for tissue engineering and regenerative therapy for the treatment of dental pulp damage [108]. The tooth is formed by sequential reciprocal interactions between epithelial cells originating from mesenchymal cells and surface ectoderm cells from cranial neural crest. iPSCs can be differentiated into neural crest-like cells, which in turn can be differentiated into odontogenic mesenchymal cells, odontoblast progenitors, and odontoblasts suitable for transplantation [109].

Periodontal disease is an important health problem that ultimately leads to tooth loss. An alternative to the existing artificially manufactured tooth replacements is the generation of complete or partial biological replacements, consisting of living periodontal tissues. Animal and human iPSCs and iPSC-derived MSCs transplanted to animals with a model of molar defects demonstrate periodontal tissue regeneration. Transplanted neural crest-like cells derived from iPSCs have also been shown to form a well-organized vascularized dentin-pulp complex and calcified tooth-like structures, demonstrating the feasibility for iPSCs use in dental tissue regeneration [108]. However, extensive research is still needed to advance the iPSC technology into dental applications.

3.10 Nephrology and urology

Renal failure is one of the most common causes of mortality and morbidity in the world. An iPSC-based cell therapy may offer an alternative therapeutic approach to kidney transplantation.

iPSCs have been successfully differentiated into nephrogenic intermediate mesoderm and renal progenitor cells [110, 111]. These cells pose the ability to differentiate into multiple cell types that constitute the adult kidney, such as metanephric mesenchyme cells, metanephric stromal cells, nephric duct, ureteric bud cells, proximal tubular cells, mature glomerular podocytes, and other types of cells capable of forming renal tubule-like structures [110]. iPSC-derived kidney organoids have been created containing multiple cell types and mimicking nephrogenesis that have a potential for regenerative medicine and personalized therapy [112]. iPSC-derived cells are also used to recellularize decellularized kidney scaffolds as an approach to bioengineering human replacement kidneys [113]. iPSCs generated from patients with specific kidney disorders have also been used in disease modeling [114]. The iPSC technology is now widely used to model kidney diseases and perform drug screening. However, no clinical trials have been approved to date.

3.11 Pulmonology

Lung transplantation remains the only treatment for many severe lung diseases. The use of iPSCs may be an effective strategy for developing patient-specific cells for lung cell therapy and lung tissue engineering as an alternative to whole-organ transplantation. iPSC-based models of lung diseases can also help to better understand lung pathologies and identify new therapeutic approaches.

Human alveolar and airway epithelial and basal cells have been successfully generated from iPSCs [115, 116]. Airway basal cells, in particular, can give rise to other airway lineages, such as secretory and ciliated cells, and can restore airway functionality [117].

Human iPSC-derived type II alveolar epithelial cells are capable of repopulating acellular lung matrices prepared from rat and human adult lungs, adhering and proliferating to form alveolar structures as a 3D lung tissue model of the distal lung regions [118]. Recently, scaffold-free structures for airway regeneration were also created using 3D bioprinting and a combination of human native chondrocytes, MSCs, and iPSC-derived endothelial cells [119].

Thus, significant progress has been achieved in deriving alveolar epithelial cells from iPSCs. However the complexity of lung tissue prevents rapid development and clinical translation of iPSC-based therapies for lung diseases.

3.12 Hepatology

Currently, liver transplants represent the only way to treat patients suffering from terminal liver failure. However, liver transplantation is associated with numerous problems, such as graft failure. As an alternative to the donor liver, human iPSCs may provide a promising source of hepatocytes for autologous cell therapy. iPSCs have been differentiated into hepatocytes [120]. Since the function of hepatocytes depends on their position in the liver globule, methods have been developed to generate iPSC-derived hepatocytes with zone-specific hepatic properties [121].

In addition to hepatocytes, the liver parenchyma also consists of many types of nonparenchymal cells, which are essential for maintaining hepatic metabolic activity and other functions. Human iPSC-derived hepatic progenitors have been differentiated into multiple liver cell types and produce functional liver models [122]. *In vitro* liver models are crucial for the study of liver diseases and development of effective therapies. Since liver transplantation is contingent on organ availability and other constraints, transplantable iPSC-derived cells and vascularized 3D organoids capable of repopulating and restoring liver functions have been developed as an alternative [123].

Hepatocyte transplantation is one of the most attractive approaches for the treatment of liver failure [124], and patient-specific iPSC-derived hepatic cells are expected to be used for patient-specific transplantations. The transplantation of hepatocyte-like cells differentiated from genetically corrected iPSCs has also been shown to ameliorate inherited liver diseases in a mouse model [125]. While iPSC-derived hepatocytes have not yet been translated into clinical trials, these cells are currently being used for screening drug hepatoxicity [126].

3.13 Gastroenterology

The gastrointestinal tract is one of the largest and most active systems, which not only breaks down and absorbs macromolecules from the lumen but also functions as an endocrine organ that regulates digestion and metabolism. The gastrointestinal system requires integrated neuronal, lymphatic, immune, and vascular tissues to function properly. The iPSC technology provides unique opportunities for modeling human diseases and novel therapeutic approaches in regenerative gastroenterology. Human iPSC-derived intestinal and gastric cells, as well as generated *in vitro* human organoids, may facilitate drug screening and modeling of gastrointestinal diseases. On the other hand, intestinal cell models can be widely used to study drug absorption and metabolism.

iPSCs can be differentiated into various types of intestinal cells and can even form multicell type intestinal tissue [127]. Among other cell types, human iPSCs have been differentiated into gastric epithelial cells and acid-secreting parietal cells [128], mature exocrine pancreatic cells [129], as well as directed along the gastric endocrine cell fate path [127]. Human iPSCs can also be efficiently differentiated into neural crest SCs and various subtypes of mature enteric neurons [130]. Enterocytes derived from human iPSCs have been used as a model system for predicting the pharmacokinetics of the human intestine and drug absorption and metabolism [131, 132].

The developed cell culture protocols allow for the derivation of self-organizing multicellular intestinal organoids from iPSCs, which resemble *in vivo* intestinal crypts [133].

Such organoids, which are composed of various intestinal cells, represent a physiologically relevant *in vitro* model for basic studies of intestinal development and pathophysiology, as well as a tool in personalized regenerative medicine and drug development. Bioengineering the intestine on a vascularized native scaffold can also be a promising approach for intestinal regeneration in patients with intestinal failure [134]. Intestinal 3D organoids derived from human primary digestive samples are currently being tested in a clinical trial to treat ulcerative colitis at Rennes University Hospitals (NCT05294107). The successful completion of this study will pave way for approval to use iPSC-derived intestinal organoids for the treatment of intestinal diseases in clinical trials.

3.14 Metabolic disorders

Several major types of diabetes are caused by the destruction and decrease in the number of functional insulin-producing β-cells. Therefore, the generation of functional insulin-secreting pancreatic β-cells represents an important goal for the treatment of various types of diabetes. Functional insulin-secreting pancreatic β-cells have been successfully generated from healthy human iPSCs [135], providing an important cell source for personalized drug screening and cell transplantation therapy in diabetes. Human iPSC-derived β-cells exhibit many of the properties of functional pancreatic β-cells, such as expression of specific transcription factors and the presence of mature endocrine secretory granules [136].

Brown adipocytes are promising cell targets for the treatment of obesity and type 2 diabetes due to their ability to actively drain and oxidize circulating glucose and triglycerides, which can prevent hyperglycemia and hypertriglyceridemia. Because of the scarcity of brown adipocytes in adults, iPSCs may be an important potential source of these cells and their progenitors for transplantation. Human iPSCs have been successfully differentiated into adipocytes [137].

MODY is a monogenic autosomal dominant disease caused by a mutation in one of the specific genes. Various mutations in each of these genes affect pancreatic β-cells, resulting in their dysfunction and diabetes development. However, despite extensive research, the mechanism by which the mutant MODY gene results in monogenic diabetes is not yet clear [138]. iPSCs have been generated from patients with different types of MODY to establish a human-based model for studying the molecular manifestations and mechanisms of these diseases [139].

iPSCs harboring disease-specific gene mutations have also been generated from somatic cells of patients with a few types of lipodystrophy and other inherited metabolic disorders for studying the pathogenesis of these diseases [140, 141].

While iPSC-based therapies for metabolic diseases are still being developed, the progress achieved by Melton's group in using human ESCs to derive functional allogeneic insulin-producing β-cells [142] has created a strong foundation for the use of iPSCs in treating type 1 diabetes. Vertex Pharmaceuticals, Inc. is currently testing the glucose-responsive allogeneic β-cells generated from human ESCs in combination with immunosuppressive therapy in Phase I clinical trial (NCT04786262). The preliminary results of this clinical trial show engraftment and functionality of implanted hESC-derived β-cells [143].

3.15 Gland regeneration

Hypofunction of salivary glands causes various life-disrupting effects. With no satisfactory therapy available, the therapeutic and regenerative potential of iPSCs has been explored for the treatment of salivary gland dysfunction [144].

Inflammatory and degenerative changes in the lacrimal gland often lead to the development of severe dry eye syndrome, a complex disease resulting in visual acuity disruption. Currently, only palliative treatments for this disease exist. iPSCs have been used for developing a therapy for lacrimal gland tissue injuries [145].

The thymus plays a significant role in the establishment of immunological self-tolerance and is required for the generation of T cell-mediated immunity. Thyroid progenitors, thymic epithelial cells, and thymic organoids derived from iPSCs can completely regenerate the thymus *in vivo* and demonstrate the potential for regenerative therapy in patients with immunodeficiency and hypothyroidism [146]. Despite the progress in generating iPSC-derived thymic cells, a poor understanding of thymus biology impedes the clinical translation of these cells.

The mammary gland is a primary target for carcinogenesis, and regenerative therapy for damaged mammary gland tissues is the best way to restore breast functions. iPSCs have been successfully reprogrammed into mammary SCs [147]. In addition, human mammary-like organoids have been produced from iPSCs. Such organoids have been shown to regenerate mammary glands upon transplantation [148]. To date, no clinical trials have been approved for the use of iPSCs in gland regeneration.

4. Conclusion

Despite tremendous progress achieved in the iPSC field, broad applications of iPSC-based therapies will take time to establish. Nevertheless, considerable advances have been made in deriving iPSCs from patients, differentiating them into tissues of interest, and using them as a platform for studying the mechanisms of diseases. The development of iPSC-based therapies is just emerging and only a limited number of clinical studies using the transplantation of iPSC-derived cells have been initiated to date. However, clinical studies related to the iPSC technology are not limited to the studies described above. Other studies worth mentioning include the successful use of autologous iPSC-derived platelets for the treatment of aplastic anemia [149] and the derivation of an off-the-shelf allogeneic chimeric antigen receptor (CAR) T-cell therapy targeting B-cell malignancies developed by Fate Therapeutics (NCT04629729). An iPSC-derived, off-the-shelf, CAR natural killer (NK) cell therapy is currently being tested in a Phase I clinic trial for refractory B-cell lymphoma (NCT 04245722) and is showing promising therapeutic efficacy [150]. Other studies have been reviewed by Kim et al. [99].

Even though iPSCs hold great potential in the field of regenerative medicine and personalized medicine, a number of challenges hinder the widespread clinical applications of these cells. These challenges include the safety of methodologies for the generation, gene correction, and differentiation of iPSCs and the high cost associated with the technology.

The first major challenge is reprogramming and gene correction methods that are known to introduce undesired genetic modifications into the patient genome. Other limitations include the heterogeneity of iPSC lines that impairs the consistency of differentiation during manufacturing of iPSC-based cell products. Establishing selection criteria for iPSC-derived cells, such as cell-specific markers, proliferation rate, lifespan, and genomic analyses, help minimize the variability of iPSC-derived cells. Just like ESCs, iPSCs are also predisposed to forming teratomas when undifferentiated. The available cell purification technologies often do not guarantee the complete eradication of undifferentiated iPSCs during manufacturing. Although many problems concerning the clinical applications of iPSCs still remain, iPSC-based therapies have a tremendous therapeutic potential for many diseases that are difficult to treat otherwise.

Acknowledgements

We are grateful for funding support from the National Institutes of Health (R21 AR074642 and U01AR075932). We also thank Epidermolysis Bullosa (EB) Research Partnership, the EB Medical Research Foundation, the Cure EB Charity, Dystrophic Epidermolysis Bullosa Research Association (DEBRA) International, and the Gates Frontiers Fund.

Conflict of interest

The authors declare no conflict of interest.

Author details

Igor Kizub, Andrii Rozhok and Ganna Bilousova*
Department of Dermatology, Gates Center for Regenerative Medicine, University of Colorado School of Medicine, Anschutz Medical Campus, Aurora, CO, USA

*Address all correspondence to: ganna.bilousova@cuanschutz.edu

References

[1] Takahashi K, Yamanaka S. Induction of pluripotent stem cells from mouse embryonic and adult fibroblast cultures by defined factors. Cell. 2006;**126**(4):663-676. DOI: 10.1016/j.cell.2006.07.024

[2] Okita K, Ichisaka T, Yamanaka S. Generation of germline-competent induced pluripotent stem cells. Nature. 2007;**448**(7151):313-317. DOI: 10.1038/nature05934

[3] Takahashi K, Tanabe K, Ohnuki M, Narita M, Ichisaka T, Tomoda K, et al. Induction of pluripotent stem cells from adult human fibroblasts by defined factors. Cell. 2007;**131**(5):861-872. DOI: 10.1016/j.cell.2007.11.019

[4] Wernig M, Meissner A, Foreman R, Brambrink T, Ku M, Hochedlinger K, et al. In vitro reprogramming of fibroblasts into a pluripotent ES-cell-like state. Nature. 2007;**448**(7151):318-324. DOI: 10.1038/nature05944

[5] Yu J, Vodyanik MA, Smuga-Otto K, Antosiewicz-Bourget J, Frane JL, Tian S, et al. Induced pluripotent stem cell lines derived from human somatic cells. Science. 2007;**318**(5858):1917-1920. DOI: 10.1126/science.1151526

[6] Gepstein L. Derivation and potential applications of human embryonic stem cells. Circulation Research. 2002;**91**(10):866-876. DOI: 10.1161/01.res.0000041435.95082.84

[7] Zomer HD, Vidane AS, Goncalves NN, Ambrosio CE. Mesenchymal and induced pluripotent stem cells: General insights and clinical perspectives. Stem Cells Cloning. 2015;**8**:125-134. DOI: 10.2147/SCCAA.S88036

[8] Kimbrel EA, Lanza R. Pluripotent stem cells: The last 10 years. Regenerative Medicine. 2016;**11**(8):831-847. DOI: 10.2217/rme-2016-0117

[9] Mitalipova M, Palmarini G. Isolation and characterization of human embryonic stem cells. Methods in Molecular Biology. 2006;**331**:55-76. DOI: 10.1385/1-59745-046-4:55

[10] Takahashi K. Cellular reprogramming--lowering gravity on Waddington's epigenetic landscape. Journal of Cell Science 2012;125(Pt 11):2553-2560. DOI: 10.1242/jcs.084822

[11] Briggs R, King TJ. Transplantation of living nuclei from blastula cells into enucleated Frogs' eggs. Proceedings of the National Academy of Sciences of the United States of America. 1952;**38**(5):455-463. DOI: 10.1073/pnas.38.5.455

[12] Gurdon JB, Elsdale TR, Fischberg M. Sexually mature individuals of Xenopus laevis from the transplantation of single somatic nuclei. Nature. 1958;**182**(4627):64-65. DOI: 10.1038/182064a0

[13] Wakayama T, Perry AC, Zuccotti M, Johnson KR, Yanagimachi R. Full-term development of mice from enucleated oocytes injected with cumulus cell nuclei. Nature. 1998;**394**(6691):369-374. DOI: 10.1038/28615

[14] Ambrosi DJ, Rasmussen TP. Reprogramming mediated by stem cell fusion. Journal of Cellular and Molecular Medicine. 2005;**9**(2):320-330. DOI: 10.1111/j.1582-4934.2005.tb00358.x

[15] Wilmut I, Schnieke AE, McWhir J, Kind AJ, Campbell KH. Viable offspring

derived from fetal and adult mammalian cells. Nature. 1997;**385**(6619):810-813. DOI: 10.1038/385810a0

[16] Niwa H, Miyazaki J, Smith AG. Quantitative expression of Oct-3/4 defines differentiation, dedifferentiation or self-renewal of ES cells. Nature Genetics. 2000;**24**(4):372-376. DOI: 10.1038/74199

[17] Ben-Shushan E, Thompson JR, Gudas LJ, Bergman Y. Rex-1, a gene encoding a transcription factor expressed in the early embryo, is regulated via Oct-3/4 and Oct-6 binding to an octamer site and a novel protein, Rox-1, binding to an adjacent site. Molecular and Cellular Biology. 1998;**18**(4):1866-1878. DOI: 10.1128/MCB.18.4.1866

[18] Johansson H, Simonsson S. Core transcription factors, Oct4, Sox2 and Nanog, individually form complexes with nucleophosmin (Npm1) to control embryonic stem (ES) cell fate determination. Aging (Albany NY). 2010;**2**(11):815-822. DOI: 10.18632/aging.100222

[19] Adhikary S, Eilers M. Transcriptional regulation and transformation by Myc proteins. Nature Reviews. Molecular Cell Biology. 2005;**6**(8):635-645. DOI: 10.1038/nrm1703

[20] Fernandez PC, Frank SR, Wang L, Schroeder M, Liu S, Greene J, et al. Genomic targets of the human c-Myc protein. Genes & Development. 2003;**17**(9):1115-1129. DOI: 10.1101/gad.1067003

[21] Rowland BD, Bernards R, Peeper DS. The KLF4 tumour suppressor is a transcriptional repressor of p53 that acts as a context-dependent oncogene. Nature Cell Biology. 2005;7(11):1074-1082. DOI: 10.1038/ncb1314

[22] Evans PM, Zhang W, Chen X, Yang J, Bhakat KK, Liu C. Kruppel-like factor 4 is acetylated by p300 and regulates gene transcription via modulation of histone acetylation. The Journal of Biological Chemistry. 2007;**282**(47):33994-34002. DOI: 10.1074/jbc.M701847200

[23] Nakagawa M, Koyanagi M, Tanabe K, Takahashi K, Ichisaka T, Aoi T, et al. Generation of induced pluripotent stem cells without Myc from mouse and human fibroblasts. Nature Biotechnology. 2008;**26**(1):101-106. DOI: 10.1038/nbt1374

[24] Kim JB, Sebastiano V, Wu G, Arauzo-Bravo MJ, Sasse P, Gentile L, et al. Oct4-induced pluripotency in adult neural stem cells. Cell. 2009;**136**(3):411-419. DOI: 10.1016/j.cell.2009.01.023

[25] Huangfu D, Osafune K, Maehr R, Guo W, Eijkelenboom A, Chen S, et al. Induction of pluripotent stem cells from primary human fibroblasts with only Oct4 and Sox2. Nature Biotechnology. 2008;**26**(11):1269-1275. DOI: 10.1038/nbt.1502

[26] Chou BK, Mali P, Huang X, Ye Z, Dowey SN, Resar LM, et al. Efficient human iPS cell derivation by a non-integrating plasmid from blood cells with unique epigenetic and gene expression signatures. Cell Research. 2011;**21**(3):518-529. DOI: 10.1038/cr.2011.12

[27] Utikal J, Polo JM, Stadtfeld M, Maherali N, Kulalert W, Walsh RM, et al. Immortalization eliminates a roadblock during cellular reprogramming into iPS cells. Nature. 2009;**460**(7259):1145-1148. DOI: 10.1038/nature08285

[28] Yang DH, Moss EG. Temporally regulated expression of Lin-28 in diverse tissues of the developing mouse. Gene Expression Patterns. 2003;**3**(6):719-726. DOI: 10.1016/s1567-133x(03)00140-6

[29] Meir YJ, Li G. Somatic Reprogramming-Above and Beyond Pluripotency. Cells. 2021;**10**:11. DOI: 10.3390/cells10112888

[30] Okita K, Nakagawa M, Hyenjong H, Ichisaka T, Yamanaka S. Generation of mouse induced pluripotent stem cells without viral vectors. Science. 2008;**322**(5903):949-953. DOI: 10.1126/science.1164270

[31] Stadtfeld M, Nagaya M, Utikal J, Weir G, Hochedlinger K. Induced pluripotent stem cells generated without viral integration. Science. 2008;**322**(5903):945-949. DOI: 10.1126/science.1162494

[32] Fusaki N, Ban H, Nishiyama A, Saeki K, Hasegawa M. Efficient induction of transgene-free human pluripotent stem cells using a vector based on Sendai virus, an RNA virus that does not integrate into the host genome. Proceedings of the Japan Academy. Series B, Physical and Biological Sciences. 2009;**85**(8):348-362. DOI: 10.2183/pjab.85.348

[33] Narsinh KH, Jia F, Robbins RC, Kay MA, Longaker MT, Wu JC. Generation of adult human induced pluripotent stem cells using nonviral minicircle DNA vectors. Nature Protocols. 2011;**6**(1):78-88. DOI: 10.1038/nprot.2010.173

[34] Wang W, Lin C, Lu D, Ning Z, Cox T, Melvin D, et al. Chromosomal transposition of PiggyBac in mouse embryonic stem cells. Proceedings of the National Academy of Sciences of the United States of America. 2008;**105**(27):9290-9295. DOI: 10.1073/pnas.0801017105

[35] Rao MS, Malik N. Assessing iPSC reprogramming methods for their suitability in translational medicine. Journal of Cellular Biochemistry. 2012;**113**(10):3061-3068. DOI: 10.1002/jcb.24183

[36] Guan J, Wang G, Wang J, Zhang Z, Fu Y, Cheng L, et al. Chemical reprogramming of human somatic cells to pluripotent stem cells. Nature. 2022;**605**(7909):325-331. DOI: 10.1038/s41586-022-04593-5

[37] Kim D, Kim CH, Moon JI, Chung YG, Chang MY, Han BS, et al. Generation of human induced pluripotent stem cells by direct delivery of reprogramming proteins. Cell Stem Cell. 2009;**4**(6):472-476. DOI: 10.1016/j.stem.2009.05.005

[38] Warren L, Manos PD, Ahfeldt T, Loh YH, Li H, Lau F, et al. Highly efficient reprogramming to pluripotency and directed differentiation of human cells with synthetic modified mRNA. Cell Stem Cell. 2010;**7**(5):618-630. DOI: 10.1016/j.stem.2010.08.012

[39] Judson RL, Babiarz JE, Venere M, Blelloch R. Embryonic stem cell-specific microRNAs promote induced pluripotency. Nature Biotechnology. 2009;**27**(5):459-461. DOI: 10.1038/nbt.1535

[40] Kogut I, McCarthy SM, Pavlova M, Astling DP, Chen X, Jakimenko A, et al. High-efficiency RNA-based reprogramming of human primary fibroblasts. Nature Communications. 2018;**9**(1):745. DOI: 10.1038/s41467-018-03190-3

[41] Raab S, Klingenstein M, Liebau S, Linta L. A comparative view on human somatic cell sources for iPSC generation. Stem Cells International. 2014;**2014**:768391. DOI: 10.1155/2014/768391

[42] Polo JM, Anderssen E, Walsh RM, Schwarz BA, Nefzger CM, Lim SM, et al.

A molecular roadmap of reprogramming somatic cells into iPS cells. Cell. 2012;**151**(7):1617-1632. DOI: 10.1016/j.cell.2012.11.039

[43] Boyer LA, Lee TI, Cole MF, Johnstone SE, Levine SS, Zucker JP, et al. Core transcriptional regulatory circuitry in human embryonic stem cells. Cell. 2005;**122**(6):947-956. DOI: 10.1016/j.cell.2005.08.020

[44] Papapetrou EP, Tomishima MJ, Chambers SM, Mica Y, Reed E, Menon J, et al. Stoichiometric and temporal requirements of Oct4, Sox2, Klf4, and c-Myc expression for efficient human iPSC induction and differentiation. Proceedings of the National Academy of Sciences of the United States of America. 2009;**106**(31):12759-12764. DOI: 10.1073/pnas.0904825106

[45] Chan EM, Ratanasirintrawoot S, Park IH, Manos PD, Loh YH, Huo H, et al. Live cell imaging distinguishes bona fide human iPS cells from partially reprogrammed cells. Nature Biotechnology. 2009;**27**(11):1033-1037. DOI: 10.1038/nbt.1580

[46] Aasen T, Raya A, Barrero MJ, Garreta E, Consiglio A, Gonzalez F, et al. Efficient and rapid generation of induced pluripotent stem cells from human keratinocytes. Nature Biotechnology. 2008;**26**(11):1276-1284. DOI: 10.1038/nbt.1503

[47] Doi A, Park IH, Wen B, Murakami P, Aryee MJ, Irizarry R, et al. Differential methylation of tissue- and cancer-specific CpG island shores distinguishes human induced pluripotent stem cells, embryonic stem cells and fibroblasts. Nature Genetics. 2009;**41**(12):1350-1353. DOI: 10.1038/ng.471

[48] Kim K, Zhao R, Doi A, Ng K, Unternaehrer J, Cahan P, et al. Donor cell type can influence the epigenome and differentiation potential of human induced pluripotent stem cells. Nature Biotechnology. 2011;**29**(12):1117-1119. DOI: 10.1038/nbt.2052

[49] Bilousova G, Chen J, Roop DR. Differentiation of mouse induced pluripotent stem cells into a multipotent keratinocyte lineage. The Journal of Investigative Dermatology. 2011;**131**(4):857-864. DOI: 10.1038/jid.2010.364

[50] Ohta S, Imaizumi Y, Okada Y, Akamatsu W, Kuwahara R, Ohyama M, et al. Generation of human melanocytes from induced pluripotent stem cells. PLoS One. 2011;**6**(1):e16182. DOI: 10.1371/journal.pone.0016182

[51] Hewitt KJ, Shamis Y, Hayman RB, Margvelashvili M, Dong S, Carlson MW, et al. Epigenetic and phenotypic profile of fibroblasts derived from induced pluripotent stem cells. PLoS One. 2011;**6**(2):e17128. DOI: 10.1371/journal.pone.0017128

[52] Veraitch O, Kobayashi T, Imaizumi Y, Akamatsu W, Sasaki T, Yamanaka S, et al. Human induced pluripotent stem cell-derived ectodermal precursor cells contribute to hair follicle morphogenesis in vivo. The Journal of Investigative Dermatology. 2013;**133**(6):1479-1488. DOI: 10.1038/jid.2013.7

[53] Ebner-Peking P, Krisch L, Wolf M, Hochmann S, Hoog A, Vari B, et al. Self-assembly of differentiated progenitor cells facilitates spheroid human skin organoid formation and planar skin regeneration. Theranostics. 2021;**11**(17):8430-8447. DOI: 10.7150/thno.59661

[54] Sebastiano V, Zhen HH, Haddad B, Bashkirova E, Melo SP, Wang P, et al. Human COL7A1-corrected induced

pluripotent stem cells for the treatment of recessive dystrophic epidermolysis bullosa. Science Translational Medicine. 2014;**6**(264):264ra163. DOI: 10.1126/scitranslmed.3009540

[55] Toyoshima KE, Ogawa M, Tsuji T. Regeneration of a bioengineered 3D integumentary organ system from iPS cells. Nature Protocols. 2019;**14**(5):1323-1338. DOI: 10.1038/s41596-019-0124-z

[56] Lee J, Rabbani CC, Gao H, Steinhart MR, Woodruff BM, Pflum ZE, et al. Hair-bearing human skin generated entirely from pluripotent stem cells. Nature. 2020;**582**(7812):399-404. DOI: 10.1038/s41586-020-2352-3

[57] Fine JD. Inherited epidermolysis bullosa. Orphanet Journal of Rare Diseases. 2010;**5**:12. DOI: 10.1186/1750-1172-5-12

[58] Lapasset L, Milhavet O, Prieur A, Besnard E, Babled A, Ait-Hamou N, et al. Rejuvenating senescent and centenarian human cells by reprogramming through the pluripotent state. Genes & Development. 2011;**25**(21):2248-2253. DOI: 10.1101/gad.173922.111

[59] Madrid M, Sumen C, Aivio S, Saklayen N. Autologous induced pluripotent stem cell-based cell therapies: Promise, progress, and challenges. Current Protocols. 2021;**1**(3):e88. DOI: 10.1002/cpz1.88

[60] Bajpai VK, Mistriotis P, Loh YH, Daley GQ, Andreadis ST. Functional vascular smooth muscle cells derived from human induced pluripotent stem cells via mesenchymal stem cell intermediates. Cardiovascular Research. 2012;**96**(3):391-400. DOI: 10.1093/cvr/cvs253

[61] Orlova VV, Drabsch Y, Freund C, Petrus-Reurer S, van den Hil FE, Muenthaisong S, et al. Functionality of endothelial cells and pericytes from human pluripotent stem cells demonstrated in cultured vascular plexus and zebrafish xenografts. Arteriosclerosis, Thrombosis, and Vascular Biology. 2014;**34**(1):177-186. DOI: 10.1161/ATVBAHA.113.302598

[62] Prasain N, Lee MR, Vemula S, Meador JL, Yoshimoto M, Ferkowicz MJ, et al. Differentiation of human pluripotent stem cells to cells similar to cord-blood endothelial colony-forming cells. Nature Biotechnology. 2014;**32**(11):1151-1157. DOI: 10.1038/nbt.3048

[63] Clayton ZE, Tan RP, Miravet MM, Lennartsson K, Cooke JP, Bursill CA, et al. Induced pluripotent stem cell-derived endothelial cells promote angiogenesis and accelerate wound closure in a murine excisional wound healing model. Bioscience Reports. 2018;**38**(4). DOI: 10.1042/BSR20180563

[64] Kachamakova-Trojanowska N, Stepniewski J, Dulak J. Human iPSCs-derived endothelial cells with mutation in HNF1A as a model of maturity-onset diabetes of the young. Cell. 2019;**8**(11):1440. DOI: 10.3390/cells8111440

[65] Samuel R, Daheron L, Liao S, Vardam T, Kamoun WS, Batista A, et al. Generation of functionally competent and durable engineered blood vessels from human induced pluripotent stem cells. Proceedings of the National Academy of Sciences of the United States of America. 2013;**110**(31):12774-12779. DOI: 10.1073/pnas.1310675110

[66] Songstad AE, Wiley LA, Duong K, Kaalberg E, Flamme-Wiese MJ, Cranston CM, et al. Generating iPSC-derived choroidal endothelial cells to study age-related macular degeneration.

Investigative Ophthalmology & Visual Science. 2015;**56**(13):8258-8267. DOI: 10.1167/iovs.15-17073

[67] Cho H, Macklin BL, Lin YY, Zhou L, Lai MJ, Lee G, et al. iPSC-derived endothelial cell response to hypoxia via SDF1a/CXCR4 axis facilitates incorporation to revascularize ischemic retina. JCI Insight. 2020;**5**(6):e131828. DOI: 10.1172/jci.insight.131828

[68] Sances S, Ho R, Vatine G, West D, Laperle A, Meyer A, et al. Human iPSC-derived endothelial cells and microengineered organ-Chip enhance neuronal development. Stem Cell Reports. 2018;**10**(4):1222-1236. DOI: 10.1016/j.stemcr.2018.02.012

[69] Nakayama KH, Joshi PA, Lai ES, Gujar P, Joubert LM, Chen B, et al. Bilayered vascular graft derived from human induced pluripotent stem cells with biomimetic structure and function. Regenerative Medicine. 2015;**10**(6):745-755. DOI: 10.2217/rme.15.45

[70] Generali M, Casanova EA, Kehl D, Wanner D, Hoerstrup SP, Cinelli P, et al. Autologous endothelialized small-caliber vascular grafts engineered from blood-derived induced pluripotent stem cells. Acta Biomaterialia. 2019;**97**:333-343. DOI: 10.1016/j.actbio.2019.07.032

[71] Sayed N, Liu C, Ameen M, Himmati F, Zhang JZ, Khanamiri S, et al. Clinical trial in a dish using iPSCs shows lovastatin improves endothelial dysfunction and cellular cross-talk in LMNA cardiomyopathy. Science Translational Medicine. 2020;**12**(554):eaax9276. DOI: 10.1126/scitranslmed.aax9276

[72] Dubois NC, Craft AM, Sharma P, Elliott DA, Stanley EG, Elefanty AG, et al. SIRPA is a specific cell-surface marker for isolating cardiomyocytes derived from human pluripotent stem cells. Nature Biotechnology. 2011;**29**(11):1011-1018. DOI: 10.1038/nbt.2005

[73] Ward MC, Gilad Y. A generally conserved response to hypoxia in iPSC-derived cardiomyocytes from humans and chimpanzees. eLife. 2019;**8**:8. DOI: 10.7554/eLife.42374

[74] Wu H, Yang H, Rhee JW, Zhang JZ, Lam CK, Sallam K, et al. Modelling diastolic dysfunction in induced pluripotent stem cell-derived cardiomyocytes from hypertrophic cardiomyopathy patients. European Heart Journal. 2019;**40**(45):3685-3695. DOI: 10.1093/eurheartj/ehz326

[75] Drawnel FM, Boccardo S, Prummer M, Delobel F, Graff A, Weber M, et al. Disease modeling and phenotypic drug screening for diabetic cardiomyopathy using human induced pluripotent stem cells. Cell Reports. 2014;**9**(3):810-821. DOI: 10.1016/j.celrep.2014.09.055

[76] Giacomelli E, Bellin M, Sala L, van Meer BJ, Tertoolen LG, Orlova VV, et al. Three-dimensional cardiac microtissues composed of cardiomyocytes and endothelial cells co-differentiated from human pluripotent stem cells. Development. 2017;**144**(6):1008-1017. DOI: 10.1242/dev.143438

[77] Pecha S, Yorgan K, Rohl M, Geertz B, Hansen A, Weinberger F, et al. Human iPS cell-derived engineered heart tissue does not affect ventricular arrhythmias in a Guinea pig cryo-injury model. Scientific Reports. 2019;**9**(1):9831. DOI: 10.1038/s41598-019-46409-z

[78] Jiwlawat S, Lynch E, Glaser J, Smit-Oistad I, Jeffrey J, Van Dyke JM, et al. Differentiation and sarcomere formation in skeletal myocytes

directly prepared from human induced pluripotent stem cells using a sphere-based culture. Differentiation. 2017;**96**:70-81. DOI: 10.1016/j.diff.2017.07.004

[79] Quattrocelli M, Swinnen M, Giacomazzi G, Camps J, Barthelemy I, Ceccarelli G, et al. Mesodermal iPSC-derived progenitor cells functionally regenerate cardiac and skeletal muscle. The Journal of Clinical Investigation. 2015;**125**(12):4463-4482. DOI: 10.1172/JCI82735

[80] Song KY, Guo XM, Wang HQ, Zhang L, Huang SY, Huo YC, et al. MBNL1 reverses the proliferation defect of skeletal muscle satellite cells in myotonic dystrophy type 1 by inhibiting autophagy via the mTOR pathway. Cell Death & Disease. 2020;**11**(7):545. DOI: 10.1038/s41419-020-02756-8

[81] Sun C, Serra C, Lee G, Wagner KR. Stem cell-based therapies for Duchenne muscular dystrophy. Experimental Neurology. 2020;**323**:113086. DOI: 10.1016/j.expneurol.2019.113086

[82] Wernig M, Zhao JP, Pruszak J, Hedlund E, Fu D, Soldner F, et al. Neurons derived from reprogrammed fibroblasts functionally integrate into the fetal brain and improve symptoms of rats with Parkinson's disease. Proceedings of the National Academy of Sciences of the United States of America. 2008;**105**(15):5856-5861. DOI: 10.1073/pnas.0801677105

[83] Lippmann ES, Al-Ahmad A, Azarin SM, Palecek SP, Shusta EV. A retinoic acid-enhanced, multicellular human blood-brain barrier model derived from stem cell sources. Scientific Reports. 2014;**4**:4160. DOI: 10.1038/srep04160

[84] Canfield SG, Stebbins MJ, Faubion MG, Gastfriend BD, Palecek SP, Shusta EV. An isogenic neurovascular unit model comprised of human induced pluripotent stem cell-derived brain microvascular endothelial cells, pericytes, astrocytes, and neurons. Fluids Barriers CNS. 2019;**16**(1):25. DOI: 10.1186/s12987-019-0145-6

[85] Schweitzer JS, Song B, Herrington TM, Park TY, Lee N, Ko S, et al. Personalized iPSC-derived dopamine progenitor cells for Parkinson's disease. The New England Journal of Medicine. 2020;**382**(20):1926-1932. DOI: 10.1056/NEJMoa1915872

[86] Takahashi J. iPS cell-based therapy for Parkinson's disease: A Kyoto trial. Regenerative Therapy. 2020;**13**:18-22. DOI: 10.1016/j.reth.2020.06.002

[87] Loring JF. Autologous induced pluripotent stem cell-derived neurons to treat Parkinson's disease. Stem Cells and Development. 2018;**27**(14):958-959. DOI: 10.1089/scd.2018.0107

[88] Lahlou H, Lopez-Juarez A, Fontbonne A, Nivet E, Zine A. Modeling human early otic sensory cell development with induced pluripotent stem cells. PLoS One. 2018;**13**(6):e0198954. DOI: 10.1371/journal.pone.0198954

[89] Takeda H, Hosoya M, Fujioka M, Saegusa C, Saeki T, Miwa T, et al. Engraftment of human pluripotent stem cell-derived progenitors in the inner ear of prenatal mice. Scientific Reports. 2018;**8**(1):1941. DOI: 10.1038/s41598-018-20277-5

[90] Fukunaga I, Fujimoto A, Hatakeyama K, Kurebayashi N, Ikeda K, Kamiya K. Generation of functional CX26-gap-junction-plaque-forming cells with spontaneous Ca(2+) transients via a gap junction characteristic of developing cochlea. Current Protocols in Stem Cell

Biology. 2019;**51**(1):e100. DOI: 10.1002/cpsc.100

[91] Gunewardene N, Crombie D, Dottori M, Nayagam BA. Innervation of Cochlear hair cells by human induced pluripotent stem cell-derived neurons In vitro. Stem Cells International. 2016;**2016**:1781202. DOI: 10.1155/2016/1781202

[92] Hayashi R, Ishikawa Y, Sasamoto Y, Katori R, Nomura N, Ichikawa T, et al. Co-ordinated ocular development from human iPS cells and recovery of corneal function. Nature. 2016;**531**(7594):376-380. DOI: 10.1038/nature17000

[93] Shibata S, Hayashi R, Kudo Y, Okubo T, Imaizumi T, Katayama T, et al. Cell-type-specific adhesiveness and proliferation propensity on Laminin isoforms enable purification of iPSC-derived corneal epithelium. Stem Cell Reports. 2020;**14**(4):663-676. DOI: 10.1016/j.stemcr.2020.02.008

[94] Parameswaran S, Balasubramanian S, Babai N, Qiu F, Eudy JD, Thoreson WB, et al. Induced pluripotent stem cells generate both retinal ganglion cells and photoreceptors: Therapeutic implications in degenerative changes in glaucoma and age-related macular degeneration. Stem Cells. 2010;**28**(4):695-703. DOI: 10.1002/stem.320

[95] Zhu J, Reynolds J, Garcia T, Cifuentes H, Chew S, Zeng X, et al. Generation of transplantable retinal photoreceptors from a current good manufacturing practice-manufactured human induced pluripotent stem cell line. Stem Cells Translational Medicine. 2018;**7**(2):210-219. DOI: 10.1002/sctm.17-0205

[96] Uyama H, Mandai M, Takahashi M. Stem-cell-based therapies for retinal degenerative diseases: Current challenges in the establishment of new treatment strategies. Development, Growth & Differentiation. 2021;**63**(1):59-71. DOI: 10.1111/dgd.12704

[97] Yang JM, Chung S, Yun K, Kim B, So S, Kang S, et al. Long-term effects of human induced pluripotent stem cell-derived retinal cell transplantation in Pde6b knockout rats. Experimental & Molecular Medicine. 2021;**53**(4):631-642. DOI: 10.1038/s12276-021-00588-w

[98] Takagi S, Mandai M, Gocho K, Hirami Y, Yamamoto M, Fujihara M, et al. Evaluation of transplanted autologous induced pluripotent stem cell-derived retinal pigment epithelium in exudative age-related macular degeneration. Ophthalmol Retina. 2019;**3**(10):850-859. DOI: 10.1016/j.oret.2019.04.021

[99] Kim JY, Nam Y, Rim YA, Ju JH. Review of the current trends in clinical trials involving induced pluripotent stem cells. Stem Cell Reviews and Reports. 2022;**18**(1):142-154. DOI: 10.1007/s12015-021-10262-3

[100] Bilousova G, Jun du H, King KB, De Langhe S, Chick WS, Torchia EC, et al. Osteoblasts derived from induced pluripotent stem cells form calcified structures in scaffolds both in vitro and in vivo. Stem Cells. 2011;**29**(2):206-216. DOI: 10.1002/stem.566

[101] Hosseini FS, Soleimanifar F, Khojasteh A, Ardeshirylajimi A. Promoting osteogenic differentiation of human-induced pluripotent stem cells by releasing Wnt/beta-catenin signaling activator from the nanofibers. Journal of Cellular Biochemistry. 2019;**120**(4):6339-6346. DOI: 10.1002/jcb.27921

[102] Ye JH, Xu YJ, Gao J, Yan SG, Zhao J, Tu Q, et al. Critical-size calvarial bone defects healing in a mouse model with silk scaffolds and SATB2-modified iPSCs.

Biomaterials. 2011;**32**(22):5065-5076. DOI: 10.1016/j.biomaterials.2011.03.053

[103] Ardeshirylajimi A, Dinarvand P, Seyedjafari E, Langroudi L, Adegani FJ, Soleimani M. Enhanced reconstruction of rat calvarial defects achieved by plasma-treated electrospun scaffolds and induced pluripotent stem cells. Cell and Tissue Research. 2013;**354**(3):849-860. DOI: 10.1007/s00441-013-1693-8

[104] Dicks A, Wu CL, Steward N, Adkar SS, Gersbach CA, Guilak F. Prospective isolation of chondroprogenitors from human iPSCs based on cell surface markers identified using a CRISPR-Cas9-generated reporter. Stem Cell Research & Therapy. 2020;**11**(1):66. DOI: 10.1186/s13287-020-01597-8

[105] Tang R, Jing L, Willard VP, Wu CL, Guilak F, Chen J, et al. Differentiation of human induced pluripotent stem cells into nucleus pulposus-like cells. Stem Cell Research & Therapy. 2018;**9**(1):61. DOI: 10.1186/s13287-018-0797-1

[106] Bloor AJC, Patel A, Griffin JE, Gilleece MH, Radia R, Yeung DT, et al. Production, safety and efficacy of iPSC-derived mesenchymal stromal cells in acute steroid-resistant graft versus host disease: A phase I, multicenter, open-label, dose-escalation study. Nature Medicine. 2020;**26**(11):1720-1725. DOI: 10.1038/s41591-020-1050-x

[107] Zhang J, Chen M, Liao J, Chang C, Liu Y, Padhiar AA, et al. Induced pluripotent stem cell-derived mesenchymal stem cells hold lower heterogeneity and great promise in biological research and clinical applications. Frontiers in Cell and Development Biology. 2021;**9**:716907. DOI: 10.3389/fcell.2021.716907

[108] Malhotra N. Induced pluripotent stem (iPS) cells in dentistry: A review. International Journal of Stem Cells. 2016;**9**(2):176-185. DOI: 10.15283/ijsc16029

[109] Seki D, Takeshita N, Oyanagi T, Sasaki S, Takano I, Hasegawa M, et al. Differentiation of odontoblast-like cells from mouse induced pluripotent stem cells by Pax9 and Bmp4 transfection. Stem Cells Translational Medicine. 2015;**4**(9):993-997. DOI: 10.5966/sctm.2014-0292

[110] Araoka T, Mae S, Kurose Y, Uesugi M, Ohta A, Yamanaka S, et al. Efficient and rapid induction of human iPSCs/ESCs into nephrogenic intermediate mesoderm using small molecule-based differentiation methods. PLoS One. 2014;**9**(1):e84881. DOI: 10.1371/journal.pone.0084881

[111] Imberti B, Tomasoni S, Ciampi O, Pezzotta A, Derosas M, Xinaris C, et al. Renal progenitors derived from human iPSCs engraft and restore function in a mouse model of acute kidney injury. Scientific Reports. 2015;**5**:8826. DOI: 10.1038/srep08826

[112] Takasato M, Er PX, Chiu HS, Maier B, Baillie GJ, Ferguson C, et al. Kidney organoids from human iPS cells contain multiple lineages and model human nephrogenesis. Nature. 2015;**526**(7574):564-568. DOI: 10.1038/nature15695

[113] Figliuzzi M, Bonandrini B, Remuzzi A. Decellularized kidney matrix as functional material for whole organ tissue engineering. Journal of Applied Biomaterial and Functional Material. 2017;**15**(4):e326-ee33. DOI: 10.5301/jabfm.5000393

[114] Shimizu T, Mae SI, Araoka T, Okita K, Hotta A, Yamagata K, et al. A novel ADPKD model using kidney organoids derived from disease-specific

human iPSCs. Biochemical and Biophysical Research Communications. 2020;**529**(4):1186-1194. DOI: 10.1016/j.bbrc.2020.06.141

[115] Abo KM, Ma L, Matte T, Huang J, Alysandratos KD, Werder RB, et al. Human iPSC-derived alveolar and airway epithelial cells can be cultured at air-liquid interface and express SARS-CoV-2 host factors. bioRxiv. 2020. DOI: 10.1101/2020.06.03.132639

[116] Hawkins FJ, Suzuki S, Beermann ML, Barilla C, Wang R, Villacorta-Martin C, et al. Derivation of airway basal stem cells from human pluripotent stem cells. Cell Stem Cell. 2021;**28**(1):79-95 e8. DOI: 10.1016/j.stem.2020.09.017

[117] Rock JR, Onaitis MW, Rawlins EL, Lu Y, Clark CP, Xue Y, et al. Basal cells as stem cells of the mouse trachea and human airway epithelium. Proceedings of the National Academy of Sciences of the United States of America. 2009;**106**(31):12771-12775. DOI: 10.1073/pnas.0906850106

[118] Ghaedi M, Calle EA, Mendez JJ, Gard AL, Balestrini J, Booth A, et al. Human iPS cell-derived alveolar epithelium repopulates lung extracellular matrix. The Journal of Clinical Investigation. 2013;**123**(11):4950-4962. DOI: 10.1172/JCI68793

[119] Taniguchi D, Matsumoto K, Machino R, Takeoka Y, Elgalad A, Taura Y, et al. Human lung microvascular endothelial cells as potential alternatives to human umbilical vein endothelial cells in bio-3D-printed trachea-like structures. Tissue & Cell. 2020;**63**:101321. DOI: 10.1016/j.tice.2019.101321

[120] Si-Tayeb K, Noto FK, Nagaoka M, Li J, Battle MA, Duris C, et al. Highly efficient generation of human hepatocyte-like cells from induced pluripotent stem cells. Hepatology. 2010;**51**(1):297-305. DOI: 10.1002/hep.23354

[121] Mitani S, Takayama K, Nagamoto Y, Imagawa K, Sakurai F, Tachibana M, et al. Human ESC/iPSC-derived hepatocyte-like cells achieve zone-specific hepatic properties by modulation of WNT signaling. Molecular Therapy. 2017;**25**(6):1420-1433. DOI: 10.1016/j.ymthe.2017.04.006

[122] Koui Y, Kido T, Ito T, Oyama H, Chen SW, Katou Y, et al. An In vitro human liver model by iPSC-derived parenchymal and non-parenchymal cells. Stem Cell Reports. 2017;**9**(2):490-498. DOI: 10.1016/j.stemcr.2017.06.010

[123] Takebe T, Sekine K, Enomura M, Koike H, Kimura M, Ogaeri T, et al. Vascularized and functional human liver from an iPSC-derived organ bud transplant. Nature. 2013;**499**(7459):481-484. DOI: 10.1038/nature12271

[124] Nagamoto Y, Takayama K, Ohashi K, Okamoto R, Sakurai F, Tachibana M, et al. Transplantation of a human iPSC-derived hepatocyte sheet increases survival in mice with acute liver failure. Journal of Hepatology. 2016;**64**(5):1068-1075. DOI: 10.1016/j.jhep.2016.01.004

[125] Sin YY, Ballantyne LL, Richmond CR, Funk CD. Transplantation of gene-edited hepatocyte-like cells modestly improves survival of Arginase-1-deficient mice. Molecular Therapy and Nucleic Acids. 2018;**10**:122-130. DOI: 10.1016/j.omtn.2017.11.012

[126] Jin M, Yi X, Liao W, Chen Q, Yang W, Li Y, et al. Advancements in stem cell-derived hepatocyte-like cell models for hepatotoxicity testing. Stem Cell Research & Therapy. 2021;**12**(1):84. DOI: 10.1186/s13287-021-02152-9

[127] Spence JR, Mayhew CN, Rankin SA, Kuhar MF, Vallance JE, Tolle K, et al. Directed differentiation of human pluripotent stem cells into intestinal tissue in vitro. Nature. 2011;**470**(7332):105-109. DOI: 10.1038/nature09691

[128] DeLaForest A, Quryshi AF, Frolkis TS, Franklin OD, Battle MA. GATA4 is required for budding morphogenesis of posterior foregut endoderm in a model of human stomach development. Frontiers in Medicine (Lausanne). 2020;7:44. DOI: 10.3389/fmed.2020.00044

[129] Ito K, Matsuura K, Mihara Y, Sakamoto Y, Hasegawa K, Kokudo N, et al. Delivery of pancreatic digestive enzymes into the gastrointestinal tract by pancreatic exocrine tissue transplant. Scientific Reports. 2019;**9**(1):5922. DOI: 10.1038/s41598-019-42362-z

[130] Workman MJ, Mahe MM, Trisno S, Poling HM, Watson CL, Sundaram N, et al. Engineered human pluripotent-stem-cell-derived intestinal tissues with a functional enteric nervous system. Nature Medicine. 2017;**23**(1):49-59. DOI: 10.1038/nm.4233

[131] Iwao T, Kodama N, Kondo Y, Kabeya T, Nakamura K, Horikawa T, et al. Generation of enterocyte-like cells with pharmacokinetic functions from human induced pluripotent stem cells using small-molecule compounds. Drug Metabolism and Disposition. 2015;**43**(4):603-610. DOI: 10.1124/dmd.114.062604

[132] Ozawa T, Takayama K, Okamoto R, Negoro R, Sakurai F, Tachibana M, et al. Generation of enterocyte-like cells from human induced pluripotent stem cells for drug absorption and metabolism studies in human small intestine. Scientific Reports. 2015;**5**:16479. DOI: 10.1038/srep16479

[133] McCracken KW, Howell JC, Wells JM, Spence JR. Generating human intestinal tissue from pluripotent stem cells in vitro. Nature Protocols. 2011;**6**(12):1920-1928. DOI: 10.1038/nprot.2011.410

[134] Kitano K, Schwartz DM, Zhou H, Gilpin SE, Wojtkiewicz GR, Ren X, et al. Bioengineering of functional human induced pluripotent stem cell-derived intestinal grafts. Nature Communications. 2017;**8**(1):765. DOI: 10.1038/s41467-017-00779-y

[135] Enderami SE, Kehtari M, Abazari MF, Ghoraeian P, Nouri Aleagha M, Soleimanifar F, et al. Generation of insulin-producing cells from human induced pluripotent stem cells on PLLA/PVA nanofiber scaffold. Artif Cells Nanomed Biotechnol. 2018;**46**(sup1):1062-1069. DOI: 10.1080/21691401.2018.1443466

[136] Rajaei B, Shamsara M, Sanati MH. In vitro generation of glucose-responsive insulin-secreting cells from PDX1-overexpressing human-induced pluripotent stem cell derived from diabetic patient. ASAIO Journal. 2018;**64**(6):819-826. DOI: 10.1097/MAT.0000000000000728

[137] Mohsen-Kanson T, Hafner AL, Wdziekonski B, Takashima Y, Villageois P, Carriere A, et al. Differentiation of human induced pluripotent stem cells into brown and white adipocytes: Role of Pax3. Stem Cells. 2014;**32**(6):1459-1467. DOI: 10.1002/stem.1607

[138] Bonnefond A, Philippe J, Durand E, Dechaume A, Huyvaert M, Montagne L, et al. Whole-exome sequencing and high throughput genotyping identified KCNJ11 as the thirteenth MODY gene. PLoS One. 2012;7(6):e37423. DOI: 10.1371/journal.pone.0037423

[139] Griscelli F, Ezanno H, Soubeyrand M, Feraud O, Oudrhiri N, Bonnefond A, et al. Generation of an induced pluripotent stem cell (iPSC) line from a patient with maturity-onset diabetes of the young type 3 (MODY3) carrying a hepatocyte nuclear factor 1-alpha (HNF1A) mutation. Stem Cell Research. 2018;**29**:56-59. DOI: 10.1016/j.scr.2018.02.017

[140] Friesen M, Cowan CA. FPLD2 LMNA mutation R482W dysregulates iPSC-derived adipocyte function and lipid metabolism. Biochemical and Biophysical Research Communications. 2018;**495**(1):254-260. DOI: 10.1016/j.bbrc.2017.11.008

[141] Matsubara M, Kanda H, Imamura H, Inoue M, Noguchi M, Hosoda K, et al. Analysis of mitochondrial function in human induced pluripotent stem cells from patients with mitochondrial diabetes due to the A3243G mutation. Scientific Reports. 2018;**8**(1):949. DOI: 10.1038/s41598-018-19264-7

[142] Melton D. The promise of stem cell-derived islet replacement therapy. Diabetologia. 2021;**64**(5):1030-1036. DOI: 10.1007/s00125-020-05367-2

[143] Jones PM, Persaud SJ. Beta-cell replacement therapy for type 1 diabetes: Closer and closer. Diabetic Medicine. 2022;**39**(6):e14834. DOI: 10.1111/dme.14834

[144] Ono H, Obana A, Usami Y, Sakai M, Nohara K, Egusa H, et al. Regenerating salivary glands in the microenvironment of induced pluripotent stem cells. BioMed Research International. 2015;**2015**:293570. DOI: 10.1155/2015/293570

[145] Zhang Y, Deng C, Qian J, Zhang M, Li X. Improvement of radiotherapy-induced lacrimal gland injury by induced pluripotent stem cell-derived conditioned medium via MDK and inhibition of the p38/JNK pathway. International Journal of Molecular Sciences. 2014;**15**(10):18407-18421. DOI: 10.3390/ijms151018407

[146] Provin N, Giraud M. Differentiation of pluripotent stem cells into Thymic epithelial cells and generation of Thymic organoids: Applications for therapeutic strategies against APECED. Frontiers in Immunology. 2022;**13**:930963. DOI: 10.3389/fimmu.2022.930963

[147] Li Y, Hong N, Zhang A, Chen W, Wang RH, Xu XL, et al. Enhancing mammary differentiation by overcoming lineage-specific epigenetic modification and signature gene expression of fibroblast-derived iPSCs. Cell Death & Disease. 2014;**5**:e1550. DOI: 10.1038/cddis.2014.499

[148] Dai X, Wang X, Yang C, Huang M, Zhou Z, Qu Y, et al. Human fibroblasts facilitate the generation of iPSCs-derived mammary-like organoids. Stem Cell Research & Therapy. 2022;**13**(1):377. DOI: 10.1186/s13287-022-03023-7

[149] Sugimoto N, Kanda J, Nakamura S, Kitano T, Hishizawa M, Kondo T, et al. iPLAT1: The first-in-human clinical trial of iPSC-derived platelets as a phase 1 autologous transfusion study. Blood. 2022;**140**:2398-2402. DOI: 10.1182/blood.2022017296

[150] Sabbah M, Jondreville L, Lacan C, Norol F, Vieillard V, Roos-Weil D, et al. CAR-NK cells: A chimeric Hope or a promising therapy? Cancers (Basel). 2022;**14**(15):3839. DOI: 10.3390/cancers14153839

Chapter 14

Stem Cell Therapy and Its Products Such as Exosomes: Modern Regenerative Medicine Approach

Leila Dehghani, Amir Hossein Kheirkhah, Arsalan Jalili, Arman Saadati Partan, Habib Nikukar and Fatemeh Sadeghian-Nodoushan

Abstract

Regenerative Medicine is a developing and multidisciplinary field of science that uses tissue engineering, biology, and cell or cell-free therapy to regenerate cells, tissues, and organs to restore their impaired or lost function. Regenerative medicine uses a new element linked to stem cells, which call exosomes, introduces it to the healthcare market. Exosomes are present in almost all body fluids, such as synovial fluid and blood. Exosomes and microvesicles are very efficient mediators of cell-to-cell communication by transferring their specific cargo to recipient cells. Furthermore, the modification of extracellular vesicles is possible that can become an excellent choice for drug delivery systems and vaccines. Isolation of exosomes for their use as therapeutic, research, or diagnostic agents for a specific type of disease is of particular importance. Five techniques have been used to isolate exosomes from different sources, including ultracentrifugation-based, size-based, immunoassay, exosome sedimentation, and microfluidic techniques. The use of exosomes in medicine has many applications, including in Bone and cartilage, dental, immune system, liver, kidney, skeletal muscle, nervous, heart systems, skin and wound, microbial and infectious, and also in cancers. This chapter focuses on stem cells, especially exosomes, as novel approaches in disease treatment and regenerative medicine.

Keywords: stem cell therapy, exosome, regenerative medicine, disease treatment, organ

1. Introduction

Regenerative medicine is a developing and multidisciplinary field of science that uses tissue engineering, biology, surgery, and cell or cell-free therapy to repair and regenerate cells, tissues, and organs to restore their impaired or lost function [18]. Cell-based treatment methods, especially stem cell therapy have been recognized for many years as the main methods in regenerative medicine for their characteristics, such as easy isolation, self-renewal, multidirectional differentiation, immunomodulatory function, and stimulation of tissue regeneration. During the last decade, it was discovered that most of the therapeutic effect of Stem cells is due to different paracrine factors such

IntechOpen

as exosomes. So consequently, cell-free treatment was introduced as a novel approach in regenerative medicine. Although exosomes do not have cell therapy-associated complications such as tumor formation, transplant rejection in the host, and the formation of ectopic tissue, stem cells can differentiate into different types of tissues, which is their main advantage over paracrine factors. This section focuses on stem cells, especially exosomes, as novel approaches in disease treatment and regenerative medicine.

2. Stem cells in clinic, advantages, limitations

Stem cells (SCs) have been extensively known for their reparative actions. There is enormous global anticipation for stem cell-based therapies that are safe and effective. Numerous pre-clinical studies represent encouraging results on the therapeutic potential of different stem cell types, such as tissue-derived stem cells.

SCs are classified into two broad categories according to their differentiation capacity and tissue of origin. Based on stem cell hierarchy, SCs are classified into totipotent, pluripotent, multipotent, or unipotent cells, depending on their cluster of differentiation [1].

If we want to compare adult stem cells, embryonic stem cells, and induced pluripotent stem cells, the positive points of adult stem cells are the ability of transdifferentiation and reprogramming of these cells which is possible but is not well studied, being less likely to be rejected if used in transplants, and successful results have already been demonstrated in various clinical applications. On the other hand, there are some concerns about them such as limitations on the differentiation ability of ASCs which is still uncertain; they are currently thought to be multi or unipotent, being not able to grow for long periods of time in culture, usually, a very small number in each tissue making them difficult to find and purify. There is no technology available to generate large quantities of stem cells in culture, and no major ethical concerns have been raised [2].

Embryonic stem cells possess remarkable properties, including their ability to maintain and grow in culture for extended periods of one year or more. Well-established protocols for their culture maintenance are available, making them a promising research subject. With their pluripotency, they can generate a wide range of cell types, making them an important tool for understanding the process of development. By further studying embryonic stem cells, we can gain a deeper understanding of developmental processes. Also, there are some limitations, such as the inefficient process to generate ESC lines, being unsure whether they would be rejected if used in transplants, therapies using ESC avenues are mainly new, and much more research and testing are needed. Also if they are used directly from the ESC undifferentiated culture prep for tissue transplants, they can cause tumors (teratomas) or cancer development, and finally, the ethical concerns as the embryo is destroyed to acquire the inner cell mass, and the risks for female donors [3].

Also, positive points about induced pluripotent stem cells are that abundant somatic cells of the donor can be used for therapeutic approaches, concerns about histocompatibility mismatch are avoided, they are beneficial for drug development and developmental studies, information learned from the "reprogramming" process may be transferable for *in vivo* therapies to reprogram damaged or diseased cells/tissues.

Furthermore, there are some limitations such as methods for reproducibility and maintenance, differentiated tissues are not specific, viruses are currently used to introduce embryonic genes and have been shown to cause cancers in mouse studies.

Moreover, as an ethical concern, it should be noted that iPS cells can become embryos if exposed to the right conditions [4].

Although stem cell transplantation has good efficacy, weak immunogenic potential, and high multi-potential differentiation, there are some concerns about that, such as safety considerations in terms of tumorigenicity and transmission of infection, tight regulations, short shelf life, and high cost associated with strict production, transport, and storage conditions. To overcome these challenges, CM's induction of SCs in their native niche to stimulate the regeneration process is a promising cell-free approach [5]. Despite the advantages of this method, such as Immuno-compatibility, improved safety compared with stem cell transplantation, and feasibility of mass production, some limitations should be considered, such as limited therapeutic efficacy due to low concentration of paracrine factors, difficulty in obtaining the CM with a consistent composition, short half-lives of paracrine factors, and requiring frequent administrations with large doses [6]. Stromal Vascular Fraction (SVF) treatment efficacy likely depends upon several patient and treatment-specific characteristics, including the severity and cause of hair loss, treatment frequency, preparation methods, and adjunctive therapies [7]. Literature reviews propose that all kinds of cell types will have the therapeutic application with the potency of regenerative therapy. Different types of cells include many diagnostic and therapy factors. Differential potency of SC such a neurogenesis, synaptogenesis, vasculogenesis, myogenesis, oligodendrogenesis, axonal connectivity, myelin formation, etc.

Although the exact mechanisms by which SCs perform are still unidentified, recent documents have proposed they might be associated with their contents, such as exosomes [8]. Extracellular vesicles (EV) are lipid bilayer-enclosed and small (40–1000 nm) vesicles secreted into biological fluids. EVs are highly heterogeneous in the context of contents, size, and membrane composition, depending on the source of origin. So far, three main categories of EVs have been identified, including exosomes, apoptotic bodies, and cellular microparticles/microvesicles/ ectosomes [8]. EVs have been identified as vital components in intercellular communication and information transfer to other cells, affecting both the recipient and parent cells' physiological and pathological functions. Also, the roles of EVs in cancer and autoimmune disease have been suggested in some research [9].

3. Exosomes: properties and applications

Extracellular vesicles (EVs) are small lipid particles secreted from all human cell types, both healthy and malignant. They can be released either directly from the plasma membrane or upon fusion among multivesicular bodies (MVBs) and the plasma membrane. Based on their size, origin, and cargo heterogeneity (i.e., DNA, proteins, various types of RNAs), EVs have been classified into several groups, such as exosomes, microvesicles, apoptotic bodies, and other vesicle types [10].

Two scientists named Pan and Johnston first defined intercellular communication by exosomes in 1983. They discovered that during the maturation of sheep reticulocytes into erythrocytes, transferrin receptors were enclosed in nanovesicles of endosomal origin. During these years, scientists considered the term exosome for these nanovesicles, which are between 30 and 150 nm [11].

Exosomes are present in almost all body fluids, such as synovial fluid and blood. Exosomes and microvesicles are very efficient mediators of cell-to-cell communication by transferring their specific cargo to recipient cells. For example, exosomes are involved in the delivery of genetic materials causing epigenetic modifications in the target cells, antigen transfer to dendritic cells (DCs) for cross-presentation to T cells, extracellular matrix remodeling, and several signaling pathways [12, 13].

The most crucial feature of EVS, including exosomes, is to be loaded as delivery systems and vaccines because it can be easily loaded with different molecules, such as drugs, antibodies, miRNAs, and siRNAs, especially in anti-tumor treatments, resulting in more specific and efficient systems compared to the carried molecules alone. Furthermore, the modification of EVs is possible that can become an excellent choice for drug delivery systems and vaccines [14].

Exosomes enter the cell through pinocytosis, endocytosis, or direct fusion with the plasma membrane. Today, stem cells are used significantly in regenerative medicine treatment protocols due to their ability and capacity to differentiate into various cell lines. It has also been proven that the ability to heal and regenerate stem cells is due to exosomes secreted from them, which act in paracrine [15, 16]. Similar to cancer, exosomes act as a double-edged sword due to their ability to carry and deliver molecules to target cells in infectious diseases. Exosomes play a crucial role in the pathogenesis of infection but also trigger immune responses to confer protection against pathogens [17]. In general, the advantages of using exosomes in regenerative medicine include the following:

1. The risks associated with treatment with exosomes derived from different cells are relatively lower compared to other cell therapy methods in which the cells themselves are used, which created a project in the field of cell therapy called Cell-free therapy, which reduces the risks of rejection of transplanted cells due to the reaction of the host's immune system against the transplant.

2. Due to the possibility of direct communication of exosomes with the target cells and the ability to target, they are used as a preferred method over other methods, which have less toxicity and faster cleaning in the body.

3. In vivo preclinical studies have shown that targeting exosomes to specific cells can reduce transported molecules' concentration and save materials and costs.

4. Structure and biogenesis

The biogenesis of exosomes starts with endocytosis, and the cargo enters the primary endosome membrane by budding. Endocytosis can be dependent or independent of clathrin protein. The early endosome enters the late endosome phase, which has a spherical shape and is located close to the nucleus [18]. The budding of the cargo into the lumen of the endosome causes the formation of intraluminal vesicles (ILVs) with sizes of 30–150 nm, called Multivesicle bodies (MVBs). These multivesicular bodies (MVBs) may fuse with the cell membrane and release their intraluminal vesicles (ILVs), which are exosomes, outside the cell, or they may fuse with the lysosomal membrane to degrade their contents [19]. The membrane components of exosomes that have been identified so far include: lipid rafts containing sphingomyelin, cholesterol, ceramide, phosphatidylserine, and more than 4000 exosomal proteins [20].

Common proteins in all exosomes include transfer proteins such as annexin, Rab GTPase, proteins related to the biogenesis of exosomes such as Alix, TSG101, actin, myosin, and cofilin, as well as tetraspanins such as CD9, CD63, CD81, CD82, CD151, and MHC classes one and two [21, 22]. Sometimes, on the membrane of exosomes, there are glycoproteins related to targeting lysosomes called Lamp1 or Lamp2, as well as integrins and heat shock proteins such as HSP90 and HSP70. Usually, integrins and tetraspanins play the roles of adhesion and targeting [23, 24].

5. Separation methods

Isolation of exosomes for their therapeutic use, research, or diagnostic agents for a specific type of disease is of particular importance. With the rapid progress of science and technology, techniques for isolating exosomes in a high-quality and high-purity form have been expanded in large quantities. Each technique uses a particular feature of exosomes, such as size, shape, or surface proteins, for their separation. Five techniques have been used to isolate exosomes from different sources, including ultracentrifugation-based, size-based, immunoassay, exosome sedimentation, and microfluidic techniques. For the investigation of the quality of isolated exosomes, several optical and non-optical techniques have been developed to check the size, shape, and quantity of chemical components [25, 26].

Ultracentrifuge: There are usually two types of ultracentrifuge: analytical and preliminary. Analytical ultracentrifuge is used to research the physicochemical properties of particles and molecular interactions of polymeric materials [27]. The preliminary ultracentrifuge is used as the gold standard for the isolation of exosomes because its use is simple and does not require particular expertise. Of course, it is fast and cheap. Separation of exosomes by the type of differential usually includes a series of centrifugation cycles with different centrifugal forces, the duration of which is different from other components based on density and size [28]. A purification step is performed at the beginning of the separation of exosomes from human plasma and serum to get rid of large biological particles. Of course, protease inhibitors are added to the sample to prevent the destruction of exosome membrane proteins [29, 30]. There are two types of density gradient ultracentrifuge: isopycnic and moving zone. The use of density gradient ultracentrifuge to separate extracellular vesicles such as exosomes has received much attention. In the density gradient ultracentrifuge, different densities of the substance are created in the tube, which usually decreases from the bottom to the top, so the exosomes with different densities are placed in a different part of the tubes based on the force exerted on them during the centrifugation [31, 32]. In moving zone ultracentrifuge, samples containing exosomes are placed in a narrow area above the density gradient of the environment, which has a lower density than any of the substances dissolved in the sample, unlike isopycnic ultracentrifuge, which separates only based on density [33, 34].

Based on size: Ultrafiltration is one of the most popular methods of separating exosomes based on size [35]. Methods such as western blotting or electron microscopic methods are used to confirm the successful isolation of exosomes. For cell-free samples such as urine, serum, spinal fluid, and cell fluid culture media, kits based on filtration separation have been developed [27, 36]. Sequential filtration is performed to separate exosomes from the solution on the cell culture medium. Initially, filtration with 100 nm filters is used to separate floating cells, and large cell debris, but components with a size larger than 100 nm but flexible are possible [37]. In this method, in the chromatography column, the stationary phase is a porous substance and is used to separate particles and macromolecules that are smaller in size than exosomes. These substances enter the pores, and when washing the column, the exosomes are separated earlier, which are finally separated by western blotting of the isolated exosomes [38].

Immunoaffinity capture: For example, the Enzyme-linked immunosorbent assay or ELISA method is usually used to isolate exosomes in body fluids such as serum or urine. The results of absorption values indicate surface biomarkers produced on the membrane of exosomes. In another method, the surface characteristics of exosomes are evaluated by immunoprecipitation methods based on microplates by ultracentrifugation. This method's more accurate results are obtained with a smaller sample, which shows its superiority over the ultracentrifugation method [28]. It also depends on the quality of the exosomes and the environment in which they are located [35, 39]. Examples of these diagnostic kits are used to isolate exosomes from the plasma of acute myeloid leukemia (AML) patients that contain abundant amounts of CD34 on their membrane, and separation is done by magnetic beads coated with the target antibody of this marker [28, 40].

To increase the capacity of immunoaffinity capture, the mass spectrometry method is used along with the immunoassay method. For exosome isolation, antibodies on highly porous silica micropipettes that are integrally immobilized are used to isolate CD9-containing exosomes. Another example demonstrating the combination of immunoaffinity trapping with other exosome isolation methods is a magnetically activated cell sorter that uses epithelial cell adhesion molecules to purify isolated tumor exosomes. It uses plasma samples of lung cancer using the method (SEC) [41, 42].

Based on sedimentation: Exosomes can be precipitated in biological fluids by changing the solutions containing exosomes or by creating a particular type of dispersion in these solutions. For this purpose, special polymers like polyethylene glycol (PEG) can be used [27]. Typically, the incubation of samples with precipitating substances such as PEG, which have a molecular weight of up to 8000 D, for one night at 4°C causes exosomes to precipitate at a low speed by centrifugation or filtration. This method is an easy separation method that does not require special tools [43]. Several exosome precipitation kits have been commercially developed and available that isolate exosomes from body fluids such as plasma, serum, urine, and spinal fluid. The urinary exosomes isolated with these kits have been proven to be quantitatively higher than in ultracentrifugation [44]. The main weakness of the polymers' exosome deposition method is co-precipitation with other non-exosome components, such as proteins and other substances [28].

Separation based on microfluidics: Rapid advances in microfluidic technology have led to the development and manufacture of devices for the fast and high-yield separation of exosomes; applying such devices savings the use of materials, reagents, and time. An example of such a device is an acoustic nano-filter that uses ultrasound waves to separate the constituents of a sample based on size and density. Larger particles experience more wave pressure and move faster toward the pressure nodes embedded in the device [45]. To take advantage of the size difference between exosomes and other extracellular vesicles, Wang and colleagues have tailored porous nanowire structures on the micropillars of a microfluidic device, which preferentially transport exosomes between 40 nm and 400 nm in diameter. Proteins and other cellular components are filtered out, and the entrapped exosomes are isolated by dissolving the porous silicon nanowires in PBS buffer [46]. To increase the specificity and introduction of exosomes, Chen and his colleagues tried for the first time to integrate the immunoaffinity trapping technique with the microfluidic chip, which is similar to the immunoassay methods of exosome isolation based on the specific interaction between exosome membrane proteins and their proven antibodies on a chip. Based on this method, commercial Exochips have been designed, as shown in **Table 1**, to isolate exosomes. These exochips are immunochips with anti-CD63 function present on the membrane of most exosomes, and a special fluorescent dye called carbocyanine (Dio) stain exosomes are also used in these special chips [47].

6. Application of exosomes in regenerative medicine

The use of exosomes in Medicine has many applications, including in drug delivery to transfer a specific drug or therapeutic molecule, as well as in regenerative medicine and cell therapy, because nanovesicles are made of cell membranes, which are effective due to their genuineness. They have high and typical safety, and also their presence in all body fluids makes them reliable candidates for diagnosing methods (**Figure 1**).

The role of exosomes in diagnosing and treating various diseases lies in their function as carriers of intracellular communication signals. Recent exosome investigations

Methods	Specification	Advantage and disadvantage
Ultracentrifuge	Using a very strong centrifugal force of up to 1,000,000 G [30] To separate exosomes, isopycnic or moving zone techniques are preferred [34].	Advantages: Accessible and easy to use Disadvantages: Differential ultracentrifugation is typically linked to contamination and exosome loss because some exosomes in heterogeneous mixtures overlap with other particles in terms of size.
Based on size	Ultrafiltration Size exclusion chromatography (SEC) [35]	Advantages: works faster than ultracentrifugation and does not require special equipment or expertise. Disadvantages: it is usually repeated two more times [36]
Immunoaffinity capture	Exosomes have been used for immune targeting based on antibody affinity to an antigen or receptor affinity to a ligand with the goal of isolation. The first characteristic and special feature of the membrane of exosomes is a protein called CD63, which is abundantly expressed on the membrane of human exosomes [28].	Advantages: yields better RNAs than ultracentrifugation. Immune trapping is preferable to ultracentrifugation for isolating colon cancer exosomes [28, 40].
Sedimentation	Exosomes can precipitate in biological fluids by altering the solutions that contain them or by introducing a specific kind of dispersion into them. Specialized polymers like polyethylene glycol (PEG) can be utilized for this purpose [27].	Advantage: It does not call any specialized equipment. The co-precipitation of other non-exosome components, such as proteins and other substances, is the mechanism of exosome deposition of the polymers' main drawback [43].
Microfluidics	An acoustic nano-filter separates sample constituents by size and density using ultrasonic waves. Larger particles feel higher wave pressure and migrate faster to device pressure nodes [45].	Advantage: can reduce the use of materials, reagents, and time. Disadvantage: expensive to build

Table 1.
Summery of exosomes separation methods.

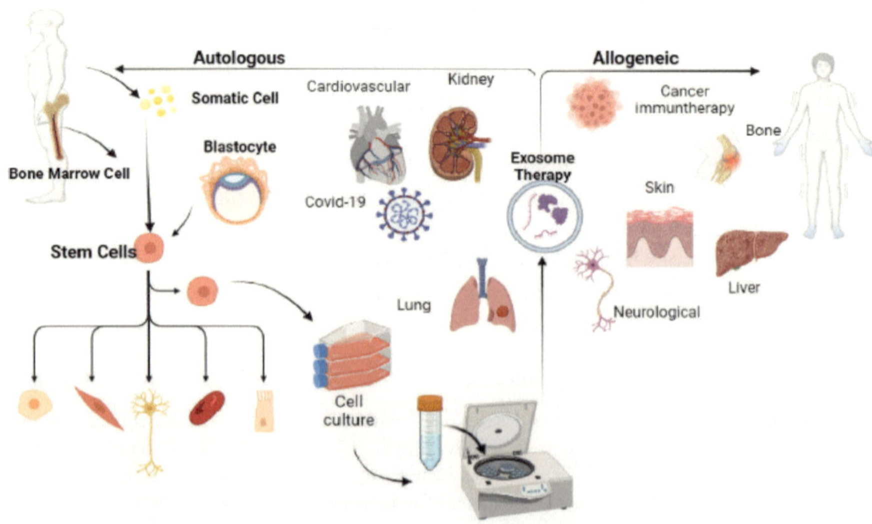

Figure 1.
Exosomes and its application in regenerative medicine.

focused on exosomes derived from humans and plants. Based on registered clinical trials (https://clinicaltrials.gov/), exosomes are used in different issues as biomarkers for cancer diagnosis like, lung, breast, colorectal, thyroid, exosome-therapy, drug delivery, and vaccines. In the case of exosome therapy, 35 studies have been registered which 22 clinical trials used MSCs-derived exosomes and most of them have been used in infectious diseases. Although some stem cell therapies have been approved for blood and immune diseases, but there is no approval for exosome therapy and more clinical trials are needed, yet.

7. Bone and cartilage tissue

Clinical studies on laboratory animals have shown the therapeutic effect of mesenchymal stem cells (MSCs) in healing cartilage injuries. The reason for these therapeutic effects is the secretion of various exosomes by these cells. It has been found that intra-articular injection of exosomes derived from mesenchymal stem cells improved osteoporotic defects in model rats. Most articular cartilages have a limited capacity to heal after injury. As challenging surgeries should be done to replace the joint, there is always a fear of rejection of the graft in this treatment [48, 49]. Recently, mesenchymal stem cells have been used to treat these defects. However, using mature mesenchymal stem cells as multipotent stem cells may not be able to optimally repair these damages because, usually, with the increasing age of the donor cells, the ability of these cells to self-regenerate and multiply and differentiate decreases dramatically [50].

Another theory suggests using pluripotent stem cells as permanent sources in cell therapy. Although these cells can differentiate into various cell lines clinically, they may cause tumors and teratoma, and they cannot be considered a reliable treatment source. So an alternative method is to use exosomes derived from these types of cells [51].

In a study by Zang et al., osteoporotic defects were created in two groups of model mice, one as a sample and the other as a control of the same size. For 8 weeks, 100 μL of exosomes derived from mesenchymal stem cells were injected intra-articularly into the PBS control group and the sample group (the environmental and nutritional conditions of both groups were the same). To observe the progress and effectiveness of the treatment, tissue evaluation was performed. It was observed that four cartilage defects were completely healed within 6 weeks, and hyaline and collagen type 2 were created entirely. In one case, tissue repair was done by fibrosis [48]. In the field of bone regeneration, most of the attention is on exosomes derived from mesenchymal stem cells due to their ability to influence and interact with the bone microenvironment at different levels of regeneration.

At different levels of regeneration, cells such as osteoblasts, osteoclasts, osteocytes, chondrocytes, and endothelial cells are associated with different repair mechanisms [52]. Exosomes derived from MSCs can generally distinguish osteoblasts by transmitting various miRNA cargoes. Exosomes produced by stimulated cells can connect to the extracellular matrix and induce the distinction of osteoblasts through regional interaction with them. In addition, considering the importance of the interaction between osteoclasts and osteoblasts in bone homeostasis, specific targeting molecules to inhibit or induce osteogenesis, such as miR-214-3p, which is secreted from osteoclasts, are of great interest [48, 53]. Another study identified the pathways by which miRNAs in exosomes regulate osteoblastic differentiation.

WNT and PI3K/AKT pathways directly affect the induction of bone formation. One of the essential miRNAs in this field is miR-27A-3P, which affects different signaling pathways in bone, such as TGFβ, BMP, and WNT, which influence proliferation genes involved in bone distinction, such as STAB2, DLX2, OSX, and Runx2 [52, 54]. According to various research, the bone regeneration capacity of exosomes derived from MSCs depends on the type of tissue from which they are isolated. For example, exosomes derived from MSCs isolated from human adipose tissue can increase the speed of healing and regeneration of bone defects [55].

8. Dental tissue

Recently, studies have been conducted on miR133b derived from certain exosomes of dental dentin cells, which regulate apoptosis in tooth development. This miRNA induces apoptosis in the primary mesenchyme of the dental tissue's upper part and causes the dental tissue's proper formation in the laboratory environment. Such studies highlight exosomes' critical role in signaling growth, differentiation, and regeneration of oral, facial, and cranial tissue [56]. Exosomes derived from adipose tissue stem cells have tremendous healing effects in treating oral and dental diseases. They have opened a clear horizon for dental treatments without the need for surgery. Exosomes may be isolated from mesenchymal stem cells of dental tissues, including dental pulp stem cells (DPSCs). These are multipotent cells that can directly contribute to the regeneration of dental pulp, bone, muscle, nerve, blood vessels, and even the liver [57, 58]. DPCS-derived exosomes play an important role in regenerative medicine. Research has confirmed that these exosomes in primary cell culture and animal models have a modulatory role and support the immune system and anti-apoptotic activity, similar to MSC-derived. They have the unique ability to regenerate dental pulps. Periodontal ligament cells (PDLCs), known as old sources of multipotent stem cells, are other cells in tooth regeneration. One of these cells' unique features is maintaining the ability of self-regeneration when transplanted [59]. Although extensive research has not been done on the effects of exosomes derived from dental ligament cells, there are shreds of evidence about the modulating properties of these cells. In addition, it has been found that exposure of PDLCs to lipopolysaccharide (LPS) produces exosomes that can induce polarity in proinflammatory macrophages [60, 61].

9. Modulation of the immune system

The ability of exosomes to modulate the immune system, and increase or inhibit inflammation, has introduced them as an attractive choice as therapeutic agents. Exosomes can transport different antigens, load on MHC class I and II complexes, and stimulate immune response through epitope presentation by cell-presented antigens [62, 63]. Exosomes derived from dendritic cells (DC) loaded with viral antigens can activate TCD8+ cells. Exosomes secreted from cells infected by bacterial and viral antigens can stimulate the release of macrophages and determine the activity of T cells. Similarly, after being made inside dendritic cells, cancer-specific epitopes can stimulate the activity of cytotoxic T cells against cancer cell antigens [64]. Exosomes derived from regulatory T cells (Treg) play the role of modulating and sometimes suppressing the immune system. TCD4+, CD25+, and Foxp3+ cells can activate TCRs or receptors on T cells. Exosomes produced by this type of T cells are quantitatively more than other T cells. Treg-derived exosomes can reduce the release of inflammatory cytokines such as IL2 and IFNγ [65]. The suppressive nature of Treg exosomes has been attributed to CD73 ectoenzymes, and the loss of CD73 in Treg exosomes reverses its natural suppressive property. It has been found that the contents of exosomes that move between Treg cells and T effector cells (Teffs) contain miRNAs such as Let-7d, Let-7b, and miR155, which indicate the modulating and inhibitory function of these exosomes [65, 66]. It has been reported that using exosomes derived from cancer cells as a vaccine for chronic myeloid leukemia (CML) patients increases the power of cytotoxic T cells against CML cells. In addition, exosomes derived from mesenchymal stem cells present cancer epitopes on their membrane, which can stimulate the activity of antibody secretion by B cells and Th1 memory cells [67]. Exosomes can be used to transfer molecules that modulate the immune system, such as vitamin D derivatives, which play a role in regulating the immune system in osteoporosis, or as vectors for gene transfer of anti-inflammatory molecules to reduce the damage caused by osteoporosis. Immunotherapy based on exosomes has numerous advantages over cellular immunotherapy because its production is of higher quality, a safer method, and they are more stable and less toxic [68, 69].

10. Liver tissue

The special features of mesenchymal stem cells, such as multipotency and self-regeneration, have been used as promising tools for treating liver diseases. According to the figure, exosomes derived from these cells can regenerate various tissues, including the liver, in damaged models [70]. Several studies have been conducted on the therapeutic effects of MSC-derived exosomes in mice models of liver fibrosis. Carbon tetrachloride (CCL4) was used to cause this damage. It was determined that exosomes derived from MSCs isolated from umbilical cord blood have improved liver fibrosis by inhibiting the epithelial-mesenchymal transition of hepatocytes and increasing collagen production [71, 72]. It has been found that exosomes significantly restore the activity of the liver aspartate aminotransferase enzyme and inhibit the smad/TGFβ signaling pathway by inactivating phosphorylation and increasing the production of type 1 and 3 collagen [73]. Another study showed that hepatic mesenchymal stem cells could release exosomes containing miR-125b, which are transported between these cells and target cells, such as stellate cells that respond to the Hedgehog (Hh) signaling pathway and heal the fibrosis caused by CCL4 damage in mouse models by inhibiting Hh pathway signaling which preventing SMO protein expression [74]. It has been reported that exosomes derived from adipose-derived MSCs (AD-MSC) contain miR-122, which affects hepatocytes and regulates the expression of specific genes such as P4HA1 and IGF1R, which are effective in collagen production and increase the speed of liver fibrosis treatment [75]. Exosomes derived from AD-MSC can significantly reduce the level of alanine and aspartate aminotransferase and concanavalin A, as well as the serum level of pro-inflammatory cytokines such as TNFα, INFγ, IL6, IL8, and also reduces IL1β, which causes severe liver inflammation. Exosomes derived from MSCs can also improve the acute liver damage caused by acetaminophen or H_2O_2 by affecting the genes of anti-apoptotic proteins such as BCL-XL and transcription activators such as STAT3 and increasing their expression [70, 76].

11. Renal tissue

Exosomes secreted by bone marrow mesenchymal stem cells can enhance the growth of cisplatin-damaged proximal tubule epithelial cells by horizontal transfer of IGF-1 receptor mRNA. It has also been shown that exosomes derived from human umbilical cord blood mesenchymal stem cells can improve acute kidney injury by inhibiting kidney oxidative stress and apoptosis, increasing kidney epithelial cells' growth [77, 78]. A scientist named Borges discovered that by placing renal tubule epithelial cells in hypoxia condition, these cells release TGF-β1 mRNA-rich exosomes into the culture medium, which can activate fibroblasts to initiate a fibrotic remodeling response [79]. Burger investigated the therapeutic potential of colony-forming cells derived from umbilical cord blood in acute kidney ischemic injury models, which showed the therapeutic abilities of these cells in treating this type of injury due to miR-486-containing exosomes. 5p is derived from these cells, which can repair this damage by targeting the PTEN gene [80]. Research has also been done on the therapeutic abilities of exosomes derived from stem cells isolated from the urine; which has shown that the damage caused by streptozotocin-induced renal damage models by weekly injection of urine-derived stem cell (USCs) exosomes can inhibit apoptosis and increase survival and vessel regeneration [81].

12. Skeletal muscle tissue

Recently, the use of mesenchymal stem cell secretomes, especially exosomes derived from mesenchymal stem cells (MSCs), for skeletal muscle regeneration has been researched. *In vivo* studies have shown that exosomes derived from MSCs can increase the speed of muscle regeneration by increasing angiogenesis and reducing muscle fibrosis. Concerning skeletal muscle injuries, researchers have discovered miRNAs with anti-apoptotic activity, such as miR-21, and myogenic activities, such as miR-1, miR-133, miR-206, and miR-494, which were able to reduce these types of injuries in mouse models [82, 83]. Choi and colleagues found that exosomes derived from human skeletal myoblasts (hSKMs) during myotube differentiation could induce myogenesis response in hASCs. Experiments on skeletal muscle injury model mice confirm that using hSKMs-derived exosomes can accelerate skeletal muscle regeneration by reducing the collagen deposition rate and increasing myofibrils' regeneration in injured muscles. According to studies, exosomes derived from MSCs regenerate skeletal muscles by strengthening myogenesis and angiogenesis; at least part of these effects are caused by miRNAs such as miR-494 [82, 84].

13. Nervous system

Exosomes have also been investigated to improve regenerative medicine's central and peripheral nerve systems. They can cross the blood-brain barrier as moderators of inflammatory responses and regeneration of nerve damage. Nervous system injuries are very debilitating for patients and often cause severe skeletal muscle disorders, and the management and recovery of these injuries are complicated and unresolved. Peripheral nervous system damage causes inflammation, loss of neuron function, and destruction, resulting in cell death. Today, we know that exosomes derived from MSCs support nerve growth by stimulating the secretion of growth factors needed to support and stimulate Schwann cells, which play an essential role in myelin production [85, 86]. Exosomes derived from MSCs can significantly induce repair of the nervous system by miR-133b, which is modified with lentiviral expression vectors and determine the overexpression or silencing of miR-133b and thus cause the regeneration of neurons [87]. Recently, it has been found that in debilitating diseases such as Parkinson's and Alzheimer's, neurons release exosomes containing α synuclein and β amyloid, respectively. These exosomes can play a role in the nucleation and physical release of these aggregated proteins that cause these diseases [88]. In the research, it has been proven that exosomes can be used as biomarkers of brain damage, for example, exosomes containing miR-9 and miR-124 isolated from blood as biomarkers are used to diagnose acute ischemic stroke (AIS) and also evaluate the amount or degree of damage caused by this ischemia [89]. The effects of exosomes in the regeneration of neurons and the nervous system have also been proven, for example, exosomes derived from oligodendrocytes stimulated with glutamate can increase neurons' survival in hypoxia conditions without glucose. Exosomes derived from bone marrow tissue stem cells can significantly increase the survival of retinal ganglion cells (RGCs) and the regeneration of their axons [90–92]. Usually, after nerve damage, Schwann cells differentiate and grow and direct axons to their target tissue. It has been found that the exosomes derived from these cells inhibit the activity of RhoA, a GTPase that can cause axons to lengthen and repair them [93, 94].

14. Heart muscle tissue

The protective effects of exosomes in myocardial ischemia re-injury models are being investigated. Scientists have shown that exosomes isolated from cardiosphere-derived cells (CDCs), when injected into mice model of ischemia, can inhibit apoptosis and induce the growth of heart cells. It has been found that these beneficial effects are due to the richness of these exosomes in miR-146a [95]. During another study, it was determined that exosomes secreted from bone marrow mesenchymal stem cells stimulate the formation of umbilical cord vein endothelial cell tubes and inhibit the production of T cells in vitro. In addition, the severity reduces infarction [96].

According to other research on exosomes derived from mouse embryonic stem cells, these exosomes can restore the internal function of the heart after myocardial infarction. With further research, the researchers found that this is due to miR-294 in these exosomes, which are transferred to cardiac progenitor cells [97, 98]. Another research group in China investigated the protective effects of mesenchymal stem cells derived from human umbilical cord blood on acute myocardial infarction (AMI) animal models and noticed the effects of these exosomes in protecting myocardial cells. Apoptosis and increased angiogenesis in the damaged area. They found that these effects are related to the modulation of BCL-2 pro-apoptotic protein family gene expression [99]. Certain exosomes in the heart's pericardial fluid improve the survival, growth, and communication of endothelial cells in the culture medium. It restores the angiogenic capacity of endothelial cells, and even these exosomes improve blood flow and angiogenesis after ischemic injuries in model mice. Further research has shown that these exosomes contain miR let-7b to carry out this process [100].

15. Skin

Angiogenesis is essential in various physiological processes, including wound healing and skin tissue regeneration. Scientists found that exosomes secreted by mesenchymal stem cells derived from adipose tissue can significantly stimulate the angiogenesis of endothelial cells *in vitro* and *in vivo*. Further research showed that these exosomes transfer miR-125a to endothelial cells, which decreases the expression of inhibitory proteins Delta-like 4 (DLL4) that inhibit angiogenesis [101, 102]. Burns is one of the most common skin injuries that significantly intensify inflammatory reactions by increasing the level of factors such as TNFα, Il-1β, and decreasing the level of IL-10. The scientists found that using exosomes derived from umbilical cord blood stem cells successfully reduced the inflammatory reactions caused by burns, and further research revealed that this effect is due to the presence of miR-181c in these exosomes, which reduces pain and severe inflammation caused by burns by reducing TLR4 signaling [103]. It has been determined that exosomes derived from umbilical cord blood endothelial progenitor cells can heal diabetic wounds in rat models. Microarray analysis has shown that exosomes significantly increase the expression of some genes. They change the ERK1/2 signaling pathway, which is very important in the healing and regeneration of these wounds. Studies by Guo and colleagues have shown that platelet-rich plasma-derived exosomes can effectively induce the proliferation and migration of endothelial cells and fibroblasts to increase angiogenesis and regeneration and repair severe skin wounds. Exosomes can also control skin regeneration bipolarly. They can also prevent scarring caused by burn healing and collagen deposition through the induction of phosphorylation that inhibits the WNT/*β*catenin YAP pathway [84, 104].

16. Cancer

In the mid-2000s, the first results from clinical trials on exosomes as a treatment for cancer were published. Exosomes derived from dendritic cells (DEXs) are potential targets for cancer therapeutic strategy. DEXs can directly catalyze the transfer of peptide-MHC complexes from their membrane surface to the T cell membrane surface (cross-dressing). Moreover, DEXs can indirectly stimulate T cell responses via cross-dressing with dendritic cells or via exosome uptake and processing, following the peptide-MHC complex presentation to T cells. DEXs can also induce activation and proliferation of NK cells by establishing interaction of the NKG2D ligand on DEXs with NKG2D receptors on the NK cell membrane [105]. In 2005, two phase I clinical trials using DEX vaccines were performed. The first trial reported the use of DEXs loaded with HLA-restricted melanoma-associated antigen (MAGE) peptides, which were infused into patients with HLA A2+ non-small cell lung cancer (NSCLC); the second trial reported the use of DEXs derived from DCs pulsed with MAGE and inoculated them to conduct immunization of melanoma patients [106]. Exosomes exhibit features for application as adjuvant carriers, such as optimal size, biocompatibility, stability in the systemic circulation, and target-specific delivery. Recently, an exosome-based adjuvant delivery system was developed using genetically modified murine melanoma B16BL6 cells. The exosomes derived from these cells containing CpG DNA were injected three times at 3-day intervals and successfully induced immunostimulatory signals in mice 7 days after the last immunization [107, 108]. In a recent study by Shi et al., a vaccine with exosomes derived from IFNγ-modified RM-1 prostate cancer cells under a vaccination regimen of four injections (on days 0, 4, 8, and 12) decreased the number of Tregs. It reduced the metastatic tumor rate in C57BL male mice with lung metastasis [109]. Recently, a non-randomized phase I/II clinical trial showed promising results with a vaccine designed using exosomes derived from DCs pulsed with SART1, a biomarker of esophagus squamous cell carcinoma. Pulsed DCs obtained from patients could generate exosomes that were well tolerated and induced antigen-specific CTLs in seven patients. One patient in this study remained stable for 20 months after DEXs therapy, although he developed lung metastasis after the stable period. The other six patients had progressive disease and died up to 10 months post-vaccination. These findings indicate that developing personalized exosome-based immunotherapy is feasible, although challenging [110]. A phase I clinical trial reported using exosomes derived from ascites (AEXs) in combination with granulocyte-macrophage colony-stimulating factor (GM-CSF) as immunotherapy for colorectal cancer. Injection of AEXs for colorectal cancer was safe and well tolerated by all patients during the four weekly doses administered [111].

17. Applied exosomes on scaffolds

The short tissue retention of exosomes after *in vivo* implantation is still a significant challenge in clinical applications. Scaffold encapsulation of exosomes can enable continuous delivery in the injured environment, thereby improving the therapeutic effect. Researchers have developed several methods to deliver exosomes to the post-infarct environment sustainably. For example, exosomes isolated from cardiomyocyte-derived induced pluripotent stem cells encapsulated in hydrogel patches were directly delivered to the hearts of infarcted rats [112]. The exosome patches demonstrated prolonged exosome release and promoted recovery of the ejection fraction, prevented cardiomyocyte hypertrophy, alleviated the ischemic injury, and promoted heart recovery. Another study loaded endothelial progenitor cell-derived exosomes into a shear-thinning gel to achieve precise administration and sustained delivery [113]. In a rat model of myocardial infarction, the exosome hydrogels enhanced angiogenesis and myocardial hemodynamics around the infarct. The cell-free scaffold material improved the effects of exosome-mediated myocardial therapy. In another study, exosomes isolated from human umbilical cord-derived

MSCs were encapsulated in functional peptide hydrogels to increase their stability and provide sustained release. The exosome hydrogels protected cardiomyocytes from oxidative stress induced by H_2O_2, which improved cardiac function in a rat myocardial infarction model. These studies provide practical and effective methods for exosome-laden scaffolds in myocardial regeneration [114]. Exosome-laden scaffolds are most widely used for skin repair.

Several findings indicate that combining bioactive scaffold materials with the controlled release of exosomes heals skin wounds. For example, exosomes isolated from human umbilical cord-derived MSCs encapsulated in polyvinylalcohol (PVA)/Alg nanohydrogels were used to heal diabetic wounds. The PVA/Alg nanohydrogel promoted cell proliferation, migration, angiogenesis, enhanced the efficacy of exosomes, and accelerated the healing of diabetic wounds. In another study, exosomes were loaded in a novel injectable bioactive hydrogel called FHE. Exosomes isolated from adipose-derived MSCs were loaded into FHE hydrogel through electrostatic interactions with poly-ε-L-lysine. The exosome hydrogel promoted angiogenesis, cell proliferation, and granulation tissue formation at the wound site and accelerated the healing of diabetic wounds and skin regeneration [115, 116]. In another study, methylcellulose-chitosan hydrogels loaded with exosomes isolated from placenta-derived MSCs were shown to heal diabetic wounds and form new tissues similar to natural skin. Similarly, chitosan/silk hydrogels with swelling and moisturizing capabilities loaded with exosomes isolated from gingival-derived MSCs promoted collagen epithelial regeneration and angiogenesis and accelerated the healing of diabetic skin defects [117, 118]. Chitosan scaffolds have also been shown to provide controlled release of exosomes isolated from synovium-derived MSCs, which accelerated wound healing by increasing the formation of granulation tissue and angiogenesis [118, 119]. Modified exosomes also have the potential to stimulate bone regeneration. For example, miR-375 was enriched in exosomes by overexpression in parental cells. The exosomes were loaded into a hydrogel and injected into a rat skull defect model. The exosomes were continuously released into the wound, which enhanced bone regeneration [120]. Liu et al. developed a photoinduced imine functional group cross-linking hydrogel glue to generate a decellularized tissue patch for cartilage regeneration. The patch retained stem cell-derived exosomes in the cartilage for a long time. In addition, the exosome-laden scaffold integrated with the natural cartilage matrix induced cell migration in the cartilage defect and promoted the repair and regeneration of articular cartilage [121]. Another study constructed a cell-free bone tissue engineering system by combining poly (lactic-co-glycolic acid) (PLGA)/polydopamine (pDA) scaffolds and exosomes isolated from human adipose-derived stem cells. The exosomes were slowly and continuously released from the scaffold, which promoted the migration of MSCs and significantly enhanced bone regeneration [112].

18. Exosome for microbial diseases

MSCs express various types of anti-microbial peptides and proteins (AMPs). Some of them are known for anti-bacterial properties, such as cathelicidin LL-37), β-defensin-2 (BD-2), hepcidin, and Lipocalin-2 (Lcn2) [122]. It is suggested in recent studies that MSCs can improve bacterial infection in preclinical models by AMPs. Therefore, MSCs can enhance the innate immune response against bacteria. It is suggested in recent studies that MSCs can improve bacterial infection in preclinical models by AMPs. Therefore, MSCs can enhance the innate immune response against bacteria [123].

In a study by Yagi et al. in 2020, the anti-microbial activity of human Adipose-derived MSCs (AD-MSCs) on *Staphylococcus aureus* was assessed. The findings showed that human AD-MSCs conditioned medium significantly prevented the growth of *S. aureus*. The results also showed the critical anti-microbial activity of cathelicidin LL-37 in AD-MSCs [124]. A previous study also showed that the anti-microbial activity of BM-MSCs against the growth of Gram-negative (*Escherichia coli* and *Pseudomonas aeruginosa*) and Gram-positive (*S. aureus*) bacteria was mediated by LL-37 [125]. On the other hand, human umbilical cord blood-derived MSCs attenuated acute lung injury through *E. coli* infection in mice. The results demonstrated that MSCs secreted BD-2 through the TLR-4 signaling pathway and mediated the anti-microbial effects [126].

Moreover, menstrual-derived MSCs secreting hepcidin in synergy with antibiotics in sepsis were responsible for bacterial clearance. MSCs also secret different growth factors, such as the Keratinocyte growth factor (KGF), to exert anti-bacterial activity. In the research performed by Lee et al. on an *E. coli* infection model in an *ex vivo* perfused human lung, BM-MSCs improved alveolar fluid bacterial clearance and mitigated inflammation [127, 128]. The metabolomics and immunomodulatory effects of MSCs-MVs are performed by enhancing the intracellular ATP levels in injured alveolar epithelial cells and reducing the secretion of inflammatory cytokines, including tumor necrosis factor-alpha (TNF-α) in human monocytes. It should be considered that MSC-MVs expressed Cyclooxygenase2 (COX2) mRNA. COX2 is the crucial enzyme in prostaglandin E2 (PGE2) synthesis that is a critical factor for transforming the polarization of M1 into M2 macrophages. As articles suggest, the enhancement in PGE2 secretion by monocytes following the transfer of COX2 mRNA from MSC-MVs to these cells caused the phenotype switch from M1 to M2. MSC-MVs, by direct transfer of KGF or indirectly by activating monocytes, mitigated lung inflammation, cytokine permeability, bacterial growth, and improved survival. This therapeutic effect of MVs was abrogated by KFG neutralizing antibody, proposing a possible mechanism for the anti-bacterial effect of MSC-MVs [129, 130]. As the anti-bacterial effect of KGF in MSCs was previously reported [128], these studies supported the hypothesis that MVs can partly conserve the anti-microbial effects of parent cells, by using growth factors, including KGF. Cell-Exo can overcome the shortage of stem cells to treat microbial and other infectious diseases and provide a new generation in medical science from cellular to acellular therapy.

Both intact and engineered exosomes have been applied, and their efficacy on various infectious diseases has been assessed in preclinical studies and limited clinical trials. Although exosomes perform part of their antimicrobial activity by directly transferring mRNA, miRNA, and protein cargos, their beneficial effects are mainly indirectly applied through reprogramming immune cells and activating innate and adaptive immune responses.

Author details

Leila Dehghani[1,2], Amir Hossein Kheirkhah[3], Arsalan Jalili[4,5], Arman Saadati Partan[5], Habib Nikukar[6]* and Fatemeh Sadeghian-Nodoushan[7]*

1 Department of Applied Cell Sciences, Isfahan University of Medical Sciences, Isfahan, Iran

2 Department of Tissue Engineering and Applied Cell Sciences, School of Advanced Technologies in Medicine, Tehran University of Medical Sciences, Tehran, Iran

3 Department of Tissue Engineering and Applied Cell Science, School of Medicine, Qom University of Medical Sciences, Qom, Iran

4 Department of Applied Cell Sciences, Faculty of Basic Sciences and Advanced Medical Technologies, Royan Institute, ACECR, Tehran, Iran

5 Parvaz Research Ideas Supporter Institute, Tehran, Iran

6 Medical Nanotechnology and Tissue Engineering Research Center, Yazd Reproductive Sciences Institute, Shahid Sadoughi University of Medical Sciences, Yazd, Iran

7 Research and Clinical Center for Infertility, Yazd Reproductive Sciences Institute, Shahid Sadoughi University of Medical Sciences, Yazd, Iran

*Address all correspondence to: habibnik@gmail.com and fsadeghian@gmail.com

References

[1] Ntege EH, Sunami H, Shimizu Y. Advances in regenerative therapy: A review of the literature and future directions. Regenerative Therapy. 2020;**14**:136-153

[2] Ratajczak M, Zuba-Surma E, Kucia M, Poniewierska A, Suszynska M, Ratajczak J. Pluripotent and multipotent stem cells in adult tissues. Advances in Medical Sciences. 2012;**57**(1):1-17

[3] Ebert AD, Shelley BC, Hurley AM, Onorati M, Castiglioni V, Patitucci TN, et al. EZ spheres: A stable and expandable culture system for the generation of pre-rosette multipotent stem cells from human ESCs and iPSCs. Stem Cell Research. 2013;**10**(3):417-427

[4] Taylor CJ, Bolton EM, Bradley JA. Immunological considerations for embryonic and induced pluripotent stem cell banking. Philosophical Transactions of the Royal Society, B: Biological Sciences. 2011;**366**(1575):2312-2322

[5] Copland IB, Galipeau J. Death and inflammation following somatic cell transplantation. Seminars in Immunopathology. 2011;**33**(6):535-550

[6] de Abreu RC, Fernandes H, da Costa Martins PA, Sahoo S, Emanueli C, Ferreira L. Native and bioengineered extracellular vesicles for cardiovascular therapeutics. Nature Reviews. Cardiology. 2020;**17**(11):685-697

[7] Kang BY, Li AW, Lee MH, Wong CC, Naseer N, Ibrahim SA, et al. The safety and efficacy of autologous adipose-derived stromal vascular fraction for nonscarring alopecia: A systematic review. Archives of Dermatological Research. 2021;**2021**:1-8

[8] Kalluri R, LeBleu VS. The biology, function, and biomedical applications of exosomes. Science. 2020;**367**(6478):eaau6977

[9] LeClaire MJ. Biophysical Characterization of Cancer-Derived Cells and Extracellular Vesicles. Los Angeles: University of California; 2021

[10] Théry C, Witwer KW, Aikawa E, Alcaraz MJ, Anderson JD, Andriantsitohaina R, et al. Minimal information for studies of extracellular vesicles 2018 (MISEV2018): A position statement of the International Society for Extracellular Vesicles and update of the MISEV2014 guidelines. Journal of Extracellular Vesicles. 2018;**7**(1):1535750

[11] Pan B-T, Johnstone RM. Fate of the transferrin receptor during maturation of sheep reticulocytes in vitro: Selective externalization of the receptor. Cell. 1983;**33**(3):967-978

[12] Raposo G, Stoorvogel W. Extracellular vesicles: Exosomes, microvesicles, and friends. The Journal of Cell Biology. 2013;**200**(4):373-383

[13] Van Niel G, d'Angelo G, Raposo G. Shedding light on the cell biology of extracellular vesicles. Nature Reviews. Molecular Cell Biology. 2018;**19**(4):213-228

[14] Zhao X, Wu D, Ma X, Wang J, Hou W, Zhang W. Exosomes as drug carriers for cancer therapy and challenges regarding exosome uptake. Biomedicine & Pharmacotherapy. 2020;**128**:110237

[15] Hu G, Drescher KM, Chen X-M. Exosomal miRNAs: Biological properties and therapeutic potential. Frontiers in Genetics. 2012;**3**:56

[16] Schäfer R, Spohn G, Baer PC. Mesenchymal stem/stromal cells in regenerative medicine: Can preconditioning strategies improve therapeutic efficacy. Transfusion Medicine and Hemotherapy. 2016;**43**(4):256-267

[17] Schorey JS, Harding CV. Extracellular vesicles and infectious diseases: New complexity to an old story. The Journal of clinical investigation. 2016;**126**(4):1181-1189

[18] Keller S, Sanderson MP, Stoeck A, Altevogt P. Exosomes: From biogenesis and secretion to biological function. Immunology Letters. 2006;**107**(2):102-108

[19] Skotland T, Sandvig K, Llorente A. Lipids in exosomes: Current knowledge and the way forward. Progress in Lipid Research. 2017;**66**:30-41

[20] Théry C, Zitvogel L, Amigorena S. Exosomes: Composition, biogenesis and function. Nature Reviews. Immunology. 2002;**2**(8):569-579

[21] Rana S, Yue S, Stadel D, Zöller M. Toward tailored exosomes: The exosomal tetraspanin web contributes to target cell selection. The International Journal of Biochemistry & Cell Biology. 2012;**44**(9):1574-1584

[22] Vabulas RM, Raychaudhuri S, Hayer-Hartl M, Hartl FU. Protein folding in the cytoplasm and the heat shock response. Cold Spring Harbor Perspectives in Biology. 2010;**2**(12):a004390

[23] Andreu Z, Yáñez-Mó M. Tetraspanins in extracellular vesicle formation and function. Frontiers in Immunology. 2014;**5**:442

[24] Conde-Vancells J, Rodriguez-Suarez E, Embade N, Gil D, Matthiesen R, Valle M, et al. Characterization and comprehensive proteome profiling of exosomes secreted by hepatocytes. Journal of Proteome Research. 2008;**7**(12):5157-5166

[25] Khatun Z, Bhat A, Sharma S, Sharma A. Elucidating diversity of exosomes: Biophysical and molecular characterization methods. Nanomedicine. 2016;**11**(17):2359-2377

[26] Zeringer E, Barta T, Li M, Vlassov AV. Strategies for isolation of exosomes. Cold Spring Harbor Protocols. 2015;**2015**(4):top074476

[27] Zarovni N, Corrado A, Guazzi P, Zocco D, Lari E, Radano G, et al. Integrated isolation and quantitative analysis of exosome shuttled proteins and nucleic acids using immunocapture approaches. Methods. 2015;**87**:46-58

[28] Rechavi O, Erlich Y, Amram H, Flomenblit L, Karginov FV, Goldstein I, et al. Cell contact-dependent acquisition of cellular and viral nonautonomously encoded small RNAs. Genes & Development. 2009;**23**(16):1971-1979

[29] Yamashita T, Takahashi Y, Nishikawa M, Takakura Y. Effect of exosome isolation methods on physicochemical properties of exosomes and clearance of exosomes from the blood circulation. European Journal of Pharmaceutics and Biopharmaceutics. 2016;**98**:1-8

[30] Tauro BJ, Greening DW, Mathias RA, Ji H, Mathivanan S, Scott AM, et al. Comparison of ultracentrifugation, density gradient separation, and immunoaffinity capture methods for isolating human colon cancer cell line LIM1863-derived exosomes. Methods. 2012;**56**(2):293-304

[31] Miranda KC, Bond DT, Levin JZ, Adiconis X, Sivachenko A, Russ C, et al.

Massively parallel sequencing of human urinary exosome/microvesicle RNA reveals a predominance of non-coding RNA. PLoS One. 2014;**9**(5):e96094

[32] Grapp M, Wrede A, Schweizer M, Hüwel S, Galla H-J, Snaidero N, et al. Choroid plexus transcytosis and exosome shuttling deliver folate into brain parenchyma. Nature Communications. 2013;**4**(1):1-13

[33] Booth AM, Fang Y, Fallon JK, Yang J-M, Hildreth JE, Gould SJ. Exosomes and HIV Gag bud from endosome-like domains of the T cell plasma membrane. The Journal of Cell Biology. 2006;**172**(6):923-935

[34] Quintana JF, Makepeace BL, Babayan SA, Ivens A, Pfarr KM, Blaxter M, et al. Extracellular Onchocerca-derived small RNAs in host nodules and blood. Parasites & Vectors. 2015;**8**(1):1-11

[35] Li P, Kaslan M, Lee SH, Yao J, Gao Z. Progress in exosome isolation techniques. Theranostics. 2017;**7**(3):789

[36] Channavajjhala SK, Rossato M, Morandini F, Castagna A, Pizzolo F, Bazzoni F, et al. Optimizing the purification and analysis of miRNAs from urinary exosomes. Clinical Chemistry and Laboratory Medicine (CCLM). 2014;**52**(3):345-354

[37] Feng Y, Huang W, Wani M, Yu X, Ashraf M. Ischemic preconditioning potentiates the protective effect of stem cells through secretion of exosomes by targeting Mecp2 via miR-22. PLoS One. 2014;**9**(2):e88685

[38] D-k K, Nishida H, An SY, Shetty AK, Bartosh TJ, Prockop DJ. Chromatographically isolated CD63+ CD81+ extracellular vesicles from mesenchymal stromal cells rescue cognitive impairments after TBI. Proceedings of the National Academy of Sciences. 2016;**113**(1):170-175

[39] Hong CS, Muller L, Boyiadzis M, Whiteside TL. Isolation and characterization of CD34+ blast-derived exosomes in acute myeloid leukemia. PLoS One. 2014;**9**(8):e103310

[40] Ueda K, Ishikawa N, Tatsuguchi A, Saichi N, Fujii R, Nakagawa H. Antibody-coupled monolithic silica microtips for highthroughput molecular profiling of circulating exosomes. Scientific Reports. 2014;**4**(1):1-9

[41] Rabinowits G, Gerçel-Taylor C, Day JM, Taylor DD, Kloecker GH. Exosomal microRNA: A diagnostic marker for lung cancer. Clinical Lung Cancer. 2009;**10**(1):42-46

[42] Batrakova EV, Kim MS. Using exosomes, naturally-equipped nanocarriers, for drug delivery. Journal of Controlled Release. 2015;**219**:396-405

[43] Alvarez ML, Khosroheidari M, Ravi RK, DiStefano JK. Comparison of protein, microRNA, and mRNA yields using different methods of urinary exosome isolation for the discovery of kidney disease biomarkers. Kidney International. 2012;**82**(9):1024-1032

[44] Hou R, Li Y, Sui Z, Yuan H, Yang K, Liang Z, et al. Advances in exosome isolation methods and their applications in proteomic analysis of biological samples. Analytical and Bioanalytical Chemistry. 2019;**411**(21):5351-5361

[45] Davies RT, Kim J, Jang SC, Choi E-J, Gho YS, Park J. Microfluidic filtration system to isolate extracellular vesicles from blood. Lab on a Chip. 2012;**12**(24):5202-5210

[46] Wang Z, Wu H-j, Fine D, Schmulen J, Hu Y, Godin B, et al. Ciliated micropillars

for the microfluidic-based isolation of nanoscale lipid vesicles. Lab on a Chip. 2013;**13**(15):2879-2882

[47] Kanwar SS, Dunlay CJ, Simeone DM, Nagrath S. Microfluidic device (ExoChip) for on-chip isolation, quantification and characterization of circulating exosomes. Lab on a Chip. 2014;**14**(11):1891-1900

[48] Zhang S, Chu W, Lai R, Lim S, Hui J, Toh W. Exosomes derived from human embryonic mesenchymal stem cells promote osteochondral regeneration. Osteoarthritis and Cartilage. 2016;**24**(12):2135-2140

[49] Jiang YZ, Zhang SF, Qi YY, Wang LL, Ouyang HW. Cell transplantation for articular cartilage defects: Principles of past, present, and future practice. Cell Transplantation. 2011;**20**(5):593-607

[50] Wall A, Board T. Mesenchymal cell-based repair of large full thickness defects of articular cartilage. Classic Papers in Orthopaedics: Springer. 2014;**2014**:441-443

[51] Charalambous CP. Cell origin and differentiation in the repair of full-thickness defects of articular cartilage. Classic Papers in Orthopaedics: Springer. 2014;**2014**:377-379

[52] Toh WS, Foldager CB, Pei M, Hui JHP. Advances in mesenchymal stem cell-based strategies for cartilage repair and regeneration. Stem Cell Reviews and Reports. 2014;**10**(5):686-696

[53] Li D, Liu J, Guo B, Liang C, Dang L, Lu C, et al. Osteoclast-derived exosomal miR-214-3p inhibits osteoblastic bone formation. Nature Communications. 2016;**7**(1):1-16

[54] Chen Y, Xue K, Zhang X, Zheng Z, Liu K. Exosomes derived from mature chondrocytes facilitate subcutaneous stable ectopic chondrogenesis of cartilage progenitor cells. Stem Cell Research & Therapy. 2018;**9**(1):1-14

[55] Li W, Liu Y, Zhang P, Tang Y, Zhou M, Jiang W, et al. Tissue-engineered bone immobilized with human adipose stem cells-derived exosomes promotes bone regeneration. ACS Applied Materials & Interfaces. 2018;**10**(6):5240-5254

[56] Chew JRJ, Chuah SJ, Teo KYW, Zhang S, Lai RC, Fu JH, et al. Mesenchymal stem cell exosomes enhance periodontal ligament cell functions and promote periodontal regeneration. Acta Biomaterialia. 2019;**89**:252-264

[57] Gandia C, Arminan A, García-Verdugo JM, Lledó E, Ruiz A, Miñana MD, et al. Human dental pulp stem cells improve left ventricular function, induce angiogenesis, and reduce infarct size in rats with acute myocardial infarction. Stem Cells. 2008;**26**(3):638-645

[58] Gronthos S, Mankani M, Brahim J, Robey PG, Shi S. Postnatal human dental pulp stem cells (DPSCs) in vitro and in vivo. Proceedings of the National Academy of Sciences. 2000;**97**(25):13625-13630

[59] Zhu W, Liang M. Periodontal ligament stem cells: Current status, concerns, and future prospects. Stem Cells International. 2015;**2015**:972313

[60] Rajan TS, Giacoppo S, Trubiani O, Diomede F, Piattelli A, Bramanti P, et al. Conditioned medium of periodontal ligament mesenchymal stem cells exert anti-inflammatory effects in lipopolysaccharide-activated mouse motoneurons. Experimental Cell Research. 2016;**349**(1):152-161

[61] Kang H, Lee M-J, Park SJ, Lee M-S. Lipopolysaccharide-preconditioned

periodontal ligament stem cells induce M1 polarization of macrophages through extracellular vesicles. International Journal of Molecular Sciences. 2018;**19**(12):3843

[62] Admyre C, Johansson SM, Paulie S, Gabrielsson S. Direct exosome stimulation of peripheral humanT cells detected by ELISPOT. European Journal of Immunology. 2006;**36**(7):1772-1781

[63] Robbins PD, Morelli AE. Regulation of immune responses by extracellular vesicles. Nature Reviews. Immunology. 2014;**14**(3):195-208

[64] Bhatnagar S, Schorey JS. Exosomes released from infected macrophages contain Mycobacterium avium glycopeptidolipids and are proinflammatory. The Journal of Biological Chemistry. 2007;**282**(35):25779-25789

[65] Yu D, Rao S, Tsai LM, Lee SK, He Y, Sutcliffe EL, et al. The transcriptional repressor Bcl-6 directs T follicular helper cell lineage commitment. Immunity. 2009;**31**(3):457-468

[66] Smyth LA, Ratnasothy K, Tsang JY, Boardman D, Warley A, Lechler R, et al. CD73 expression on extracellular vesicles derived from CD4+ CD25+ Foxp3+ T cells contributes to their regulatory function. European Journal of Immunology. 2013;**43**(9):2430-2440

[67] Qazi KR, Gehrmann U, Domange Jordö E, Karlsson MC, Gabrielsson S. Antigen-loaded exosomes alone induce Th1-type memory through a B cell–dependent mechanism. Blood, The Journal of the American Society of Hematology. 2009;**113**(12):2673-2683

[68] Bellavia D, Costa V, De Luca A, Maglio M, Pagani S, Fini M, et al. Vitamin D level between calcium-phosphorus homeostasis and immune system: New perspective in osteoporosis. Current Osteoporosis Reports. 2016;**2016**:1-12

[69] Tran T-H, Mattheolabakis G, Aldawsari H, Amiji M. Exosomes as nanocarriers for immunotherapy of cancer and inflammatory diseases. Clinical Immunology. 2015;**160**(1):46-58

[70] Tan CY, Lai RC, Wong W, Dan YY, Lim S-K, Ho HK. Mesenchymal stem cell-derived exosomes promote hepatic regeneration in drug-induced liver injury models. Stem Cell Research & Therapy. 2014;**5**(3):1-14

[71] Fiore EJ, Mazzolini G, Aquino JB. Mesenchymal stem/stromal cells in liver fibrosis: Recent findings, old/new caveats and future perspectives. Stem Cell Reviews and Reports. 2015;**11**(4):586-597

[72] Huang B, Cheng X, Wang H, Huang W, Wang D, Zhang K, et al. Mesenchymal stem cells and their secreted molecules predominantly ameliorate fulminant hepatic failure and chronic liver fibrosis in mice respectively. Journal of Translational Medicine. 2016;**14**(1):1-12

[73] Li T, Yan Y, Wang B, Qian H, Zhang X, Shen L, et al. Exosomes derived from human umbilical cord mesenchymal stem cells alleviate liver fibrosis. Stem Cells and Development. 2013;**22**(6):845-854

[74] Hyun J, Wang S, Kim J, Kim GJ, Jung Y. MicroRNA125b-mediated Hedgehog signaling influences liver regeneration by chorionic plate-derived mesenchymal stem cells. Scientific Reports. 2015;**5**(1):1-15

[75] Li J, Ghazwani M, Zhang Y, Lu J, Li J, Fan J, et al. miR-122 regulates collagen production via targeting hepatic stellate cells and suppressing P4HA1 expression. Journal of Hepatology. 2013;**58**(3):522-528

[76] Fu J, Zhang H, Zhuang Y, Liu H, Shi Q, Li D, et al. The role of N-acetyltransferase 8 in mesenchymal stem cell-based therapy for liver ischemia/reperfusion injury in rats. PLoS One. 2014;**9**(7):e103355

[77] Tomasoni S, Longaretti L, Rota C, Morigi M, Conti S, Gotti E, et al. Transfer of growth factor receptor mRNA via exosomes unravels the regenerative effect of mesenchymal stem cells. Stem Cells and Development. 2013;**22**(5):772-780

[78] Zhou Y, Xu H, Xu W, Wang B, Wu H, Tao Y, et al. Exosomes released by human umbilical cord mesenchymal stem cells protect against cisplatin-induced renal oxidative stress and apoptosis in vivo and in vitro. Stem Cell Research & Therapy. 2013;**4**(2):1-13

[79] Borges FT, Melo SA, Özdemir BC, Kato N, Revuelta I, Miller CA, et al. TGF-β1–containing exosomes from injured epithelial cells activate fibroblasts to initiate tissue regenerative responses and fibrosis. Journal of the American Society of Nephrology. 2013;**24**(3):385-392

[80] Viñas JL, Burger D, Zimpelmann J, Haneef R, Knoll W, Campbell P, et al. Transfer of microRNA-486-5p from human endothelial colony forming cell–derived exosomes reduces ischemic kidney injury. Kidney International. 2016;**90**(6):1238-1250

[81] Jiang Z-z, Liu Y-m, Niu X, Yin J-y, Hu B, Guo S-c, et al. Exosomes secreted by human urine-derived stem cells could prevent kidney complications from type I diabetes in rats. Stem Cell Research & Therapy. 2016;**7**(1):1-13

[82] Nakamura Y, Miyaki S, Ishitobi H, Matsuyama S, Nakasa T, Kamei N, et al. Mesenchymal-stem-cell-derived exosomes accelerate skeletal muscle regeneration. FEBS Letters. 2015;**589**(11):1257-1265

[83] Nakasa T, Ishikawa M, Shi M, Shibuya H, Adachi N, Ochi M. Acceleration of muscle regeneration by local injection of muscle-specific microRNAs in rat skeletal muscle injury model. Journal of Cellular and Molecular Medicine. 2010;**14**(10):2495-2505

[84] Guo S-C, Tao S-C, Yin W-J, Qi X, Yuan T, Zhang C-Q. Exosomes derived from platelet-rich plasma promote the re-epithelization of chronic cutaneous wounds via activation of YAP in a diabetic rat model. Theranostics. 2017;**7**(1):81

[85] Oliveira JT, Bittencourt-Navarrete RE, de Almeida FM, Tonda-Turo C, Martinez AMB, Franca JG. Enhancement of median nerve regeneration by mesenchymal stem cells engraftment in an absorbable conduit: Improvement of peripheral nerve morphology with enlargement of somatosensory cortical representation. Frontiers in Neuroanatomy. 2014;**8**:111

[86] Cartarozzi LP, Spejo AB, Ferreira RS Jr, Barraviera B, Duek E, Carvalho JL, et al. Mesenchymal stem cells engrafted in a fibrin scaffold stimulate Schwann cell reactivity and axonal regeneration following sciatic nerve tubulization. Brain Research Bulletin. 2015;**112**:14-24

[87] Xin H, Li Y, Buller B, Katakowski M, Zhang Y, Wang X, et al. Exosome-mediated transfer of miR-133b from multipotent mesenchymal stromal cells to neural cells contributes to neurite outgrowth. Stem Cells. 2012;**30**(7):1556-1564

[88] Yuyama K, Sun H, Usuki S, Sakai S, Hanamatsu H, Mioka T, et al. A potential function for neuronal exosomes: Sequestering intracerebral amyloid-β peptide. FEBS Letters. 2015;**589**(1):84-88

[89] Ji Q, Ji Y, Peng J, Zhou X, Chen X, Zhao H, et al. Increased brain-specific

MiR-9 and MiR-124 in the serum exosomes of acute ischemic stroke patients. PLoS One. 2016;**11**(9):e0163645

[90] Fröhlich D, Kuo WP, Frühbeis C, Sun J-J, Zehendner CM, Luhmann HJ, et al. Multifaceted effects of oligodendroglial exosomes on neurons: Impact on neuronal firing rate, signal transduction and gene regulation. Philosophical Transactions of the Royal Society, B: Biological Sciences. 2014;**369**(1652):20130510

[91] Takeda YS, Xu Q. Neuronal differentiation of human mesenchymal stem cells using exosomes derived from differentiating neuronal cells. PLoS One. 2015;**10**(8):e0135111

[92] Mead B, Tomarev S. Bone marrow-derived mesenchymal stem cells-derived exosomes promote survival of retinal ganglion cells through miRNA-dependent mechanisms. Stem Cells Translational Medicine. 2017;**6**(4):1273-1285

[93] Lopez-Verrilli MA, Picou F, Court FA. Schwann cell-derived exosomes enhance axonal regeneration in the peripheral nervous system. Glia. 2013;**61**(11):1795-1806

[94] Selmaj I, Mycko MP, Raine CS, Selmaj KW. The role of exosomes in CNS inflammation and their involvement in multiple sclerosis. Journal of Neuroimmunology. 2017;**306**:1-10

[95] Ibrahim AG-E, Cheng K, Marbán E. Exosomes as critical agents of cardiac regeneration triggered by cell therapy. Stem Cell Reports. 2014;**2**(5):606-619

[96] Teng X, Chen L, Chen W, Yang J, Yang Z, Shen Z. Mesenchymal stem cell-derived exosomes improve the microenvironment of infarcted myocardium contributing to angiogenesis and anti-inflammation. Cellular Physiology and Biochemistry. 2015;**37**(6):2415-2424

[97] Khan M, Nickoloff E, Abramova T, Johnson J, Verma SK, Krishnamurthy P, et al. Embryonic stem cell–derived exosomes promote endogenous repair mechanisms and enhance cardiac function following myocardial infarction. Circulation Research. 2015;**117**(1):52-64

[98] Zhao Y, Sun X, Cao W, Ma J, Sun L, Qian H, et al. Exosomes derived from human umbilical cord mesenchymal stem cells relieve acute myocardial ischemic injury. Stem Cells International. 2015;**2016**:761643

[99] Vicencio JM, Yellon DM, Sivaraman V, Das D, Boi-Doku C, Arjun S, et al. Plasma exosomes protect the myocardium from ischemia-reperfusion injury. Journal of the American College of Cardiology. 2015;**65**(15):1525-1536

[100] Beltrami C, Besnier M, Shantikumar S, Shearn AI, Rajakaruna C, Laftah A, et al. Human pericardial fluid contains exosomes enriched with cardiovascular-expressed microRNAs and promotes therapeutic angiogenesis. Molecular Therapy. 2017;**25**(3):679-693

[101] Gopal SK, Greening DW, Rai A, Chen M, Xu R, Shafiq A, et al. Extracellular vesicles: Their role in cancer biology and epithelial–mesenchymal transition. The Biochemical Journal. 2017;**474**(1):21-45

[102] Liang X, Zhang L, Wang S, Han Q, Zhao RC. Exosomes secreted by mesenchymal stem cells promote endothelial cell angiogenesis by transferring miR-125a. Journal of Cell Science. 2016;**129**(11):2182-2189

[103] Li X, Liu L, Yang J, Yu Y, Chai J, Wang L, et al. Exosome derived from human umbilical cord mesenchymal stem cell mediates MiR-181c attenuating burn-induced excessive inflammation. eBioMedicine. 2016;**8**:72-82

[104] Zhang B, Wu X, Zhang X, Sun Y, Yan Y, Shi H, et al. Human umbilical cord mesenchymal stem cell exosomes enhance angiogenesis through the Wnt4/β-catenin pathway. Stem Cells Translational Medicine. 2015;**4**(5):513-522

[105] Tkach M, Kowal J, Zucchetti AE, Enserink L, Jouve M, Lankar D, et al. Qualitative differences in T-cell activation by dendritic cell-derived extracellular vesicle subtypes. The EMBO journal. 2017;**36**(20):3012-3028

[106] Morse MA, Garst J, Osada T, Khan S, Hobeika A, Clay TM, et al. A phase I study of dexosome immunotherapy in patients with advanced non-small cell lung cancer. Journal of Translational Medicine. 2005;**3**(1):1-8

[107] Bowen WS, Svrivastava AK, Batra L, Barsoumian H, Shirwan H. Current challenges for cancer vaccine adjuvant development. Expert Review of Vaccines. 2018;**17**(3):207-215

[108] Morishita M, Takahashi Y, Matsumoto A, Nishikawa M, Takakura Y. Exosome-based tumor antigens–adjuvant co-delivery utilizing genetically engineered tumor cell-derived exosomes with immunostimulatory CpG DNA. Biomaterials. 2016;**111**:55-65

[109] Walker S, Busatto S, Pham A, Tian M, Suh A, Carson K, et al. Extracellular vesicle-based drug delivery systems for cancer treatment. Theranostics. 2019;**9**(26):8001

[110] Narita M, Kanda T, Abe T, Uchiyama T, Iwafuchi M, Zheng Z, et al. Immune responses in patients with esophageal cancer treated with SART1 peptide-pulsed dendritic cell vaccine. International Journal of Oncology. 2015;**46**(4):1699-1709

[111] Dai S, Wei D, Wu Z, Zhou X, Wei X, Huang H, et al. Phase I clinical trial of autologous ascites-derived exosomes combined with GM-CSF for colorectal cancer. Molecular Therapy : The Journal of the American Society of Gene Therapy. 2008;**16**(4):782-790

[112] Liu B, Lee BW, Nakanishi K, Villasante A, Williamson R, Metz J, et al. Cardiac recovery via extended cell-free delivery of extracellular vesicles secreted by cardiomyocytes derived from induced pluripotent stem cells. Nature Biomedical Engineering. 2018;**2**(5):293-303

[113] Chen CW, Wang LL, Zaman S, Gordon J, Arisi MF, Venkataraman CM, et al. Sustained release of endothelial progenitor cell-derived extracellular vesicles from shear-thinning hydrogels improves angiogenesis and promotes function after myocardial infarction. Cardiovascular Research. 2018;**114**(7):1029-1040

[114] Han C, Zhou J, Liang C, Liu B, Pan X, Zhang Y, et al. Human umbilical cord mesenchymal stem cell derived exosomes encapsulated in functional peptide hydrogels promote cardiac repair. Biomaterials Science. 2019;**7**(7):2920-2933

[115] Zhang Y, Zhang P, Gao X, Chang L, Chen Z, Mei X. Preparation of exosomes encapsulated nanohydrogel for accelerating wound healing of diabetic rats by promoting angiogenesis. Materials Science and Engineering: C. 2021;**120**:111671

[116] Chin JS, Chooi WH, Wang H, Ong W, Leong KW, Chew SY. Scaffold-mediated non-viral delivery platform for CRISPR/Cas9-based genome editing. Acta Biomaterialia. 2019;**90**:60-70

[117] Yao Y, Zhang H, Wang Z, Ding J, Wang S, Huang B, et al. Reactive oxygen species (ROS)-responsive biomaterials mediate tissue microenvironments and

tissue regeneration. Journal of Materials Chemistry B. 2019;7(33):5019-5037

[118] Qian Z, Bai Y, Zhou J, Li L, Na J, Fan Y, et al. A moisturizing chitosan-silk fibroin dressing with silver nanoparticles-adsorbed exosomes for repairing infected wounds. Journal of Materials Chemistry B. 2020;**8**(32):7197-7212

[119] Li M, Ke Q-F, Tao S-C, Guo S-C, Rui B-Y, Guo Y-P. Fabrication of hydroxyapatite/chitosan composite hydrogels loaded with exosomes derived from miR-126-3p overexpressed synovial mesenchymal stem cells for diabetic chronic wound healing. Journal of Materials Chemistry B. 2016;**4**(42):6830-6841

[120] Chen S, Tang Y, Liu Y, Zhang P, Lv L, Zhang X, et al. Exosomes derived from miR-375-overexpressing human adipose mesenchymal stem cells promote bone regeneration. Cell Proliferation. 2019;**52**(5):e12669

[121] Liu X, Yang Y, Li Y, Niu X, Zhao B, Wang Y, et al. Integration of stem cell-derived exosomes with in situ hydrogel glue as a promising tissue patch for articular cartilage regeneration. Nanoscale. 2017;**9**(13):4430-4438

[122] Keshtkar S, Kaviani M, Soleimanian S, Azarpira N, Asvar Z, Pakbaz S. Stem cell-derived exosome as potential therapeutics for microbial diseases. Frontiers in Microbiology. 2021;**12**:786111

[123] Chow L, Johnson V, Impastato R, Coy J, Strumpf A, Dow S. Antibacterial activity of human mesenchymal stem cells mediated directly by constitutively secreted factors and indirectly by activation of innate immune effector cells. Stem Cells Translational Medicine. 2020;**9**(2):235-249

[124] Yagi H, Chen AF, Hirsch D, Rothenberg AC, Tan J, Alexander PG, et al. Antimicrobial activity of mesenchymal stem cells against Staphylococcus aureus. Stem Cell Research & Therapy. 2020;**11**(1):1-12

[125] Krasnodembskaya A, Song Y, Fang X, Gupta N, Serikov V, Lee J-W, et al. Antibacterial effect of human mesenchymal stem cells is mediated in part from secretion of the antimicrobial peptide LL-37. Stem Cells. 2010;**28**(12):2229-2238

[126] Rezaie J, Aslan C, Ahmadi M, Zolbanin NM, Kashanchi F, Jafari R. The versatile role of exosomes in human retroviral infections: From immunopathogenesis to clinical application. Cell & Bioscience. 2021;**11**(1):1-15

[127] Alcayaga-Miranda F, Cuenca J, Martin A, Contreras L, Figueroa FE, Khoury M. Combination therapy of menstrual derived mesenchymal stem cells and antibiotics ameliorates survival in sepsis. Stem Cell Research & Therapy. 2015;**6**(1):1-13

[128] Lee JW, Krasnodembskaya A, McKenna DH, Song Y, Abbott J, Matthay MA. Therapeutic effects of human mesenchymal stem cells in ex vivo human lungs injured with live bacteria. American Journal of Respiratory and Critical Care Medicine. 2013;**187**(7):751-760

[129] Monsel A, Zhu Y-g, Gennai S, Hao Q, Hu S, Rouby J-J, et al. Therapeutic effects of human mesenchymal stem cell–derived microvesicles in severe pneumonia in mice. American Journal of Respiratory and Critical Care Medicine. 2015;**192**(3):324-336

[130] Abreu SC, Weiss DJ, Rocco PR. Extracellular vesicles derived from mesenchymal stromal cells: A therapeutic option in respiratory diseases? Stem Cell Research & Therapy. 2016;7(1):1-10